ELSEVIER ERGONOMICS BOOK SERIES

VOLUME 1

Ergonomics Guidelines and Problem Solving

ELSEVIER SCIENCE INTERNET HOMEPAGE

http://www.elsevier.nl (Europe)
http://www.elsevier.com (America)
http://www.elsevier.jp (Asia)

Consult homepage for full catalogue information on all books, journals and electronic products and services

ELSEVIER TITLES OF RELATED INTEREST

BOOKS
Cushman and Rosenberg / Human Factors in Product Design
Gale et al / Vision in Vehicle Series
Helander et al / Handbook of Human-Computer Interaction
Scott et al / Global Ergonomics
Vink et al / Human Factors in Organizational Design and Management Series

JOURNALS
Accident Analysis & Prevention
Applied Ergonomics
Cognition
Computers in Human Behaviour
Computers in Industry
Design Studies
Human Movement Science
International Journal of Industrial Ergonomics
Journal of Safety Research
Performance Evaluation
Safety Science

ELSEVIER ERGONOMICS BOOK SERIES

VOLUME 1

Ergonomics Guidelines and Problem Solving

Edited by

Anil Mital
USA

Åsa Kilbom
Sweden

Shrawan Kumar
Canada

ERGONOMICS SERIES EDITORS
A. Mital, M.M. Ayoub & K. Landau
Consulting Editor: T. Leamon

2000

ELSEVIER

Amsterdam • Lausanne • New York • Oxford • Shannon • Singapore • Tokyo

ELSEVIER SCIENCE Ltd
The Boulevard, Langford Lane
Kidlington, Oxford OX5 1GB, UK

© 2000 Elsevier Science Ltd. All rights reserved.

This work is protected under copyright by Elsevier Science, and the following terms and conditions apply to its use:

Photocopying
Single photocopies of single chapters may be made for personal use as allowed by national copyright laws. Permission of the Publisher and payment of a fee is required for all other photocopying, including multiple or systematic copying, copying for advertising or promotional purposes, resale, and all forms of document delivery. Special rates are available for educational institutions that wish to make photocopies for non-profit educational classroom use.

Permissions may be sought directly from Elsevier Science Rights & Permissions Department, PO Box 800, Oxford OX5 1DX, UK; phone: (+44) 1865 843830, fax: (+44) 1865 853333, e-mail: permissions@elsevier.co.uk. You may also contact Rights & Permissions directly through Elsevier's home page (http://www.elsevier.nl), selecting first 'Customer Support', then 'General Information', then 'Permissions Query Form'.

In the USA, users may clear permissions and make payments through the Copyright Clearance Center, Inc., 222 Rosewood Drive, Danvers, MA 01923, USA; phone: (978) 7508400, fax: (978) 7504744, and in the UK through the Copyright Licensing Agency Rapid Clearance Service (CLARCS), 90 Tottenham Court Road, London W1P 0LP, UK; phone: (+44) 171 631 5555; fax: (+44) 171 631 5500. Other countries may have a local reprographic rights agency for payments.

Derivative Works
Tables of contents may be reproduced for internal circulation, but permission of Elsevier Science is required for external resale or distribution of such material. Permission of the Publisher is required for all other derivative works, including compilations and translations.

Electronic Storage or Usage
Permission of the Publisher is required to store or use electronically any material contained in this work, including any chapter or part of a chapter.

Except as outlined above, no part of this work may be reproduced, stored in a retrieval system or transmitted in any form or by any means, electronic, mechanical, photocopying, recording or otherwise, without prior written permission of the Publisher.
Address permissions requests to: Elsevier Science Rights & Permissions Department, at the mail, fax and e-mail addresses noted above.

Notice
No responsibility is assumed by the Publisher for any injury and/or damage to persons or property as a matter of products liability, negligence or otherwise, or from any use or operation of any methods, products, instructions or ideas contained in the material herein. Because of rapid advances in the medical sciences, in particular, independent verification of diagnoses and drug dosages should be made.

First edition 2000

Library of Congress Cataloging in Publication Data
A catalog record for this title is available from the Library of Congress.

British Library Cataloguing in Publication Data
A catalogue record for this title is available from British Library.

Reprinted with revisions from the International Journal of Industrial Ergonomics,
Vol. 10, Nos. 1-2, Vol. 14, Nos 1-2 and Vol. 22, Nos. 1-2

ISBN: 0-08-043643-9

Transferred to digital 2007

PREFACE

A science, or field of science, is not truly defined until it has a set of guidelines or standards that can be referred to. This book is the outcome of three different invited symposia on ergonomics guidelines and problem solving held between 1991 and 1996. The First Invited International Symposium on Ergonomics Guidelines and Problem Solving was held in Lake Tahoe, USA in June 1991. The Second International Symposium on Ergonomics Guidelines and Problem Solving was held in Copenhagen, Denmark in June 1993. The Third International Symposium on Ergonomics Guidelines and Problem Solving was held in Zurich, Switzerland in July 1996. All symposia were sponsored by the National Institute for Occupational Safety and Health (NIOSH) of the United States Department of Health and Human Services. The first two symposia were also co-sponsored by Sweden's National Institute for Working Life (formerly, the National Institute of Occupational Health). The last symposium was further co-sponsored by the Swiss Federal Institute of Technology (ETH) and the Integrated Furniture Solutions – EckAdams of the United States.

The primary purpose of these invited symposia was to provide time-tested and applied information to ergonomics practitioners (managers, production system designers, shop supervisors, occupational safety and health professionals, union representatives, labor inspectors, production and industrial engineers, industrial hygienists, ergonomists and all those interested in workplace productivity and efficiency, and worker health and safety). The organization of these symposia also reflected the gradual realization that there is an urgent need to disseminate ergonomics "know-how". Nearly 75 international experts participated in one or more of the symposia. These experts, all invited, felt that it was important to provide useable guidelines to practitioners now, even if the information was based on limited scientific data, rather than to wait until all questions can be completely and conclusively answered. It was felt that these Guidelines would provide valuable assistance in alleviating the problems.

International experts were invited from around the World and Guidelines on various topics were discussed. The topics were chosen by the symposia organisers (Asa Kilbom and Anil Mital for the first two symposia and Anil Mital and Shrawan Kumar for the third symposium) according to the following criteria:

- need to address pervasive ergonomics problems,
- availability of reasonable research knowledge, and
- lack of up-to-date guidelines.

Experts in the areas selected were approached to write draft Guidelines. These draft Guidelines were circulated to the symposium participants and other experts for their comments ahead of the symposium. Each Guideline was also sent to several prominent workers in the field for their comments and criticisms. These comments and criticisms were utilized in revising the Guidelines. The revised Guidelines were presented at one of the symposiums and discussed by the participants. An expert in the area of the guideline was asked to coordinate the critique and consolidate the comments. The Guidelines were further revised and reviewed by various experts, with the final version appearing in three separate issues of the *International Journal of Industrial Ergonomics*. For the first time this book consolidates all these Guidelines and the scientific basis on which the Guidelines are based.

The readers and the users of these Guidelines should realize that even though the information contained herein has been subjected to extensive discussion and review and are based on available scientific data, in no way these are endorsed by the symposia sponsors. In other words, these Guidelines do not have the status of official documents representing the views of any sponsoring organization or group. These Guidelines are the product of individual researchers, who, after extensive feedback from other experts in the field have developed the Guidelines based on their review of scientific literature. Each Guideline is endorsed by a number of individuals in their personal capacity.

Each Guideline is divided into two parts. Part I contain Guidelines for practitioners and Part II provides the scientific basis or knowledge base for the Guideline. Such separation of the applied and theoretical content is designed to facilitate rapid incorporation of these Guidelines into practice.

Finally, the editors wish to express their gratitude to the authors and reviewers who lent their expertise and time in the development of these Guidelines. We sincerely hope that these Guidelines will provide users with a practical tool to enhance workplace productivity and safety, and will be conducive in preserving the health of workers.

Anil Mital	Asa Kilbom	Shrawan Kumar
Cincinnati, USA	Solna, Sweden	Edmonton, Canada

CONTENTS

Preface v

Task analysis: Part I – Guidelines for the practitioner 1
K. Landau, W. Rohmert and R. Brauchler

Task analysis: Part II – The scientific basis (knowledge base) for the guide 9
R. Brauchler and K. Landau

Allocation of functions to humans and machines in a manufacturing environment: Part I – Guidelines for the practitioner 33
A. Mital, A. Motorwala, M. Kulkarni, M. Sinclair and C. Siemieniuch

Allocation of functions to humans and machines in a manufacturing environment: Part II – The scientific basis (knowledge base) for the guide 61
A. Mital, A. Motorwala, M. Kulkarni, M. Sinclair and C. Siemieniuch

Occupational and individual risk factors for shoulder–neck complaints: Part I – Guidelines for the practitioner 79
J. Winkel and R. Westgaard

Occupational and individual risk factors for shoulder–neck complaints: Part II – The scientific basis (literature review) for the guide 83
J. Winkel and R. Westgaard

Human muscle strength definitions, measurement, and usage: Part I – Guidelines for the practitioner 103
A. Mital and S. Kumar

Human muscle strength definitions, measurement, and usage: Part II – The scientific basis (knowledge base) for the guide 123
A. Mital and S. Kumar

Repetitive work of the upper extremity: Part I – Guidelines for the practitioner 145
Å. Kilbom

Repetitive work of the upper extremity: Part II – The scientific basis (knowledge base) for the guide 151
Å. Kilbom

The reduction of slip and fall injuries: Part I – Guidelines for the practitioner 179
T.B. Leamon

The reduction of slip and fall injuries: Part II – The scientific basis (knowledge base) for the guide 183
T.B. Leamon

Job design for the aged with regard to decline in their maximal aerobic capacity: Part I – Guidelines for the practitioner 189
J. Ilmarinen

Job design for the aged with regard to decline in their maximal aerobic capacity: Part II – The scientific basis for the guide 199
J. Ilmarinen

Design, selection and use of hand tools to alleviate trauma of the upper extremities: Part I – Guidelines for the practitioner 213
A. Mital and Å. Kilbom

Design, selection and use of hand tools to alleviate trauma of the upper extremities: Part II – The scientific basis (knowledge base) for the guide 217
A. Mital and Å. Kilborn

Equipment design for maintenance: Part I – Guidelines for the practitioner 233
S.N. Imrhan

Equipment design for maintenance: Part II – The scientific basis for the guide 241
S.N. Imrhan

Designing warning signs and warning labels: Part I – Guidelines for the practitioner........... 249
M.R. Lehto

Designing warning signs and warning labels: Part II – Scientific basis for initial guidelines ... 257
M.R. Lehto

Vision at the workplace: Part I – Guidelines for the practitioner 281
S. Konz

Vision at the workplace: Part II – Knowledge base for the guide 285
S. Konz

Evaluation and control of industrial inspection: Part I – Guidelines for the practitioner 301
T.J. Gallwey

Evaluation and control of industrial inspection: Part II – The scientific basis for the guide 313
T.J. Gallwey

Evaluation and control of hot working environments: Part I – Guidelines for the practitioner.. 329
J.D. Ramsey, T.E. Bernard and F.N. Dukes-Dobos

Evaluation and control of hot working environments: Part II – The scientific basis (knowledge base) for the guide 337
T.E. Bernard, F.N. Dukes-Dobos and J.D. Ramsey

Cold stress: Part I – Guidelines for the practitioner 347
I. Holmér

Cold stress: Part II – The scientific basis (knowledge base) for the guide......................... 357
I. Holmér

Noise in the office: Part I – Guidelines for the practitioner... 367
A. Kjellberg and U. Landström

Noise in the office: Part II – The scientific basis (knowledge base) for the guide 371
A. Kjellberg and U. Landström

Work/rest: Part I – Guidelines for the practitioner 397
S. Konz

Work/rest: Part II – The scientific basis (knowledge base) for the guide 401
S. Konz

Managing stress in the workplace: Part I – Guidelines for the practitioner...................... 429
A.M. Williamson

Managing stress in the workplace: Part II – The scientific basis (knowledge base) for the guide   437
A.M. Williamson

Economic evaluation of ergonomic solutions: Part I – Guidelines for the practitioner 463
E.R. Andersson

Economic evaluation of ergonomic solutions: Part II – The scientific basis 473
E.R. Andersson

Author Index... 479

Task analysis: Part I – Guidelines for the practitioner[1,2]

Kurt Landau[a,*], Walter Rohmert[a], Regina Brauchler[b]

[a] *University of Technology, D-64287 Darmstadt, Germany*
[b] *University of Hohenheim, D-70599 Stuttgart, Germany*

1. Who is the practitioner?

The practitioner is anyone involved in analysis, evaluation, design or redesign of work places or work systems. This includes work study specialists, production engineers and other engineers working in industry, industrial designers, supervisory staff, industrial medical and safety officers, ergonomists and labour inspectors.

2. When and where should these guidelines be used?

The aim of these guidelines is to provide guidance on methods of analysing the tasks involved in work systems.

Task analysis yields data on the demands imposed on the worker by a given job and enables the elements of a work system to be identified and compared with those of other work systems. It provides information on peak stress situations that may occur and indications of how these could be eliminated or reduced by job redesign. Other types of task analysis procedure can be used to forecast the characteristics of and stresses likely to arise in new work systems that are still in the design phase.

Procedures that are capable of analysing job demands are a valuable tool in human resources management when used in combination with mental and physical aptitude tests.

Applications in the fields of research and planning, for example, in the classification of jobs and occupations/professions, in accident research or in the planning of educational and training programmes demonstrate the broad spectrum covered by task analysis procedures.

3. Definitions and terminology

The type of task analysis procedure used will depend on the aspect from which the task is defined and the terminology used. It is necessary to distinguish between organizational terminology, work study terminology and ergonomical terminology (Fig. 1).

---

* Corresponding author.

[1] The recommendations provided in this guide are based on numerous published and unpublished scientific studies and are intended to enhance worker safety and productivity. These recommendations are neither intended to replace existing standards, if any, nor should be treated as standards. Furthermore, these documents should not be construed to represent institutional policy.

The following individuals participated in the discussion of the earlier version of this guide. Their suggestions (written or verbal) were incorporated by the authors in this version: A. Aaras, Norway; J.E. Fernandez, USA; A. Freivalds, USA; T. Gallwey, Ireland; M. Jager, Germany; S. Konz, USA; S. Kumar, Canada; H. Krueger, Switzerland; A. Luttmann, Germany; A. Mital, USA; J.D. Ramsey, USA; M.-J. Wang, Taiwan.

[2] See also Brauchler et al. (1998).

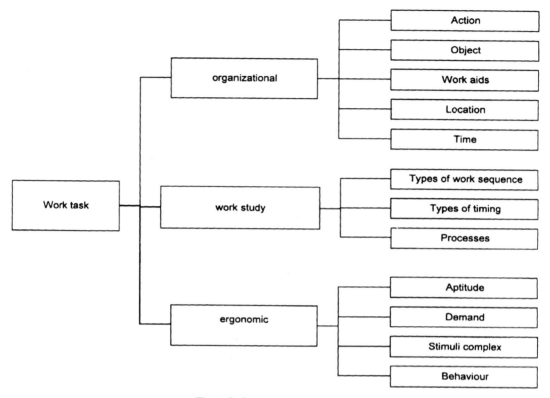

Fig. 1. Definitions and terminology.

Organizational definitions relate to the functional actions of the industrial worker. The task thus represents a set target that must be fulfilled. Under this heading tasks can be distinguished by the following characteristics:
- Each task implies a given activity, e.g. planning, analysing, preparing, assembling, etc. The work process is performed either as a (predominantly) mental exercise or (more usually) as combination of mental and physical activities.
- Each task relates to an object on which the expected activity must be performed. The object can be either inanimate (e.g. screws), animate (e.g. personnel) or mental (e.g. verification of stock lists).
- Each task also regularly involves the use of inanimate work aids that are used to perform the work process.
- Each task can be classified in both time and space, i.e. it is determined by these two fundamental existential conditions.

The core of the organizational approach to definition of the elements comprising the task is a task analysis followed by a task synthesis. The partial or subtasks that have been identified are assembled into a bundle which can be delegated to specific persons or departments.

Whilst the organizational definition classifies tasks into partial or subtasks, the work study definition frequently adopts a variable time approach. Work sequences can be classified as temporal and spatial consequences of the interaction between man/machine and the work object in either macro (days and weeks) or micro-time frames (h and min).

Whereas the aim of these classifications is merely to estimate the duration of specific work sequences, there is a second version of work study terminology for the classification of tasks into types of sequence. Types of work sequence are terms used to define the interaction of man and work object with the input of the work system, for example, "turning the work

Preparing the work system and executing the order	Equipping/executing
Worker's ability to influence the timing of the work sequence	Total/partial/none
Type of use of man and work object	In use/not in use/ factory not operating/ not definable
Method of executing order	Processing/interrupting processing/factory not operating/not definable

Fig. 2. Methods of classifying types of work sequence.

object", "trimming the work object", "removing the work object", etc.

Fig. 2 lists ways of classifying types of work sequence: A job is made up of a number of types of tasks. These can be broken down into actions. The task is thus a generic term for an associated series of actions which are normally performed in the prescribed sequence and place demands on the worker which, although similar, may vary in level.

This concept sees the work task as a complete activity performed by the worker. It represents the worker's smallest, but at the same time, most important "unit of activity".

Whereas the organizational and scientific terminologies look at the worker primarily as a production factor or a functional element, the ergonomic definition of a work task focuses on the man/work interface and regards the task as
- a description of behaviour
- an aptitude test
- a behavioural demand
- a complex of stimuli.

Whilst the first item listed above uses purely descriptive terms to define the visible and recordable actions of people at work, the second item seeks to make an indirect evaluation of the characteristics, skills, abilities and knowledge required to perform the task. The third item describes the various forms of information processing required. The fourth item is based on the assumption that a task consists of a so-called complex of stimuli and a series of instructions specifying what has to be done in the job in relation to the stimuli.

For more details of these definitions please refer to Part II of this guideline.

4. Problem identification

Planning, design and evaluation of work should be preceded by an analysis of the work tasks and the resulting demands that are standard pattern is performed in only very few cases, mainly manual jobs in industry. Instead, ad hoc procedures relating to the individual case are used. No further use is made of the data after the immediate problem has been solved. This would, in any case, be impossible because the analytical instrument is either totally or, at least, partially inapplicable outside the confines of the company that used it. This means that companies regularly "reinvent the wheel". Analytical data that could be further evaluated for general purposes is not passed on and the opportunity to further develop the discipline of ergonomics is lost. No taxonomies of tasks can be compiled and questions relating to occupational research are left unanswered.

This raises the question of whether it would be possible to develop task analysis procedures that are universally applicable. Such procedures should cover the whole spectrum from heavy physical work to mental work and they should be equally suitable for use in large and small operations and in different branches of industry.

The broader the spectrum of task analysis instruments, the less danger there is of being forced to apply suboptimal solutions in the subsequent job design. One consequence of a broad spectrum, universal approach to task analysis is, however, that the data produced is on a high level of abstraction. This means that it may be difficult for the practitioners on the ground to apply these data directly to their own problem.

5. Data collection

Task analysis procedures must fulfil the following criteria:
They must
- be based on a theoretical model that enables interpretation of the data obtained in a form which is relevant for practical purposes;
- as far as possible completely register all the tasks involved in a given work system;
- be as economical as possible in their application and in the subsequent processing and evaluation of the data;
- be standardizable;
- enable conclusions to be drawn that are quantifiable, at least on an ordinal scale, and which extend beyond a purely verbal description of the job;
- provide data on the validity of the procedure.

It is impossible in these guidelines to discuss the various types of task analysis in detail. They can, in principle, be classified as unstandardized, semi-standardized and standardized. If, when using an unstandardized task analysis procedure, reference is made to information from analyses of documents, training reports, reports prepared by individual employees actually doing the job or their superiors, job analysts, etc., this means that a qualitative task analysis will be performed in accordance with certain guidelines and then set down on paper in freely formulated form. The use of a semi-standardized procedure makes it possible to limit the analyst's discretionary powers. Such procedures include, for example, guidelines for observations and interviews, work journals and the critical incident technique. If, however, one wishes to reduce the analyst's discretionary powers to a minimum, it is necessary to resort to standardized procedures such as questionnaires, interviews, check-lists, observation interview and time studies. These produce a qualitative and quantitative record of the detailed characteristics of a given job in a fixed schematic form. All these survey techniques largely exclude the actual person doing the job, thereby concentrating on compliance with classic statistical criteria and avoiding the problem of identifying the job's subjective aspects. Evaluations by outsiders are frequently criticized as being one-sided and incomplete. If the technique of job analysis by the worker actually doing the job is used, this technique elevates the worker to the position of a person who is competent to make an ergonomic assessment of the job situation. This enables an analysis of the tasks as redefined by the individual worker. It often makes sense to use both "objective" and "subjective" task analysis procedures, because this will in any event yield additional information which would not have been obtained with a single survey technique.

6. Data analysis

From the wide variety of potential uses of task analysis data discussed above it is possible to identify several basic applications arising in virtually all fields.

1. Classification of individual work systems on the basis of predominant characteristics revealed by the task analysis and other information relating to the work system and the worker.
2. Definition of work systems or groups of work systems by their predominant characteristics.
3. Grouping of work systems on the basis of the task analysis using, e.g. cluster analysis techniques.
4. Grouping of task analysis characteristics using, e.g. factor analysis techniques.

With the help of relatively simple univariate statistical procedures task analysis data can provide answers to the sort of fundamental question listed in items 1–5 below.

The results obtained make it possible to examine questions relating, for example, to individual branches of industry, sex or educational level of the worker. Personal data relating to the individual workers may also be used.

1. How do tasks performed by men and women differ in terms of task content, work object, work tools, workplace, work environment and job organization in different branches of industry,
- different factories,
- different departments,
- different wage categories?

2. What differences are there in the task characteristics of workers with different levels of academic education and job-related training?

3. What differences are there in the task characteristics of locally born and immigrant workers?

4. What differences are there in the task characteristics of manual workers, salaried employees, executives and civil servants?

5. Which task exhibits marked similarities or differences in terms of their various stress components?

Univariate statistical analysis can also be used to interpret cluster analysis data.

The main point of interest here is the distribution of intensity frequencies of individual task characteristics and also determinations of the main trends and degree of dispersion revealed by the data in cases where it is possible to display the data on an ordinal or nominal scale.

When ordinal scales are used, the median will show the main trend and the percentiles the degree of dispersion.

The median test can be used to test hypothesis Hø by determining whether task data from two samples show the same main trend. Here all the data from both samples is tested for deviations above or below the overall median.

The Kolmogoroff-Smirnoff test can be used to ascertain whether the two samples have similar frequency distributions. If other tests, e.g. Wilcoxon's U-test or Mann and Whitney are used, it will generally be necessary to have genuine rank orders, which means that each level on the ordinal scale can only be occupied once.

Graphs showing profiles compiled from the scores obtained for groups of characteristics are an excellent way of depicting the extent or duration of stresses occurring during the performance of given tasks/activities or groups of tasks/activities. In ergonomics this is called "profile analysis", a term which may be used in a different sense in other disciplines. The scores used in the preparation of the profile analyses are obtained from the relevant items of the task analysis procedure. It is also possible to use factor analyses to prepare a job profile.

Metric scales should be used for both the classification and the summation of the characteristics revealed by the analysis procedure and when using the data from a factor analysis. If data derived from task analysis is displayed on interval or proportional scales, there is a danger that incompatible values may be added together. When calculating profile heights, the relevant scores for a task or a group of tasks can be used to calculate an approximate maximum score. The height of the profile is determined by the number of characteristics attributed to the relevant item. This means that, although the individual columns of the profile are not comparable, it is possible to compare the heights of the various columns for different jobs.

Factor analysis can be used to determine those stress factors inherent in a task, which cannot be measured or observed directly. This procedure normally involves the reduction of a mass of data to a small number of separate hypothetical dimensions or factors.

The main purpose of cluster analytical methods is to reduce the volume of data. Task groups must be formed and numerical relationships between these groups must be determined and interpreted. When classifying tasks, it is necessary to use the numerical relationships to draw conclusions that can then be used for compiling a taxonomy of human tasks. The influence of specific work content on the composition of the groups and on the numerical relationships between the groups is an important aspect for the ergonomist.

Table 1

General sphere of activity	Actual application and result that can be achieved
Ergonomic and technical design of work systems	
Documentation and diagnosis of work systems	Description and, where applicable, quantification of system elements and their characteristics, e.g. stresses occurring, design needs; establishment and checking of design priorities
	Prevention of potential problems by identifying inacceptable levels of stress
	Active reduction of stress
Product design	Description and, where applicable, quantification of impact of product characteristics when product is subsequently used by the consumer or by industry
Management organization	
Planning of corporate reorganization	Analysis of before/after job demand profiles when corporate or technical/technological changes are planned. Forecasting of shifts in demand as a result of technical change
Work process design during a given work period or shift	Task and demand sequences
Design of shifts and breaks	Planning of breaks based on the level of stresses occurring within the work system, and of work sharing during individual shifts to achieve an even spread of the types/levels of demands arising
	Reduction of absenteeism by analyses to establish causal relationships between stresses and disease
Human resources management	
Staff recruitment	Drafting of advertisements for staff based on available task analyses
Selection	Selection of applicants by comparison of demand and aptitude profiles
	Establishment of qualification/aptitude criteria
Instruction/training/retraining	Establishment of training needs/necessary training programs on the basis of the task/demand profiles of the work systems
Integration of disabled persons and other handicapped groups into the organization	Identification of work systems with task/demand profiles that are compatible with the working capacity of these groups
Qualitative human resources planning/ appointment planning	Long- and short-term comparisons of aptitude and demand profiles
Labour relations	Task analyses as a basis for negotiations between employers and staff
Wage negotiations	Review of wage/salary grades by comparison of demand profiles of standard work systems with those of work systems for which reclassification is being demand
Analysis of corporate remuneration policy	Checking of adequacy of remuneration policy within the company and in relation to industry as a whole
Research and career advice	
Job classification	Drafting of demand- or task-related job classifications and comparison of these with conventional job classifications
	Investigation of health status in given occupational groups
	Establishment of psychological and physical profiles required for given jobs or occupations
	Determination of gravity and frequency of accidents by job/occupational group
	Job classification by socio-demographic criteria
Career advice and information	Description of job spectra when giving career advice to school-leavers

7. Problems and solutions

Table 1 lists examples of problems arising in industry and possible solutions involving task analysis.

The following recommendations, based on many years of experience with the applications listed in the above table, can be given to practitioner:

- Use general, universal task analysis procedures in order to enable comparison of the data with the results obtained in different groups, corporate divisions, companies, etc.
- Use task analysis procedures with standardized instruments to improve the reliability of your data.
- If you use supplementary procedures to obtain more detailed data, do not include the results in the database of the core analysis. Data obtained from supplementary procedures can be presented in modular form.
- Use task analysis procedures producing data that can be processed into task registers. Task registers are useful tools for documenting all the available ergonomic and work safety data for a whole company.
- Select only task analysis procedures using validated, reliable, standardized and economical techniques.
- In studies that can be relevant for epidemiological research, try to use task analysis procedures providing data that can be interrelated with the workers' medical data. Epidemiologists have in past been forced to rely mainly on occupational or professional classifications and are in urgent need of task analysis data.
- Use only task analysis procedures that are supported by appropriate software. A procedure that is only available in written form is merely of theoretical interest. For practical purposes, a software tool that is being constantly maintained and updated by its developer is essential. It must be able to function as a hot line and, most importantly, it must be supported by training programs.

References

Brauchler, R., Landau, K., 1998. Task analysis: Part II – The scientific basis for the guide (knowledge base). Int. J. Ind. Ergon., 22 (1–2): 13–35 (this issue).

Task analysis: Part II – The scientific basis (knowledge base) for the guide[1]

Regina Brauchler[a,*], Kurt Landau[b]

[a] *University of Hohenheim, D-70593 Stuttgart, Germany*
[b] *University of Technology, D-64287 Darmstadt, Germany*

1. Objective of task analysis

The technological changes that are being introduced into modern life at an ever increasing speed cause constant variations in work content. Whereas the typical work tasks performed at the beginning of this century had a high physical content, this is now being increasingly replaced by work involving high mental-intellectual content, for example, monitoring and controlling automated production processes and complicated management tasks (for the basic types of work task refer to Rohmert (1972, 1983)). This means that systems directed at the analysis of human work, like "work systems", have to cover the interaction of individuals and equipment involved in the work process being performed at the workplace and within a specific organizational chemo-physical and social environment. The term "work system" is a synonym of "man-at-work system" and it also covers the term "man–machine system (MMS)" used in industrial psychology.

Attempts have been made to define the concept of the work system in both national and international standards (e.g. the German DIN 33 400) and this concept is now widely used in ergonomic research and work study. Work systems are dynamic, socio-technical, open systems. They interrelate with their environment in a material, an energetic and an informatory sense. Work systems are generally complex or ultracomplex systems that can be broken down into successive hierarchical levels and can also form an integral part of higher-level systems. They can be either homeostatic, i.e. self-regulating, or externally regulated.

Mainly for the purposes of work study, interlinked work systems operating at different hierarchical levels are described as micro- or macro-systems, depending on their extent and complexity. Fig. 1 shows the work system in the form of a basic model including the two elements "man" and "task".

---

* Corresponding author.

[1] The recommendations provided in this guide are based on numerous published and unpublished scientific studies and are intended to enhance worker safety and productivity. These recommendations are neither intended to replace existing standards, if any, nor should be treated as standards. Furthermore, these documents should not be construed to represent institutional policy.

The following individuals participated in the discussion of the earlier version of this guide. Their suggestions (written or verbal) were incorporated by the authors in this version: A. Aaras, Norway; J.E. Fernandez, USA; A. Freivalds, USA; T. Gallwey, Ireland; M. Jager, Germany; S. Konz, USA; S. Kumar, Canada; H. Krueger, Switzerland; A. Luttmann, Germany; A. Mital, USA; J.D. Ramsey, USA; M.-J. Wang, Taiwan.

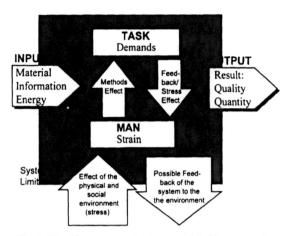

Fig. 1. Working sytem as the basic model of human work.

The model sees the worker as one of the technical elements of the work system. If the technical approach is taken to its logical conclusion, the worker or "man" is reduced to an ultracomplex, biosocial element of the system. For obvious reasons this one-sided technical approach fails to make allowance for all the relevant human factors involved. Its applications are therefore limited and its main use is in mathematical work system analyses for the simulation of work processes (Laurig, 1971).

There are various ways of analysing the structure of a work system and the interaction of its individual elements, and it is consequently incorrect to speak of work or job analysis or activity-oriented or task-related analysis in the singular as though referring to a monolithic procedure. The design and execution of work system analyses depend primarily on their theoretical basis and the object of the analysis, e.g. the work system, the job, the individual worker, etc.

An often used approach to the evaluation of job analysis procedures is explained in Part I of this guide (cf. Landau et al., 1998, this issue). Each procedure should be applicable to the full spectrum of both physical and non-physical types of work involving work contents ranging from the generation of force to the generation of information (cf. Rohmert, 1972).

A major advantage of the most recently developed task-related analysis procedures is that they can be used in the analysis of predominantly non-physical work. Whereas it was in the past only possible in this area to analyse selected, highly specialized activities like jobs or tasks in production management, it is now possible to apply these new procedures extensively to a wide range of service and management jobs.

It is not intended to include a taxonomy of job analysis procedures in this guide. The reader can refer to publications by Frieling (1975), Graf Hoyos (1974), von Pupka (1977), Rohmert et al. (1975b) and Frei (1981). For the theoretical basis and the statistical requirements of individual job analysis procedures, refer to the following publications (Fleishman, 1975; Theologus et al., 1970; Frieling, 1977; Hackman, 1970; McCormick, 1976; Morsh, 1964; Prien and Ronan, 1971; Zerga, 1943).

2. Job analysis as task analysis

This section discusses the analysis of work tasks from various aspects. From the work study and organizational aspects, the individual worker is seen as an element of the system or as a production factor.

In contrast, ergonomists and industrial psychologists are interested only in the interactions between the worker and work performed.

Work study task analysis examines technical performance or technical work functions. Complex entities like the individual job are broken down into their various elements. Work processes, the spatial and temporal interaction of the workers and the equipment with the work object (REFA, 1978) are broken down into time frames, process steps and actions (macrosections of the process) and into partial actions and steps and elements of actions (microsections of the process). A company's ability to meet agreed delivery dates is generally a matter of crucial, existential importance. This involves various factors, including the speed of its order processing, optimal use of its production capacities by reducing throughput times, efficient invoicing

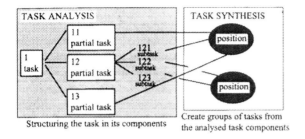

Fig. 2. Task analysis and task synthesis.

systems to prevent capital being locked up unnecessarily, etc.

Time and motion analyses are used to determine whether individual procedures should run concurrently or consecutively. The actual times recorded are converted into required or standard times and these are passed on to the worker as target figures. Work flow problems, e.g. minimization of throughput times, optimal work breaks, optimal order acceptance procedures, optimal ordering points and delivery dates for materials and optimal shift work arrangements are the main factors hindering the practical attainment of the required or standard times obtained from the time and motion analyses.

With this method of classification it is only possible to estimate the duration of sections of the process and the method is therefore too inaccurate and of little practical use for the purposes of task-related job analysis.

A second method of classification involves type coding of the sections of the process. This is described in greater detail in Part I (cf. Landau et al., this issue; Fig. 2).

In addition to time-based analyses, work study task analysis can also adopt a structural approach. For example, Kirchner and Rohmert (1973) break down the job into tasks and functions. Function in this context is used in a purely technical sense and means the achievement of the purpose of the work system.

Performance-related job inventories used mainly in the military field are examples of this type of structural task analysis which basically uses a work study approach.

Research on this type of job inventory is described, for example, by Morsh et al. (1961), and Morsh (1967, 1966). Job inventories are usually designed for a specific group of workplaces (e.g. within a company, a factory or a department) which can then be extrapolated for a larger group with the help of the "auditorium method".

As the items evaluated relate only to the conditions prevailing in the area under examination, the results, although certainly of value for that area, cannot be transposed to other groups or, if so, only to a limited degree. These procedures are unsuitable for comparison of stresses or for forecasting purposes, e.g. for estimating the consequences of technical change.

Organizational analyses start from a company's overall task, e.g. a market-oriented production task supported by auxiliary tasks in areas like accounting, warehousing and procurement. They break down these tasks into their integral parts from the organizational point of view. This method makes it possible to characterize the task by defining the following five items or elements (Kosiol, 1973, p. 202):

1. Type of activity. The task is performed, for example, as a (predominantly) intellectual activity or (more usually) as a combination of intellectual and physical activities.
2. Work object. Every task involves a work object on which the required activity has to be performed. The object can be either material or immaterial.
3. Physical adjuncts required. Every task involves the use of physical aids (materials or equipment) for the performance of the work process.
4. Location.
5. Time factor.

Organizational task analysis first establishes the task structure, i.e. a list of partial tasks and subtasks (these are defined as partial tasks that cannot be broken down any further). The degree of break-down will depend partly on the planned degree of work sharing and partly on the size of the relevant company. The scope of the subtasks should be reduced to a level which makes it possible to assign them to a single worker (cf. Fig. 2). This will involve answers to the following specific

questions:
- What action/process has to be performed?
- What work object is involved?
- What tools have to be used to perform the task? What purpose does this partial task/subtask serve?
- Is this partial task actually necessary? Can it be dispensed with?
- Does it make sense to repeat this partial task in all the procedures involved?
- When and how often do specific partial tasks have to be performed?

The task analysis should reveal the following information:
- all the basic tasks involved,
- the necessary sequence of these basic tasks in cases where this is not arbitrary,
- the identification of those basic tasks that can be performed in parallel.

Fig. 2 shows the principle of a task analysis.

Task analysis is followed by task synthesis to create groups of tasks from the analysed task components. These groups of tasks can then be allocated to specific persons or departments.

The following criteria can be used to integrate a series of tasks into a job:
- Tasks involving similar actions are brought together.
- Tasks performed on similar products or groups of products are brought together.
- Jobs are created by bringing together various planning or control tasks.
- Tasks are brought together with the aim of optimizing the use of tangible assets (e.g. an automatic production plant).
- Tailoring jobs (e.g. management jobs or jobs for handicapped persons) to fit specific employees.
- Tasks are brought together on a regional basis.

In contrast to the procedures discussed above, which see the individual worker as an element in the system or as a production factor, *ergonomic and industrial psychological* analysis procedures concentrate on the aptitude, the personal requirements and the attitude/behaviour of the individual worker. When the aim of the analysis is to investigate the tasks involved in a job, it is not always possible to separate the results of the analysis from the individual worker whose specific aptitudes inevitably influence the nature of the task and its performance. This point will be reverted to and discussed in greater detail in the next section dealing with the stress–strain concept (cf. Fig. 3).

The psychological analysis procedures address themselves to four possible definitions of the task in terms of its interaction with the individual worker (Hackman, 1970; Wheaton, 1968; Graf Hoyos, 1974; Frieling, 1975; Graf Hoyos and Frieling, 1977, etc.):

1. *Task as defined by the type of actions performed* (*Fine*, 1967; *Rabideau*, 1964). This is a purely descriptive procedure under which the visible and recordable actions of the worker are noted under the relevant items.

The fact that it is difficult to automate an operator's cognitive processes has placed renewed emphasis on the importance of the human component in advanced manufacturing systems. Whereas traditional task analysis procedures have made significant contributions towards improving productivity in cases where the major elements of the task are observable, their usefulness is limited to manual procedures and they are considerably less effective in the analysis of cognitive activities. In recent years, a start has been made on the development of methods of analysis for cognitive tasks using a combination of procedures from various disciplines (Koubek et al., 1994). Other new methods of analysis for cognitive tasks include Baber (1994), van der Schaaf (1993), Roth (1992), Leplat (1990), and Rasmussen (1990).

These do not attempt to evaluate physiological or psychological processes within the individual or the skills and abilities required.

2. *Task as defined by the aptitude required* (*e.g. Theologus et al., 1970; Fleishman et al., 1970; Fleishman, 1975*). Tasks are defined and rated by their aptitude requirements, i.e. personal characteristics, abilities, skills and knowledge. Expert ratings are used to specify the aptitude required and factor analysis is extremely useful here for summarizing the aptitude requirements. It is not possible to use this type of procedure for evaluating the actions of workers whilst actually performing the job.

3. *Task as defined by the behaviour required from the worker* (Miller, 1971). This approach analyses tasks primarily by their individual information processing functions. A total of 24 information processing functions is classified. This classification takes account of the fact that informative-intellectual work content is tending to replace energetic-effective work content in the modern world. These 24 functions are derived from the six determinants of the task: sequence, activity, information processing, control, execution status and tools. Shepherd (1993) has developed a method of dealing with the problem of specifying information requirements within a processing plant. This method translates task analysis into a series of standard task elements from which standard sets of information, called "sub-goal templates", can be derived.

4. *Task defined as a complex of stimuli* (Hackman, 1969, 1970). The task is viewed in isolation from the worker and the complex of stimuli associated with it is analysed to obtain data on the information required for the performance of the task. This makes it possible for tasks to be assigned to individual workers or groups either by an external person or by the individual or the group him/itself. This approach defines a task as a complex of stimuli and a set of instructions specifying how to respond to these stimuli. These instructions specify the operations to be performed by the worker(s) in response to the stimuli and/or the objectives to be achieved (Hackman, 1970, p. 210 in the translation by Graf Hoyos, 1974).

Individual elements of these four task definitions are often extracted and combined for the purpose of performing specific types of task analysis. It is, for example, possible to use a combination of technological/technical, functional (Kirchner and Rohmert, 1973) and information processing (Miller, 1971) approaches when evaluating a work task. In this case, the work task is seen as a phenomenon that is scientifically explicable against the background of the work system (i.e. work process, work object, equipment and materials) and the requirements needed from the worker (personal characteristics, abilities and skills).

When recording the exact job requirements, it is important to identify not only the relationships between the various activities within the company but also the general positioning of the job within the overall work process. Kirchner and Rohmert (1972) suggest that this should be done stepwise as follows:
- general definition of the job,
- definition of the basic task involved in the job (action, work object),
- definition of the partial tasks involved in the job,
- definition of the individual functions involved in the job (e.g. insertion, assembly),
- definition of the special demands imposed by the individual functions,
- definition of the special demands imposed by the overall job structure and its position in the work process.

The following questions need to be examined in detail:
- What aptitude does the worker performing the job possess? What status does he/she have within the corporate hierarchy?
- What are the minimum qualifications and experience required from the workers? Does the worker possess the aptitude for the task? The following are examples of demands arising:
  - nature of action (dynamic muscular work)
  - part of body used (hand-arm region)
  - dimensions of action (turning)
  - accuracy of action (adjustment)
  - speed of action
  - resistance occurring (reactions of work object)
  - disturbing environmental factors.
- Are the workers capable of coping with present and future demands?
- What additional qualifications are needed (further training)?
- Does the company have a staff development program?

One particularly important feature of this approach is that the model of a work system can be used to define the job in terms of its impact on the worker instead of treating the worker merely as one of the elements of the work system.

3. Task analysis and strain

If task analysis is seen as an analysis of stress determinants, it can be assumed that it will be

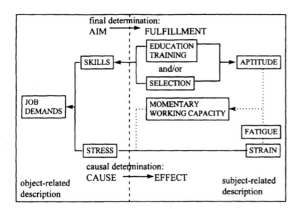

Fig. 3. Stress–strain concept (cf. Rohmert et al., 1975b).

possible to make a quantitative evaluation of stress factors (generally rated on an ordinal scale) by duration, intensity, sequence, overlap and time of occurrence within a work shift (cf. Laurig, 1977). If it is also claimed that the analysis procedure will provide information on the strains resulting from the stress patterns, this implies that the procedure is capable of:
- producing repeatable qualitative and quantitative analyses of the strains arising (with the exception of emotional strains) (cf. Luczak, 1975),
- allocating psychological or physical strain to selected items qualitatively,
- rating specific items for the psychological or physical strains produced by them,
- helping to make quantitative evaluations of strains (ratings on a set scale) based on the results of physical or psychological examinations of the workers involved.

This type of strain-oriented job analysis naturally goes beyond the "objective section" of the stress–strain concept of human work (cf. Fig. 3; see, for example, Rohmert et al., 1975b; Luczak, 1975), in which the work task is part of the "objective section". In this theoretical model the objective section consists of the work task itself and the environmental conditions under which the task has to be performed.

The combination of these elements results in job-specific and situation-specific demands on the worker, which in turn determine the degrees of energetic-effective intensity and informative-intellectual difficulty involved in the work.

The definition of the stress–strain relationships applying in this objective section was used in the development of the AET job analysis procedure (Arbeitswissenschaftliche Erhebungsverfahren zur Tätigkeitsanalyse, Landau et al., 1975; Rohmert et al., 1975a; Landau, 1978a,b; Rohmert and Landau, 1979).

Landau et al. (1990) have examined the feasibility of using various task analysis procedures to identify an industrial micro-epidemiology. This comparison shows that a task analysis procedure aimed at the development of preventive measures must examine both the ergonomic and the psychological aspects of the tasks. As a worker's behaviour must be viewed in relation to factors like subjective job perception, redefinition and regulation to external fluences, objective stress analyses must be complemented by subjective assessments of the working conditions. Subjective perception of work tasks needs to be investigated regularly by industrial medical staff during their routine reviews of preventive measures.

For the purposes of forecasting work-related health damage on the basis of data obtained from task analyses, the stress–strain concept (cf. Rohmert et al., 1975b) can be expanded to cover general lifestyle, e.g. diet, habits, drugs, etc., and also the worker's social situation and general attitudes, i.e. his redefinition of his lifestyle and his social situation. However, the high degree of mobility between occupations and workplaces during the course of a person's working life will make it necessary to dynamize the concept and to develop a system which determines stresses at both the existing and previous workplaces.

In summary, a model must be so designed that it can demonstrate the hypothetical association between work tasks and disease. It must take account of:
- the current situation in the person's working and private life,
- the external factors influencing the worker both as a person and an employee (objective influences). These include the stresses arising from the work tasks and from the working environment,

and also leisure stresses from outside the working environment.
- the worker's redefinition of the job situation. This will be strongly influenced by lifestyle (e.g. diet and habits) and the social situation (attainments, aptitudes, skills and ability).
- the dynamization of the model – it must take into account previous situations applying in the person's working and private life.

The use of this multi-causal model to prove the hypothetical relationship between those determinants of the work tasks causing stresses, strains and disease and also to predict work-related diseases is discussed in detail by Brauchler (cf. Brauchler, 1992; Brauchler and Landau, 1989, 1992, 1991; Landau et al., 1990; Brauchler et al., 1990). Experiences in creating a knowledge base on questions relating to micro-epidemiology are described by Landau and Brauchler (1994).

Job analysis procedures using broad-spectrum variables (Frieling, 1975; McCormick, 1976) which enable simultaneous evaluation of the stress potentials inherent in tasks, demands, environmental conditions and skills required – like the AET – constitute an adequate theoretical basis for ergonomic/human factors task analysis.

Hacker (1973) defines job demands as the general personal performance requirements imposed on a worker for the correct execution of the tasks involved in the job. These job demands are supplemented by environmental demands to produce the total demands imposed on the worker. If any demand produced by work is classified as stress, as Kirchner (1986) stated, work itself must be stress (cf. Greiner and Leitner, 1989) "...because work not demanding human effort is not imaginable. If two, until now, little differentiated aspects of work are considered separately, the implied equation of work and stress can be avoided:
- job requirements, for which a person uses work capacity in order to attain a certain goal (for the area of mental requirements the "Instruments for the Assessment of Regulation Requirements in Industrial Work" (VERA) can be applied, see Volpert et al., 1983);
- job stress (regulation hindrances) which increases the difficulty of reaching the defined goal causing an unnecessary additional expenditure of energy" (cf. Greiner and Leitner, 1989).

However, task analysis procedures cannot and should not be used to evaluate interindividual variances in job performance. Hackman (1970) uses "redefinition" to enlarge the objective aspect of the work task. The worker must first understand the work task and be willing to accept it and capable of coping with the demands imposed by it. The objective aspect of the work task can then be redefined to make allowance for the worker's experience and intellectual powers, the term "experience" including not only the person's experience in performing that particular task, but also any previous experience (Graf Hoyos, 1974).

Subjective task analysis procedures examine the differences between individual workers, as revealed by their subjective perception of the job and their job performance. These subjective procedures are only capable of analysing limited parts of the psychological processes involved, because they depend heavily on the individual's ability to perceive and describe work processes which varies widely from worker to worker (Gablenz-Kolakovic et al., 1981).

It should be noted that task analysis procedures using broad-spectrum variables and claiming to be universally applicable are totally unsuitable for subjective job analysis and are not intended for this purpose.

4. Task analysis methods and procedures

4.1. Review of existing procedures

It is necessary to distinguish between task analysis procedures or tools on the one hand and the methods or techniques used to implement these procedures on the other.

Whereas the term "procedure" covers the whole theoretical basis, including the procedure's objectives, possible applications and the statistical tests used, the term "method" or "technique" refers solely to the way in which it is implemented. Thus, there will usually be several equally valid methods of implementation, for example, interviews, observation interviews, self- recording, etc.

Predominantly demand-oriented task analysis procedures using a work study approach are regarded as irrelevant in the present context and are therefore not included. The reader can refer to the literature (DIN 33 407; Nutzhorn, 1964; REFA, 1977 etc.).

Also excluded are those subjective job analysis procedures used for analyses of job satisfaction and appraisals of a superior's management skills. Here again, reference can be made to the available literature (Benninghaus, 1981; Bowers and Franklin, 1977; Bruggemann, 1976; Celluci and de Vries,

Table 1
Selected task analysis procedures (Landau et al., 1990)

Name	Title	Author/year	Aim of procedure
PAQ	Postition Analysis Questionnaire	McCormick et al. (1969)	Broad spectrum procedure for the classification of different jobs.
FAA	Job Analysis Questionnaire	Frieling and Hoyos (1978)	Classification is made by comparison of items. Used mainly for obtaining similarity data for different jobs based on analysis of fixed items. Primarily action-oriented procedures.
	Standard Job Card	Arendt and Uhlemann (1974)	Primary data source for all job-related information. Planning aid for action to improve working conditions and for organizational purposes.
AET	Ergonomic Job Analysis Procedure	Landau et al. (1975)	Broadspectrum job and stress analysis procedure for identifying weak points. Used in job design, structuring and evaluation for identification of industrial medical risks and for estimating the consequences of automation.
EBA	Ergonomic Evaluation of Work Systems	Schmidtke (1975)	Description of the technical components and environmental factors affecting work systems with subsequent evaluation of their functional and beneficial values. Assessments of the associated health risk. The work systems' operational and functional safety and the feasibility of the planned functional and performance targets.
	Profiles de Postes	Services des conditions de travail de la Région Nationale des Usines Renault (1976)	Identification of design weak points mainly in factor jobs, Numerical evaluation of weak points.
OWAS	Ovako Working Posture Analysis System	Heinsalmi (1978)	Improvement of working conditions, tools and environment by identifying and eliminating injurious postures.
	Job Hygiene with Professiography Analysis Procedure	Häublein et al. (1979)	Job analysis from the health aspect for evaluation of stresses and identification of causal relationships between the characteristic elements of an occupation and health.
	Procedure for defining relationships between jobs Evaluation of Work Systems	Janes (1980)	For obtaining information on actions and elements of actions mediating process-specific and non-process-specific qualifications. Identification of relationships between different jobs and job flexibility.

Table 1 Continued

Name	Title	Author/year	Aim of procedure
	Analysis of Disease Types	Georg et al. (1981)	Link-up of occupational and health insurance data. Survey of work sphere to analyse working conditions.
BEAT	Industrial sociological questionnaire for the identification of job characteristics	Volpert et al. (1981)	Comparison of various jobs and before/after comparison to evaluate personality gains and to identify possible approaches to job redesign.
TBS	Evaluation and design of progressive work content, Job Evaluation System	Baars et al. (1981); Hacker et al. (1983)	Analysis and evaluation of personality gain resulting from job-related tasks. Positioning of the task in the sharing of functions between man and machine and job sharing. Job analysis relating to cognitive and manual activities. Identification of learning requirements.
ASA	Work System Analysis	Darmstädter and Nohl (1982)	Survey of actual working conditions and on which to base a workers' health care program. Main aims: job design, job sequence design, job environment design.
FSD	Questionnaire on Safety Diagnosis	Bernhardt and Hoyos (1987)	Identification of elements and constellations within a work system relating specifically to safety.
SQD	Safety Diagnosis Questionnaire	Ruppert and Hoyos (1989)	Development of psychological aids to motivation and training for safety work and safe job design.
VERA	Procedure for Identifying Regulation Requirements in Jobs	Volpert et al. (1983)	Determination of thought and planning processes (regulation requirements for the performance of task) to identify jobs in need of redesign and to evaluate personally gain obtained from specific jobs.
	Job Load and Hazard Analysis	Mattila (1985)	Identification of safety and health risks at work.
TAI I	Job Analysis Inventory	Frieling et al. (1984)	Identification of stressors resulting from technical and organizational conditions.
TAI II	Job Analysis Invetory	Facaoaru and Frieling (1985, 1986)	Analysis of informatory content of jobs, determination of duration and density of informatory stresses, i.e. qualitative and quantative analysis. Operationalization of intellectual freedom and freedom in decision-taking. Characterization of organization and technical conditions applying in jobs.
	Method for Monitoring Psychological Stress at Work	Elo (1986, 1989)	Evaluation of psychological stresses by observation. Used in the analysis of the most frequently occurring "underload" and "overload" situations in industrial jobs.
ISTA	Instrument for Analysing Stress Foci	Dunckel and Semmer (1987)	Identification of stresses foci in industrial jobs.

Table 1 Continued

Name	Title	Author/year	Aim of procedure
RHIA	Regulation Hindrances in Jobs	Greiner et al. (1987)	Illustration of action structures, identification of hindrances to regulation, i.e. external influences hindering attainment of work target and consequently demanding additional effort. Evaluation of technical and organizational changes and concepts.
ATAA	Analysis of Job Structures and Design in Automated Processes	Wächter et al. (1989)	Check-list for job design, preliminary analyses in connection with investment decisions.

1978; Fischer and Lück, 1972; Fittkau-Garthe and Fittkau, 1971; Hackman and Oldham, 1974; IG-Metall, 1979; Kern and Schumann, 1970; Lynch, 1974; Martin et al., 1980; Müller-Böling, 1978; Neuberger and Allerbeck, 1978; Plath and Richter, 1976; Sims et al., 1976; Smith et al., 1969; Staehle et al., 1981; Turner and Lawrence, 1965; Ulich, 1981; Udris, 1977, 1981; White, 1975, etc.).

Broad surveys of job and task analysis procedures have been published by Landau and Rohmert (1989), Frei (1981), Frieling (1975), Hennecke (1976), Graf Hoyos and Frieling (1977), Jones et al. (1953), Karg and Staehle (1982), Kenton (1979), Neunert (1979), Prien and Ronan (1971), Rohmert et al. (1975a) and other authors.

For a review of the various techniques that can be used to implement a job analysis procedure, refer to Part I of this guide (cf. Landau et al., this volume).

Table 1 lists task analysis procedures and documents their objectives (cf. Landau et al., 1990). No attempt has been made to evaluate the extent to which the procedures meet qualitative job analysis criteria, e.g. theoretical validation, economy, quantifiability, standardizability, teachability and reliability. Although the survey makes no claim to completeness, care has been taken to include as many variants as possible of task-oriented job analysis. The procedures are listed in the order of the year of their publication.

Individual procedures have been selected from the list given in Table 1. These are then described and in some cases compared. It would be impossible for space reasons to describe all the procedures listed in Table 1 in this publication and the reader is therefore requested to refer in the relevant cases to the original literature quoted in the bibliography.

4.2. *Ergonomic task analysis based on stimulus-reaction models*

The procedures designed by McCormick et al. (1969), Landau et al. (1975) and Frieling and Hoyos (1978) all comply to a large extent with the criteria formulated by Frieling and Hoyos, which were explained at the beginning of the article, whereas the procedures listed above are based on stimulus-reaction models derived mainly from neo-behaviouristic thinking (Hull, 1952; Spence, 1948; Skinner, 1958).

The Ergonomic Job Analysis Procedure (AET) (Landau et al., 1975; Rohmert et al., 1975a; Landau, 1978a,b; Rohmert and Landau, 1979) will be examined here in greater detail as an example of this group of analysis procedures.

The accents of the theoretical concepts (cf. Section 3) are highlighted by the basic structure of the AET, which shows three parts:

A. Analysis of the man at work system

B. Task analysis
C. Demand analysis.

Observation interviews performed by trained analysts are used to collect the data required for the various AET items. As the items contained in the demand analysis might present difficulties with insufficient knowledge of ergonomics, "activity scales" are given as additional aids to classification. These activity scales are based on data obtained in previous investigations and list an ascending scale for typical activities. It can be assumed that there will be an approximate correlation between ascending degrees of stress and the intensity of the resulting strain (see Fig. 4).

AET data can be used to obtain answers to fundamental questions arising in nearly all potential fields of application:
1. The AET codings can be used to characterize a work system.
2. The AET codings can be used to describe work systems or groups of work systems.
3. The characteristics revealed by all or most of the AET codings can be used to classify a work system.
4. Conversely, work systems can be grouped together according to common AET characteristics.

Examples of results obtained in various studies are given below. These are intended to give the practitioner guidance on the evaluation and interpretation of AET data, especially for the types of evaluation listet under items 2 and 3 above.

The type of evaluation described under 2 above is called profile analysis. The scores obtained for the various items can be presented in the form of profiles that demonstrate graphically the extent or duration of stress occuring during performance of specific activities or groups of activities. Fig. 5 shows examples of job profiles obtained from AET analyses.

The vertical plane shows the types of demand and the horizontal scale shows the maximum AET classification in percent. The upper bar represents the female jobs, the lower bar the male jobs. The analysis shows that the most important tasks for males involve operating, controlling, supervising, planning, organizing, and analysing. The main tasks performed by females are checking, and also

---

Stress by Static Work

The term static work implies a long-term (>4 sec) muscular effort which does not result in a movement of the body (contrary to dynamic work). Static work is therefore not measurable in the mechanical sense.
During static work, a muscular effort can take place not only due to the exertion of an external force but also because of the effort required to bear the weight of the body extremities.

205 D   Finger-Hand-Forearm
        Characteristic: Muscular effort without support of body weight
        Example: Seizing and maintaining objects of work, operating a keyboard

D
0
↓   Filing
        Works management
        Assembling
        Secretarial work using mechanical typewriter

Fig. 4. Example of an AET Item (Landau and Rohmert, 1979).

---

a variety of general, people-oriented service tasks. The tasks are divided into the stereotyped patterns of "typically male" and "typically female". Males perform more (complex) operating, controlling, and assembly tasks where they are required to plan and organize their own work, while females are employed in industry mainly for simple checking activities and also for people-oriented services approximating to their role as mother or housewife.

The jobs occupied by males were exposed to far higher levels of physical or chemical stress from the working environment. This applies both to factors like illumination, climatic conditions,

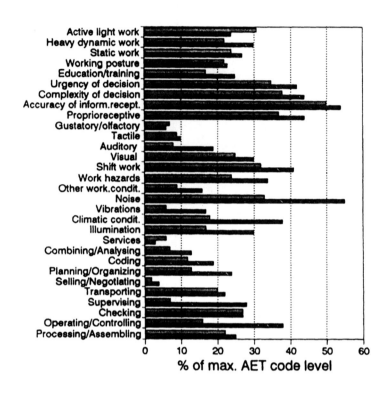

Fig. 5. Analysis of job tasks and demands for 2838 males and 866 females (basis: AET items 1–216) (Landau and Rohmert, 1992).

vibration, and noise and also to other environmental influences like noxious materials. The work hazards, including the frequency or probability of a work accident or an occupational disease, are rated higher for the male jobs.

The male jobs also show a higher level of demands in both the organization of working time (shift work), the sequence of operations, and overall planning. As shift work by females is severely restricted by law, it is largely a male preserve in industry, and the demands in this respect are therefore much higher for men.

Closer investigation shows that information reception and information processing place substantially higher degrees of certain types of demand on the male than on the female jobs. This led to a higher classification against the AET criteria for demands involving the reception of visual information, of auditory information, and proprioception. Similar levels for both male and female jobs were registered only for information reception via the senses of smell, taste, touch and temperature sensitivity of the skin. There are only very slight differences between the sexes in the demands for

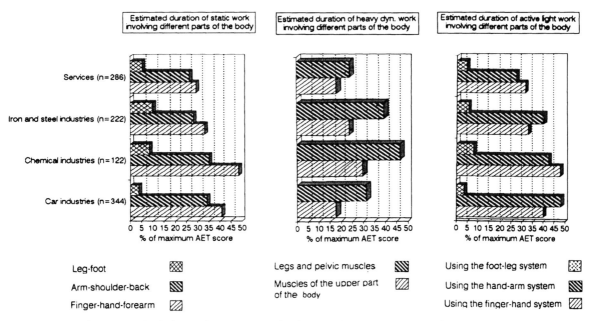

Fig. 6. Estimated duration of physical work forms. An example of task analysis results obtained with AET (Landau and Rohmert, 1992)

accuracy of information reception. The male jobs involve tasks of greater complexity and, in some cases, greater critical stress. The level of knowledge required in male jobs is rated higher than in female jobs.

The proportion of physical work demands is very similar for both sexes. They are identical for static handling work and only insignificantly higher for males in the case of static holding work. In male jobs, longer periods of the shift are devoted to heavy dynamic work and, in female jobs, to active light work. This corresponds to the role expectations in the division of work between men and women in industry, i.e. heavy dynamic work for men and active light work involving monotonous procedures for women.

Cluster analysis is used to depict similarities in body postures revealed by the AET items. This enables the identification of similarities between subgroups in different samples (cf. the type of evaluation described under 3 above). These subgroups may, for example, be linked by the type of tasks predominating in the jobs, the sector of industry in which the jobs are performed or the sex of the workers normally performing the relevant jobs. For further information on the use of cluster analysis (classification, numerical taxonomy) as a method of breaking down a large number of prima facie unrelated objects (in this case jobs) into smaller, homogeneous and practically relevant categories or groups exhibiting similarities or factual relationships, refer to Landau (1978a,b). This publication describes the application of cluster analytical models, techniques and algorithms to ergonomic data. The "dissection" of the results of a cluster analysis makes it possible to identify the job clusters at a given hierarchical level. In cases where jobs are grouped by sectors of industry, the mean duration of the observed postures per shift is shown graphically. As this involves the calculation of mean arithmetical values from data rated on an ordinal scale, it is advisable to interpret the results with caution (Fig. 6).

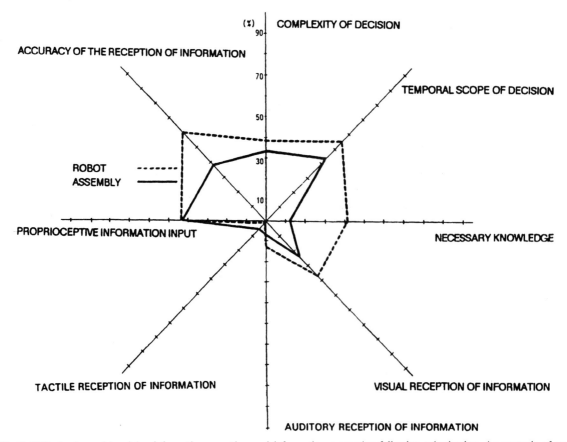

Fig. 7. Shifts in demand involving information reception and information processing following robotization. An example of task analysis results obtained with AET (Landau and Rohmert, 1992). n1, solid line; 67 traditional assembly jobs. n2, dashed line; 8 jobs in mechanized assembly.

Comparing different industries, Landau and Rohmert (1992) estimated for example, the percentage of shift time involving static work (Fig. 6). The jobs in the iron and steel industry received the highest rating for static work, followed by the chemical industry, the automotive industry, and the services sector in that order. Static work mainly involves the use of the finger/hand/forearm region or the arm/shoulder/back region. Static work using the leg or foot region is of only minor importance in all the industries covered by this study.

The chemical industry has the highest percentages of heavy dynamic work and active light work. This is followed by the iron and steel industry in the case of heavy dynamic work, and by the automotive industry in the case of active light work. The percentages of heavy dynamic work and active light work are lowest in the services sector (Fig. 6).

Heavy dynamic work can involve either the arms and upper body muscles or the legs and pelvic muscles. Heavy dynamic stress on the legs and pelvic muscles was caused by walking, climbing, etc., in some cases with loads. Walking and climbing are still important work factors in the chemical industry, followed by the iron and steel industry, the automotive industy and the service sector in that order.

Active light work in the chemical industry is involving mainly the finger/hand system. In the automotive industry there are more gross motor activities using the hand/arm system. The foot/leg system is not used to any significant degree in any of the industries investigated (Fig. 6).

The AET is capable of analysing exceptionally high stress levels affecting specific body organs in man-at-work systems and also of quantifying tasks and demands at the workplace and, by extension, identifying shifts in demands such as:
- cessation or addition of specific types of demand
- changes in intensity of one or more types of demand
- changes in duration of specific types of demand
- changes in time spread of specific types of demand
- changes in association between different types of demand.

Using the AET data, Landau and Rohmert (1992) compared eight jobs in mechanized assembly with 67 traditional assembly jobs in the automotive industry (cf. Fig. 7). The small size of the sample populations makes it necessary to interpret the present results with caution.

Visual information reception increases from 24% to 37% of the maximum AET score. There is an even higher increase, from 39% to 60% of the maimum score, in the requirement for accuracy of information reception. Proprioceptive information reception remains almost unchanged, while demands involving information reception by touch and via the thermosensors of the skin have been eliminated, because gripping actions are now carried out by the mechanized assembly. Demands involving auditory information reception, especially in cases where problems are starting to develop, increase because of the high noise level in the work environment.

Demands relating to information processing produce an increase in qualification requirements. The amount of knowledge required increases from 12% to 41% of the maximum AET score. Decision complexity shows a small increase of 5%. However, the urgency of decisions rises from 43% to 55% because of the highly integrated assembly processes in which defects in the equipment have to be rectifed without delay.

In conclusion, it can be stated that technological change per se does not necessarily bring an improvement in working conditions or a more humane work pattern. Industrial robots actually have both positive and negative effects for the workers. Their advantage is merely that they offer the option of a new production process better adapted to the needs of the workers. Whether this option and the option of improving the qualifications of the workers is exercised, depends very much on the attitudes of management and engineering departments.

4.3. Psychological task analysis based on action theory ideas

The procedures developed by Hacker (Baars et al., 1981; Hacker et al., 1983), Volpert et al. (1983) and Greiner et al. (1987) are based solely on action theory ideas. Psychological analysis procedures are generally designed to analyse the mental processes controlling work activities.

Typical examples of this kind of procedures are the "TBS – Job Evaluation System" (Hacker et al., 1983) and the "VERA – Procedure for Identifying Regulation Requirements in Jobs" (Volpert et al., 1983). These procedures provide information on the extent to which a given job and the demands made by it can be designed to enable the achievement of work-related goals and strategies that is taylor-made for the person actually performing that job. Psychological procedures are suitable for analysing the degree of cognitive control in a task but not for analysing physical job components.

These procedures are designed around the technique of the observation interview in which answers are obtained to extremely detailed questions (items, characteristics) by observations at the workplace and by questioning the worker, his immediate superior and his colleagues. They are thus standardizable and can be passed on to a large group of users by means of simple training courses.

Although these job analysis procedures involve a relatively high degree of technical input, they can nevertheless be rated as economical because the

B 10	Occurence of planning phases

Planning phases are defined as phases during which the worker has to think before performing the task or a part of it, e.g. about the sequence of actions, materials required etc.

These planning phases may often consist of only short pauses but, in some cases, test runs etc. may be necessary.

Routine preparations following a prescribed sequence or defined in the relevant working instructions are not planning phases.

Do planning phases occur in the task under observation?

(1) Do not occur.

(2) Occur in exceptional cases.

(3) Occur not only in exceptional cases.

Fig. 8. Example of a VERA item (Volpert et al., 1983).

Job planning by the worker himself
After initial training:

(0) Planning is not possible, visualization of future action not necessary.
(1) Planning is not possible, visualization of future action is necessary.
(2) Although planning is possible, the task can be correctly performed without planning.
(3) Planning of timing of actions involved in parts of the task (point in time, sequence) is essential for correct performance of the task. The work content is standardized.
(4) Planning of timing and content of some of the actions involved is essential for correct performance of the task.
(5) Planning of timing and content of subtasks is essential for correct performance of the whole task.
(6) Planning of timing and content of a complex of tasks is essential for correct performance of the task.
(7) Preparation of plans (strategies) that can be adapted to changing circumstances is essential at task or subtask level for correct performance of the task.

Fig. 9. Example of a TBS item (Hacker et al., 1983).

database obtained can be used for a wide variety of evaluations (cf. Landau et al., Part I, this issue).

VERA (Volpert et al., 1983) identifies the needs for regulation in specific work tasks. VERA investigates the worker's actual tasks and not the worker himself. The job, as defined by the working conditions within the company, is the object of the analysis. Fig. 8 shows an example of a VERA item.

The results obtainable with VERA are in the form of verbal definitions of levels of regulation. These show the action required to maintain or improve the existing degree of regulation. There are five levels of regulation:

Level 5: Introducing new work processes.
Level 4: Coordinating several work processes.
Level 3: Planning of subsidiary objectives.
Level 2: Action planning.
Level 1: Sensorimotor regulation.

VERA microanalysis (Moldaschl and Weber, 1986) subdivides a complete work task into a series of subtasks, each of which is analysed separately (Weber and Oesterreich, 1989). The assessment of regulation requirements in individual tasks or subtasks is a useful extension of the conventional VERA procedure. If the VERA microanalysis is carried out in more detail, it can be used to specify the documents, tools, materials and machines needed for the performance of a given job.

The TBS also uses observation interviews to make an objective job analysis aimed at designing jobs with a more progressive, personality-developing work content. The TBS can identify work contents providing the necessary conditions for or creating obstacles to the maintenance and consolidation of the worker's mental health. Fig. 9 shows an example of a TBS item.

TBS data can be presented either in the form of separate profiles for each of the characteristics investigated or, alternatively, as a comparison of different jobs.

A comparison of the various elements of the TBS (Baars et al., 1981), the AET (Landau and Rohmert, 1979) and VERA (Volpert et al., 1983) has been published by Brauchler (1992).

4.4. Analysis of safety and postural aspects

Sections 4.2 and 4.3 describe the main task analysis procedures. We will now take a look at examples of two types of work analysis procedures that

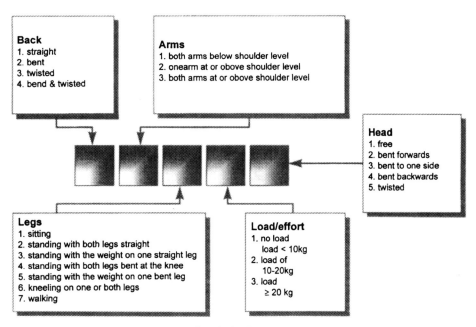

Fig. 10. OWAS-code.

are frequently used in conjunction with task analysis procedures.

Although the procedures described above and also the TAI (Frieling et al., 1984) investigate safety or exposure to danger as one element of job tasks, there is one procedure specially designed for safety analysis.

The Safety Diagnosis Questionnaire (SDQ; Hoyos and Bernhardt, 1987; Ruppert and Hoyos, 1989) contains the groupings listed below and the list gives examples of the items investigated (cf. Ruppert and Hoyos, 1989):
1. Work organization (shift work)
2. Hazards and danger sources (Moving parts, electricity)
3. Perception and operation of hazard signals (importance)
4. Prediction and rating of hazards (importance)
5. Planning and prevention (attendance at first aid courses)
6. Actions (mixing of hazardous liquids)
7. Cooperation and communication (unusual machine noise).

The SDQ contains 276 items and enables a safety expert to examine the work situation and determine whether the worker is capable of complying with norms of safe behaviour both routinely and under unfavourable conditions.

Finally, we wish to mention a procedure that is very suitable for use as a complement to ergonomical task analyses and especially for identifying specific job redesign measures.

The Working Postures Analysing System (OWAS; Karhu et al., 1977; Heinsalmi, 1978; Stoffert, 1985) is used for mapping and classifying working postures, with the aim of developing working methods that are less damaging to health. The OWAS includes 84 so-called typical working postures, 36 additional postures and 5 positions of the head. When other criteria like external load or force exerted in performance of the job are taken into account, the total number of combinations comes to 360. All postures are a combination of back, arms, legs, head and load (see Fig. 10).

The data obtained are converted into four action categories according to the estimated strain caused. One recent publication used OWAS is an investigation of aircraft turnaround (cf. Landau et al., 1995; Landau and Wendt, 1995). Although various mechanical loading and unloading aids are available, 40% of the loading and stowing work has to be performed manually. The OWAS procedure was used to collect body posture data.

In summary, it can be stated that there is no task analysis procedure satisfying all conceivable requirements. The user has to decide in each individual case whether his problems are mainly of an ergonomical or a psychological nature and whether, after the initial data has been obtained, it would be better to switch to other more appropriate analysis procedures.

5. Conclusions

Task analysis can be reduced to two fundamentally different types of procedure. The first type involves recording and classifying the main elements of the work system, the tasks involved and the demands imposed on the basis of a stimulus organism response (SOR) paradigm. The second type evaluates the subjective perception of the work by the worker himself. These two types are not mutually exclusive. On the contrary, they are complementary in that the redefinition process (the worker's subjective perception) is not directly observable, whereas the tangible elements of the tasks and the stress factors recorded by objective task analysis procedures like PAQ, AET (cf. Table 1) are directly observable. The optimal solution is to select specific elements from both approaches and to use the tools associated with each type of procedure either simultaneously or successively to analyse a given work system.

Task analysis procedures are often criticized because of their unsuitability for a wide range of applications. The problem is that the development of universally applicable task analysis procedures is a costly business. The optimal solution is to design the procedure as universal as possible within the financial limitations imposed, universality being defined in this case as suitability for the interpretation and generalization of data obtained from either different investigations, different sectors of industry or different companies. The contribution from Landau and Brauchler (Part I, this issue) deals in greater detail with the application of task analysis procedures.

A comprehensive task analysis system can provide information on the undesirable components of a given job, better appreciation of the psychological stresses arising in the job and a greater understanding of the tasks and requirements involved.

One frequent disadvantage of broad-spectrum analytical procedures is that the results obtained are at a high supersign level which makes direct identification of the area of application impossible. Another problem is that a strain-related interpretation of task analysis data quickly reveals the limitations of the observation interview. Stress-related interpretation is based on the fallacious assumption that all individuals behave alike in work situations which are classified as similar (Frei, 1981), irrespective of whether the individual is being confronted with the relevant work situation for the first time or whether he performs it routinely. Moreover, task analysis procedures which endeavour to operate on an individual level cannot strictly be regarded as "objective", because the behaviour of the individual within the work system is inevitably included in the analysis. In this case the analyst will not, or at least only partly, be able to ignore the observed behaviour of the individual and use the data to code in abstract job behaviour.

In some cases the task analysis is too complex and it is impossible to record all the items necessary for the evaluation of work tasks and stress factors. It may be necessary to use additional procedures, for example, in the form of supplements to the basic method of analysis.

The main aim of applied ergonomics and preventive medicine in industry is the early identification of work-related deterioration before serious health damage occurs, thereby minimizing the human and social costs caused by degenerative diseases and the resulting disabilities. It is necessary to collect all available data on potential health risks at the workplace and to evaluate these to produce a database that can be used to provide answers to ergonomic, technical and epidemiological questions.

Consequently, the main aim of further developments of task analysis procedures should be to maintain or even increase the number of items examined and to achieve maximum cost-effectiveness by applying the database obtained from successive investigations as a forecasting and indicative aid in the widest possible range of work situations and to achieve maximum cost-effectiveness by evaluating and interpreting the data.

References

Arendt, M., Uhlemann, K., 1974. Vorschlag zur Anwendung einer einheitlichen Arbeitsplatzkarte. In: Arbeit und Arbeitsrecht, vol. 29, Verlag Die Wirtschaft, Berlin.

Baars, A., Hacker, W., Hartmann, W., Iwanowa, A., Richter, P., Wolff, E., 1981. Psychologische Arbeitsanalysen zur Erfassung der Persönlichkeitsförderlichkeit von Arbeitsinhalten. In: Frei, F., Ulich, E. (Eds.), Beiträge zur psychologischen Arbeitsanalyse. Huber Verlag, Bern.

Baber, C., 1994. Task analysis for error identification: a methodology for designing error tolerant consumer products. Ergonomics 37 (11), 1923–1941.

Benninghaus, 1981. Zwischenbericht über das Forschungsprojekt "Merkmale und Auswirkungen beruflicher Tätigkeit 1980". Technische Universität Berlin, Institut für Soziologie, unpublished.

Bernhardt, U., Graf Hoyos, C., 1987. Fragebogen zur Sicherheits-diagnose (FDS) - Konzeption, Erprobung und Anwendungs-möglichkeiten eines verhaltensorientierten Instruments zur Sicherheitsanalyse von Arbeitssystemen. In: Sonntag, Kh. (Ed.), Arbeitsanalyse und Techniventwicklung. Wirtschaftsverlag Bachem, Köln.

Bowers, D.G., Franklin, J.L., 1977. Data Based Organizational Change (Survey-Guided Development) Michigan, 1975 La Lolla, CA.

Brauchler, R., 1992. Schädigungsanalyse als Basis eines epidemiologischen Frühwarn-systems. Dissertation, Universität Hohenheim, 1991. Verlag Dr. Kovàc, Hamburg.

Brauchler, R., Brauchler, W., Landau, K., 1990. ESES – A methodological approach to predict job-related diseases. In: Noro, K., Brown, O. Jr. (Eds.), Human Factors in Organizational Design and Management – III. Elsevier, North-Holland, pp. S203–206.

Brauchler, R., Landau, K., 1989. Epidemiological analysis of workload data using ergonomic data bases. In: Recent Developements in Job Analysis, Proceedings of an International Symposium on Job Analysis. University of Hohenheim, 14–15 March 1989. Taylor and Francis, London, pp. 143–154.

Brauchler, R., Landau, K., 1991. Setting up an epidemiological early-warning system – first steps of implementation. In: Quéinnec, Y., Daniellou, F. (Eds.), Designing for Everyone. Taylor and Francis, London, pp. 640–642.

Brauchler, R., Landau, K., 1992. Implementation of an epidemiological early-warning system by using rule induction algorithms. In: Karwowski, W., Mattila, M. (Eds.), Proceedings of the International Conference on "Computer Aided Ergonomics and Safety", Tampere, Finnland, pp. 249–254.

Bruggemann, 1976. Zur empirischen Untersuchung verschiedener Formen der Arbeitszufriedenheit. Zeitschrift für Arbeitswissenschaft, 30(2): 71 ff.

Celluci, A.J., Vries, de, D.L., 1978. Measuring managerial satisfaction: a manual for the MJS. Technical report No. 11 des Center for Creative Leadership, Greensboro, NC.

Darmstädter, P., Nohl, J., 1982. Erweiterung der DIN 33407 (Arbeitsanalyse, Merkmale) z.B. mit einer Arbeitsplatzkarte als Element einer Komplexanalyse – Möglichkeit des Einsatzes als Arbeitnehmerorientierte Analyse. Abschlußbericht, Uni Hannover

Dunckel, H., Semmer, N., 1987. Streßbezogene Arbeitsanalyse: Ein Instrument zur Abschätzung von Belastungsschwerpunkten in Industriebetrieben. In: Sonntag, K.-H. (Ed.), Arbeitsanalyse und Technikentwicklung. Wirtschaftsverlag Bachem, Köln.

DIN 33400. Gestalten von Arbeitssystemen nach arbeitswissenschaftlichen Erkenntnissen. Beuth Vertrieb GmbH, Berlin/Köln.

DIN 33407. Arbeitsanalyse. Merkmale. Dezember 1980.

Elo, A.-L., 1986. Assessment of Psychic Stress Factors at Work. Institute of Occupational Health, Helsinki.

Elo, A.-L., 1989. Method for monitoring psychic stress factors by occupational health personnel. In: Landau, K., Rohmert, W. (Eds.), Recent Developments in Job Analysis. Taylor and Francis, London, pp. 81–90.

Facaoaru, C., Frieling, E., 1985. Verfahren zur Ermittlung informatorischer Belastungen Teil 1: Theoretische und konzeptionelle Grundlagen. ZfA 39 (11), 2.

Facaoaru, C., Frieling, E., 1986. Verfahren zur Ermittlung informatorischer Belastungen Teil 2: Aufbau und Darstellung eines Verfahrensentwurfs. ZfA 40, 2.

Fine, S., 1967. Matching job requirements and worker qualifications. In: Fleishman, E. (Ed.), Studies in Personnel and Industrial Psychology. The Dorsey Press, Homewood (III.).

Fischer, L., Lück, H.F., 1972. Entwicklung einer Skala zur Messung von Arbeitszufriedenheit (SAZ). Psychologie und Praxis, 16: 64ff.

Fittkau-Garthe, H., Fittkau, B., 1971. Fragebogen zur Vorgesetzten-Verhaltensbeschreibung (F-V-V-B), Verlag für Psychologie, Göttingen.

Fleishman, E.A., 1975. Taxonomic problems in human performance research. In: Singleton, W.T., Spurgeon, P. (Eds.), Measurement of Human Resources. Taylor and Francis Ltd., London, pp. 49-72.

Fleishman, E.A., Teichner, W.H., Stephenson, R.W., 1970. Development of a Taxonomy of Human Performance: a review of the second years progress. Technical Report No. 2, American Institute for Research, Washington, D.C.

Frei, F., 1981. Psychologische Arbeitsanalyse - Eine Einführung zum Thema. In: Frei, F., Ulrich, E. (Eds.), Beiträge zur psychologischen Arbeitsanalyse. Huber Verlag, Bern.

Frieling, E., 1975. Psychologische Arbeitsanakyse. Verlag W. Kohlhammer, Stuttgart.

Frieling, E., 1977. Occupational analysis; some details of an illustrative German project. International Review Applied Psychology 26 (2), 77-85.

Frieling, E., Hoyos, C.G., 1978. Fragebogen zur Arbeitsanalyse - FAA. Hans Huber Verlag, Bern.

Frieling, E., Kannheiser, W., Facaoaru, C., Wöcherl, H., Därholt, E., 1984. Entwurf eines theoriegeleiteten standardisierten verhaltenswissenschaftlichen Verfahrens zur Tätigkeitsanalyse. In: Endbericht des HDA Projektes 01 HA 029. ZA-TAP-0015. Münche Verlag, Oldenburg.

Gablenz-Kolakovic, S., Krogoll, T., Oesterreich, R., Volpert, W., 1981. Subjektive oder objektive Arbeitsanalyse? Zeitschrift für Arbeitswissenschaft 35, 217-220.

Georg, A., Stuppardt, R., Zoike, E., 1981. Krankheit und Arbeitsbedingte Belastungen (Band 1), Voraussetzungen, Schwerpunkte und erste Ergebnisse. Bundesverband der Betriebskrankenkassen, Essen.

Greiner, B., Leitner, K., 1989. Assessment of job stress - the RHIA.Instrument. In: Recent Developements in Job Analysis, Proceedings of an International Symposium on Job Analysis. University of Hohenheim, 14-15 March, 1989. Taylor and Francis, London, pp. 53-66.

Greiner, B., Leitner, K., Weber, W.-G., Hennes, K., Volpert, W., 1987. RHIA - Ein Verfahren zur Erfassung psychischer Belastung. In: Sonntag, Kh. (Ed.), Arbeitsanalyse und Technikentwicklung. Wirtschaftsverlag Bachem, Köln.

Hacker, W., 1973. Allgemeine Arbeits- und Ingenieurpsychologie. VEB Deutscher Verlag der Wissenschaften, Berlin.

Hacker, W., Iwanowa, I., Richter, P., 1983. Das Tätigkeitsbewertungssystem - TBS. Psychodiagnostisches Zentrum, Berlin.

Hackman, J.R., 1969. Toward understanding the role of tasks in behavioral research. Acta Psychologica 31, 97-128.

Hackman, J.R., 1970. Tasks and performance in research on stress. In: McGrath, J.E. (Ed.), Social and Psychological Factors in Stress. Holt, Rinehart and Winston, New York, pp. 202-237.

Hackman, J.R., Oldham, G.R., 1974. The job diagnostic survey: an instrument for the diagnosis of jobs and evaluation of job redesign projects. Technical Report No. 4, Department of Administrative Sciences, Yale University, New Haven.

Häublein, H.-G., Heuchert, G., Schulz, G., Blau, E., 1979. Methodische Anleitung zur arbeitshygienischen Professiographie. Forschungsverband Arbeitsmedizin der DDR, Berlin.

Heinsalmi, P., 1978. Method to measure working posture heads at working sites (OWAS). In: Corlett, N., Wilson, J., Manenica, J. (Eds.), The Ergonomics of Working Postures. Taylor and Francis, London.

Hennecke, A., 1976. Neuere Verfahren der Anforderungsvermittlung durch Arbeitsanalyse. Mitteilungen des IfaA A 64, 3-19.

Hoyos, G., 1974. Arbeitspsychologie. Verlag W. Kohlhammer GmbH, Stuttgart.

Hoyos, G.C., Frieling, E., 1977. Die Methodik der Arbeits- und Berufsanalyse. In: Seifert, K.H. (Ed.), Handbuch der Berufspsychologie. Verlag für Psychologie Dr. C.J. Hogrefe, Göttingen.

Hoyos, G.C., Bernhardt, U., 1987. Fragebogen zur Sicherheitsdiagnose (FDS) - Konzeption, Erprobung und Anwendungsmöglichkeiten eines verhaltensorientierten Instruments zur Sicherheitsanalyse von Arbeitssystemen. In: Sonntag, Kh. (Ed.), Arbeitsanalyse und Technikentwicklung. Wirtschaftsverlag Bachem, Köln.

Hull, C.L., 1952. A Behavior System. Yale University Press, New Haven.

IG-Metall Bezirksleitung Stuttgart, 1979. Werktage müssen menschlicher werden! Stuttgart.

Janes, A., 1980. Verfahren zur Beschreibung der Verwandtschaft zwischen Tätigkeiten. Ref. am 26. Kongreß der GfA in Hamburg am 06.03.1980.

Jones, M.H., Hulbert, S.F., Haase, R.H., 1953. A survey of the literature of job analysis of technical positions. Personnel Psychology 6, 173-194.

Karg, P.W., Staehle, W.H., 1982. Analyse der Arbeitssituation. Haufe Verlag, Freiburg.

Karhu, O., Kansi, P., Kuorinka, I., 1977. Correcting working postures in industry: a practical method for analysis. Applied Ergonomics 8, 199-201.

Kenton, E., 1979. Job analysis methodology. National Technical Information Service. Springfield, VA.

Kern, H., Schumann, M., 1970. Industriearbeit und Arbeiterbewußtsein. Europäische Verlagsanstalt.

Kirchner, J.-H., 1986. Belastungen und Beanspruchungen. Zeitschrift Arbeitswissenschaft 40, 69-74.

Kirchner, J.-H., Rohmert, W., 1973. Problemanalyse zur Erarbeitung eines arbeitswissenschaftlichen Instrumentariums für Tätigkeitsanalysen. In: Bundesinstitut für Berufsbildungsforschung (Hrsg.), Arbeitswissenschaftliche Studien zur Berufsbildungsforschung. Gebr. Jänecke Verlag, Hannover, pp. 7–48.

Kosiol, E., 1973. Aufgabenanalyse. In: Grochla, E. (Ed.), Handwörterbuch der Organisation. Poeschel Verlag, Stuttgart.

Koubek, R.J., Salvendy, G., Noland, S., 1994. The use of protocol analyses for determining ability requirements for personnel selection on a computer-based task. Ergonomics 37 (11), 1787–1800.

Landau, K., 1978a. Das Arbeitswissenschaftliche Erhebungsverfahren zur Tätigkeitsanalyse - AET. Dissertation, TH Darmstadt.

Landau, K., 1978b. Das Arbeitswissenschaftliche Erhebungsverfahren zur Tätigkeitsanalyse - AET im Vergleich zu Verfahren der analytischen Arbeitsbewertung. Fortschrittliche Betriebsführung 27 (1), 33–38.

Landau, K., Luczak, H., Rohmert, W., 1975. Arbeitswissenschftliche Erhebungsbogen zur Tätigkeitsanalyse. In: Rohmert, W., Rutenfranz, J. (Eds.), Arbeitswissenschaftliche Beurteilung der Belastung und Beanspruchung an unterschiedlichen Arbeitsplätzen. Der Bundesminister für Arbeit und Sozialordnung, Bonn.

Landau, K., Rohmert, W., 1979. Das Arbeitswissenschaftliche Erhebungsverfahren zur Tätigkeitsanalyse (AET) - Merkmalheft. Verlag Hans Huber, Bern.

Landau, K., Rohmert, W., 1989. Recent Developments in Job Analysis, Proceedings of an International Symposium on Job Analysis. University of Hohenheim, 14–15 March, 1989. Taylor and Francis, London, pp. 1–24.

Landau, K., Rohmert, W., 1992. Evaluation of worker workload in flexible manufacturing industry. International Journal of Human Factors in Manufacturing 2, 369–388.

Landau, K., Brauchler, R., Brauchler, W., Balle, W., Blankenstein, U., 1990. Eignung arbeitsanalytischer Verfahrensweisen zur Prognose möglicher arbeitsbedingter Schädigungen. In: Schriftreihe der Bundesanstalt für Arbeitsschutz und Unfallforschung. Wirtschaftsverlag NW, Dortmund.

Landau, K., Brauchler, R., 1994. Disease prediction using ergonomic knowledge bases. In: Proceedings of the 12th triennial congress of the International Ergonomics Association, vol. 2, Ergonomics in Occupational Health and Safety, Toronto, pp. S. 159–160

Landau, K., Wendt, G., Mirsching, Chr., Schühle, A., Pressel, G., 1995. Untersuchung der Körperhaltungen bei der Flugzeugbeladung mit der OWAS-Methode. Zeitschrift für Arbeitswissenschaft 49 (2), 84–89.

Landau, K., Wendt, G., 1995. Body postures involved in loading work – an investigation using the OWAS method. In: Institut de recherche en santé et en sécurité du travail du Quebec (Eds.). PREMUS 95, Book of Abstracts, Montreal, pp. 271–273.

Landu, K., Rohmert, W., Brauchler, R., 1998. Task analysis: Part I: Guidelines for the practitioner. International Journal of Industrial Ergonomics 22(1–2): 3–11 (this issue).

Laurig, W., 1971. Simulationsmethoden zum Abschätzen des Erholzeitbedarfs. Werkstatt und Betrieb 104, 263–267.

Laurig, W., 1977. Der Arbeitsinhalt als ergonomische Fragestellung. Dortmund: Vortrag gehalten auf der Internationalen Tagung der Sozialakademie vom 20.–23.06.1977.

Leplat, J., 1990. Relations between tsk and activity: elements for elaborating a framework for error analysis. Ergonomics 33 (10/11), 1389–1402.

Luczak, H., 1975. Untersuchungen informatorischer Belastung und Beanspruchung des Menschen. Fortschritts-Berichte der VDI-Zeitschriften, Reihe 10, 2. VDI-Verlag, Düsseldorf.

Lynch, B.P., 1974. An empirical assessment of Perrow's technology construct. Administrative Science Quarterly, 338.

McCormick, E.J., Jeanneret, P.R., Mechaman, R.C., 1969. The development and background of the position analysis questionnaire (PQA). Report No. 5, Occupational Research Center, Purdue University.

McCormick, E., 1976. Job and task analysis. In: Dunette, M.D. (Ed.), Handbook of Industrial and Organizational Psychology. Rand McNally College Publishing Company, Chicago, pp. 651–696.

Martin, W., Ackermann, E., Udris, H., Oegerli, K., 1980. Monotonie in der Industrie. Huber, Bern.

Mattila, M.K., 1985. Job load and hazard analysis: A method for the analysis of workplace conditions for occupational health care. British Journal of Industrial Medicine 42, S656–666.

Miller, R.B., 1971. Development of a Taxonomy of Human Performance. American Institutes for Research, Washington.

Moldaschl, M., Weber, W.-G., 1986. Prospective Arbeitsplatzbewertung an flexiblen Fertigungssystemen. Psychologische Analyse von Arbeitsorganisation, Qualifikation und belastung. In: Volpert, W., Oesterreich, R. (Eds.), Forschungen zum Handeln in Arbeit und Alltag, Band 1. Technische Universität, Berlin.

Morsh, J.E., 1967. The analysis of jobs - use of the task inventory method of job analysis. In: Fleishman, E. (Ed.), Studies in Personnel and Industrial Psychology. The Dorsey Press, Homewood (IL.).

Morsh, J.E., 1964. Job analysis in the United States Air Force, Personnel Psychology 17.

Morsh, J.E., 1966. Impact of the computer on job analysis in the United States Air Force. PRL-TR-66-19, Personnel Research Laboratory, Aerospace Medical Division, Air Force Systems Command, Lackland (TX).

Morsh, J.E., Madden, J.M., Christal, R.E., 1961. Job analysis in the United States Air Force Technical Report WADDTR-61-113, Lackland Air Force Base, TX, USA.

Müller-Böling, D., 1978. Arbeitszufriedenheit bei automatisierter Datenverarbeitung. München/Wien.

Neuberger, O., Allerbeck, M., 1978. Messung und Analyse von Arbeitszufriedenheit, Bern, Stuttgart.

Neunert, J., 1979. Der Anfang einer Verständigung? Zusammenfassung der Podiumsdiskussion "Arbeitsanalyse" auf dem Frühjahrskongreß der Gesellschaft für Arbeitswissenschaft in Wien, 2-4 Mai 1979. Angewandte Arbeitswissenschaft 82, 57-61.

Nutzhorn, H., 1964, Leitfaden der Arbeitsanalyse. Verlag für Wissenschaft, Bad Harzburg, Wirtschaft, Technik.

Plath, H.E., Richter, P., 1976. Erfassung von Beeinträchtigung durch Belastungswirkungen. Monotonie und psychische Sättigung. Sozialistische Arbeitswissenschaft 20 (1), 27-37.

Prien, E.P., Ronan, W.W., 1971. Job analysis. A review of research findings. Personnel Psychology 24, 371-396.

Pupka von, M., 1977. Anforderungsgerechtes menschliches Verhalten bei Transporttätigkeit - Anforderungs- und Eignungsprofile. Bundesanstalt für Arbeitsschutz und Unfallforschung, Forschungsbericht Nr. 158. Dortmund, Wirtschaftsverlag NW.

Rabideau, G., 1964. Field measurement of human performance in man-machine-systems. Human Factors 6, 663-672.

Rasmussen, J., 1990. The role of error in organizing behaviour. Ergonomics 33 (10/11), 1185-1199.

REFA (Hrsg.), 1977. Methodenlehre des Arbeitsstudiums. Teil 4 - Anforderungsermittlung (Arbeitsbewertung). Carl Hanser Verlag, München.

REFA (Hrsg.), 1978. Methodenlehre des Arbeitsstudiums. Teil 2 - Datenermittlung. Carl Hanser Verlag, München.

Rohmert, W., 1972. Aufgaben und Inhalt der Arbeitswissenschaft. Die berufsbildende Schule 24 (1), 3-14.

Rohmert, W., 1983. Formen menschlicher Arbeit. In: Rohmert, W., Rutenfranz, J. (Eds.), Praktische Arbeitsphysiologie. Thieme Verlag, Stuttgart.

Rohmert, W., Luczak, H., Landau, K., 1975a. Arbeitswissenschaftlicher Erhebungsbogen zur Tätigkeitsanalyse - AET. Zeitschrift für Arbeitswissenschaft 29 (4), 199-207.

Rohmert, W., Rutenfranz, J., Luczak, H., Landau, K., Wucherpfennig, D., 1975b. Arbeitswissenschaftliche Beurteilung der Belastung und Beanspruchung an unterschiedlichen industriellen Arbeitsplätzen. In: Rohmert, W., Rutenfranz, J. (Eds.), Arbeitswissenschaftliche Beurteilung der Belastung und Beanspruchung an unterschiedlichen industriellen Arbeitsplätzen. Der Bundesminister für Arbeit und Sozialordnung, Bonn, pp. 15-250.

Rohmert, W., Landau, K., 1979. Das Arbeitswissenschaftliche Erhebungsverfahren zur Tätigkeitsanalyse (AET). Handbuch und Merkmalheft. Hans Huber Verlag, Bern.

Roth, E.M., 1992. Cognitive simulation as a tool for cognitive task analysis. Ergonomics 35 (10), 1163-1198.

Ruppert, F., Hoyos, Graf.v., C., 1989. Safety diagnosis in industrial plants: concepts and preliminary results. In: Landau, K., Rohmert, W. (Eds.), Recent Developments in Job Analysis. Taylor and Francis, London, pp. 167-178.

Schaaf van der, T., 1993. Developing and using cognitive task typologies. Ergonomics 36 (11), 1439-1444.

Schmidtke, H., 1975. Ergonomische Bewertung von Arbeitssystemen - Entwurf eines Verfahrens. Hanser Verlag, München.

Services des conditions de travail de la Région Nationale, 1976. Profils de Postes. Usines Renault.

Shepherd, A., 1993. An approach to information requirements specification for process control tasks. Ergonomics 36 (11), 1425-1437.

Sims, H.D., Szlagyi, A.D., Keller, R.T., 1976. The measurement of job characteristics. Academy of Management Journal, 195.

Skinner, B.F., 1958. Science and Human Behavior. MacMillan, New York.

Smith, P.C., Kendall, L.M., Hulin, C.L., 1969. The Measurement of Satisfaction in Work and Retirement, Chicago.

Spence, K.W., 1948. The methods and postulates of "behaviorism". Psychology Review 55, S67-78.

Staehle, W.H., Hattke, W., Sydow, J., 1981. Die Arbeit an Datensichtgeräten aus der Sicht der Betroffenen. DBWDepot-Papier 81-5-1, Stuttgart, Poeschel.

Stoffert, G., 1985. Analyse und Einstufung von Körperhaltungen bei der Arbeit nach der OWAS-Methode. Zeitschrift für Arbeitswissenschaft 39 (1), 31-39.

Theologus, G.C., Romashko, T., Fleishman, E.A., 1970. Development of a taxonomy of human performance: a feasibility study of ability dimensions for classifying human taks. American Institutes for Research, AD 705 672, Pittsburgh.

Turner, A.N., Lawrence, P.R., 1965. Industrial Jobs and the Worker, Boston.

Udris, I., 1977. Fragebogen zur Arbeitsbeanspruchung; unpublished working papers. Lehrstuhl für Arbeits- und Betriebspsychologie der ETH, Zürich.

Udris, I., 1981. Redefinition als Problem der Arbeitsanalyse. In: Frei, F., Ulich, E. (Eds.), Beiträge zur psychologischen Arbeitsanalyse. Huber Verlag, Bern.

Ulich, E., 1981. Subjektive Tätigkeitsanalyse als Voraussetzung autonomieorientierter Arbeitsgestaltung. In: Frei, F., Ulich, E. (Eds.), Beiträge zur psychologischen Arbeitsanalyse. Huber Verlag, Bern.

Volpert, W., Oesterreich, R., Gablenz-Kolakovic, S., Krogoll, T., Resch, M., 1983. Verfahren zur Ermittlung von Regulationserfordernissen in der Arbeitstätigkeit (VERA). Verlag TÜV Rheinland, Köln.

Volpert, W., Ludborzs, B., Muster, M., 1981. Lernrelevante Aspekte in der Aufgabenstruktur von Arbeitstätigkeiten - Probleme und Möglichkeiten der Analyse. In: Frei, F., Ulich, E. (Eds.), Beiträge zur psychologischen Arbeitsanalyse. Huber Verlag, Bern.

Wächter, H., Modrow-Thiel, B., Schmitz, G., 1989. Entwicklung eines Verfahrens zur Analyse von Tätigkeitsstrukturen und zur vorausschauenden Arbeitsgestaltung bei Automatisierung (ATAA), Endbericht des HdA-Projektes, Trier.

Weber, W., Oesterreich, R., 1989. VERA Microanalysis : applied to a flexible manufacturing system. In: Landau, K., Rohmert, W. (Eds.), Recent Developements in Job Analysis. Taylor and Francis, London, pp. 91–100.

Wheaton, G., 1968. Development of a taxonomy of human performance: a review of classif. American Institutes for Research, Pittsburgh, PA.

White, G.C., 1975. Job design and work organization, diagnosis and measurement. Research Unit, Paper No. 4, Department of Employment, London.

Zerga, J.E., 1943. Job analysis – a resume and bibliography. Journal of Applied Psychology 27, 249–267.

Allocation of functions to humans and machines in a manufacturing environment: Part I – Guidelines for the practitioner

Anil Mital [a,*], Arif Motorwala [a], Mangesh Kulkarni [a], Murray Sinclair [b], Carys Siemieniuch [b]

[a] *Ergonomics and Engineering Controls Research Laboratory, Industrial Engineering, University of Cincinnati, Cincinnati, OH 45221-0116, USA*
[b] *HUSAT Research Institute, Loughborough University of Technology, Elms Grove, Loughborough, Leicestershire LE11-1RG, UK*

1. Audience

This guide is intended for practitioners. The practitioner in the context of this guide is defined as an individual who is responsible for deciding which of the functions in the manufacturing environment should be assigned to humans and which to machines (including all kinds of automated equipment). This person may be the product or process designer, process engineer, manufacturing or production engineer, mechanical or industrial engineer, task analyst, ergonomist, safety engineer, or a professional responsible for deciding who (human or machine) performs a specific manufacturing function.

2. Problem identification

Modern manufacturing requires a close association between the humans and machines. The determination of which of the various functions, necessary in order to convert raw materials into finished goods, are allocated to humans and which to machines must be based on a systematic consideration of the design of the entire manufacturing system; both physical and human elements of the system must be considered. Currently, the assignment of functions is carried out on an ad hoc basis, rather than through scientific deliberations.

A manufacturing system may be basically classified into three basic categories: (1) pre-dominantly manual manufacturing systems (all func-

---

* The recommendations provided in this guide are based on numerous published and unpublished scientific studies and are intended to enhance worker safety and productivity. These recommendations are neither intended to replace existing standards, if any, nor should be treated as standards. Furthermore, this document should not be construed to represent institutional policy.

The following individuals participated in the discussion of the earlier version of this guide. Their suggestions (written or verbal) were incorporated by the authors in this version: Arne Aaras, *Norway*; Fred Aghazadeh, *USA*; Roland Andersson, *Sweden*; Jan Dul, *The Netherlands*; Jeffrey Fernandez, *USA*; Ingvar Holmér, *Sweden*; Matthias Jäger, *Germany*; Åsa Kilbom, *Sweden*; Anders Kjellberg, *Sweden*; Olli Korhonen, *Finland*; Helmut Krueger, *Switzerland*; Shrawan Kumar, *Canada*; Ulf Landström, *Sweden*; Tom Leamon, *USA*; Ruth Nielsen, *Denmark*, Jerry Ramsey, *USA*; Rolf Westgaard, *Norway*; Ann Williamson, *Australia*; Jørgen Winkel, *Sweden*; Pia Zätterström, *Sweden*. The guide was also reviewed in depth by several anonymous reviewers.
* Corresponding author.

tions require human intervention in some form or the other), (2) hybrid manufacturing systems (some functions require human intervention while others are completely automated), and (3) automated manufacturing systems (all functions are automated). Even though examples of all three kinds of manufacturing systems can be found in industry, it is the hybrid system that has become most pervasive. And it is the hybrid system that raises the basic question "How should functions be allocated to humans and machines?".

All basic functions in a hybrid manufacturing environment can be grouped under one of the three classes: (1) functions which can only be performed by humans, such as high level decision-making, (2) functions that can only be per-

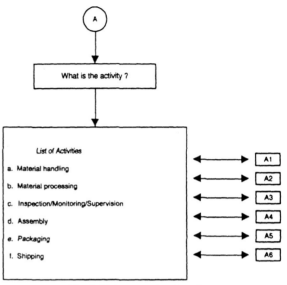

Fig. 2. Main operation flow chart.

formed by machines, such as water-jet cutting, and (3) functions that both humans and machines can perform. It is this last class of functions that poses a major problem for the practitioner. A sub-optimal decision could lead to inefficiency, loss in productivity, ineffectiveness and unnecessary costs and injuries and accidents. If a function can be performed by both humans and machines, it is imperative that the function be thoroughly and systematically analyzed prior to deciding whether it is best that a human performs it or a machine.

There are three other issues which we must consider. These are: (a) the scope of the function allocation problem, (b) the immediacy of the problem, and (c) what happens once the function allocation has been carried out. The first two are inter-related, as will be seen. The third discusses the place of function allocation in the overall design of jobs and organizational structures.

Firstly, we discuss the scope of function allocation. At one extreme a company may be considering the problem of designing a new manufacturing facility on a greenfield site in a foreign country; at the other extreme, the problem might be the revised layout of a manufacturing cell to accommodate a new machine or addition of a new production line, with consideration of the

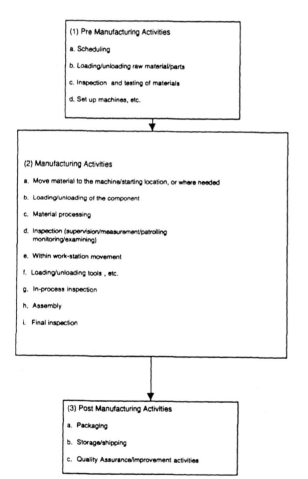

Fig. 1. Activities performed in manufacturing a product.

changed tasks and changed operating environment that might ensue. Problems of the scope of the former are rare, and represent a step-change in the organization's operations; problems of the latter scope are far more common, and are the ones that represent gradual, incremental improvements to processes that maintain competitiveness in the ever-changing markets of today. For two reasons, these guidelines consider the latter class of problems; they are frequent, and the process of function allocation is better understood than for the first class of problem.

Secondly, we discuss the immediacy of the problem. Like most design problems, the design of manufacturing processes is an exercise in design on a broad front, with progress being made through the problem space by incremental approximations towards an acceptable answer. These incremental approximations require the gradual fixing of parameters and constraints, until the answer is discovered. In the case of problems of large scope, these parameters and constraints are fluid, and tend to be non-stationary. Furthermore, the solution can take years to be

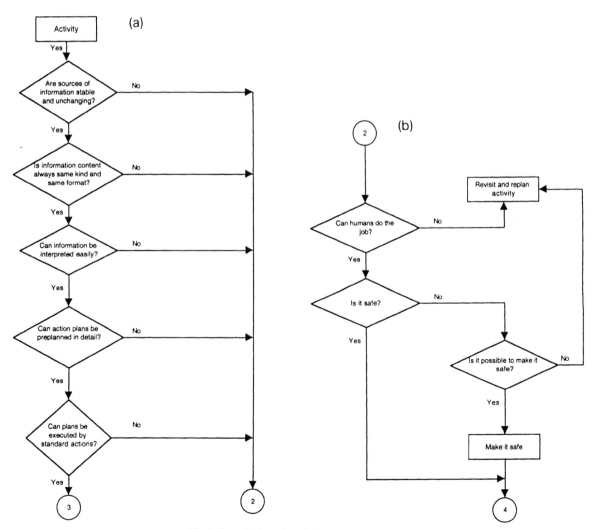

Fig. 3. Generic flow chart for function allocation.

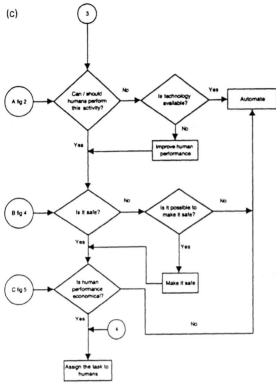

Fig. 3 (continued).

finalized and implemented. On the other hand, the problems of smaller scope tend to have fixed boundaries and constraints, with relatively few unknown (and unrecognized) variables. Their time frame is limited, and the environment can be considered to be stationary. Because of the attributes of the latter class, the process of function allocation is better understood, and more easily applied; the former is a particularly difficult problem for which adequate approaches have yet to be defined.

Thirdly, we discuss the role of function allocation. This is the first step in the design of jobs. What function allocation does is to assign activities or tasks to humans or machines. The flow charts in these guidelines are intended to help the practitioner to arrive at these assignment decisions. However, in real life there is always a gray area, where the task or activity could be performed by either humans or by a machine, and there is little to be gained by either choice.

This is where issues of job design enter; the next stage in the redesign of organizational structures. Many more variables enter at this stage, all aimed at devising jobs which are meaningful to the people who have to perform them, which provide a sense of achievement, which make minimum demands on the company's resources for control and communications, and which enable the company to attain cost-effective production. Principles such as compatibility of process, localized control of operational variances, minimum necessary change, minimum training requirement, etc. will enter at this stage. The application of these principles will necessitate some rearrangement of the task allocation decisions, particularly those that fall in the 'grey area', to provide good jobs for people. However, discussion of this process is not part of the allocation of functions, and we address this issue no further; future guidelines will be devoted to this particular issue. What these guidelines achieve is to partition the activities of the company into two classes, such that the processes of job design, and subsequently human–machine interface design, can be undertaken.

Summarizing this discussion, we concentrate on the frequent problem of allocating functions for shopfloor problems within a manufacturing facility.

3. Information gathering

The activities involving conversion of raw materials into useful products may be divided into: pre-manufacturing activities, manufacturing activities (actual raw material conversion), and post manufacturing activities (Fig. 1). These manufacturing activities can be grouped under the following broad categories, as shown in Fig. 2:
(a) Material handling,
(b) Material processing (machining or material conversion),
(c) Inspection/Monitoring/Supervision,
(d) Assembly,
(e) Packaging,
(f) Shipping, and
(g) Improvement activities.

It should be noted immediately that the last of these classes of activities is essentially a human function. Necessarily, it involves changing the operating environment, and it requires 'envisioning the future'. Both of these attributes preclude this activity from being carried out by automation (though obviously humans responsible for this can be assisted by simulation tools, etc.).

For each of the first six activities, independent analysis must be carried out before optimal assignment of functions can be determined. A set of mandatory, generic questions must be answered (as outlined in Fig. 3). These include:
- Requirements of complex decision making, experience to efficiently perform the task: If a task requires complex decision making and experience, then it must be assigned to humans (due to present technological limitations).

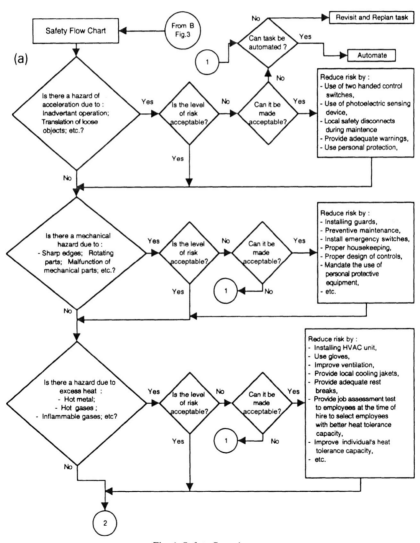

Fig. 4. Safety flow chart.

However, with further developments in the field of artificial intelligence, the automated option may become a viable (if perhaps a costly) choice in the future.
- Physical ability of humans to perform the task: A detailed assessment of the capabilities/limitations of humans versus those of the automated equipment for each of the activities listed in Fig. 2 must be performed (Figs. 6–11). If it is determined that humans are capable of performing the task then further analysis must be performed; if humans are determined incapable of performing the function, the function must be allocated to automated equipment. An

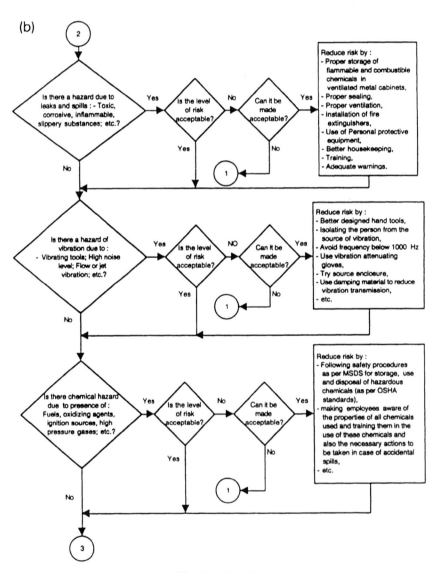

Fig. 4 (continued).

important decision that the company must take is whether in this process the company should consider humans in general, or those that work for the company. If labor turnover is fairly rapid, or seasonal, it would be sensible to consider all humans. If the staff are relatively permanent, it would be sensible to consider only those working in the company and likely to be affected by the project.

- Safety considerations: If it is determined that both the human and automated options can be used to perform the task, safety of the human operator must be considered. A detailed safety analysis must be carried out to determine the severity and likelihood of potential injury to humans (Fig. 4). If performing the function is determined to be unsafe for humans but technology is available to automate the function, the function must be assigned to automated equipment; otherwise the job cannot be performed as specified. In this case, the functions to be undertaken should be reanalyzed and respecified; it is always possible to devise more than one way to solve a problem.

- Economic considerations: If it is established that there are no safety hazards, and both the human and automated options are still viable, the decision must be based on systematic economic analysis (Fig. 5). Standard engineering economy techniques, such as economic evaluation of alternatives, should be used.

The flow chart in Fig. 3 displays the generic process of function allocation, with more detailed exposition of parts of this in Fig. 4 (safety analy-

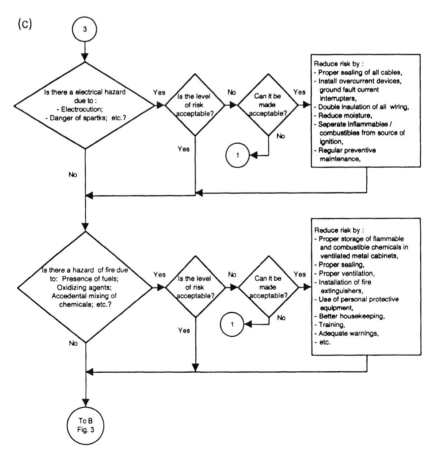

Fig. 4 (continued).

sis) and Fig. 5 (economic analysis). Figs. 6 to 11 display flow charts for each of the activities in Fig. 2, and represent adaptation of the generic flow chart in Fig. 3 for the activity in question.

4. Analysis and decision-making

4.1. Complex decision analysis

The particular attributes that humans bring to the workplace are the abilities to perceive and interpret information, to think, and to act, in the context of a variable environment. Automated systems are still unable to cope efficiently where these abilities cannot be specified in advance, and are unlikely to be competent for some time to come. Consequently, if any of these abilities are required in order to perform some function, and the ability cannot be specified in advance, then necessarily the function must be allocated to humans.

Questions that need to be considered for each of the activities identified in Fig. 2 include:

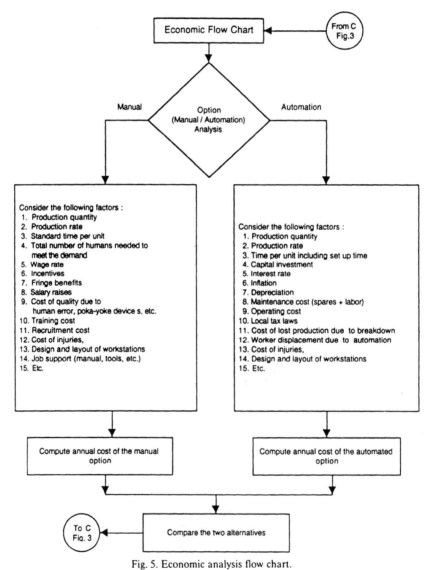

Fig. 5. Economic analysis flow chart.

(a) Are the sources of information for carrying out this function easily identified and unchanging (e.g. do the raw materials always arrive in the same place, with the same orientation, and at the same time)?

(b) Is the content of information from the sources always of the same kind and in the same format (e.g. cell status information is always given the same variables, with a fixed range of known categories)?

(c) Can the meaning of the information be understood without interpretation (e.g. little interpretation is required if the burner should be turned down when the needle on the temperature gauge is in the red region)?

(d) Can action plans be determined in detail beforehand (e.g. how to start a machine; how to check that the machine is operating correctly, etc.)?

(e) Can the action plans be carried out using

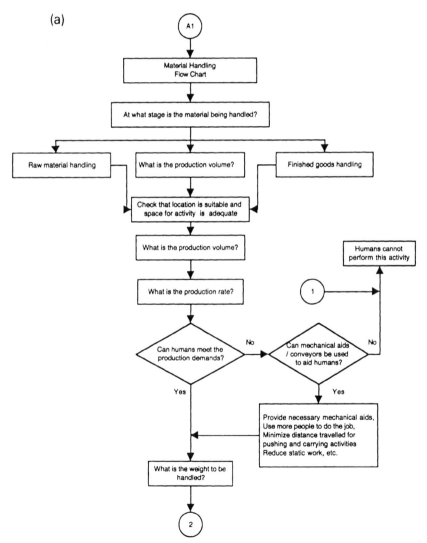

Fig. 6. Material handling flow chart.

standard actions (e.g. "Pick up part A using overgrasp, and place in holder B, oriented as shown in diagram C")?

(f) Is past experience irrelevant to carrying out activities (c) and (d) above?

The answer 'No' to any of these questions indicates that the function should be allocated to humans in the first instance. Note that this does not mean that a human must perform the function unaided; merely that a human must be in direct, realtime and online control of the function. It is likely that the human will require job support (e.g. manuals, tools, computer applications) in order to ensure safe, efficient, and error-free performance of the function; these requirements can be determined after the allocation of functions has been completed. It is possible that as the design progresses, these decisions can be reviewed.

4.2. Safety analysis

For each of the activities identified in Fig. 2, a detailed safety analysis must be performed. The safety analysis must incorporate investigation of all of the following hazards (not in any particular order):

(1) Hazards due to acceleration (inadvertent operation, etc.),
(2) Mechanical hazards (high speed rotating parts, sharp edges, etc.),
(3) Environmental hazards (lighting, noise, pollution, heat, vibration, dust, gases, etc.),
(4) Chemical hazards (leaks, spills, accidental contact with toxic chemicals, etc.),
(5) Potential hazard of fire and explosion due to the presence of fuels, strong oxidizing agents, etc.
(6) Electrical hazards.

Fig. 4 outlines the major risks involved in a manufacturing facility, and provides a list of actions that can reduce the hazard by reducing the risk of accidents. Each of the hazards listed above must be considered individually and the level of risk associated with each hazard must be determined. If the level of risk is beyond the acceptable limit (exceeds existing standards, recommendations, levels considered safe, etc.), an attempt must be made to reduce risk as suggested in Fig. 4. If the level of risk to humans cannot be reduced to fall within acceptable (safe) levels, the activity must be performed by automated equipment.

4.3. Economic analysis

In the last several years, industries in general and manufacturing industries in particular have experienced a substantial growth in automation. The need to produce goods economically and, by doing so, compete worldwide is the major reason that is cited most frequently to justify such moves.

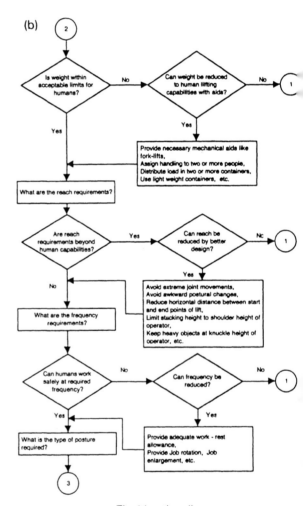

Fig. 6 (continued).

There are, however, other reasons, such as the fear of obsolescence, and the desire to follow the latest trend. In fact, many decisions to automate functions are made for these reasons rather than economical or technical necessities. The results often are unreasonable expectations from the changeover, and an economic disaster. It is critical to realize that just because the technology is available, it is not sufficient reason to automate. The final justification must be based on economics. When introducing automated equipment in the work place, a large amount of capital is generally at stake and one does not have the luxury to say "It did not work out, let us revert back to manual methods".

However, economic analysis must be preceded by thorough analyses of the capabilities and limitations of both humans and the automated equipment and the safety issues involving manual and automated options. If, even after replanning, a function is determined to be unsafe for humans, it must be allocated to the automated equipment irrespective of the outcome of the economic analysis. (In fact, it is very unlikely that the manual option will turn out to be economically attractive once the injury and associated costs are included.) Fig. 5 includes a list of factors that must be considered in making a detailed economic analysis for both manual and automated options.

4.3.1. Sample cost computations

The following are some examples of simplified cost calculations. For detailed economic analysis procedures and cost quantification, the user is

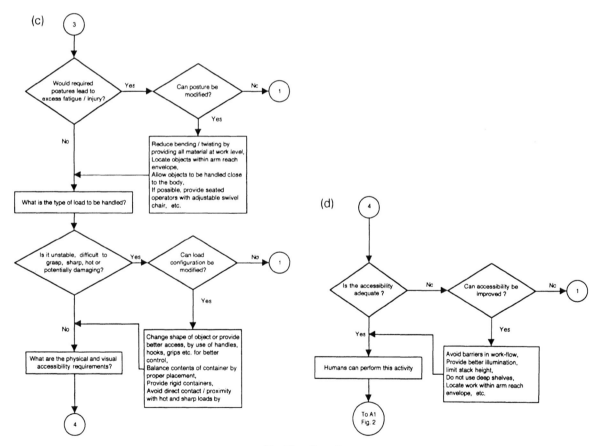

Fig. 6 (continued).

referred to a book on engineering economy or the economic analysis guide that appeared in a previous issue of the *International Journal of Industrial Ergonomics* (Volume 10, numbers 1-2, pages 161-178).

(i) Manual option

Workstation cost:
- Determine the cost and quantity of each kind of equipment required (ex., gravity chutes, finished parts bin, work bench, stools, chairs, etc.). In estimating total annual capital cost, interest rate, inflation, etc., should be considered. The annual capital cost of the equipment, without interest rate and inflation considerations, is:

= equipment capital cost ($)/life of the equipment (yrs).

Labor cost:
- Determine normal time of operation (based on time study, MTM, etc.) for all direct and indirect manual activities.

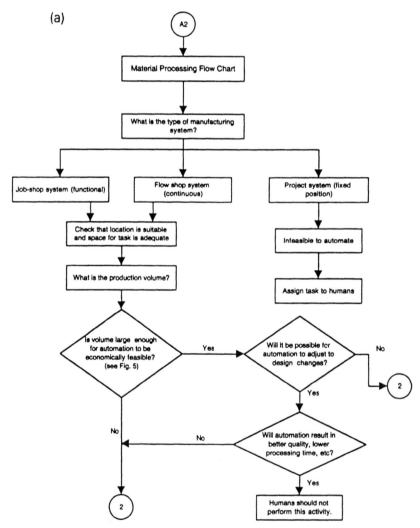

Fig. 7. Material processing flow chart.

- Add necessary allowances (personal, delay, and fatigue) to compute standard time per unit.
- Based on estimated (forecast) production demand, calculate total direct and indirect standard labor hours per year:

  = standard time (hrs/unit) * production demand (units/yr).

- Calculate labor cost ($/yr) based on the hourly rate:

  = total labor hours/yr * hourly rate ($/hr) * (1 + necessary allowances for insurance, FICA, medical expenses, etc.).

Equipment installation cost:
- Equipment installation cost ($)

  = setup time (hrs) * hourly rate of operation ($/hr).

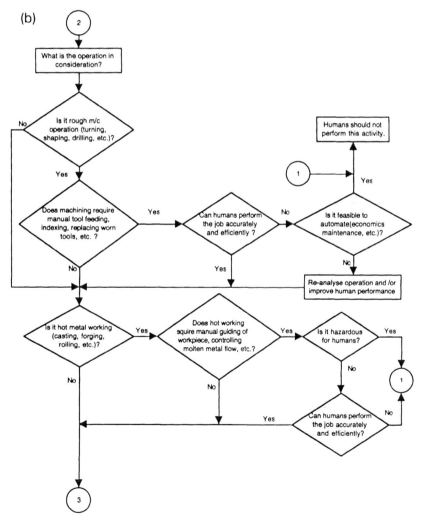

Fig. 7 (continued).

Total annual production cost:
- Total annual cost for the manual option

    = work station cost + labor cost + equipment installation cost + ...

(ii) Automated option

Workstation cost:
- Determine the cost and quantity of each kind of equipment required (ex., feeders, conveyor belts, gravity chutes, robots, finished parts bins, computer software and hardware, etc.). In estimating the total capital investment, interest rate, inflation, life of setup, etc., must be considered. The capital cost without these considerations is:

    = capital investment ($)/life of the equipment (yrs).

Annual maintenance and set-up costs:
- Estimate the downtime of the automated equipment and associated cost of lost production.
- Estimate lifetime costs necessitated by the provision of additional maintenance facilities, programming requirements, etc. Any new piece of

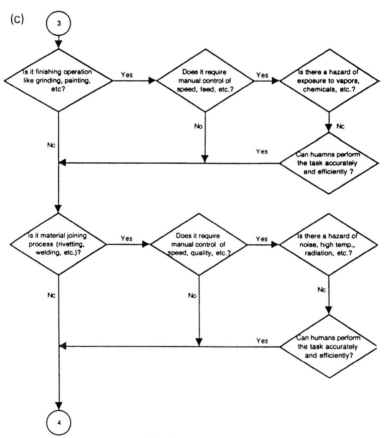

Fig. 7 (continued).

equipment will affect a number of functions or departments in the company, and the effects on these should be calculated.
- Determine the set-up cost based on the hourly rate of set-up personnel, time required for set-up, and the equipment used in setting-up.

Energy cost per year:
- Estimate the total energy cost based on the number of hours the automated equipment will be in operation and the hourly utility cost.

 = energy consumption per hour (KW/hr) * number of operating hours per year * Cost ($/KWH).

Installation cost:
- Installation cost should be computed based on the number of hours required for installation and the hourly rate of the worker.

 = installation time (hr) * labor cost ($/hr).

Total production cost:
- Total annual cost of the automated option.

 = workstation cost + maintenance and set-up costs + energy cost + installation cost + ...

The costs described above and in Fig. 5 should be calculated for both the manual and automated

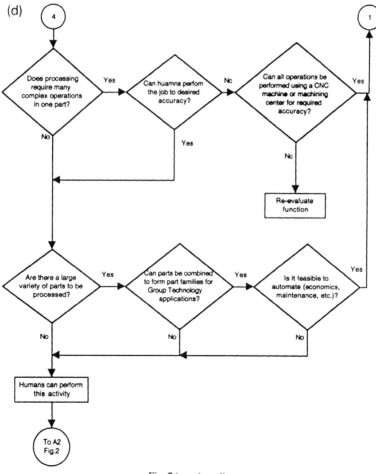

Fig. 7 (continued).

options and compared. Whichever option results in lower total annual cost should be allocated the task.

4.4. Explanation of activity-specific flow charts

The flow charts are predicated on the assumption that the company will have carried out an analysis of the functions to be performed, and that this will result in a hierarchical tree of activities. The flow charts are then applied to the activities of the tree at the most detailed level that the company feels to be appropriate.

The flow charts have the following structure. Fig. 1 lists the activities that occur on a day-to-day basis on the shopfloor. Fig. 2 is the master flow chart for the detailed activity charts in Figs. 6 to 11. Fig. 3 is a generic flow chart that shows the process of function allocation in general; Figs. 6 to 11 are activity-specific versions of this generic

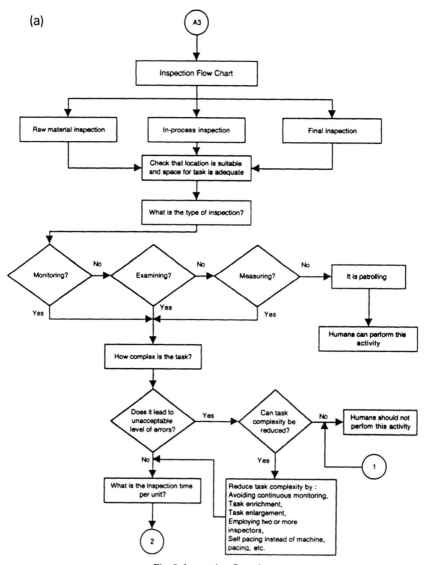

Fig. 8. Inspection flow chart.

flow chart. Fig. 4 is a generic flow chart for considering safety aspects; as such it forms a part of Fig. 3, and is cross-referenced in Figs. 6 to 11. Fig. 5 is a flow chart for considering cost aspects for function allocation; again, it is generic, and is cross-referenced from Figs. 6 to 11.

Within the flow charts, the symbols may be interpreted as follows. The sequential flow of these diagrams is shown on the page. Square boxes are information boxes for the reader. Diamonds represent decisions that the reader must make. Circles represent the flow of attention from one part of the chart to another, either on another page, or to a distant point on the same page. Arrows join the boxes to order and control the flow of attention.

The flow charts should be treated as checklists; they order the series of decisions which must be made, and they indicate the kind of decision that must be undertaken. The information necessary, and the criteria used to make the decisions have only been specified in certain cases; typically, these are case and/or situation-specific, and require the level of details (e.g. tolerance that can be achieved on a specific type of equipment or by a specific manufacturing process,

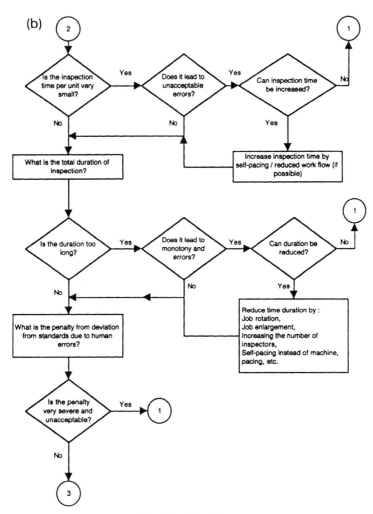

Fig. 8 (continued).

quality requirements, etc.) that cannot be included in general guidelines at this stage, without making the entire guidelines excessively large. Future efforts will be directed at expanding specific flow charts (e.g. Materials Processing flow chart) to include specific criteria to aid the user in making decisions. Further, it should be noted that in some cases the decision-making criteria are determined by legislation within a particular country or accepted good practice within the particular industry, and, therefore, cannot be included in general guidelines (e.g. the weight that can be handled safely by humans). The reader should consult the relevant agencies to obtain these criteria.

4.4.1. Material handling flow chart

Materials can be handled manually (if within human capabilities), or mechanically. Manual handling of materials, while not capital intensive,

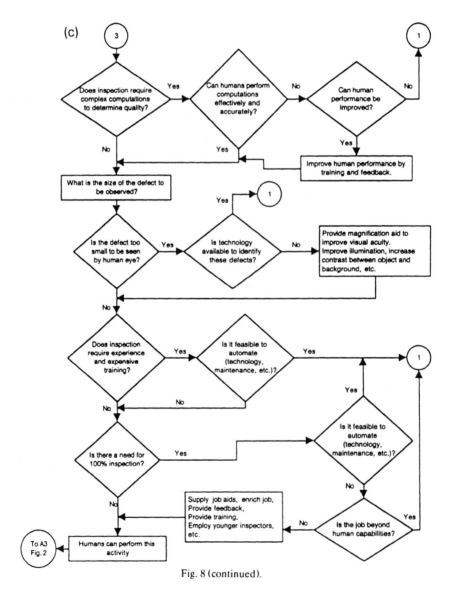

Fig. 8 (continued).

can be an expensive industrial problem. The major cost components in this case are indemnity compensation and funds spent on treatment (medical tests, treatments, medications, surgeries, etc.). Other costs may be incurred due to the imprecision and non-repeatability of human movements; initially this may introduce quality costs, and subsequently the costs of *poka-yoke* devices. The various material handling devices that are generally available for use are conveyors, cranes and hoists, and industrial trucks. These devices are, however, mainly suitable only for materials handling between work stations. Even though cranes and hoists, and devices such as lift tables are available for materials handling at the work station, these devices are often slower and more difficult to operate than manual handling methods. Much of the materials handling at the work stations, therefore, may still have to be done manually.

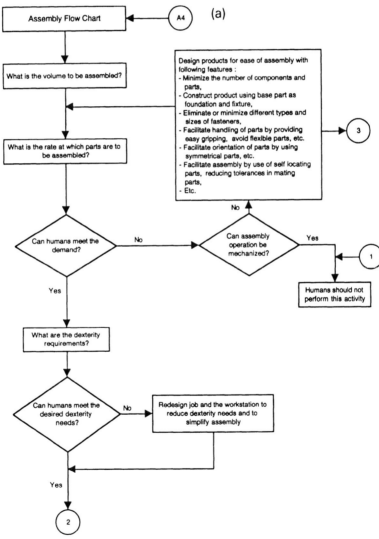

Fig. 9. Assembly flow chart.

Fig. 6 presents an outline of the decision-making process for allocating material handling functions between humans and automated equipment (note the components above – the company must decide whether to refer to humans in general, or just those in the company). The intent of the flow chart is to determine whether humans can perform the activity without a serious risk of overexertion injuries. Should they be able to, the activity should be assigned to humans. If humans are at risk, mechanical aids should be used to assist workers. In the event mechanical aids cannot be used, the task must be assigned to automated equipment. The practitioner is guided through a series of questions and suggestions. The major factors that are considered include:
– production volume and production rate,
– object weight,
– dexterity,
– reach/posture required,

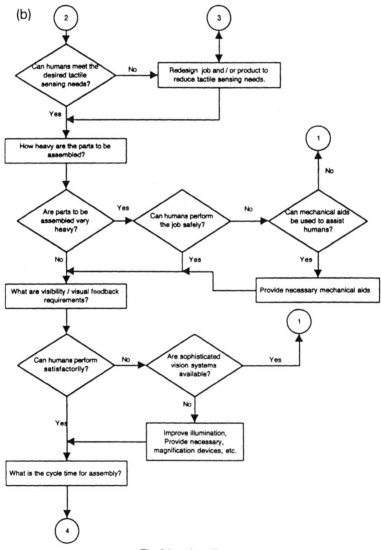

Fig. 9 (continued).

- frequency of performance and
- physical accessibility.

The intent here is to prompt the user to consider these factors for specific situations. The acceptable limits should be determined from widely available ergonomics literature on the subject. Consult the flow chart for specific recommendations to reduce the risk of injury.

4.4.2. Material processing flow chart

Manufacturing systems for processing activities can be classified into three different kinds: job shop systems (process type layout), flow shop or continuous manufacturing systems (product type layout), and project shop systems. Job shop systems are very flexible and capable of manufacturing a wide variety of product families, but are

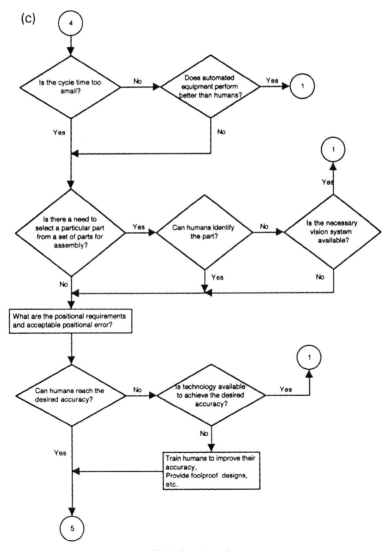

Fig. 9 (continued).

difficult to automate. The flow shop manufacturing systems are used for producing the same product, or products with few variations, in very large volumes. These systems are generally easier to automate. In the project shop systems, workers and machines must be brought to the project site for processing. These systems are almost impossible to automate.

Fig. 7 provides a generic set of prompting questions that must be answered for efficient allocation of functions to humans and machines. The following factors must be considered:
- production volume and production rate,
- flexibility in terms of adaptation to design changes,
- quality issues and processing times,
- efficiency of the system,
- processing accuracy (tolerances, etc.),
- safety of humans performing material processing.

Higher production volumes, need for precise dimensional control (lower tolerances), lower processing times, hazardous work operations, and fewer part variations favor the automated option.

4.4.3. Inspection flow chart

Inspection activities can be classified into four general categories: measurement (using instruments to make a decision about the status of an item as good or defective), monitoring (monitor-

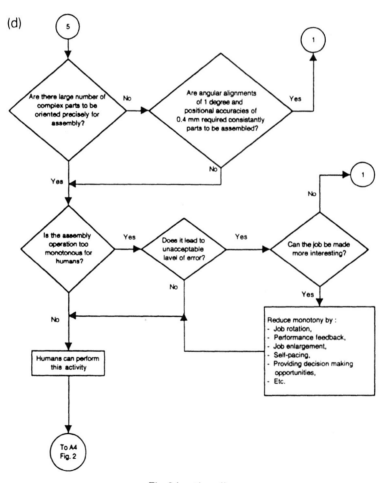

Fig. 9 (continued).

ing the process for deviations from the normal or acceptable), examining (searching for defects in the attributes of an item), and patrolling (checking and organizing others' work). With the current technology, measurements and monitoring and some classes of examining can be automated. However, patrolling still must be performed by humans. The factors that favor automated inspection are simplicity of the inspection task, high production volumes, higher quality and litigation costs if faulty parts are placed in the market, need for 100% inspection, higher inspection times, etc.

Manual inspection performance depends upon job-related factors, such as degree of machine-pacing, task complexity, defect rate, defect frequency, type and number of defects to be inspected, available inspection time; environmental

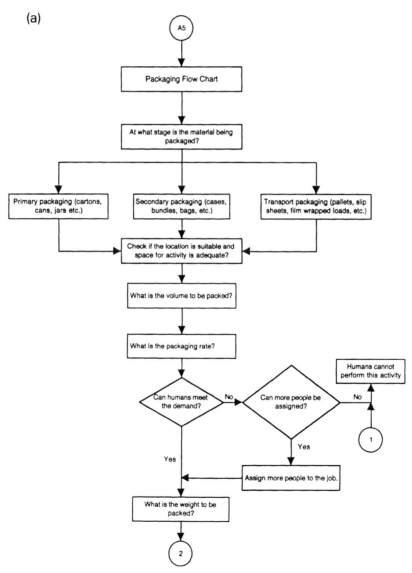

Fig. 10. Packaging flow chart.

factors, such as lighting and magnification; organizational factors, such as self inspection, multiple inspectors; motivational factors, such as feedback, pay off structure, training; and personal factors, such as individual differences, visual acuity, age and gender. In cases where human performance may not be acceptable but technological limitations may render automation impossible, attempts must be made to improve human performance.

Fig. 8 identifies the major factors that must be considered in allocating inspection function between humans and automated equipment. Note that the associated activities of reworking and/or disposal are not included here; they represent activities that would be analyzed separately.

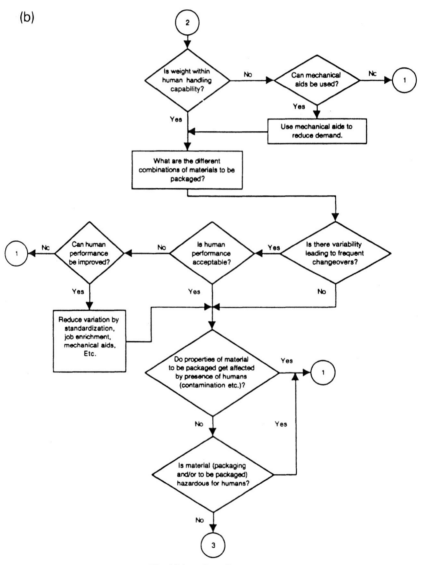

Fig. 10 (continued).

4.4.4. Assembly flow chart

Assembly of components generally requires great dexterity and visual and tactile sensing. Fig. 9 poses critical questions that must be answered in order to arrive at the optimal allocation of function. The factors that favor automated assembly are high production volume, low dexterity/visual/tactile sensing needs, long cycle time, low acceptable positional error, fewer number of complex parts to be assembled, and routine monotonous assembly operations. Assembly operations with very small cycle times (less than 1 second) may also be suitable for automation.

Unavailability of complex visual and tactile sensing aids and the need to manufacture parts with very narrow tolerances and feeding them at precise assembly locations and orientations are major obstacles to automation. In practice, products designed for the ease of manufacturing render themselves to easy assembly, either manually

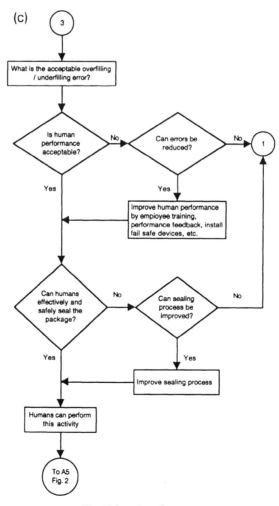

Fig. 10 (continued).

or with the aid of automated equipment (robots). In such cases, humans may be faster than the automated equipment.

4.4.5. Packaging flow chart

The decision to allocate packaging function to either humans or automated equipment should be based on the following considerations:
- volume to be packed,
- weight to be packed,
- combination of shapes and sizes of material to be packed, and
- acceptable overfilling/underfilling error.

The decision-making process for allocation of packaging function between humans and automated equipment has been outlined in Fig. 10. Higher volumes, heavy and hazardous materials, smaller variability in shapes and sizes of material to be packed, higher over-filling and under-filling error on the part of humans favor the automated option.

4.4.6. Shipping flow chart

Storage and shipping are major cost centers in total manufacturing. Materials handling is the major activity involved in shipping (palletizing, loading/unloading, and order-picking). Even though cranes and hoists are available for han-

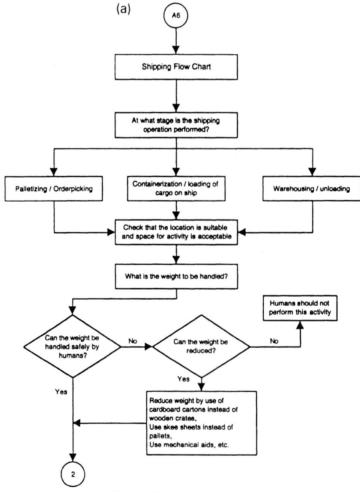

Fig. 11. Shipping flow chart.

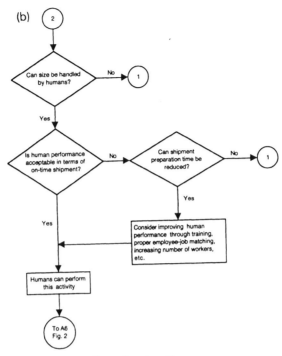

Fig. 11 (continued).

dling activities at the work station, these are often too slow for certain kinds of handling activities, such as palletizing and stacking. As a result, much of the material handling and loading of pallets, etc., may still have to be performed manually.

Technological limitations, a major reason for limited shipping function automation, have been overcome in recent years. High capital investment and excessive maintenance costs, however, may render the automated equipment unsuitable for consideration.

Fig. 11 outlines a step-by-step approach to function allocation for shipping operations. The factors that should be given consideration include:

– weight (tonnage) to be handled,
– the types of objects (shapes and sizes of boxes to be handled),
– availability of technology,
– acceptable performance in terms of on-time delivery,
– injury to humans and associated costs.

Reference

Mital, A., Motorwala, A., Kulkarni, M., Sinclair, M. and Siemieniuch, C., 1994. Allocation of functions to humans and machines in a manufacturing environment: Part II – The scientific basis (knowledge base) for the guide. International Journal of Industrial Ergonomics, 14(1–2): 33–49 (this issue).

Allocation of functions to humans and machines in a manufacturing environment: Part II – The scientific basis (knowledge base) for the guide

Anil Mital [a,*], Arif Motorwala [a], Mangesh Kulkarni [a], Murray Sinclair [b], Carys Siemieniuch [b]

[a] *Ergonomics and Engineering Controls Research Laboratory, Industrial Engineering, University of Cincinnati, Cincinnati, OH 45221-0116, USA*
[b] *HUSAT Research Institute, Loughborough University of Technology, Elms Grove, Loughborough, Leicestershire LE11-1RG, UK*

1. Problem overview

In modern manufacturing, automated equipment and computer applications are sharing the work environment with humans at an ever-increasing pace. Functions that in the past were routinely performed by humans are now being regularly assigned to machines. One does not have to venture far to encounter examples in the areas of material processing, handling, assembly, etc., nor in the more 'office' jobs of scheduling, shopfloor management, etc. This rush to automate manufacturing functions has been and is being justified through the general belief that since people have great difficulty in providing the required quality, uniformity, cleanliness, reliability, repeatability, ability to log events, and, furthermore since remarkable human performance is not the norm, the manual alternative is no longer a feasible option. For instance, in Japan, even one assembly error in 250,000 manual operations is not considered good enough human performance (Sata, 1986). Advocates of complete automation believe that all manufacturing activities, such as metal cutting, materials handling, assembly, and inspection, which are routinely handled by people, must be raised to a level that would allow machines to perform them. However, the fact is that fully automated factories are not yet viable for technical and economic reasons except in a very few special cases (Yamashita, 1987). Furthermore, for cybernetic reasons, it is likely that full automation will be a suboptimal solution for manufacturing organizations for the foreseeable future compared to hybrid systems combining people, machines and computers in effective partnership. The importance of people as components for control and innovation in manufacturing systems is recognized worldwide (Brodner, 1985, 1987; Corbett et al., 1991; Hockley, 1990; Hammer, 1992; Grant et al., 1991; Gill, 1990; Kidd, 1992; Kearney, 1989; Kohler, 1988; Nonaka, 1991; PA Consultants, 1989; Porter, 1990; Bishop and Schofield, 1989; Senehi et al., 1991; Sinclair, 1986, 1992; Wobbe, 1992; Yoshikawa, 1992).

The cybernetics case is as follows; the Law of Requisite Variety (Ashby, 1962a,b; Brehmer, 1988) states that for any system to remain under control, the controller of that system must be able

---

* Corresponding author.

to absorb the entire range of inputs that may affect the system (i.e. the system or the control process must be at least as complex in its behavior as the system it is trying to control). Given that a manufacturing organization is an open system, i.e. is affected by its physical, commercial, legal and social environment as well as its own environment, then the control subsystem must be able to remain in a stable state. Quite apart from the nature of such inputs (most of which exist in human-compatible form, but many of which could be made computer-compatible), there is the content of this input. Within the organization, the decisions to be made vary in importance, and hence the type of information required for decision-making also varies. The information content varies from very low level of detail (e.g. "The tool has engaged the workpiece") to very high level of detail (e.g. "The recent government legislation carries the following implications"). It is difficult to see how any automated system could deal with high level information of the latter sort, and make the necessary deductions and other inferences without an enormous investment in background knowledge stores. Furthermore, the unexpected nature of some of the inputs ("Behold, a hurricane has blown the roof in") will require human-like intelligent control procedures, unlikely to be available in the foreseeable future, to be generated from automated systems.

Secondly, there are the system improvement, system monitoring, and maintenance roles to be performed within manufacturing units. Again, these roles require human-type intelligence, and frequently great manipulative skills for their performance. It is difficult to see how such roles could be performed automatically.

It should be evident, therefore, that people will be necessary in manufacturing plants for a long time to come. There can be no doubt that many of the functions currently being performed by humans will be taken over by automation, and indeed in many cases this is to be welcomed (in general, these functions are those that are fairly well proceduralized, require little creative input, and permit algorithmic analysis). The functions that are not automated will be those requiring cognitive skills of a high order: design, planning, monitoring, exception-handling, and so on, and will still be performed by humans.

While it is easy to comprehend the need for large capital investment associated with the automated option, the technical reasons are not always evident. The prime requirements for automating any function are the availability of a model of the activities necessary for that function, the ability to quantify that model, and a clear understanding of the associated information and control requirements. One cannot simply automate functions by mimicking what people do (Nevins and Whitney, 1989). According to Nevins and Whitney, there are several reasons why automation purely/largely based on what people do does not work well:

(1) In many cases we really do not know exactly what people are doing.
(2) We are limited by technology to build machines that are comparable to human capabilities, particularly in flexibility and decision-making, and mimicking their techniques.
(3) People have unique means – sensing capability for instance.
(4) People are often too innovative or resourceful and improve processes in ways machines cannot perform – mating parts that do not quite meet the tolerances; finding ways to prevent errors or inefficiencies in process (the *poka yoke* attribute).

Some manufacturing researchers and practitioners, in fact, openly admit that human is the most versatile worker and products and systems redesigned with this fact in mind can tremendously reduce cost while improving quality (Coleman, 1988).

However, because complete automation is not yet a practical alternative it does not necessarily mean that no functions can be automated or all functions must be performed manually (Mital et al., 1991). Clearly, there are some functions that should be performed by machines because of:

(1) design accuracy and tolerance requirements,
(2) the nature of the activity is such that it can not be performed by humans (e.g. water jet cutting, laser drilling),
(3) speed and high production volume requirements,

(4) size, force, weight and volume requirements (e.g. materials handling),
(5) hazardous nature of the work (e.g. welding, painting),
(6) special requirements (e.g. prevent contamination), etc.

Equally, there are some activities that should be performed by humans because of:
(1) information-acquisition and decision-making needs (e.g. supervision, some forms of inspection),
(2) higher level skill needs (e.g. programming),
(3) specialized manipulation, dexterity, and sensing needs (e.g. maintenance),
(4) space limitations (e.g., work that must be done in narrow and confined spaces),
(5) situations involving poor reliability equipment or where equipment failure could be catastrophic,
(6) activities for which technology is lacking (soil remediation), etc.

This leaves a very large number of activities that could be performed either by machines or people. These activities must be reviewed carefully and then assigned either to automated equipment or humans in a systematic manner for smooth and trouble-free operation of a manufacturing organization. For instance, consider assembly of small and varied parts with cycle times of 1–3 s. The situation does not lend itself to hard automation because of the variability required. Both humans and robots may be able to provide the necessary flexibility to accomplish the task. While humans are both *capable* and *flexible*, existing robots only have the flexibility. For robots to be capable, they must have the necessary intelligence (Brady et al., 1984) and must be able to sense the working situation to see, touch, feel pressure, and sense their own movement; acquire knowledge and judgement to carry out the task properly and act according to the knowhow of the skilled operator (for example, clean parts that may have accidently acquired some swart); and communicate with other humans. Therefore, if the task requires both flexibility and capability, it must be assigned to humans; otherwise both humans and robots can perform it together with the final task assignment based on other considerations (safety and economics).

However, this example illustrates one of the major problems in function allocation; while it might be sensible to allocate the function to humans initially, this may not be a sensible decision when aspects such as job design (the next level of consideration in function allocation) are included. It is difficult to see how any job having a major function that can also be performed by a robot could be considered to be either intrinsically rewarding or fulfilling; nor is it easy to see how any degree of personal improvement is possible in such a job. It is far more likely that there will be a strong sense of monotony in the job, which is usually associated with computer-productive attitudes and rapid labor turnover. At this point in the allocation of functions, one usually asks the question, "Well, why do we do things this way?", and return to the drawing board to see if the function or process could be redesigned. This multi-level approach to function allocation is discussed later.

Some of the activities that may generally fall under the category of activities that can be performed by both humans and machines are:
(1) assembly of parts and subassemblies,
(2) routine on-line inspection,
(3) packaging and shipping,
(4) palletizing and stacking,
(5) materials handling,
(6) sorting, etc.

In order to resolve the dilemma of human or machine, a functional database of all the operations performed in a manufacturing unit needs to be created and specific procedures and guidelines should be established and followed for efficient and effective allocation of functions between humans and machines. The first part of this requirement can be achieved by using standard techniques such as $IDEF_0$, or CORE. It is the second part of this requirement that this paper addresses.

The assignment of tasks to humans and machines must be based on in-depth analysis of:
– information and competence requirements of the activity,

- capabilities and limitations of humans and the automated equipment,
- availability of technology,
- safety and comfort,
- the design principles for "good" jobs, and
- economics.

2. Scope of the problem

Manufacturing systems are considered essential by most nations for the creation of wealth. Wealth, in turn, dictates the standard of living. Today, the majority of countries in the world engage in some form of manufacturing (narrowly defined as conversion of raw materials into finished goods). It should be noted that no nation in the developed world has been able to achieve this status without a manufacturing base that comprises at least 20% of Gross Domestic Product, and which provides at least 30% of traded goods between nations (see, for example, Australian Manufacturing Council, 1990; Office of Technology Assessment, 1991; Institution of Electrical Engineers, 1992).

Historically, manufacturing was considered as the application of manual skills acquired by on-site and informal training. It has also been considered a necessary evil that must be carried out in order to undertake more meaningful (wealth creating) business activities. In recent years, the need to compete in the world market, to improve industrial productivity, the fear of intellectual stagnation and loss of creative competitive edge have forced the manufacture of products that represent quality and excellent value for money. The ease, quickness, and economy of the manufacture of quality products have become the yardsticks that measure the efficiency and effectiveness of a manufacturing outfit. Automation, partial (hybrid) or total, was widely believed to provide the answer to the woes of yesteryears' manual manufacturing by providing high machine utilization, superior quality and more accurate and reliable components, lower labor costs, lower manufacturing lead times, and faster response to market demands (Groover, 1986, 1987). There is hardly any manufacturing organization in the industrialized world that has not automated at least some of its functions. The debate is no longer whether to automate or not automate, but to what extent.

As indicated earlier, complete automation is, at present, technically and economically not a viable option. Those advocating complete automation, and forging ahead with it, have experienced major setbacks and disasters. In fact, some organizations have had such a shock that it may be some time before automation figures highly in their strategic planning. It is now widely recognized that: (a) automation is not the answer to all problems, (b) if applied properly, it can be very effective, (c) local and regional economic and production factors play a major role in the success of automation, and (d) blindly following others (countries and manufacturers) is certain to lead to economic disaster (Sinclair, 1986, 1992; PA Consultants, 1989; Nonaka, 1991; Office of Technology Assessment, 1991; Institution of Electrical Engineers, 1992; Hoskins, 1992; Kidd, 1992; Jain, 1993; Kumar, 1993).

Since neither extreme – manual manufacturing or complete automation – is desirable, the problem of assigning functions (tasks, activities) to humans and machines has assumed an ever greater significance. It is one of the major issues that has yet to be resolved satisfactorily, even though there has been much discussion of it over the past forty years. For the practitioner, function allocation considerations have been limited due to:
- a lack of systematic and step-by-step approach to the function allocation decision-making process,
- a lack of generic guidelines which integrate all criteria for function allocation,
- a lack of systematic and concise data for addressing issues, such as the capability and limitations of humans and automated equipment, and under what circumstances one option is preferable over the other,
- a lack of methodology of symbiotic agents (ergonomists and manufacturing engineers) to integrate human and machine behaviors,
- a lack of unified theory addressing domain issues, such as roles and authorities, etc.

- a lack of integration of other decision-making criteria into this process so that decision is not made in isolation.
- a lack of easily usable tools to simulate different configurations of humans and machines.

These issues are discussed below.

3. Background (Literature review)

The function allocation problem has been in existence (perhaps) ever since the beginning of the industrial revolution. Over the years, it has drawn the attention of many researchers. Modern efforts to resolve the problem may be said to have begun with the efforts of Fitts (1951). He prepared a list of capabilities in which humans and machines excelled (Table 1), enabling some comparisons to be made between them. This list has been used frequently in providing general guidance to solving the function allocation problem.

While the "Fitts List" undoubtedly provided some direction for future thinking, it has had little impact on engineering design practices due to its overly general, non-quantitative, and incompatible criteria. However, the assumption that functions will be performed exclusively by humans or machines does have some merit in the context of modern manufacturing. In the '60s, '70s, '80s and early '90s, more elaborate lists emerged, as did criticisms of the approach (Swain and Wohl, 1961; Jordan, 1963; Chapanis, 1965; Meister and Rabideau, 1965; Coburn, 1973; Doring, 1976; Rasmussen, 1976; Price et al., 1982; Pulliam et al., 1983; Price, 1985; Ghosh and Helander, 1986; Sinclair, 1986; Madni, 1988; Mital et al., 1991), but no systematic procedure evolved that was widely accepted. A fundamental criticism that was advanced many years ago (e.g. Jordan, 1963) still holds true. At its simplest, unless one can describe functions in engineering terms, it is not possible to ascertain whether a machine could perform the function; if one can describe human behavior in engineering terms, it is usually possible to design a machine to do the job better. But many functions cannot be completely specified in engineering terms; this implies that those micro-functions which cannot be described in engineering terms should be allocated to humans, with the rest allocated to machines. And it is the lack of clear criteria which has led to the manufacturing systems of today.

A further problem is that most efforts were directed at non-manufacturing activities and, from an application standpoint, added to the frustration since no specific function allocation procedures emerged. The only approaches that received any great degree of recognition were those of Meister and Rabideau (1965), in which they attempted to show the process. But the comments of practitioners are perhaps the most telling:

"A rather stark conclusion emerges: there is no adequate systematic methodology in existence for allocating functions (in this case, test and checkout functions) between man and machine. This lack, in fact, is probably the central problem in human factors today.... It is interesting to note that ten years of research and applications experience have failed to bring us closer to our goal than did the landmark article by Fitts in 1951." (Swain and Wohl, 1961).

Table 1
The original Fitts list (Fitts, 1951)

Humans appear to surpass present-day machines with respect to the following:
1. Ability to detect small amounts of visual or acoustic energy.
2. Ability to perceive patterns of light or sound.
3. Ability to improvise and use flexible procedures.
4. Ability to store very large amounts of information for long periods and to recall relevant facts at the appropriate time.
5. Ability to reason inductively.
6. Ability to exercise judgement.

Present-day machines appear to surpass humans with respect to the following:
1. Ability to respond quickly to control signals, and to apply great force smoothly and precisely.
2. Ability to perform repetitive, routine tasks.
3. Ability to store information briefly and then to erase it completely.
4. Ability to reason deductively, including computational ability.
5. Ability to handle highly complex operations, i.e., to do many different things at once.

"Many textbooks accounts of this process assume the systems engineer goes through a deliberate cataloguing of functions, followed by a careful and reasoned weighing of alternatives: 'Should a man do this, or should a machine do it?' This rarely happens. The nature of systems engineering and the economics of modern life are such that the engineer tries to mechanize or automate every function that can be mechanized or automated. What cannot be mechanized is left for human operators to do. This method of attack, counter to what has been written by many human factors specialists, does have a considerable amount of logic behind it. In general, machines can be made to do a great many things faster, more reliably, and with fewer errors than people can do them. These considerations and the high cost of human labor make it reasonable to mechanize everything that can be mechanized. This means, however, that one important job of the engineering psychologist is to ensure that the jobs left over for human operators are within human capabilities. If they are not, the jobs must be redesigned. Men are much more flexible than machines, and it is this flexibility that gives the systems engineer some freedom in his work. A primary task of the engineering psychologist is to ensure that the limits of man's flexibility are not exceeded simply because machines cannot be made to do everything." (Chapanis, 1965).

"Because of the importance of these [allocation of] functions decisions, one would hope for some guidelines to aid the systems designer in allocating specific functions to human beings versus physical machine components ... [But there are problems with the guidelines approach, and as a result] ... The allocation process must rely on expert judgement as the final means for making meaningful decisions; allocation of functions is as much an art as a science." (Sanders and McCormick, 1993).

Sad words, indeed. In spite of these shortcomings, research work on task allocation has permitted certain general inferences:
- function allocation cannot be accomplished by a formula – for example, the rules which may apply in one situation may be irrelevant in another (i.e. the rules have limited scope);
- function allocation is not a one-shot decision – the final allocation of functions depends on considering activities at the levels of the tasks, the conflation of tasks into jobs, the relationships of jobs within a workgroup, and the likely rate of change of the manufacturing process itself (i.e. higher level strategies);
- function allocation can be systematized – it is clear that there are a number of sequential steps that can be taken;
- humans and machines can be both poor or good at certain tasks;
- the use of analogies can facilitate the function allocation process;
- function allocation has to be targeted to a specific time frame;
- function allocation needs multi-disciplinary involvement to be effective;
- function allocation varies dramatically based on whether the task is perceptual, cognitive or psychomotor;
- function allocation must not only be based on the capabilities and limitations of the humans and machines, but also on sound economic analysis;
- human and machine performances are not always antithetical;
- function allocation must consider future technology advances, within the given time frame;
- Wherever possible, for those cases in which either machines or people could perform the function, the system should be designed so that both can perform the function (so that people can delegate the function to machines, or can take over the function when circumstances demand it). Where it is not possible, the function should be allocated to people, or to machines, whichever is "better", according to the criteria being used.

Many peripheral studies also have been able to provide some guidance concerning the criteria and approaches that may be used to determine how the decisions regarding function allocation should be made. Some of the promising criteria and approaches are:
- performance: comparing the time required by machines and humans to complete the task and assigning the function on the basis of superior performance,
- job and skill analysis approach: comparing the capabilities and limitations of the humans and machines, with particular attention to knowledge, skills, information sources and channels involved,
- systematic procedural approach: to include all relevant factors,
- economics: allocating the function to least expensive option, and

– safety: allocating the function to equipment if the human safety concerns are not alleviated.

3.1. Comparison of human and automated equipment (robot) performance

Several studies carried out over the last fifteen years have shown that significant performance differences between humans and robots exist. While robots are faster when carrying out activities such as welding, when it comes to assembly, certain materials processing activities or materials handling, humans outperform robots. Paul and Nof (1979) performed a study to determine the difference between robot and human operator performances in assembling a water pump. The Methods-Time Measurement (MTM) method (Maynard et al., 1948) and Robot-Time and Motion method (RTM – Paul and Nof, 1979) were applied to determine the performance times of humans and robot, respectively. It was observed that the robot required more than eight times the human performance time. Wygant (1986) compared the actual total time taken by robots and humans for drilling a 1.27 cm deep hole in a casting. The time, estimated by ROBOT MOST (based on Maynard Operating Sequence Technique known as MOST – Zandin, 1980), was 0.22 min for humans and 0.50 min for robots. The robots, thus, took 127% more time than humans to complete the task. Mital et al. (1988), Mital and Mahajan (1989) and Mital (1991) also compared assembly time taken by robots and humans for a variety of products using MTM and RTM predetermined time systems. In every case, humans outperformed robots. Mital (1992a), using direct time study, observed that robots took almost 645% more time than humans for performing palletizing and stacking tasks. A major reason for this large difference, and those observed in other studies, is that while humans learn on the job, machines do not. A machine's performance does not vary much but is expected to deteriorate as it ages. Of course, other criteria such as reliability and precision enter into this decision, but as several of the studies mentioned above indicate, if the functions are designed with *poka-yoke* principles in mind, and design for manufacturability is practiced, human performance can often match robot performance in particular circumstances.

3.2. Job and skills analysis approach to task allocation

One particular approach to job and skills analysis for function allocation was suggested by Nof et al. (1980). This approach was directed towards materials handling tasks and to certain fabrication and assembly tasks (particularly those which could be described as "materials handling with controlled collisions"). This approach involves two sequential steps:
(1) Robot-man chart (RMC) and
(2) Job and skills analysis (JSA)

RMC provides a means of identifying jobs which can be performed by either robots or humans. Once it is determined that a robot can perform the job, JSA is conducted.

RMCs are similar to the traditional man–machine charts used in the field of motion and time study and compare the relative abilities of humans and current industrial robots in the following broad categories:
(1) Action and manipulation: body dimensions, strength and power, consistency, reaction speed, and self diagnosis,
(2) Brain and control: computational capability, memory, intelligence, reasoning, and signal processing,
(3) Energy and utilities: power requirements, utilities, fatigue, down time, life expectancy and energy efficiency,
(4) Interface: sensing and inter-operator communication,
(5) Miscellaneous factors: environmental constraints, brain muscle combination, training, and social and psychological needs,
(6) Individual differences.

The JSA method is based on the concept that a task can be broken down into elements with their time and skill requirements. The time required by robots to complete the assigned task is determined using actual time study. In a case study reported by Nof et al. (1980), it was observed that the use of a camera reduced the task time for the robot by about 30%. The method, however, is highly biased towards the automated option as once it is determined that the robots can perform the task, the task is assigned to the robot without any due consideration to other factors, such as human performance, that may affect the decision.

The JSA method uses the tabular (listing) method to compare the differences between humans and robots (physical structure, motion control and intelligence). These tables offer a basic comparison and can adjust to the technological variation, but cannot decide the allocation of those functions that both humans and robots can perform. This is as it should be; essentially, this is a management decision, depending on management's assessment of the (typically uncertain) future of the products and the process, and the need to cope with variability and flexibility.

Genaidy and Gupta (1992), based on published studies, attempted to combine the work measurement and the JSA described above to come up with an intuitive procedure that can be used to assign a physically demanding activity to humans and robots. The following procedure was proposed:

(1) determine job demands,
(2) determine robot capabilities (using RMC procedure),
(3) if job demands do not match robot capabilities, then find another alternative; otherwise proceed to the next step,
(4) if the job conditions are unsafe, select robots; otherwise proceed to the next step,
(5) determine human capabilities,
(6) compare human and robot performances (using RTM, MTM or time study),
(7) conduct economic analysis and
(8) depending upon the outcome of economic analysis select robot or human for the task.

The approach reflects the thinking of researchers working in manual materials handling, but lacks depth and does not provide any specific information about solving the function allocation problem.

So far, we have only discussed the job and skills approach as it applies to a fine-grain analysis of a manufacturing facility. However, this approach does not extend to the longer-term, more large-scale, discrete step-change problems such as are involved in the complete rehabilitation of a production line, or even the creation of a new, greenfield manufacturing facility. This is a problem of a different magnitude, but one in which the allocation of functions is even more important. The approaches above, when applied to problems of this magnitude, prove to be insufficient. Firstly, they depend on having information which typically is not available when the function allocation decisions have to be made (e.g. "Should we aim for a heavily-automated plant staffed by relatively few professional engineers?"; "Should we aim for a limited, incremental improvement over what we have now, using the same people with the same skills?"; "Where between these two extremes should we aim?"). Secondly, the time required to perform the analysis in the detail required is unacceptable.

In response to this, other approaches based on a jobs and skills analysis have been suggested. One of these is based on the arrangement of functions from a strongly human-oriented viewpoint, with the criterion of producing "good" jobs, making best use of human skills (e.g. Engstrom, 1991; Ehn, 1988). Factories have been designed on these principles (e.g. the Volvo plants at Uddevalla in Sweden). However, in recent years these facilities have been closed, though whether this is due to local economic conditions, or to less economic production compared to the best plants elsewhere is not known.

Another approach which has been applied in Europe is Methods GRAI (Doumeingts, 1984; Roboam et al., 1989), which is based on $IDEF_0$ functional representations, followed by a cost engineering approach. However, this approach does not pay much attention to human capabilities, tending to leave those jobs which cannot be performed well by machines to humans.

A third approach based on jobs and skills analysis which offers promise is based on several different strands of thought, all concerned with identifying the knowledge and skills required to operate an organizational structure, and to define jobs (Andersson et al., 1979; Badaracco, 1990; Majchrzak, 1988; Bainbridge and Quintanilla,

1989; Prahalad and Hamel, 1990; Rasmussen, 1979, 1980; Shipley, 1990; Wirstad, 1979, 1988). Independently, all these authors have discussed the configuration of knowledge and information in the organization, and it is but a short step to use these classifications to determine the allocation of functions (see, for example, Siemieniuch and Sinclair, 1993). However, this approach is only gradually emerging from laboratories, and it will be some time before it could be considered to be accepted.

The above approaches are not discussed further in this set of guidelines; they will be the subject of a later set that will appear in this journal sometime in the future, and have been mentioned here for cross-reference. In this set, we concentrate on the fine-grain allocation of functions; it is by far the most common form.

3.3. Systematic procedural approach to task allocation

Mital et al. (1991) presented a basic framework for function allocation in an FMS environment. The paper attempted to categorize functions in a way that helps in allocating these functions to machines and human operators. It was pointed out that certain functional requirements can only be satisfied by assigning them to humans while other functional requirements can be met only by assigning them to machines. Functions that can be performed either by humans or machines can be allocated only after a thorough consideration of all influencing factors. Often, this would necessitate a rigorous quantitative analysis of these factors.

For instance, consider material handling activities. The factors that favor manual materials handling are:
(1) low investment,
(2) low wage rates (low operating cost),
(3) movement and handling flexibility,
(4) ability to operate in narrow and confined spaces,
(5) longer availability.

The factors that discourage the use of manual materials handling are:
(1) high volume of materials to be handled,
(2) awkward loads (e.g. bulky, poor gripping points, weight, etc.),
(3) worker injuries and associated costs,
(4) special needs (e.g. contamination control, hazardous materials),
(5) high wages.

The factors favoring automated materials handling are:
(1) low interest rates,
(2) storage space needs,
(3) reduction in operating costs, particularly with the operation of a second or third shift,
(4) ability to reach high and deep locations,
(5) ability to optimize the utilization of the storage space,
(6) ability to work in inhospitable environments.

The factors that discourage the use of automated materials handling equipment are:
(1) high initial investment,
(2) specialized maintenance requirements,
(3) equipment reliability,
(4) high interest rates,
(5) inflexibility.

Since the influence of these factors will vary from situation-to-situation and from time-to-time, a detailed consideration of each factor is warranted. The feedback from the analysis will determine the allocation of function (manual or automated materials handling).

A systems approach to allocating tasks between humans and robots was provided by Ghosh and Helander (1986). An inventory of common human and robotic tasks in a manufacturing environment was developed. The following steps were considered to represent a systems approach to designing human robotic interaction:
(1) inventory of anticipated common tasks in manufacturing,
(2) design of products to be manufactured,
(3) allocation of tasks between humans and robots,
(4) iterative improvement in product design.

The basic human tasks that were listed include programming, maintenance, supervision, materials handling, assembly, quality control, and leftover tasks. The basic robot tasks include materials handling, load/unload, surface cleaning and coating, assembly, and inspection. The allocation of functions between human and robots was based purely on the capabilities and limitations of both options in a particular situation. The discussion is general and qualitative in terms and the inventory of tasks is not very comprehensive. The study, however, provides an elementary framework for function allocation and can be further developed.

Price (1985) suggested a five-step approach to function allocation:

Step 1:
To prepare for design organizing teams, clarify requirements and plan a design documentation base.

Step 2:
To identify functions, define each necessary function in terms of its inputs and outputs, and identify a set of accessory functions.

Step 3:
Hypothesize design solutions. This is the major step in the design cycle, the step at which a lot of interaction must take place between engineering, allocation, and human factors decisions. For each function, an engineering hypothesis must be made. If a good engineering solution cannot be found, then the functions must be redefined. Based on the engineering hypothesis, an allocation hypothesis must be made. Revisions must be made to the engineering hypothesis and design hypothesis until a mutually consistent set of hypotheses is determined.

Step 4:
The allocation of function hypothesis must be tested and thoroughly evaluated.

Step 5:
The design cycle must be iterated to correct errors and optimize design to acceptable level of detail.

In addition to spelling out the procedure, four rules have also been suggested that can be applied to develop the allocation hypothesis as a part of Step 3. These rules are based on the early work of Price and Tabachinik (1968), updated by Pulliam et al. (1983):

(1) Mandatory allocation: There are mandatory reasons for allocating a function to humans or machines. For example, safety considerations, legal requirements, engineering limitations, etc.
(2) Balance of value: Estimate the relative goodness of human and machine technology for performing the intended function.
(3) Utilitarian and cost-based allocation: A rigorous economic analysis must be performed, the relative cost of human and machine performance must be computed and decision of task allocation must be based on sound economic analysis.
(4) Allocation of function for affective or cognitive support: The cognitive needs of humans must also be considered in function allocation.

3.4. Use of expert systems for task allocation

An entirely different approach to task allocation was proposed by Madni (1988), wherein a network-based environment was presented as a solution to the problem. "HUMANE" (Human–Machine Allocation Network-Based Environment) is intended to be used by designers to postulate and evaluate function allocation options. The software consists of:

(1) Task modeling: the task is modelled using simulation, and the model is then encoded using modified petri nets. The appropriate information is elicited from the database and knowledge bases. The knowledge bases have information collected from diverse sources, such as literature on human performance, laboratory experimental findings, expert opinion, etc.
(2) Function allocation option generation and evaluation: In this phase, based on user input, the system provides a set of options for function allocation.
(3) Analysis of the options: HUMANE provides advice on function allocation, and the technical feasibility and cost–benefit analysis of a particular option.

HUMANE, however, is not very easy to use as the software has been written in Zestalisp and runs on the Symbolics 3600 family of computers. It also requires computer literacy on the part of the user and is not very generic in function allocation. Furthermore, no practical application is provided.

3.5. Economic considerations in function allocation

Considerable efforts have been directed towards assigning task to humans and automated equipment based on economics. Mital and Vinayagamoorthy (1987) examined the economic feasibility of a robot installation in a metal industry from two viewpoints:
(1) the company's viewpoint and
(2) the government's viewpoint.

The results indicated that the robot installation, though desirable from the company's viewpoint, is not always attractive to the government, especially if it leads to permanent replacement of displaced employees. However, when the displaced employees are relocated with or without retraining, the robot installation becomes economically desirable from both viewpoints. A slight increase in GNP of about $(19*10^{-11}\%)$ was computed for the case study with worker displacement. However, based on the nationwide trend towards robot installations, it was pointed out that by 1990, the net effect would be a decline of approximately 2% in the GNP. If the workers are relocated, then an increase of $31.06*10^{-11}\%$ was observed.

Mital and George (1989) presented an economic analysis of selecting a product (computer printer) assembly method from the following three alternatives:
(1) manual,
(2) hybrid (part manual and part automatic), and
(3) automatic.

It was shown that if certain factors change, such as the equipment reliability, the chosen alternative, even though economically desirable initially, results in higher unit production costs and, therefore, becomes undesirable. Hence, all factors, such as equipment reliability, that are ignored in conventional economic analysis must be considered.

A logical sequence of steps that must be followed for automation project evaluation should include:

(1) Precost phase
(a) Determination of alternative manufacturing methods
(b) Technical feasibility study
(c) Selection of jobs to automate
(d) Non-economic considerations
(e) Data acquisition and operation analysis
   (i) Projected production volume
   (ii) Parts throughput requirements
   (iii) Projected daily production hours
   (iv) Automated equipment utilization
   (v) Capacity volume requirements
(f) Decision concerning future applications

(2) Cost analysis phase
This phase focuses on detailed cost analysis for investment justification and includes 5 general steps:
(a) Period evaluation, depreciation and tax data requirements
(b) Project cost analysis:
   (i) Labor considerations
   (ii) Acquisition and start up cost
   (iii) Operating expenses
(c) Evaluation techniques:
   (i) Minimum cost rule
   (ii) Payback method
   (iii) Net present worth method (NPW)
   (iv) Rate of return method
(d) Additional economic considerations:
   (i) Effect of interest rates
   (ii) Performance gaps
   (iii) Effects of recession and inflation
(e) Recommendation based on analysis results

Mital et al. (1988), Mital and Mahajan (1989), Mital (1991, 1992b) demonstrated the effects of various production and market factors on economic decision-making as it relates to product assembly methods. In particular, the influence of the following factors on the cost of manual and flexible automated assembly methods was considered:
(1) Performance time,
(2) Production quantity,
(3) Wage rate,
(4) Interest rate and
(5) Robot reliability.

Several DFA (design for assembly) products were considered and it was conclusively proved that designing products for assembly are always cost effective. The study, however, raised considerable doubt regarding the belief that the flexible automated assembly will always lead to increase in productivity. The disadvantages of the flexible automated assembly methods are primarily created by the slow performance of robots, their low reliability, high interest rates, and, recently in the United States, declining wages. In general, it was found that low interest rates and high labor costs tend to favor the flexible automated assembly method. High interest rates, low wages, low equipment reliability and ability to learn on the job favor the manual assembly method.

Function allocation based on economic analysis in one country may not always be the same as that based on economic analysis in another country. Mital (1991) compared the cost of manual and flexible assembly methods in four different countries (Japan, UK, USA, and Germany). It was shown that while robotic assembly methods may be economically more desirable in one country, the manual mode may be economically more desirable in other countries. The interest rate, production volume, labor costs, and human performance were identified as the major factors that influence the decision to choose a particular assembly method. In general it was found that the manual assembly was more economical in the U.K. and the U.S.A while the flexible assembly option was cheaper in Japan. The major reason for this difference was attributed to low interest rates in Japan. Higher hourly rates also favor automation. Relatively smaller hourly wages and higher interest rates in western countries tend to favor manual assembly.

Mital (1992a) showed that automation of manual material handling activities may not always be a realistic option to control overexertion injuries. For automation to be possible, it must meet two conditions:
(1) automation must be technically feasible and
(2) it must be within economic means of the company.

The case study of the palletizing problem discussed by the author is one of the many cases that lacked a practical solution. Recently, an expert system based algorithm was developed that could solve the problem. An economic analysis was performed to determine the cost–benefit of state of the art technology over the manual option. It was observed that in spite of the technical advances, humans were still faster than robots in performing complex pallet-loading tasks. The manual alternative was not only faster, it was economical as well. The manual alternative was found to take only 21% of the time it took the robot to the load the pallet. The manual alternative also resulted in annual savings of $268,313 for the whole set-up.

3.6. Safety considerations in function allocation

One of the major reasons for introducing automated equipment in the work-place is to avoid exposing humans to hazardous working environments (e.g. handling heavy parts, hot metal slabs, poisonous, radioactive, and other hazardous chemical atmospheres). If the task is unsafe for humans, or has a high risk of injury, regardless of the cost it must be assigned to the automated equipment.

The typical applications of robots in industry include welding, spray painting, grinding, material handling, assembly, and warehousing. One of the major concerns is that automation, instead of isolating humans from working in such a hazardous environment, may have put them at a greater risk. According to Reader (1986), the most important aspect of automation is not that it replaces people, but that it drastically changes the workplace environment. Humans, the most important components in that environment, must learn to work with the robots in order for automation to succeed in producing tangible benefits. This may put humans at a greater risk; a robot might collide with a worker or some other equipment and cause personal injury or material damage (as is discussed below). It should be remembered that while most robots are safe in normal performance, and usually are enclosed, it is when maintenance occurs that most of the safeguards are disabled, and the maintainers are in close proximity to the robot.

When considering worker safety, attention should be given not only to the individual ma-

chines or robots, but also to their interactions arising from processing needs. Jiang and Cheng (1992) proposed a Six Severity Level Design (SSLD) concept for the safe design of a manufacturing cell. The principles used in SSLD design philosophy include:
(1) The design should be based on the knowledge of the manufacturing process (Linger, 1984),
(2) Safety measures should be provided for system failure, human mistake, and misuse of the safety devices (Linger, 1984; Barnett and Switalski, 1988),
(3) The control action should depend on the severity of the risk,
(4) The guarding technique should consider both individual machines and the interactions among the machines,
(5) Warning/alarming devices should be used first for personnel awareness, and
(6) Guarding devices should be provided to isolate hazard sources from the personnel (control actions should be taken according to the level of severity of the danger).

In general, robots should be kept in areas where operators do not need to go, and where they normally cannot go. That is much easier to arrange in some installations than in others, and in any case maintenance staff must go into the robot area. In such cases, the dangers of robotic work cell must be identified (Hartley, 1983). The control system should be such that when setting the robot to do a certain job, the operator should not be able to operate the robot at fast speeds – 10% of the maximum speed is a suitable limit. The control system should also guide the operator in reprogramming, to prevent him from making errors. The system must be designed so that in the event of a failure, the robot will not make unexpected movements. Also, the emergency stop button should be the biggest and easiest button to use on the teachbox. In calculating the real cost of a robot and the time taken to install one and get it running efficiently, the safety equipment and training must be taken into account.

Mital and George (1988), and George and Mital (1989) pointed out that safety of personnel in automated factories is a major concern. In an automated factory environment, with a large number of automated components in operation simultaneously, safety becomes critical. This is especially true in the "bring up" stages of automated systems, such as robot assembly lines, when manufacturing personnel might inadvertently step into the work envelope of a robot during application, teaching or debugging. It is critical that personnel involved in maintenance, installation and operation understand the unpredictability and capabilities of an automated device, such as a moving robot arm, in case of a malfunction such as an encoder or resolver breakdown. Thus, a safety awareness needs to be created and emphasized repeatedly. The key to automation safety is to understand the most effective way for protection and still have an easy access to the automated work area for intervention and maintenance purposes.

Some of the safety measures suggested by the authors include:
(1) Hard hats for all personnel entering the manufacturing area for protection from material flow on overhead conveyors.
(2) Safety mats, safety shields and light curtains strategically placed to protect personnel from robot arms, mechanical fixtures and feeders, shuttles and moving material handling devices, such as conveyors. A key switch can be provided to override these devices for maintenance purposes.
(3) Emergency power-off buttons and emergency power-off chords installed at strategic locations to deenergize moving devices. (It is advisable to have an operator monitor the maintenance technician during the period he works within the work envelope. The operator should have easy access to the emergency power-off button during the period.)
(4) Strobe lights can be used to indicate the work-station status:

Color of strobe	Conditions
Amber	Parts low condition of feeders, magazines.
Blue	Parts out condition of feeders, magazines.
Red	Malfunctions, such as part jams, problems in product infeed, outfeed, improper

	clamping, etc. Operator intervention required.
Green	Maintenance in progress.
White	Station operational.

It should also be realized that safety cannot be compromised to accelerate installation and production schedules, as an equipment-related injury could have serious implications and cause setbacks on production schedules and employee and management morale.

These, as illustrated by the literature review, have been isolated efforts to tackle the problem of function allocation. There is a need to combine all these guidelines to develop a generic scheme of function allocation for use by practitioners in the actual manufacturing environment.

4. After the allocation of functions

It is critically important to realize that the allocation of functions by itself does not provide jobs for humans to undertake. Allocation of functions is exactly what it says; it is only the allocation of functions to humans or machines. After this must follow the grouping of the human functions into jobs. This is not a straightforward process. Frequently, it may involve returning to the allocation of functions decisions and reconsidering them, in order that "good" jobs (i.e. with the right attributes) are achieved. Essentially, three steps are involved here: firstly, grouping the functions to create jobs for individuals; secondly, ensuring that these jobs are appropriately defined for the workgroup involved (and if not, regrouping the functions); and thirdly, ensuring that the workload for these jobs is satisfactory (and if not, regrouping the functions again). There is an extensive literature on the subject of job design, dealing with these three steps. However, a discussion of this copious literature is outside the scope of this guideline.

5. Need for further research

The literature review clearly shows the dearth of studies dealing with the problem of function allocation in a manufacturing environment. Those studies that have investigated this problem have generally assumed that functions will be performed by humans or machines alone. There are numerous situations that may require automated equipment to assist the human operator. For instance, consider assembling small parts with cycle times of 1–3 s. In such cases, even though humans have both the capability and flexibility, it may not be desirable for them to exercise their flexibility, for example to reduce cumulative trauma disorders of the upper extremities. This is a situation that requires automated equipment (robot) to perform the bulk of the work while the human supervises it. Information on the interaction when people are working with automated equipment, safety when human operators have to intervene (for example, when a part is dropped or does not fit), and reliability of the parallel system is needed.

There is also the problem of allocation of functions in major investment decisions; the renovation of entire production lines, or the design of greenfield production facilities. Essentially, there is nothing available to deal with this situation, yet the need in the current manufacturing environment of global competition is great and increasing.

The decision-making flow charts based on prompting questions, provided in Part I of this guide, should be expanded by including quantitative decision-making criteria. For instance, men should not handle objects weighing in excess of 27 kg; women, in excess of 20 kg (Mital et al., 1993). Similar criteria should be included for other activities such as assembly and inspection. For assembly activities, number of components could be the basis for the decision-making criterion. In case of inspection activities, the design or model of the overlay could form such basis.

References

Andersson, H., Back, P. and Wirstad, J., 1979. Job analysis for training design and evaluation – Description of a job analysis method for process industries. Report No. 6. Ergonomrad, Karlstad, Sweden.

Australian Manufacturing Council, 1990. The global challenge – Australian manufacturing in the 1990s. World Trade Centre, Melbourne 3005, Australia.

Ashby, W.R., 1962a. Introduction to Cybernetics. Heinemann, London.

Ashby, W.R., 1962b. Principles of self-organising systems. In: E.H. von Foerster and G.W. Zopf (Eds.), Principles of Self-organising Systems. Pergamon, New York.

Badaracco, J., 1990. The Knowledge Link. Harvard Business School Press, Cambridge, MA.

Bainbridge, L.S. and Quintanilla, S.A.R., 1989. Developing Skills with Information Technology. John Wiley, Chichester.

Barnett, R.L. and Switalski, W.G., 1988. Principles of human safety. Safety Brief, 5: 1–15.

Bishop, T. and Schofield, N., 1989. Unlocking the potential of CIM. PA Consulting Services Ltd., Bowater House East, 68 Knightsbridge, London SW1X 7LJ, U.K.

Brady, M., Gerhard, L.A. and Davidson, H.F., 1984. Robotics and Artificial Intelligence. Springer-Verlag, Berlin.

Brehmer, B., 1988. Organisation for decision-making in complex systems. In: L.P. Goodstein, H.B. Anderson and S.E. Olsen (Eds.), Tasks, Errors and Mental Models. Taylor & Francis. London.

Brodner, P., 1985. Skill-based production: The superior concept to the unmanned factory. In: H.J. Bullinger and H. Warnecke (Eds.), Towards the Factory of the Future. Springer-Verlag, Stuttgart.

Brodner, P., 1987. Strategic options for new production systems: Computer and human integrated manufacturing. CEC-FAST, Brussels, Belgium.

Chapanis, A., 1965. On the allocation of functions between men and machines. Occupational Psychology, 39: 1–11.

Coburn, R., 1973. Human Engineering Guide to Ship System Development. Naval Electronics Laboratory Center, San Diego, CA.

Coleman, J.R., 1988. Design for assembly: Users speak out. Assembly Engineering. July.

Corbett, J.M., Rasmussen, L.B. and Rauner, F., 1991. Crossing the Border. Springer, Berlin.

Doring, B., 1976. Analytical methods in man–machine system development. In: K.-F. Kraiss and J. Moraal (Eds.), Introduction to Human Engineering. Verlag TUV Rheinland GmbH, Cologne.

Doumeingts, G., 1984. Methode GRAI: Méthode de conception des systèmes en productique. Ph.D. Thesis, Laboratoire GRAI, Université de Bordeaux, France.

Ehn, P., 1988. The Work Oriented Design of Computer Artifacts. Arbetsmiljö, Arbelistraum, Stockholm.

Engstrom, T., 1991. Future assembly work – Natural grouping. In: Y. Quiennec and F. Daniellou (Eds.), Designing for Everyone – Proceedings of the XIth Congress of the International Ergonomics Association. Vol. 2. Taylor & Francis, London, pp. 1317–1319.

Fitts, P.M., 1951. Human engineering for an effective air-navigation and traffic-control system. National Research Council, Washington, D.C.

Genaidy, A.M. and Gupta, T., 1992. Robot and human performance evaluation. In: M. Rahimi and W. Karwowski (Eds.), Human–Robot Interaction. Taylor & Francis, London, pp. 4–15.

George, L.J. and Mital, A., 1989. Role of robotics in the factory of the future. Robotics and Autonomous Systems, 4: 309–326.

Ghosh, B.K. and Helander, M.G., 1986. A systems approach to task allocation of humans robot interaction in manufacturing. Journal of Manufacturing Systems, 5: 41–49.

Gill, K.S., 1990. Summary of human centred research in Europe. Research paper, SEAKE Centre, Brighton Polytechnic, U.K.

Grant, R.M., Krishnan, R., Shani, A.B. and Baer, R., 1991. Appropriate manufacturing technology: A strategic approach. Sloan Management Review, 33: 43–54.

Groover, M.P., 1986. Industrial Robotics. McGraw-Hill, New York.

Groover, M.P., 1987. Automation, Production Systems, and Computer Integrated Manufacturing. Prentice-Hall, New York.

Hammer, H., 1992. The economics of flexible manufacturing systems contingent upon operating and service personnel. European Journal of Production Engineering, 16: 38–41.

Hartley, J., 1983. Robots At Work: A Practical Guide For Engineers and Managers. IFS Ltd., United Kingdom/North-Holland, Amsterdam.

Hockley, J., 1990. Manufacturing strategies (editorial). Computer-Integrated Manufacturing Systems, 4: 59.

Hoskins, J., 1992. Personal communication with A. Mital. Lexmark, Lexington, Kentucky.

Institution of Electrical Engineers, 1992. UK manufacturing – A survey of surveys and a compendium of remedies. Public Affairs Board, U.K.

Jain, S., 1993. Personal communication with A. Mital. General Electric Company, Cincinnati, Ohio.

Jiang, B.C. and Cheng, O.S.H., 1992. Six severity level design for robotic cell safety. In: M. Rahimi and W. Karwowski (Eds.), Human–Robot Interaction. Taylor & Francis, London, pp. 161–181.

Jordan, N., 1963. Allocation of functions between man and machines in automated systems. Journal of Applied Psychology, 47: 161–165.

Kearney, A.T., 1989. Computer Integrated Manufacturing: Competitive Advantage or Technological Dead End? A.T. Kearney, London.

Kidd, P., 1992. Organisation, people and technology in European manufacturing. Report No. DG XII/Monitor/FAST, Contract No. MOFA-0004-D (MB), Commission of the European Communities, Cheshire Henbury Research & Consultancy, Macclesfield SK11 8UW, U.K.

Kohler, E., 1988. Technology and the improvement of working and living conditions. Paper presented at the conference on Joint Design Technology, Organisation and People Growth, RSO Milan, Italy.

Kumar, A., 1993. Personal communication with A. Mital. Johnson Controls, Ann Arbor, MI.

Linger, M., 1984. How to design safety systems for human protection in robot applications. Paper presented at the 14th International Symposium on Industrial Robots, Gothenburg, Sweden.

Madni, A.M., 1988. HUMANE: A designer's assistant for modelling and evaluating function allocation options. In: W. Karwowski, H.R. Parsaei and M.R. Wilhelm (Eds.), Ergonomics of Hybrid Automated Systems I. Elsevier Science Publishers, Amsterdam, pp. 291–302.

Maynard, H.B., Stegemerten, G.J. and Schwab, J.L., 1948. Methods–Time Measurement. McGraw-Hill, New York.

Majchrzak, A., 1988. The human infrastructure impact statement (HIIS). Computer Integrated Manufacturing Systems, 1: 95–102.

Meister, D. and Rabideau, G.F., 1965. Human Factors Evaluation in System Development. John Wiley, New York.

Mital, A., 1991. Manual versus flexible assembly: A cross comparison of performance and cost in four different countries. In: M. Pridham and C. O'Brien (Eds.), Production Research: Approaching the 21st Century. Taylor & Francis, London, pp. 481–488.

Mital, A., 1992a. Comparison of manual and automated palletizing of mixed size and weight box containers. International Journal of Industrial Ergonomics, 9: 65–74.

Mital, A., 1992b. Economics of flexible assembly automation: Influence of production and market factors. In: H.R. Parsaei and A. Mital (Eds.), Economics of Advanced Manufacturing Systems. Chapman & Hall, London, pp. 45–72.

Mital, A. and George, L.J., 1988. Human issues in automated (hybrid) factories. In: F. Aghazadeh (Ed.), Trends in Ergonomics/Human Factors V. North-Holland, Amsterdam, pp. 373–378.

Mital, A. and George, L.J., 1989. Economic feasibility of a product line assembly: A case study. The Engineering Economist, 35: 25–38.

Mital, A. and Mahajan, A., 1989. Impact of production volume, wage and interest rates on economic decision making: The case of automated assembly. Proceedings of the Conference of the Society for Integrated Manufacturing Conference, Institute of Industrial Engineers, Norcross, Georgia, pp. 558–563.

Mital, A. and Vinayagamoorthy, R., 1987. Case study: Economic feasibility of a robot installation. The Engineering Economist, 32: 173–196.

Mital, A., Anand, S. and Parsaei, H.R., 1991. Assignment of functions to machines and humans in an FMS environment. Proceedings of the Production Engineering Division, ASME Winter Annual Meeting, Paper No. 91-WA-Prod-2.

Mital, A., Mahajan, A. and Brown, M.L., 1988. A comparison of manual and automated assembly methods. Proceedings of the IIE Integrated Systems Conference. Institute of Industrial Engineers, Norcross, Georgia, pp. 206–211.

Mital, A., Nicholson, A.S. and Ayoub, M.M., 1993. A Guide to Manual Materials Handling. Taylor & Francis, London.

Nevins, J.L. and Whitney, D.E., 1989. Concurrent Design of Products and Processes: A Strategy for the Next Generation in Manufacturing. McGraw-Hill, New York.

Nof, S.Y., Knight, J.L. and Salvendy, G., 1980. Effective utilization of industrial robots – A job and skills analysis approach. Transactions of the American Institute of Industrial Engineers, 12: 216–225.

Nonaka, I., 1991. The knowledge-creating company. Harvard Business Review, 69: 96–104.

Office of Technology Assessment, U.S.A., 1984. Computerised manufacturing automation. Library of Congress Report No. 84-601053.

PA Consultants, 1989. Manufacturing into the late 1990s. Report No. AMR/pm 4A&B to Dept. of Trade and Industry, H.M. Government, PA Consulting Group, Bowater House East, 68 Knightsbridge, London SW1X 7LJ, U.K.

Paul, R.P. and Nof, S., 1979. Work methods measurement – A comparison between robot and human task performance. International Journal of Production Research, 17: 277–303.

Porter, M.E., 1990. The competitive advantage of nations. Harvard Business Review, 90: 73–90.

Prahalad, C.K. and Hamel, G., 1990. The core competence of the corporation. Harvard Business Review, 68: 79–91.

Price, H.E., 1985. The allocation of functions in systems. Human Factors, 27: 33–45.

Price, H.E. and Tabachinik, B.J., 1968. A descriptive model for determining optimal human performances in systems, Vol. III: An approach for determining the optimal role of man and allocation of functions in an aerospace system. Report No. NASA CR-878, Serendipity Associates, Chatsworth, California.

Price, H.E., Maisano, R.E. and Van Cott, H.P., 1982. The allocation of functions in man–machine systems: A perspective and literature review. Report No. NUREG-CR-2623, Oak Ridge National Laboratories, Oak Ridge, Tennessee.

Pulliam, R., Price, H.E., Bongarra, J., Sawyer, C.R. and Kisner, R., 1983. A methodology for allocating NPP control functions to humans or automatic control. Technical Report No. NUREG-CR-3331, U.S. Nuclear Regulatory Commission, Oak Ridge National Laboratory, Oak Ridge, Tennessee.

Rasmussen, J., 1976. Outlines of a hybrid model of the process plant operator. In: T.B. Sheridan and G. Johannsen (Eds.), Monitoring Behavior and Supervisory Control, Plenum Press, New York.

Rasmussen, J., 1979. On the structure of knowledge – A morphology of mental models in man–machine system context. Report No. Riso-M-2192, Riso National Laboratory, Roskilde, Denmark.

Rasmussen, J., 1980. Some trends in man–machine interface design for industrial process plants. Report No. Riso-M-2228, Riso National Laboratory, Roskilde, Denmark.

Reader, D.E., 1986. Human factors in automation. In: P.M. Strubhar (Ed.), Working Safely With Industrial Robots. Robotics International, Society of Manufacturing Engineers, Dearborn, Michigan, pp. 46–48.

Roboam, M., Zanettin, M. and Pun, L., 1989. GRAI-IDEFO-MERISE (GIM): Integrated methodology to analyse and design manufacturing systems. Computer Integrated Manufacturing Systems, 2: 82–98.

Sanders, M.S. and McCormick, E.J., 1993. Human Factors in Engineering and Design. McGraw Hill, New York.

Sata, T., 1986. Development of flexible manufacturing systems in Japan. Paper presented at the First Japan-USA Symposium on Flexible Automation, Japan Association of Automatic Control Engineers, Osaka, Japan.

Senehi, M.K., Barkmeyer, E., Luce, M., Ray, S., Wallace, E.K. and Wallace, S., 1991. Manufacturing systems integration: Initial architecture document (NISTIA 4682), National Institute of Standards and Technology, Gaithersburg, MD.

Shipley, P., 1990. The analysis of organisations as a conceptual tool for ergonomic practitioners. In: J.R. Wilson and E.N. Corlett (Eds.), Evaluation of Human Work: A Practical Ergonomics Methodology. Taylor & Francis, London.

Siemieniuch, C.E. and Sinclair, M.A., 1993. Systems design and concurrent engineering – An organisational perspective. In: F.A. Stowell, D. West and J.G. Howell (Eds.), Systems Science: Addressing Global Issues, Proceedings of the IIIrd UK Conference on Systems. Paisley, Scotland, pp. 525–530.

Sinclair, M.A., 1986. Ergonomic aspects of the automated factory. Ergonomics, 29: 1507–1523.

Sinclair, M.A., 1992. Human factors, design for manufacturability, and the computer-integrated manufacturing enterprise. In: M. Helander and M. Nagamachi (Eds.), Human Factors and Design for Manufacturability. Taylor & Francis, London, pp. 127–146.

Swain, A.D. and Wohl, J.G., 1961. Factors affecting degree of automation in test and checkout equipment. Dunlap & Associates, Stamford, CT.

Wirstad, J., 1979. On the allocation of functions between human and machine – Analysis of concept and its implication for control system ergonomics. Report No. 13, Ergonomrad, Karlstad, Sweden.

Wirstad, J., 1988. On knowledge structures for process operators. In: L.P. Goodstein, H.B. Anderson and S.E. Olsen (Eds.), Tasks, Errors and Mental Models. Taylor & Francis, London, pp. 50–69.

Wobbe, W., 1992. What are anthropocentric production systems? Why are they a strategic issue for Europe? Final Report No. EUR 13968 EN, DG XII/Monitor/FAST, Commission of the European Communities, Brussels, Belgium.

Wygant, R.M., 1986. Robots vs. humans in performing industrial tasks: A comparison of capabilities and limitations. Unpublished Doctoral Dissertation, University of Houston, Houston, TX.

Yamashita, T., 1987. The interaction between man and robot in high-technology industries. In: K. Noro (Ed.), Occupational Health and Safety in Automation and Robotics. Taylor & Francis, London, pp. 139–142.

Yoshikawa, H., 1992. The intelligent manufacturing systems proposal. In: C. O'Brien, P. Mac Conaill and W. van Puymbroek (Eds.), Computer Integrated Manufacturing (CIME Applications and Benefit). Proceedings of VIIIth CIM-Europe Annual Conference (CEC Publication No. EUR 14350 EN), Birmingham, U.K.

Zandin, K.B., 1980. MOST Work Measurement System. Marcel Dekker, New York.

Occupational and individual risk factors for shoulder-neck complaints: Part I – Guidelines for the practitioner *

Jørgen Winkel [a] and Rolf Westgaard [b]

[a] *National Institute of Occupational Health, Department of Physiology, Division of Applied Work Physiology, S-171 84 Solna, Sweden*
[b] *The Norwegian Institue of Technology, Division of Organization and Work Science, N-7034 Trondheim-NTH, Norway*

1. Introduction

The present guidelines are solely based on physical exposure/chronic effect and individual factor/chronic effect relationships documented in the literature (see Part II). Additional guidelines may be given based on physical exposure/acute response relationships as soon as the literature has been reviewed.

The audience of these guidelines comprises all with the responsibility for or possibility to influence occupational shoulder-neck exposure; e.g. O & S personnel, production engineers, safety controllers and management. The guidelines may be used when (1) planning new work tasks, (2) doubt occurs regarding acceptability of occupational shoulder-neck exposure, (3) shoulder-neck disorders occur.

2. General concepts and suggestions

2.1

A description of physical exposure (physical work load) should comprise estimates of exposure levels, exposure repetitiveness and exposure durations (see Part II, section 5.1). 'Variation pattern' is defined as the interaction between exposure levels and repetitiveness. 'Monotony' denotes prolonged exposure comprising short and similar cycles.

2.2

With the exception of extreme physical exposures, which may be classified as accidents, physical activity as such is *not* injurious. However, it may damage the body if the *exposure time* is too long or too short! Thus, lack of physical activity may cause negative physiological effects in the body and this should be prevented by proper training. For most occupational exposures, however, the critical factor is too *long* exposure time. Accordingly, physical exposure should primarily

---

Correspondence to: Jørgen Winkel, National Institute of Occupational Health, Department of Physiology, Division of Applied Work Physiology, S-171 84 Solna, Sweden.

* The recommendations provided in this guide are based on numerous published and unpublished scientific studies and are intended to enhance worker safety and productivity. These recommendations are neither intended to replace existing standards, if any, nor should be treated as standards. Furthermore, this document should not be construed to represent institutional policy.

The following individuals participated in the discussion of the earlier version of this guide. Their suggestions (written or verbal) were incorporated by the authors in this version: Roland Andersson, *Sweden*; Alvah Bittner, *USA*; Peter Buckle, *UK*; Jan Dul, *The Netherlands*; Bahador Ghahramani, *USA*; Juhani Ilmarinen, *Finland*; Sheik Imrhan, *USA*; Asa Kilbom, *Sweden*; Stephan Konz, *USA*; Shrawan Kumar, *Canada*; Tom Leamon, *USA*; Mark Lehto, *USA*; William Marras, *USA*; Barbara McPhee, *Australia*; James Miller, *USA*; Anil Mital, *USA*; Don Morelli, *USA*; Maurice Oxenburgh, *Australia*; Jerry Purswell, *USA*; Jorma Saari, *Finland*; W. (Tom) Singleton, *UK*; Juhani Smolander, *Finland*; Terry Stobbe, *USA*. The guide was also reviewed in depth by several anonymous reviewers.

be regulated by time limits, provided the design of the workstation and hand tools is optimal.

2.3

Minimal time limits may apply to physical training to avoid negative effects of physical inactivity. Numerous guidelines are given in textbooks on physical training.

2.4

Maximal time limits may apply to physical exposures in working life. Exposure time should be reduced when the exposure level and/or the monotony is increased.

2.5

Introduction of time limits requires introduction of new work tasks to avoid non-productive periods. The exposure pattern of the new work tasks should differ from the former.

3. Line of action

3.1. Classify the work task according to exposure level (cf. Part II, section 5.4)

Group 1: 'low' exposure level. Definition of the group: The workstation design is 'good' according to standard ergonomic textbooks. Continue to 3.3.1.
Group 2: 'medium' exposure level. Definition of the group: Work with elevated shoulders and/or abducted/flexed upper arm(s) and/or flexed/extended neck. If tools are used, their contribution to the biomechanical load in the shoulder-neck is marginal compared to the contribution from head, arm and hand. Continue to 3.2.
Group 3: 'high' exposure level. Definition of the group: Exertion of large forces in the shoulder/neck, e.g. handling heavy tools with the upper arm(s) deviating from vertical position. Continue to 3.2.

3.2. Workstation design

Optimize workstation design if possible. For details, consult major ergonomic textbooks. Hence, continue to 3.3.2 (group 2) and 3.3.3 (group 3).

3.3. Regulation on exposure time

3.3.1. Group 1 ('low' exposure level)
In monotonous work tasks a reduction of exposure time from full-time (8 hours) to half-time work may cause only a marginal reduction of the risk for developing work-related shoulder-neck complaints. A more pronounced effect may be obtained if the exposure time is reduced to less than 4 hours/day.

In general, a time limit should be further reduced when one or more of the following factors are particularly striking (individual risk factors are excluded, but may be deduced from the review):
- monotonous work task,
- poor psychosocial work environment (low work control, high psychosocial job demands),
- high work/production intensity,
- lack of breaks during working hours apart from the ordinary lunch break (about 30 minutes) and 2 coffee breaks (about 15 minutes each),
- lack of alternative tasks offering a different exposure pattern during working hours.

Remember that future research may show that additional suspected risk factors may play important roles in the development of shoulder-neck complaints. Continue to 3.4.

3.3.2. Group 2 ('medium' exposure level)
In some situations it may not be possible to obtain a 'good' workstation design; e.g. a fixture may not allow the operator to lower working height to elbow height. Such work tasks belong to group 2 and the exposure times should be shorter compared to those given for group 1. Continuous work of one hour or less may in some cases be reasonable. In particular, the above mentioned modifying and confounding factors should be considered (see 3.3.1). Continue to 3.4.

3.3.3. Group 3 ('high' exposure level)
The exposure times should be shorter compared to those of group 2. In particular, the above-mentioned modifying and confounding factors should be considered (see 3.3.1). Continue to 3.4.

3.4. Group 1 + 2 + 3

An acceptable physical exposure for the shoulder-neck may be obtained by combining work tasks belonging to different exposure groups. Although a given work task may not be acceptable to perform for more than e.g. one hour at a time it may well be acceptable to repeat the task one or several times during the day.

Reference

Winkel, J. and Westgaard, R., 1992. Occupational and individual risk factors for shoulder-neck complaints: Part II – The scientific basis (literature review) for the guide. International Journal of Industrial Ergonomics, 10: 85–104 (this issue).

Occupational and individual risk factors for shoulder-neck complaints: Part II – The scientific basis (literature review) for the guide

Jørgen Winkel [a] and Rolf Westgaard [b]

[a] *National Institute of Occupational Health, Department of Physiology, Division of Applied Work Physiology, S-171 84 Solna, Sweden*
[b] *The Norwegian Institute of Technology, Division of Organization and Work Science, N-7034 Trondheim-NTH, Norway*

1. Problem description

Disorders in the neck and upper limbs were pointed out as an occupational disease as early as the beginning of the 18th century (Ramazzini, 1713). In the 1950's the complaints were recognized as an increasing problem in Japan, but it was only discussed in relevant orthopedic diagnoses; little attention was paid to a possible work-related pathogenesis. In 1973 a committee under the Japanese Association of Industrial Health proposed to call the syndrome 'Occupational Cervicobrachial Disorder' (OCD), which now has become a widely used expression. In addition, 'Repetitive Strain Injury' (RSI), 'Cumulative Trauma Disorder' (CTD) and 'Upper Limb Disorder' (ULD) are commonly used terms in Australia, US and England respectively.

However, not until the 1970's were these complaints emphasized in the international scientific literature as potentially work-related (e.g. Fergusson, 1971). In 1979 a whole issue of the *Scandinavian Journal of Work, Environment & Health* dealt with the problem of OCD (1979). At the beginning of the 1980's OCD gained further international attention by the 'International workshop on occupational neck and upper limb disorder due to constrained work' in Japan (Aoyama et al., 1982). Since then the scientific literature has increased dramatically; for references see the following review papers: Hagberg, 1987; Wallace and Buckle, 1987; Gleerup Madsen, 1990; Stock, 1991.

Reports from all over the world describe complaints in the locomotor system as a dominating health problem in working life (e.g. Grazier et al., 1984; National Occupational Health and Safety Commission, Australia, 1986; Rasmussen et al., 1988; Statistics Sweden, 1991; Socialstyrelsen redovisar, 1987). Despite the widespread documentation of the international importance of OCD in terms of prevalence and economy, its etiology still remains highly controversial. Thus, Hadler (1990) recently stated that "... science is having difficulty demonstrating the damage that is to be feared according to the CTD concept". There are several reasons for the confusion in the scientific literature; one is the lack of pathophysiological explanations linking exposure to disorder in the locomotor system; another may be the low quality of most studies (e.g. Hadler, 1989).

However, in a meta-analysis of previous published studies Hagberg and Wegman (1987) calculated aetiological fractions (proportion of exposed cases attributable to exposure) of 0.5–0.9 for shoulder-neck disorders in a variety of jobs. This suggests that ergonomic interventions often may be effective. In contrast, a prevalent viewpoint in many countries is that musculoskeletal disorders are mainly 'epidemic hysteria' (Sirois, 1974) and 'mass psychogenic illness in organizations' (Colligan and Murphy, 1979). (For further references see Molin and Nilsson, 1990).

OCD comprises the shoulder-neck as well as the arm/hand. The prevalent occupational expo-

---

Correspondence to: Jørgen Winkel, National Institute of Occupational Health, Department of Physiology, Division of Applied Work Physiology, S-171 84 Solna, Sweden.

sure patterns are poorly described. However, the few existing papers on this subject indicate marked differences between shoulder-neck and arm/hand exposures. As risk factors (and guidelines) partly are expressed in terms of exposure, these two body regions should be treated separately.

The purpose of this paper is to review the literature regarding risk factors for shoulder-neck disorders based on physical exposure/chronic effect and individual factor/chronic effect relationships.

2. Information retrieval

2.1. Data bases

Searches were carried out on computer-based bibliographic databases: Arbline, Cisilo, Mbline, Medline and Nioshtic for the years between 1980 and 1991 (February). The following key words were used: occupational diseases, musculoskeletal diseases, neck, shoulder. Different combinations of the key words were tested and they were allowed to occur in the title as well as in the

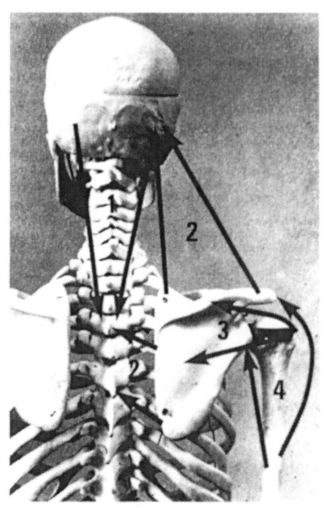

Fig. 1. Some main force actions of the four functional muscle groups in the shoulder-neck region: the neck muscles (1), the stabilizers of the scapula (2), the rotator cuff muscles (3) and the primary movers of the arm (4).

abstract. Additional references earlier than 1980 were gathered by studying reference lists. From February to October 1991 the main relevant journals have been reviewed at the Swedish main library for literature on work environment issues. In combination with the present authors' own files we arrived at the final pool of references.

2.2. Selection of references

In general, the references had to fulfil the following criteria to be included as a contribution to the guidelines:
- original full-length paper (with a few exceptions proceedings have been excluded),
- contained an estimate of exposure as well as chronic effect (complaints),
- a 'contrast' was created in the exposure and/or the individual variables in order to study a possible effect on the outcome.

Studies on exposure to vibration and pathophysiological mechanisms which may link exposure to injuries have been excluded.

The selected literature represents a considerable variation in quality. Most of the papers suffer from incomplete controlling for confounders. However, only papers with serious deficiencies have been excluded.

The present list of references is by no means complete. However, in our opinion, inclusion of additional literature would not make any further contribution to the guidelines.

3. Anatomy

There are four functional groups of muscles in the shoulder-neck (figure 1). In addition, there are joints and passive tissues (e.g. bones, ligaments and bursae).

3.1. Neck

The bony part of the neck is the cervical spine and the occipital bone. The main group of neck muscles (splenius, longissimus, semispinalis) are located underneath the trapezius and levator scapula. They arise from the lateral and dorsal processes of the cervical and upper thoracic spinal segments, and insert on the occipital bone of the head. The function is to counteract a forward flexion of the head when activated bilaterally, and to participate in the generation of a turning movement of the head (splenius).

3.2. Shoulder

The shoulder girdle, consisting of the collar bone (clavicle) and the shoulder blade (scapula) provides the bony link between the arm and the trunk. The three joints of the shoulder girdle are constructed to allow a wide range of arm movement, at the expense of stability. This is compensated by the rotator cuff muscles acting as dynamic ligaments.

Three main groups of muscles in the shoulder participate in the control of arm movement: (a) trapezius, levator scapula, rhomboid, serratus anterior, which arise from the main skeleton and insert on the scapula, moving and stabilizing this structure; (b) the rotator cuff muscles: teres minor, infraspinatus, supraspinatus, subscapularis, which arise from the scapula and insert on the tuberculum, stabilizing the glenohumeral joint; (c) primary movers of the upper arm: e.g. biceps, deltoid, triceps, which arise from the clavicle and scapula and insert onto the humerus.

4. Exposure-effect model

Guidelines should be based on relationships between exposure and effect. A simple model connecting these concepts is illustrated in figure 2. 'External exposure' describes the work task

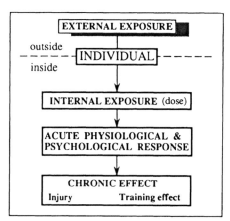

Fig. 2. Exposure-effect model.

Table 1

Summary of selected studies on risk factors for shoulder-neck disorders. The left column indicates references and the numbers (1–3) give the grading of the exposure level as estimated by us (cf. section 5.4).

+ or name of exposure/individual factor: a 'contrast' is created in the exposure or the individual variables in order to study a possible effect on the outcome.
[+] or [exposure factor]: no 'contrast' is created.
–: the exposure factor is not indicated.

Abbreviations: anthropom: static anthropometry; CK: creatinekinase concentration in blood serum; empl.: employment; endur.: endurance; env.: environment; espec.: especially; exam.: examination; exert.: exertion; ext.: extension; flex: flexion; grp.: group; h: hour; indiv.: individual; inf.: infraspinatus muscle; m.m.h.: manual material handling; mov.: movements; op.: operator, phys.: physical; pos: position; pt.: patient; qu.: questionnaire; rep.: repetitive; stat.: static; satisf.: satisfaction; sup.: supraspinatus muscle; tr.: upper trapezius muscle; trad.: traditional; train.: training; workst.: workstation; y: year.

Reference	Job title	Task	Exposure Level	Repetitive-ness	Duration	Indiv. factors	Effect	Risk factors
[1] Aronsson et al. (1988)	VDU op.	4 VDU tasks	–	–	h/day	sex	qu.	h/day, task sex
[3] Bjelle et al. (1979)	industrial worker	–	hands above shoulder	–	–	age, anthrop.	clinical exam. (chronic pain)	hand above shoulder, age
[2] Brisson et al. (1989)	garment worker	[sewing-machine operators]	–	–	y's in piecework, y's in rep. task	age, smoking, education,	interview (disability)	y's in piecework
[1+2+3] Dimberg et al. (1989a)	many	3 classes	3 classes	–	h/day	anthrop.	qu.	work task, age, mental stress, anthrop, smoking, etc.
[1+2+3] Dimberg et al. (1989b)	many	3 classes	3 classes	–	–	smoking, age, blue-collar work, etc.	sick-leave	smoking, age, blue-collar work, etc.
[1] Grieco et al. (1989)	VDU op.	6 VDU tasks	–	–	empl. time, h/day at VDU	sex, age	qu.	h/day at VDU, empl. time, age
[2] Hägg et al. (1990) (longitudinal)	assembly worker	[+]	productivity	–	empl. time	e.g. isom. strength, muscle endur., EMG fatigue, CK, personality	qu. interview clinical exam.	productivity, empl. time, personality
[1] Hedberg et al. (1981)	engine driver	[+]	workst. design	–	–	anthrop.	interview	workst. design/ anthrop.
[3] Herberts et al. (1981)	welder, control grp.: white-collar	–	3 classes	–	empl. time	age	qu., interview clinical exam.	job title
[1+2] Hünting et al. (1980)	accounting machine op.,	+	workst. design arm support	2 classes	[8.5 h/day]	anthrop.	qu.	neck angle, elbow angle

Study	Subjects	Task	Physical load measures	Other measures	—	Exposure time	Confounders / other variables	Methods	Variables analyzed
[1+2] Hünting et al. (1981)	VDU op., control grp.: trad. office work	2 VDU tasks	[strokes/h] body angles, arm support	—	—	[full-time]	—	qu. clinical exam.	job title, head angle, head rotation, arm support
[2] Hviid Andersen & Gaardboe-Poulsen (1990)	garment worker, control grp.	[sewing-machine operators]	—	—	—	empl. time	age, exercise, children, smoking personality, others	qu., clinical exam.	job title, age, children, empl. time
[1] Jeyaratnam et al. (1989)	VDU op.	—	—	—	—	h/day at VDU	marital stat., age, ethnic gr. empl. time	qu.	h/day at VDU, ethnic group, marital stat.
[2] Jonsson et al. (1988) (longitudinal)	assembly worker	2 classes	e.g. neck flex., arm elev., shoulder elev. work	micropauses (~ work technique)	—	—	e.g. isom. strength, muscle endur., productivity, spare time phys. activity, previous sick leave	qu, interview, clinical exam.	job title, task, productivity, previous sick leave, spare time phys. activity, work technique (see "exposure")
[1] Kamwendo et al. (1991)	secretary	typing	—	—	—	empl. time, h/day sitting, h/day typing	psychosocial work env., age	qu.	age, empl. time, h/day typing, psychosocial work env.
[2+3] Kilbom (1988) (longitudinal)	assembly, worker automobile worker	—	—	—	—	—	isom. strength, muscle endur.	clinical exam.	[3] isom. strength
[2] Kilbom et al. (1986)	assembly worker	—	e.g. neck flex., arm elev., shoulder elev.	e.g. no. of flex./abduc. of upper arm	—	—	work capacity, work technique (see "exposure")	qu, interview, clinical exam.	work technique
[1] Knave et al. (1985)	VDU op., control grp.	2 VDU tasks	—	—	—	y's of VDU work, h/day at VDU	sex, age	qu.	h/day at VDU, sex
[1–2] Kukkonen et al. (1983) (longitudinal, intervention)	data entry op., control grps.: data entry op., office worker	—	workst. design	health education	—	—	physiotherapy	qu., interview, clinical exam.	job title, intervention program

[Continued overleaf]

Table 1 (continued)

Reference	Job title	Task	Exposure			Indiv. factors	Effect	Risk factors
			Level	Repetitiveness	Duration			
[1+2] Kuorinka and Koskinen (1979)	assembly worker	2 classes	productivity	cycle time, mode of control	[h/day]	weight/height index, productivity, age	interview, clinical exam.	no clear risk factor
[1+2] Kvarnström (1983)	manufacturing worker	+	3 classes	10 classes	[h/day]	sex, nationality, age, type of pay, isom. strength, stature, etc.	qu., interview, clinical exam., sick leave,	sex, nationality, repetitive work, type of pay
[1] Linton & Kamwendo (1989)	medical secretary	–	–	–	[h/day]	psychosocial work env.	qu.	psychosocial work env.
[1+2] Luopajärvi et al. (1979)	packer, control grp.; shop assistant	–	+/– lifting	+/– rep.	–	[age]	clinical exam., interview	job title
[1+2+3] Maeda (1977)	many	+	–	–	–	–	qu.	job title (espec. assembly linework)
[2] Ohara et al. (1976) (longitudinal, intervention)	cash register op., control grps.	2 tasks	key pressure, productivity	rest breaks	h/day	–	qu., clinical exam.	no clear risk factor
[1–2] Ohlsson et al. (1989)	assembly worker, control grp.	–	work pace	–	empl. time	age	qu.	job title, age, empl. time, work pace
[2] Punnet et al. (1985)	garment worker, control grp.	5 main tasks	–	–	empl. time	age, nationality	qu., clinical exam..	job title, task, nationality,
[1–2] Rossignol et al. (1987)	office worker	VDU	–	–	0-7 h/day at VDU	age, VDU train., smoking, glasses	qu.	h/day at VDU
[3] Sakakibara et al. (1987)	orchard worker	2 tasks	arm elev., neck ext., [productivity]	–	[h/day]	sex	qu.	arm elev., neck ext.
[2] Sjøgaard et al. (1987)	garment worker	4 tasks	[EMG], [workst. design], productivity	no. of tasks	[empl. time]	isom. strength, muscle endur., [age], type of pay	qu.	no. of tasks, type of pay, productivity

Study	Subjects		Exposure			Confounders	Method	Effect modifiers
[2] Sokas et al. (1989)	garment worker, control grp.	–	–	–	empl.time	[age, nationality, sex]	qu.	job title
[1+2+3] Tola et al (1988)	machine op., carpenter, office worker	–	twisted/bent postures	–	[empl. time]	age, job satisf., draft	qu.	job title, age, twisted/bent postures, draft, job satisf.
[1+2+3] Ursin et al. (1988)	process worker, office worker, fireman, diver, air pilot, noise exposed subj.	–	–	–	–	many, e.g. personality, age	qu.	personality, job title
[1] Veiersted et al. (1990)	packer	[3 tasks]	EMG	EMG gaps	–	work technique (see "exposure")	qu.	work technique (see "exposure")
[3] Wells et al. (1983)	letter carrier, control grps.: gas meter reader, postal clerk	max carrying: 11 and 16 kg	–	–	[$\approx$ 5 h/day]	[age, body size, y's of service, previous heavy work]	qu.	job title, max weight,
[1+2+3] Westerling and Johnsson (1980)	several	–	heavy lift, physical exert.	–	–	age, sex, social class, Vo_2 max, stature, weight	interview, clinical exam., sick leave	age, physical heavy work.
[2] Westgaard & Aarås (1984)	assembly worker, control grp.	5 tasks	workst. design	–	empl. time	–	qu., clinical exam., sick leave,	job title, task, workst. design, empl. time
[2] Westgaard & Aarås (1985) (longitudinal, intervention)	assembly worker,	5 tasks	workst. design EMG [productivity]	–	–	–	qu., sick leave	workst. design, task
[2] Wærsted & Westgaard (1991)	garment worker	–	–	–	8 h/day + 5 h/day empl. time	age	sick leave	h/day, empl. time

independent of the operator. Examples are weight of tools, working height and duration of work task. 'Internal exposure' describes the forces in the target tissues. They are usually estimated indirectly by measuring e.g. intramuscular pressure (IMP), electrical activity of muscles (EMG) and arm position. Estimates of 'acute physiological and psychological responses' are e.g. spectral changes of EMG signals, changes in heart rate and perception of fatigue. 'Chronic effects' may either be injuries or training effects. For a given external exposure the resulting internal exposure, acute responses and chronic effects depend on the individual characteristics. The present paper concerns relationships between exposure and chronic effects ('disorders').

5. Exposure

5.1. Methods

Quantifications of exposure should consider 3 factors: level (amplitude, e.g. Newton), repetitiveness (frequency, e.g. second^{-1}, day^{-1}, year^{-1}) and duration (time, e.g. seconds, days, years) of the load (cf. Mathiassen and Winkel, 1991; table 1). 'Variation pattern' is defined as the interaction between exposure levels and repetitiveness. 'Monotony' is devoted to prolonged exposures comprising short and similar cycles. In operational situations 'level' may be estimated by recording e.g. trunk inclination and 'repetitiveness' by recording changes in load from second to second, over the day, week, month, season or occupational life. 'Duration' may refer to single tasks during the day, duration of the working day and week or total duration of working life. There is still no consensus on how all these factors should be pooled and interpreted as a dose.

Job title is the most frequently used estimate of exposure. An exposure 'contrast' is usually created by comparing the study group with a less exposed control group. This as well as additional information about exposure (e.g. postures, forces, exposure times) may be obtained by questionnaire (e.g. Wiktorin et al., 1991) or diary (Dallner et al., 1991). These methods are commonly used as a large amount of data can be collected at a low cost. However, recent studies indicate that the validity as well as the reliability may be low (Burdorf and Laan, 1991; Dallner et al., 1991; Karlqvist et al., 1991; Wigaeus-Hjelm et al., 1991; Wiktorin et al., 1991; Winkel et al., 1991a).

Check-lists and observation methods are other kinds of subjective methods which have been used. As the individuals estimating the exposure in these cases usually are experienced ergonomists, it is assumed that the validity of the collected data is higher than those obtained by questionnaires. However, little is known about this.

Highest validity is presumably obtained by using 'direct measurements', i.e. recordings of forces, postures, etc. during the working day for each individual. Heart rate and perception of fatigue are also used as direct measurements to estimate exposure although they are acute physiological and psychological responses (see figure 2 and section 4). In epidemiologic studies direct measurements imply high costs. If the exposure is only recorded during a single day, the generality of the measurements may be questioned in jobs including large day-to-day exposure variability.

In conclusion, direct exposure measurements at the workplace should be included to obtain a valid description of exposure–effect relationships. Unfortunately, this is rarely the case.

In the reviewed literature exposure levels have been estimated by e.g. job title, EMG, workstation design, productivity, posture and force recordings (see table 1). Repetitiveness has been estimated by e.g. cycle time, pause patterns and changes in postures. Duration has been estimated as e.g. hours in a given posture or in a task per day, total number of working hours per day or week and total employment time.

5.2. Neck exposure

The load on the neck is correlated to the trunk and head position. The load moment is balanced by muscle forces and tension of the passive connective tissue structures (Rizzi and Covelli, 1975 in Czech et al., 1990). Thus, flexion of the whole spine in the sitting posture increases the EMG activity in the cervical erector spinae, trapezius and thoracic erector spinae muscles (Schüldt et al., 1986). The lowest activity levels are obtained when the trunk is slightly backward inclined and the neck vertical. However, in extreme positions with the whole neck flexed, the extensor muscle activity of the cervical spine is not increased compared to the neutral upright head position although the load moment of the C7-T1 motion segment is increased 3–4 times (Harms-Ringdahl, 1986). Thus, a considerable stress is generated in the ligaments and joint capsules during extreme flexed positions of the cervical spine.

5.3. Shoulder exposure

It is not possible to use the arm/hand without stabilizing the shoulder girdle and the glenohumeral joint. Any arm movement requires continuous activation of the first two groups of the above-mentioned shoulder muscles (section 3.2, a & b). Therefore, work tasks with a demand of continuous arm movements generate load patterns with a static load component (e.g. Aarås and Westgaard, 1987; Onishi et al., 1977; Sjøgaard et al., 1987; Westgaard et al., 1986; Winkel and Oxenburgh, 1990). The load on the glenohumeral joint is transmitted to the scapula and further to the upper trapezius, which thereby acts as the principal antigravitational muscle for the arm. The relative load on the upper trapezius (in per cent of maximal voluntary contraction; %MVC) has been shown to increase linearly with the relative torque in the glenohumeral joint (in % of max torque) (Hagberg, 1981). However, recent studies seem to indicate a more complex relationship (see below).

In an optimal seated work posture, the upper trapezius static load level (Jonsson, 1982) is 2–3%MVC (Hagberg and Sundelin, 1986; Winkel and Oxenburgh, 1990). 'Optimal' for the shoulder-neck is here defined as vertical position or slightly backwards inclination of the spine, approximately vertical position of the upper arm and a work task at about elbow height without material handling (∼ e.g. data entry work). In this position the IMP is close to zero in the upper trapezius, supraspinatus and infraspinatus muscles, suggesting acceptable local blood flow in these muscles. Elevation of the shoulders while keeping the upper arms vertical (e.g. due to working surface above elbow height) may increase the load level on the upper trapezius up to about 20%MVC (Hagberg, 1987).

Deviation of the upper arm from the vertical position increases the load, particularly on the upper trapezius and the rotator cuff muscles. EMG amplitude as well as IMP increases linearly with an increase in shoulder torque (Hagberg, 1981; Järvholm et al., 1990; Jensen, 1991). However, there are large differences between muscles and between flexion and abduction. If a given shoulder torque is obtained during flexion rather than abduction the IMP of the upper trapezius and the supraspinatus is approximately halved and the EMG amplitude may be reduced by 15–45% (Järvholm, 1990; Mathiassen and Winkel, 1990). In addition, a considerable load may occur on the upper trapezius muscle due to a contralateral glenohumeral torque (Mathiassen and Winkel, 1990). The load distribution within the upper trapezius also seems to vary even when the glenohumeral torque is kept constant (Mathiassen and Winkel, 1990).

At a commonly occurring shoulder abduction of 30° (in a plane 45° to the frontal plane) with a straight elbow and without load in the hand, the IMP in the upper trapezius is only about 15 mmHg (Järvholm, 1990). The corresponding value in the supraspinatus muscle is about 80 mmHg. It has been shown that the blood flow ('acute response', see figure 2) in the supraspinatus is significantly impeded at an IMP of about 40 mmHg (Järvholm, 1990). These studies suggest that the supraspinatus is a vulnerable muscle in work situations with elevated arms.

5.4. Grouping into exposure levels

It was decided to classify the literature according to exposure level of the study groups. The aim was twofold: (a) to facilitate the step from research data to guidelines; (b) to facilitate the application of guidelines in working life. Thus, the groups should be easy to identify by safety representatives at the workplace and each of them should comprise a coherent and significant sample of exposure factors characteristic for many tasks in working life.

A Nordic research group has recently suggested a system for grouping of physical exposure into entities suitable for criteria documents (Winkel et al., 1991b). This system comprises, among other things, a grouping into 'low',

M. Trapezius pars desc.

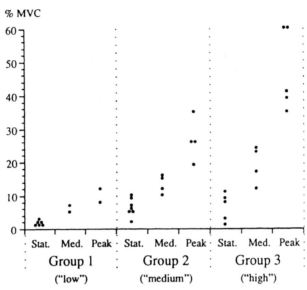

Fig. 3. Exposure levels for the upper trapezius muscle based upon EMG recordings in 15 studies (Aarås and Westgaard, 1987; Bjelle et al., 1981; Christensen, 1986; Hagberg et al., 1979; Hagberg et al., 1983; Hagberg and Sundelin, 1986; Jørgensen et al., 1989; Sjøgaard et al., 1987; Veiersted et al., 1990; Westgaard et al., 1986; Westgaard et al., 1991; Winkel and Gard, 1988; Winkel and Oxenburgh, 1990; Winkel et al., 1983a,b). Each point represents an average value obtained for the investigated group in one of these studies. See section 5.4 for definitions of exposure groups 1–3. %MVC: per cent of maximal voluntary contraction. Stat., med., peak: static, median and peak load levels according to Jonsson (1982).

'medium' and 'high' exposure levels. It was decided to test this grouping for the present guidelines. The exposure groups for the shoulder-neck region were tentatively defined and described as follows:

- 'Low' exposure level (group 1) for the shoulder-neck occurs e.g. during seated work when the workstation design is 'good' according to standard ergonomic textbooks (Chaffin and Andersson, 1984; Grandjean, 1988; Salvendy, 1987; Sanders and McCormick, 1987).
- 'Medium' exposure level (group 2) corresponds to work with elevated shoulders and/or abducted/flexed upper arm(s) and/or flexed/extended neck. If tools are used, their contribution to the biomechanical load in the shoul-

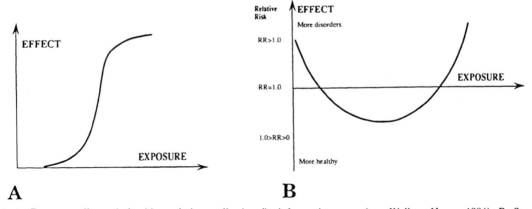

Fig. 4. A. Exposure–effect relationship as it is usually described for toxic agents (e.g. Wallace Hayes, 1984). B. Suggested exposure–effect relationship for physical exposures (modified from Winkel, 1987). RR: relative risk.

der-neck is marginal compared to the contribution from head, arm and hand.
- 'High' exposure level (group 3) corresponds to work requiring exertion of large forces in the shoulder/neck, e.g. handling heavy tools with the upper arm(s) deviating from vertical position.

Fifteen EMG studies of the upper trapezius muscle were then grouped into 'low', 'medium' and 'high' exposure levels by the present authors based on the description of the work tasks in the papers. Hence, the relative load levels on the trapezius muscle were plotted for each exposure group as illustrated in figure 3. This suggests that it is possible to classify the exposure as indicated above (Winkel et al., 1991b).

5.5. Exposure time

It should be emphasized that physical exposure as such is not injurious if the exposure time is sufficiently short. However, lack of physical exposure (inactivity) may cause negative effects on the body (figure 4) (Greenleaf, 1984; Greenleaf and Kozlowski, 1982). This may be counteracted by proper physical training which has been shown to improve a wide range of body functions (Åstrand, 1988). *Minimal exposure times* have been described to obtain such improvements (Åstrand and Rodahl, 1986). In working life the prevalent musculoskeletal disorders are considered an effect of too long exposure times (Winkel, 1987; 1989; see also section 8.2.4). Guidelines for occupational physical exposures should therefore aim at describing *maximal acceptable exposure times*.

6. Effect

6.1. Methods

In general, we have longer experience with effect measurements and in most studies they are also better described and measured in more detail than the exposures. However, most of the commonly used effect measures are far from being reliable and valid.

Chronic effects can be measured as subjective symptoms or objective clinical signs. The symptoms are most often recorded as 12-month period prevalence by using questionnaire or interview. The 'Nordic Questionnaire' (Kuorinka et al., 1987) is frequently used. It conforms with epidemiologic protocols and the reliability seems to be acceptable but the validity may be questioned (e.g. Kemmlert and Kilbom, 1988). In addition, the intention of developing a relatively simple screening questionnaire makes it difficult to assess the severity of a positive response: the quantitative description of the complaints is inadequate. A more detailed quantitative registration system has recently been published (Westgaard and Jansen, 1992a).

An interesting questionnaire variant is the diary for daily recordings of symptoms from the locomotor system (Dallner et al., 1991). This eliminates the uncertainty of remembering the occurrence of pain symptoms over long time periods, important if a quantitative description of pain symptoms is attempted. The validity of such diaries is currently being tested by several research groups.

Symptoms may lead to sick leave, which is a commonly used effect parameter. It is suspected to be influenced by e.g. physical exposures during work, sick leave compensation, fear of job loss, job satisfaction and the general psychosocial situation at the workplace. Sick leave data should be supplemented with clinical examination indicating the primary cause of the sick leave.

Shoulder-neck complaints suspected to be work-related may belong to a number of different diagnoses. Examples of commonly occurring diagnoses are (cf. Hagberg and Wegman, 1987; Viikari-Juntura, 1983; Waris, 1980):
- tension neck syndrome,
- rotator cuff tendinitis,
- thoracic outlet syndrome,
- shoulder joint osteoarthrosis,
- cervical syndrome,
- bicipital tendinitis,
- cervical spondylosis.

Unfortunately the diagnostic criteria used for the shoulder-neck disorders varies and this complicates comparisons between studies (Toomingas et al., 1991). In addition, clinical signs and subjective symptoms may show low correlation (e.g. Silverstein, 1985; Törner et al., 1990).

The outcome may also be measured as acute

physiological and psychological responses, which develops during a work task/working day (e.g. Chaffin, 1973; Jonsson, 1984; Winkel and Oxenburgh, 1990). It is presumed that an intervention causing a reduced acute response also reduces the risk of developing chronic effects (see figure 2). Unfortunately, it is still unknown which acute response parameters to use as risk indicators.

6.2. Prevalence of shoulder-neck complaints in general populations

Complaints in the shoulder-neck region may be due to occupational exposures as well as underlying medical diseases. Prevalence rates of complaints in general populations may therefore serve as a reference for evaluation of complaints

Table 2
Examples of studies showing the indicated risk factors for shoulder-neck disorders. The studies are grouped according to exposure level (see section 5.4).

Risk factors	Exposure level	
	Group 1 and 2 (low and medium)	Group 3 (high)
Individual factors		
Sex	Aronsson et al., 1988; Knave et al., 1985; Kvarnström, 1983	
Age	Grieco et al., 1989; Hviid Andersen & Gaarbo-Poulsen, 1990 (neck); Kamwendo et al., 1991; Ohlsson, 1989; Tola et al., 1988	Bjelle et al., 1979; Tola et al., 1988
Psychological factors/ personality	Hägg et al., 1990; Westgaard & Jansen, 1992b; Hviid Andersen & Gaarbo-Poulsen, 1990	Ursin et al., 1988
Isometric strength		Kilbom, 1988
Spare time phys. act.	Jonsson et al., 1988	
Ethnic group/ nationality/ language	Jeyaratnam et al., 1989; Kvarnström, 1983; Punnet et al., 1985	
Work technique/ ergonomic training	Jonsson et al., 1988; Kilbom et al., 1986; Parenmark et al., 1988;	Veiersted et al., 1990
Previous sick leave/ disorders	Jonsson et al., 1988; Westgaard & Jansen, 1992b	
Exposure level		
Workstation design/ anthropometry	Hedherg et al., 1981, Hünting et al., 1980; 1981; Sauter et al., 1991; Tola et al., 1988; van Wely, 1970; Westgaard & Aarås, 1984; 1985	Bjelle et al., 1979; Sakakibara et al., 1987
Manual material handling		Wells et al., 1983
Work pace/productivity/ type of pay	Hägg et al., 1990; Jonsson et al., 1988; Kvamström, 1983; Ohlsson et al., 1989; Sjøgaard et al., 1987	
Duration		
Empl. time/ years in piecework	Brisson et al., 1989; Grieco et al., 1989; Hägg et al., 1990; Hviid Andersen & Gaarbo-Poulsen, 1990; Ohlsson et al., 1989; Kamwendo et al., 1991; Westgaard & Aarås, 1984; Wærsted & Westgaard, 1991	
Hours per day or week	Aronsson et al., 1988; Grieco et al., 1989; Jeyaratnam et al., 1989; Kamwendo et al., 1991; Knave et al., 1985; Rossignol et al., 1987; Wærsted and Westgaard, 1991	
Repetitiveness	Karlsson et al., 1988; Kvarnström, 1983; Sjøgaard et al., 1987	
Psychosocial work environment	Kamwendo et al., 1991; Linton & Kamwendo, 1989; Theorell et al., 1991 [a]	

[a] The study comprises all three exposure groups.

in exposed groups. A study of manual workers from jobs 'without special demands for neck or shoulder activity' ($n = 2,648$) showed a 12-month period prevalence of 15% (Andersson, 1984): In a random sampled population ($n = 2,500$) aged 18–65 the 12-month period prevalence was 18% (Westerling and Jonsson, 1980). A sample of 55-year-old men and women ($n = 575$) showed a one-month period prevalence of 14% for shoulder pain (Bergenudd, 1989). Thus, a 'basic noise level' of shoulder-neck complaints always seems to occur.

7. Individual factors

The exposure–effect relationships may be influenced by individual factors (cf. figure 2). Those most frequently controlled for are sex and age; other examples are anthropometry, muscle strength, muscle endurance, smoking, psychological factors, ethnic group/nationality, productivity, spare time activity, work technique and social class (cf. table 1). Some of these factors may be obtained from registers (e.g. sex, age, productivity, ethnic group/nationality) or recorded by a

Table 3
Examples of studies failing to show the indicated risk factors for shoulder-neck disorders. The studies are grouped according to exposure level (see section 5.4).

Risk factors	Exposure level	
	Group 1 and 2 (low and medium)	Group 3 (high)
Individual factors		
Sex	Grieco et al., 1989;	Sakakibara et al., 1987
Age	Brisson et al., 1989; Jeyaratnam et al., 1989; Hviid Andersen & Gaarbo-Poulsen, 1990; Knave et al., 1985; Kuorinka & Koskinen, 1979; Kvarnström, 1983; Punnet et al., 1985; Rossignol et al., 1987; Wærsted & Westgaard, 1991; Westgaard & Jansen, 1992b	Herberts et al., 1981
Isometric strength	Hägg et al., 1990; Jonsson et al., 1988; Kvarnström, 1983; Sjøgaard et al., 1987	
Muscle endurance	Hägg et al., 1990; Jonsson et al., 1988; Sjøgaard et al., 1987	Kilbom, 1988
Spare time phys. act.	Hviid Andersen & Gaarbo-Poulsen, 1990; Westgaard & Jansen, 1992b	
Smoking	Brisson et al., 1989; Hviid Andersen & Gaarbo-Poulsen, 1990; Rossignol et al., 1987	
Work capacity	Kilbom et al., 1986	
Weight/height index	Kuorinka & Koskinen, 1979;	
Stature	Kvarnström, 1983	
Exposure level		
Work pace/productivity/ type of pay	Kuorinka & Koskinen, 1979; Ohara et al., 1976	
Duration		
Empl. time/ years in piecework	Knave et al., 1985; Punnet et al., 1985; Sjøgaard et al., 1987	Herberts et al., 1981
Hours per day or week	Ohara et al., 1976	
Repetitiveness	Ohara et al., 1976	

questionnaire (e.g. spare time activity, smoking). Others may require more effort as direct measurements have to be performed for each individual (e.g. muscle strength, work technique). Most studies only control for a few potential confounders/modifiers (cf. table 1).

8. Risk factors

Based on the literature retrieval described in section 2.1, several hundred potentially relevant studies were identified. After a critical review (cf. section 2.2), it was decided that about a hundred references would contribute significantly to the guidelines. Forty-seven of these are reviewed in tables 1–3.

8.1. Qualitative data

The significant risk factors demonstrated in each of the considered papers are indicated in the right column of table 1. In addition, the individual and exposure factors (level, repetitiveness, duration) that have been considered are listed. The inadequacy of the exposure measures in all studies is obvious.

Table 1 gives examples of job titles associated with elevated risk for shoulder-neck disorders; group 1 ('low'): VDU operator, assembly worker (some types) and cash register operator; group 2 ('medium'): garment worker, assembly worker and packer; group 3 ('high'): welder and letter carrier.

Based on the present pool of data it is possible to identify a number of risk factors (see table 2). It seems reasonable to assume that additional risk factors exist although not yet demonstrated (e.g. number of working hours per day or week, group 3 in table 2).

It should be emphasized that the significance of a factor does not necessarily imply a causal relationship. This may apply to sex, which has been shown to be a risk factor in many studies. However, several studies suggest that when proper corrections for actual sex differences in exposure are carried out, sex may no longer be a risk factor (e.g. Mergler et al., 1987). For exposure group 3 it may, however, be a significant risk factor, although we have not been able to find any proper data supporting this. Another example is muscle endurance which may be reduced due to pain and not be the cause of the complaints. This illustrates the importance of longitudinal studies. Finally, it should be emphasized that little is known about interactions between the factors (further details in section 8.2.5).

Although table 2 suggests a number of risk factors, other studies have failed to prove the occurrence of several of these (table 3). This may be due to deficiencies in the investigations or to inherent problems in studies of risk factors. Thus, many investigations fail to show age as a risk factor probably due to a 'healthy worker effect'. In addition, the same risk factors may have been estimated in different ways (see table 1), or different aspects of the risk factor may have been measured (e.g. 'personality', see Hviid Andersen and Gaardboe-Poulsen, 1990; Hägg et al., 1990).

8.2. Quantitative data

The mere demonstration of risk factors is of limited help as a basis for guidelines. In fact, most of the risk factors shown in table 2 occur in all populations and jobs in working life. What we need is a description of the exposure–effect relationship for each of these factors and information on interactions between risk factors. Unfortunately, such information is not available at present partly due to inadequate exposure measures in most studies.

8.2.1. Exposure–effect relationships based on EMG studies

Figure 3 illustrates exposure levels on the upper trapezius muscle in a variety of work tasks, each one comprising the dominating activity in an occupation. The occurrence of shoulder-neck complaints in these occupations was high, suggesting that the illustrated exposure levels in combination with exposure time and repetitiveness were unacceptable.

In *exposure group 1* (figure 3) all exposure levels ('static', 'median' and 'peak') are modest. In spite of this, the static load levels are probably unacceptable due to a long exposure time and inappropriate repetitiveness. The median and peak exposure levels are probably of minor significance.

Exposure group 2 is characterized by postures requiring elevated shoulders and/or abducted/flexed upper arm(s). Accordingly, the static exposure levels are higher in this group. The static as well as the median load levels may be critical, but the distribution of relaxation periods and exposure duration also has to be considered. The peak load levels are assumed to be of minor significance.

In *exposure group 3* all three exposure levels may be critical. This was the case for air hostesses (Winkel et al., 1983a) and cleaners (Winkel et al., 1983b). However, in crane coupling (Winkel and Gard, 1988) the peak exposure level (60% MVC) seems to be the main problem as the static level was close to zero.

8.2.2. Exposure–effect relationships based on head position measurements

Exposure group 1 and 2: A field study of 119 accounting machine operators showed a significant increase of neck pains with increasing neck flexion (Hünting et al., 1980). The daily working hours were $8\frac{1}{2}$ h and the average daily work on the accounting machine was estimated at 5–6 h. The keying speed was 8000–12000 h^{-1}. In a field study of 162 VDU operators clinical findings in the neck-shoulder area occurred more frequently with increasing average head inclination as well as head rotation during work (Hünting et al., 1981). Due to variations in daily exposure time, work intensity, etc. between the individuals, it is not possible to draw any general conclusions on the quantitative relationship between head inclination and occurrence of complaints.

8.2.3. Exposure-effect relationships based on weight carrying

Exposure group 3: A field study of letter carriers showed that an increase in the weight carried from 11 kg ($n = 92$) to 16 kg ($n = 104$) increased the occurrence of 'significant shoulder complaints' from 13 to 23% after an average exposure time of about 5 months (Wells et al., 1983). In the control groups the similar values were 7% (gas meter readers, walking but not weight carrying, $n = 76$) and postal clerks 5% (neither carrying nor walking, $n = 127$). The letter carriers as well as the meter readers walked about 5 h per day. Unfortunately, this study lacks significant details on how the load was carried and how it varied during the day. In spite of this, it still illustrates an interesting exposure-effect relationship.

8.2.4. Exposure-effect relationships based on exposure time measurements

Exposure group 1 and 2: During recent years this relationship has been investigated in a considerable number of studies. Among 1,545 clerical workers it was shown that the prevalence of shoulder and neck pain increased with the number of hours per day using VDU's (Rossignol et al., 1987). Similar results have been obtained by e.g. Aronsson et al. (1988), Grieco et al. (1989) and Jeyaratnam et al. (1989). The prevalence of shoulder-neck complaints seems to be around 50% or more when the daily time at the VDU is 6 or more hours. A reduction to about 4 hours seems to cause almost the same prevalence rate (around 50%). If the exposure time is reduced to about 2 hours per day, prevalence rates around 20–40% are reported. Thus, in order to obtain a significant reduction in occurrence of shoulder-neck complaints related to VDU work, the daily exposure time should be less than 4 hours. This is supported by a recent study of garment workers (Wærsted and Westgaard, 1991). A reduction from full-time (8 h working day, $n = 408$) to part-time work (5 h working day, $n = 210$) caused no lasting effect on sick leave due to shoulder-neck complaints. However, it should be emphasized that exposure time–effect relationships may be heavily influenced by many factors (see section 8.2.5). All of these must be considered when discussing guidelines for daily exposure time at monotonous tasks.

Exposure time may also be measured as number of years in the same job. Many studies have failed to show that this is a risk factor in monotonous tasks, e.g. garment work (e.g. Punnett et al., 1985). One possible explanation may be that workers with pain are more likely to move to another job or leave employment. Other studies of garment work show high associations between duration of employment and complaints (Brisson et al., 1989; Hviid Andersen and Gaarboe-Poulsen, 1990). Thus, more than 15 years in garment work increased the prevalence proportion ratio for chronic shoulder-neck pain by 4.8, while it was only 1.7 when the employment time was 0–7 years.

8.2.5. Exposure time / effect-relationships: interactions with other factors

The relationship between daily exposure to monotonous work and shoulder-neck complaints may be influenced by several factors; e.g. (for references, see table 2):
- Workstation design
- Work intensity
- Work technique
- Psychosocial work environment
- Psychological factors/personality
- Distribution of rest pauses and alternative tasks during the working day
- Exposure during the alternative tasks

Aronsson et al. (1988) was not able to show any clear-cut time effect on the prevalence of shoulder-neck complaints for programmers ($n = 209$). This may be due to more frequent rest pauses for the shoulder-neck as well as a higher level of control in this work task compared to data entry operators (cf. Theorell et al., 1991).

Comparison of studies presenting data on exposure time/effect relationships is complicated, as the working hours are indicated in different ways. Thus, for full-time VDU work the indicated number of h/day at the VDU may vary from 4.5 to 8 hours.

8.3. Discussion of the literature and need for further research

Only a few studies have ever been carried out with the aim of describing an exposure–effect relationship for the shoulder-neck. Common aims are to prove that a job title is related to an increased risk for shoulder-neck disorders, to create a basis for social insurance statements or to evaluate interventions. Thus, in most epidemiologic studies the different exposure factors are only quantified in terms of high/low occurrence (cf. table 1). Job title is the most frequently used exposure measure. It usually represents a specific combination of exposure factors. The risk estimate therefore becomes highly specific and valid only for the investigated work situation. Furthermore, a job title represents a wide range of exposure intensities, which contributes to the coarseness in this measure.

Thus, the scattered information on exposure–effect relationships in the scientific literature seems mainly to be due to poor design of most studies (in relation to the present needs) and lack of proper methods for quantifying exposure.

Further break-through on exposure–effect descriptions therefore requires three main lines of research:
(1) Studies on quantification of physical exposure in epidemiologic studies.
(2) Studies on the relationship between acute responses and chronic effects. Such knowledge may, among other things, increase the value of acute response measurements, thereby reducing the experimental costs.
(3) Longitudinal intervention studies in which a single-factor exposure gradient is created. In long-term studies reference groups are desirable. This design is unfortunately combined with many practical as well as ethical problems. A few have tried, as indicated in table 1, and so far the benefit has been limited.

References

Aarås, A. and Westgaard, R., 1987. Further studies of postural load and musculo-skeletal injuries of workers at an electro-mechanical assembly plant. Applied Ergonomics, 18: 211–219.

Anderson, J.A.D., 1984. Shoulder pain and tension neck and their relation to work. Scandinavian Journal of Work, Environment & Health, 10: 435–442.

Aoyama, H., Endo, T., Futatsuka, M. et al., 1982. International workshop on occupational neck and upper limb disorder due to constrained work. Journal of Human Ergology (Tokyo), 11: 3–5.

Aronsson, G., Åborg, C. and Örelius, M., 1988. Winners and losers of computerization: A study of the working conditions and health of Swedish State Employees. (In Swedish, English summary). Arbete och Hälsa 27, The Swedish National Institute of Occupational Health, Stockholm, 82 pp.

Åstrand, P.-O., 1988. From exercise physiology to preventive medicine. Annals of Clinical Research, 20: 10–17.

Åstrand, P.-O. and Rodahl, K., 1986. Textbook of work physiology. Physiological bases of exercise. McGraw-Hill Book Company, New York, 756 pp.

Bergenudd, H., 1989. Talent, occupation and discomfort. Department of Orthopaedics, Malmö General Hospital, Lund University. (Thesis).

Bjelle, A., Hagberg, M. and Michaelsson, G., 1979. Clinical and ergonomic factors in prolonged shoulder pain among industrial workers. Scandinavian Journal of Work, Environment & Health, 5: 205–210.

Bjelle, A., Hagberg, M. and Michaelsson, G., 1981. Occupational and individual factors in acute shoulder-neck disor-

ders among industrial workers. British Journal of Industrial Medicine, 38: 356–363.
Brisson, C., Vinet, A., Vézina, M. and Gingras, S., 1989. Effect of duration of employment in piecework on severe disability among female garment workers. Scandinavian Journal of Work, Environment & Health, 15: 329–334.
Burdorf, A. and Laan, J., 1991. Comparison of methods for the assessment of postural load on the back. Scandinavian Journal of Work, Environment & Health, 17: 425–429.
Chaffin, D., 1973. Localized muscle fatigue – Definition and measurement. Journal of Occupational Medicine, 15: 346–354.
Chaffin, D.B. and Andersson, G., 1984. Occupational biomechanics. John Wiley & Sons, New York, 454 pp.
Christensen, H., 1986. Muscle activity and fatigue in the shoulder muscles of assembly plant employees. Scandinavian Journal of Work, Environment & Health, 12: 582–587.
Colligan, M.J. and Murphy, L.R., 1979. Mass psychogenic illness in organizations: An overview. Journal of Occupational Psychology, 52: 77–90.
Czech, C., Czech, R., Seibt, A. and Freude, A.D., 1990. Funktionelle Veränderungen der Halswirbelsäule und Schulter-Nacken-Region bei Textilnäherinnen. Z. gesamte Hyg., 36(1): 61–64.
Dallner, M., Gamberale, F., Winkel, J. and the Stockholm-MUSIC I Study Group, 1991. Validation of a questionnaire for estimation of physical load in relation to a workplace diary. In: Y. Quéinnec and F. Daniellou (Eds.), Designing for Everyone. Proceedings of the Eleventh Congress of the International Ergonomics Association. Taylor & Francis (London), Vol. 1, pp. 233–235.
Dimberg, L., Olafsson, A., Stefansson, E., Aagaard, H., Odén, A., Andersson, G.B.J., Hagert, C.-G. and Hansson, T., 1989a. Sickness absenteeism in an engineering industry – an analysis with special reference to absence for neck and upper extremity symptoms. Scandinavian Journal of Social Medicine, 17: 77–84.
Dimberg, L., Olafsson, A., Stefansson, E., Aagaard, H., Odén, A., Andersson, G.B.J., Hansson, T. and Hagert, C.-G., 1989b. The correlation between work enviroment and the occurrence of cervicobrachial symptomes. Journal of Occupational Medicine, 31: 447–453.
Ferguson, D., 1971. An Australian study of telegraphists' cramp. British Journal of Industrial Medicine, 28: 280–285.
Gleerup Madsen, A., 1990. Besvær i nakke, skuldre og arme som følge af belastninger ved ensidigt gentaget arbejde (In Danish, English summary). The Work Environment Fund, Copenhagen, 133 pp.
Grandjean, E., 1988. Fitting the task to the man. A textbook of occupational ergonomics. Taylor & Francis, London, 363 pp.
Grazier, K.L., Holbrook, T.L., Kelsey, J.L. and Stauffer, R.N., 1984. The frequency of occurrence, impact and cost of musculoskeletal conditions in the United States. American Academy of Orthopaedic Surgeons.
Greenleaf, J.E., 1984. Physiological responses to prolonged bed rest and fluid immersion in humans. Journal of Applied Physiology: Respiration, Environment, Exercise Physiology, 57: 619–633.
Greenleaf, J.E. and Kozlowski, S., 1982. Physiological consequences of reduced physical activity during bed rest. In: R.L. Terjung (Ed.), Exercise and Sport Sciences Reviews. Franklin Inst. Press, Philadelphia, PA, 10: 84–119.
Grieco, A., Occhipinti, E. and Colombini, D., 1989. Work postures and musculo-skeletal disorders in VDT operators. Bollettino de Oculistica, suppl. 7, 99–111.
Hadler, N.M., 1989. The roles of work and of working in disorders of the upper extremity. Baillière's Clinical Rheumatology, 3(1): 121–141.
Hadler, N.M., 1990. Cumulative trauma disorders. An iatrogenic concept. Journal of Occupational Medicine, 32(1): 38–41.
Hagberg, M., 1981. On evaluation of local muscular load and fatigue by electromyography. Arbete och Hälsa 24, The Swedish National Institute of Occupational Health, Stockholm. (Thesis).
Hagberg, M., 1987. Occupational shoulder and neck disorders. The Swedish Work Environment Fund, Stockholm, 72 pp.
Hagberg, M., Jonsson, B. and Sima, S., 1979. Muskelbelastning vid spollindning. Investigation report, 01, The Swedish National Institute of Occupational Health, Stockholm. (In Swedish).
Hagberg, M., Jonsson, B., Brundin, L., Ericson, B.-E. and Örtelius, A., 1983. An epidemiological, ergonomic and electromyographic study of musculoskeletal complaints among meat-cutters. (In Swedish, English summary). Arbete och Hälsa 12, The Swedish National Institute of Occupational Health, Stockholm, 52 pp.
Hagberg, M. and Sundelin, G., 1986. Discomfort and load on the upper trapezius muscle when operating a wordprocessor. Ergonomics, 29: 1637–1645.
Hagberg, M. and Wegman, D.H., 1987. Prevalence rates and odds ratios of shoulder-neck diseases in different occupational groups. British Journal of Industrial Medicine, 44: 602–610.
Hägg, G., Suurküla, J. and Kilbom, Å., 1990. Predictors for work related shoulder/neck disorders. A longitudinal study of female assembly workers. (In Swedish, English summary). Arbete och Hälsa, 10, The Swedish National Institute of Occupational Health, Stockholm, 78 pp.
Harms-Ringdahl, K., 1986. On assessment of shoulder exercise and load-elicited pain in the cervical spine. Scandinavian Journal of Rehabilitation Medicine, suppl., 14. (Thesis).
Hedberg, G, Björkstén, M., Ouchterlony-Jonsson, E. and Jonsson, B. 1981. Rheumatic complaints among Swedish engine drivers in relation to the dimensions of the driver's cab in the Rc engine. Applied Ergonomics, 12: 93–97.
Herberts, P., Kadefors, R., Andersson, G. and Petersén, I., 1981. Shoulder pain in industry: An epidemiological study on welders. Acta Orthop. Scand., 52: 299–306.
Hünting, W., Grandjean, E. and Maeda K., 1980. Constrained postures in accounting machine operators. Applied Ergonomics, 11(3): 145–149.
Hünting, W., Läubli, Th. and Grandjean, E., 1981. Postural and visual loads at VDT workplaces. I. Constrained postures. Ergonomics, 24: 917–931.
Hvid Andersen, J. and Gaardboe-Poulsen, O., 1990. Helbredsprofil bland kvindelige beklædningsindustriarbejdere

(In Danish, English summary). The Work Environment Fund, Copenhagen, 175 pp.)

Järvholm, U., 1990. On shoulder muscle load. An experimental study of muscle pressures, EMG and blood flow. Göteborg. (Thesis).

Jensen, B.R., 1991. Isometric contractions of small muscle groups. Danish National Institute of Occupational Health and University of Copenhagen, August Krogh Institute. (Thesis).

Jeyaratnam, J., Ong, C.N., Kee, W.C., Lee, J. and Koh, D., 1989. Musculoskeletal symptoms among VDU operators. In: M.J. Smith and G. Salvendy (Eds.), Work with Computers: Organizational, Management, Stress and Health Aspects. Elsevier Science Publishers, Amsterdam, pp. 330-337.

Jonsson, B., 1982. Measurement and evaluation of local muscular strain in the shoulder during constrained work. Journal of Human Ergology (Tokyo), 11: 73-88.

Jonsson, B., 1984. Muscular fatigue and endurance. Basic research and ergonomic applications. In: M. Kumamoto (Ed.) Neural and Mechanical control of movement. Yamaguchi Shoten, Kyoto, pp. 63-76.

Jonsson, B.G., Persson, J. and Kilbom, Å., 1988. Disorders of the cervicobrachial region among female workers in the electronics industry. International Journal of Industrial Ergonomics, 3: 1-12.

Jørgensen, K., Fallentin, N. and Sidenius, B., 1989. The strain on the shoulder and neck muscles during letter sorting. International Journal of Industrial Ergonomics, 3: 243-248.

Kamwendo, K., Linton, S.J. and Moritz, U., 1991. Neck and shoulder disorders in medical secretaries. Scandinavian Journal of Rehabilitation Medicine, 23, 57-63.

Karlqvist, L., Dallner, M., Ericson, M., Fransson, C., Nygård, C.-H., Selin, K., Wigaeus Hjälm, E., Wiktorin, C., Winkel, J. and the Stockholm-MUSIC I Study Group. Validity of questions estimating manual materials handling, working postures and movements. In: Y. Quéinnec and F. Daniellou F. (Eds.), Designing for Everyone. Proceedings of the Eleventh Congress of the International Ergonomics Association. Taylor & Francis, London, Vol. 1: 251-253.

Karlsson, G.-B., Adle, B., Löfgren, B. and Sundström, I., 1988. Mikropauser minskade besvär från rörelseapparaten. Läkartidningen, 85: 3463-3463.

Kemmlert, K. and Kilbom, Å., 1988. Muskuloskeletala besvär i nacke/skuldra och samband med arbetssituation. En utvärdring med hjälp av frågeformulär och arbetsplatsbesök. (In Swedish, English summary). Arbete och Hälsa 17, The Swedish National Institute of Occupational Health, Stockholm, 42 pp.

Kilbom, Å., 1988. Isometric strength and occupational muscle disorders. European Journal of Applied Physiology and Occupational Physiology, 57: 322-326.

Kilbom, Å., Persson, J. and Jonsson, B.G., 1986. Disorders of the cervicobrachial region among female workers in the electronics industry. International Journal of Industrial Ergonomics, 1: 37-47.

Knave, B.G., Wibom, R.I., Voss, M., Hedström, L.D. and Bergqvist, U.O.V., 1985. Work with video display terminals among office employees. I. Subjective symptoms and discomfort. Scandinavian Journal of Work, Environment & Health, 11: 457-466.

Kukkonen, R., Luopajärvi, T. and Riihimäki, V., 1983. Prevention of fatigue amongst data entry operators. In: T.O. Kvålseth, T.O. (Ed.), Ergonomics of Workstation Design. Butterworths, London, chap. 3, pp. 28-34.

Kuorinka, I. and Koskinen, P., 1979. Occupational rheumatic diseases and upper limb strain in manual jobs in light mechanical industry. Scandinavian Journal of Work, Environment & Health, 5: 39-47, suppl. 3.

Kuorinka, I., Jonsson, B., Kilbom, Å., Vinterberg, H., Biering-Sørensen, F., Andersson, G. and Jørgensen, K., 1987. Standardised Nordic questionnaires for the analysis of musculoskeletal symptoms. Applied Ergonomics, 18: 233-237.

Kvarnström, S., 1983. Occurrence of musculoskeletal disorders in a manufacturing industry, with special attention to occupational shoulder disorders. Scandinavian Journal of Rehabilitation Medicine, Suppl. 8, 114 pp.

Linton, S.J. and Kamwendo, K., 1989. Risk factors in the psychosocial work environment for neck and shoulder pain in secretaries. Journal of Occupational Medicine, 31: 609-613.

Luopajärvi, T., Kuorinka, I., Virolainen, M. and Holmberg, M., 1979. Prevalence of tenosynovitis and other injuries of the upper extremities in repetitive work. Scandinavian Journal of Work, Environment & Health, 5: 48-55, suppl. 3.

Mathiassen, S.E. and Winkel, J., 1990. Electromyographic activity in the shoulder-neck region according to arm position and glenohumeral torque. European Journal of Applied Physiology and Occupational Physiology, 61: 370-379.

Mathiassen, S. E. and Winkel, J., 1991. Quantifying variation in musculoskeletal load using exposure-vs-time data. Methodological considerations and presentation of a new method. Ergonomics, 34: 1455-1468.

Mergler, D.M., Brabant, C., Vézina, N. and Messing, K., 1987. The weaker sex? Men in women's working conditions report similar health symptoms. Journal of Occupational Medicine, 29(5): 417-421.

Molin, C. and Nilsson, C.G., 1990. Psykogent smittsamma sjukdomar - Några exempel och en förklaringsmodell. Läkartidningen, 87(32-33): 2510-2511.

National Occupational Health and Safety Commission, 1986. Repetition strain injury. A report and model code of practice. Worksafe Australia. Australian Government Publishing Service, Canberra, 117 pp.

Ohara, H., Aoyama, H. and Itani, T., 1976. Health hazard among cash register operators and the effects of improved working conditions. Journal of Human Ergology (Tokyo), 5: 31-40.

Ohlsson, K., Attewell, R. and Skerfving, S., 1989. Self-reported symptoms in the neck and upper limbs of female assembly workers. Scandinavian Journal of Work, Environment & Health, 15: 75-80.

Onishi, N., Sakai, K., Itani, T. and Shindo, H., 1977. Muscle load and fatigue of film rolling workers. Journal of Human Ergology (Tokyo), 6: 179-186.

Parenmark, G., Engvall, B. and Malmkvist, A.-K., 1988. Ergonomic on-the-job training of assembly workers. Arm-neck-shoulder complaints drastically reduced amongst beginners. Applied Ergonomics, 19(2): 143–146.

Punnett, L., Robins, J.M., Wegman, D.H. and Keyserling, W.M., 1985. Soft tissue disorders in the upper limbs of female garment workers. Scandinavian Journal of Work, Environment & Health, 11: 417–425.

Rafnsson, V., Steingrimsdóttir, O.A., Ólafsson, M.H. and Sveinsdóttir, T., 1989. Muskuloskeletala besvär bland islänningar. (In Swedish, English summary). Nordisk Medicin, 104: 104–107.

Ramazzini, B., 1713. De morbis artificum. Second edition. Translated by Wright W.C. Hafner Publishing Co., New York.

Rasmussen, N.K., Groth, M.V., Bredkjær, S.R., Madsen, M. and Kamper-Jørgensen, F., 1988. Sundhed & sygelighed i Danmark 1987. En rapport fra DIKEs undersøgelse. DIKE, Copenhagen, 265 pp.

Rossignol, A.M., Morse, E.P., Summers, V.M. and Pagnotto, L.D., 1987. Video display terminal use and reported health symptoms among Massachusetts clerical workers. Journal of Occupational Medicine, 29: 112–118.

Sakakibara, H., Miyao, M., Kondo, T., Yamada, S., Nakagawa, T. and Kobayashi, F., 1987. Relation between overhead work and complaints of pear and apple orchard workers. Ergonomics, 30: 805–815.

Sanders, M.S. and McCormick E.J. (Eds.), 1987. Human Factors in Engineering and Design. McGraw-Hill Book Company, New York.

Salvendy, G. (Ed.), 1987. Handbook of Human Factors. John Wiley & Sons, New York.

Sauter, S.L., Schleifer, L.M. and Knutson, S.J., 1991. Work posture, workstation design, and musculoskeletal discomfort in a VDT data entry task. Human Factors, 33(2): 151–167.

Scandinavian Journal of Work, Environment & Health, 1979. Suppl. 3.

Schüldt, K., Ekholm, J., Harms-Ringdahl, K., Németh, G. and Arborelius, U.P., 1986. Effects of changes in sitting work posture on static neck and shoulder muscle activity. Ergonomics, 29(12): 1525–1537.

Silverstein, B., 1985. The prevalence of cumulative trauma disorders in industry. Occupational Health and Safety Engineering, University of Michigan, 131 pp.

Sirois, F., 1974. Epidemic hysteria. Acta Psychiat. Scand. (suppl.), 252: 1–46.

Sjøgaard, G., Ekner, D., Schibye, B., Simonsen, E.B., Jensen, B.R., Christiansen, J.U. and Pedersen, K.S., 1987. Skulder/nakke-besvær hos syersker. En epidemiologisk og arbejdsfysiologisk undersögelse. (In Danish, English summary). Arbejdsmiljøfondet, Copenhagen, 134 pp.

Socialstyrelsen redovisar, 1987. Att förebygga sjukdomar i rörelseorganen. Underlag till hälsopolitiskt handlingsprogram. Socialstyrelsen, Stockholm, 216 pp.

Sokas, R.K., Spiegelman, D. and Wegman, D.H., 1989. Self-reported musculoskeletal complaints among garment workers. American Journal of Industrial Medicine, 15: 197–206.

Statistics Sweden, 1991. Arbetsskador 1990. HS 10 SM 9101. (Partly in English). SCB-Publishing Unit, Örebro, Sweden, 36 pp.

Stock, S.R., 1991. Workplace ergonomic factors and the development of musculoskeletal disorders of the neck and upper limbs: A meta-analysis. American Journal of Industrial Medicine, 19: 87–107.

Theorell, T., Harms-Ringdahl, K., Ahlberg-Hultén, G. and Westin, B., 1991. Psychosocial job factors and symptoms from the locomotor system – A multicausal analysis. Scandinavian Journal of Rehabilitation Medicine, 23: 165–173.

Tola, S., Riihimäki, H., Videman, T., Viikari-Juntura, E. and Hänninen, K., 1988. Neck and shoulder symptoms among men in machine operating, dynamic physical work and sedentary work. Scandinavian Journal of Work, Environment & Health, 14: 299–305.

Toomingas, A., Németh, G., Hagberg, M. and the Stockholm MUSIC I Study Group, 1991. Besvärsförekomst, kliniska fynd samt diagnoser i nacke, skuldror och axlar i Stockholmsundersökningen. In: Hagberg, M. and Hogstedt, C. (Eds.) Stockholmsundersökningen 1. Data från en tvärsnittsundersökning av ergonomisk och psykosocial exponering samt sjuklighet och funktion i rörelseorganen. MUSIC Books, Stockholm.

Törner, M., Zetterberg, C., Hansson, T. and Lindell, V., 1990. Musculoskeletal symptoms and signs and isometric strength among fishermen. Ergonomics, 33(9): 1155–1170.

Ursin, H., Endresen, I.M. and Ursin, G., 1988. Psychological factors and self-reports of muscle pain. European Journal of Applied Physiology and Occupational Physiology, 57: 282–290.

van Wely, P., 1970. Design and disease. Applied Ergonomics, 1(5): 262–269.

Veiersted, K.B., Westgaard, R.H. and Andersen, P., 1990. Pattern of muscle activity during stereotyped work and its relation to muscle pain. International Archives of Occupational and Environmental Health, 62: 31–41.

Viikari-Juntura, E., 1983. Neck and upper limb disorders among slaughterhouse workers. An epidemiologic and clinical study. Scandinavian Journal of Environmental Health, 9: 283–290.

Wærsted, M. and Westgaard, R.H., 1991. Working hours as a risk factor in the development of musculoskeletal complaints. Ergonomics, 34(3): 265–276.

Wallace Hayes, A. (Ed.), 1984. Principles and methods of toxicology. Student edition. Raven Press, New York.

Wallace, M. and Buckle, P., 1987. Ergonomic aspects of neck and upper limb disorders. International Reviews of Ergonomics, 1: 173–200.

Waris, P., 1980. Occupational cervicobrachial syndromes. A review. Scandinavian Journal of Environment and Health, suppl. 3: 3–14.

Wells, J.A., Zipp, J.F., Schuette, P.T. and McEleney, J., 1983. Musculoskeletal disorders among letter carriers. Journal of Occupational Medicine, 25(11): 814–820.

Westerling, D. and Jonsson, B.G., 1980. Pain from the neck-shoulder region and sick leave. Scandinavian Journal of Social Medicine, 8: 131–136.

Westgaard, R.H. and Aarås, A., 1984. Postural muscle strain

as a causal factor in the development of musculo-skeletal illnesses. Applied Ergonomics, 15: 162-174.

Westgaard, R.H. and Aarås, A., 1985. The effect of improved workplace design on the development of work-related musculo-skeletal illnesses. Applied Ergonomics, 160: 91-97.

Westgaard, R.H., Wærsted, M., Jansen, T. and Aarås, A., 1986. Muscle load and illness associated with constrained body postures. In: N. Corlett and I. Manenica (Eds.) The ergonomics of working postures. Models, Methods and Cases. Taylor & Francis, London, chap. 1, pp. 5-18.

Westgaard, R.H. and Jansen, T., 1992a. Individual and work related factors associated with symptoms of musculoskeletal complaints. I. A quantitative registration system. British Journal of Industrial Medicine, 49: 147-153.

Westgaard, R.H. and Jansen, T., 1992b. Individual and work related factors associated with symptoms of musculoskeletal complaints. II. Different risk factors among sewing machine operators. British Journal of Industrial Medicine, 49: 154-162.

Wigaeus Hjelm, E., Karlqvist, L., Nygård, C.-H., Selin, T., Wiktorin, C., Winkel, J. and the Stockholm-MUSIC I Study Group, 1991. Validity of questions regarding physical activity and perceived exertion in occupational work. In: Y. Quéinnec & F. Daniellou (Eds.), Designing for Everyone. Proceedings of the Eleventh Congress of the International Ergonomics Association. Taylor & Francis (London), Vol. 1, pp. 239-241.

Wiktorin, C., Karlqvist, L., Nygård, C.-H., Winkel, J. and the Stockholm-MUSIC I Study Group, 1991. Design and reliability of a questionnaire for estimation of physical load in epidemiologic studies. In: Y. Quéinnec and F. Daniellou (Eds.), Designing for Everyone. Proceedings of the Eleventh Congress of the International Ergonomics Association. Taylor & Francis (London), 1991, Vol. 1, pp. 230-232.

Winkel, J., 1987. On the significance of physical activity in sedentary work. In: B. Knave and P.-G. Widebäck (Eds.), Work with display units. Elsevier Science Publishers (North-Holland), Amsterdam, pp. 229—36.

Winkel, J., 1989. Varför ökar belastningsskadorna? (In Swedish, English summary). Nordisk Medicin, 104(12): 324-327.

Winkel, J., Ekblom, B., Hagberg, M. and Jonsson, B., 1983a. The working environment of cleaners. Evaluation of physical strain in mopping and swabbing as a basis for job redesign. In: T.O. Kvålseth (Ed.), Ergonomics of Workstation Design. Butterworth & Co, London, chap. 4, pp. 35-44.

Winkel, J., Ekblom, B. and Tillberg, B., 1983b, Ergonomic and medical factors in shoulder/arm pain among cabin attendants as a basis for job redesign. In: H. Matsui and K. Kobayashi (Eds.), Biomechanics VIII-A. Human Kinetics Publishers, Illinois, pp. 567-573.

Winkel, J. and Oxenburgh, M., 1990. Towards optimizing physical activity in VDT/office work. In: S. Sauter, M. Dainoff and M. Smith (Eds.), Promoting Health and Productivity in the Computerized Office: Models of Successful Ergonomic Interventions. Taylor & Francis, London, pp. 94-117.

Winkel, J., Dallner, M., Ericson, M., Fransson, C., Karlqvist, L., Nygård, C.-H., Selin, T., Wigaeus Hjälm, E., Wiktorin C. and the Stockholm-MUSIC I Study Group, 1991a. Evaluation of a questionnaire for the estimation of physical load in epidemiologic studies: Study design. In: Y. Quéinnec & F. Daniellou (Eds.), Designing for Everyone. Proceedings of the Eleventh Congress of the International Ergonomics Association. Taylor & Francis (London), Vol. 1, pp. 227-229.

Winkel, J., Jørgensen, K., Ólafsdóttir, H., Sejersted, O., Sjøgaard, G., Smolander, J. and Westgaard, R., 1991b. Criteria documents for the prevention of musculo-skeletal disorders: A proposal for assessment of exposure categories. Proceedings of the Nordic Work Environment Meeting, Nyborg, Denmark, 16-18 September, pp. 43-45.

Winkel, J. and Gard, G., 1988. An EMG-study of work methods and equipment in crane coupling as a basis for job redesign. Applied Ergonomics, 19: 178-184.

Human muscle strength definitions, measurement, and usage: Part I – Guidelines for the practitioner[1]

Anil Mital[a,][*], Shrawan Kumar[b]

[a] *Ergonomics and Engineering Controls Research Laboratory, College of Engineering, University of Cincinnati, Cincinnati, OH 45221-0116, USA*
[b] *Department of Physical Therapy, University of Alberta, Edmonton, Alberta T6G 2G4, Canada*

1. Audience

This guide is intended for practitioners. The practitioner in the context of this guide is the individual who is responsible for designing work, workplace, equipment, and tools, and selecting workers for a specific task. This person may be a design engineer entrusted with designing equipment or tools, or an ergonomist, product designer, occupational health and safety professional, industrial or production engineer.

It should be explicitly understood that this guide is intended for the purpose of designing those activities and devices that require the use of muscular strengths. While human muscle strengths are an important measure of human capabilities, knowledge of these strengths alone may not be sufficient to correct an "ergonomic" problem in the workplace. It may be necessary to evaluate the working situation and correct it in order to alleviate a musculoskeletal problem. For instance, if a task requires frequent exertions, activities and devices designed on the basis of maximum strength will not alleviate the fatigue problem.

The practitioners should also keep in mind that the guidelines provided here are based on what we currently know about human strengths. It may be necessary to modify these guidelines as a result of future research and experience.

2. Need for human muscular strengths

Most occupational activities are performed through human intervention. In such cases, a person's capacity to perform mechanical work is determined by his/her ability to exert muscular strength. For instance, stronger individuals can lift heavier weights than their weaker colleagues. The demands for human strengths to accomplish physical

---

[*] Corresponding author.

[1] The recommendations provided in this guide are based on numerous published and unpublished scientific studies and are intended to enhance worker safety and productivity. These recommendations are neither intended to replace existing standards, if any, nor should be treated as standards. Furthermore, this document should not be construed to represent institutional policy.

The following individuals participated in the discussion of the earlier version of this guide. Their suggestions (written or verbal) were incorporated by the authors in this version: A. Aaras, Norway; J.E. Fernandez, U.S.A.; A. Freivalds, U.S.A.; T. Gallewey, Ireland; M. Jager, Germany; S. Konz, U.S.A.; H. Krueger, Switzerland; K. Landau, Germany; A. Luttmann, Germany; J.D. Ramsey, U.S.A.; M-J. Wang, Taiwan.

activities remain strong despite increasing automation. The nature of many tasks and work situations mandates recruitment of muscle power. For instance, when handling materials in narrow and confined spaces, transferring patients to and from their beds, or performing maintenance activities.

Muscular strengths are also necessary to exert forces and torques necessary to operate equipment and controls and sustain external loading without inflicting personal injury. Insufficient strength can lead to overloading of the muscle-tendon-bone–joint system and possible consequent injury. The relationship between insufficient strength and injury is widely admitted for tasks such as manual materials handling and those requiring the use of hand tools. Knowledge of human muscular strengths thus is necessary in designing devices which are physically activated by humans and for preventing musculoskeletal injuries.

3. Classification and definition of human muscular strengths

Any scheme of classification of human strength is neither precise nor entirely accurate. However, for the purpose of understanding and application a common approach and terminology will be useful. In this vein human muscular strengths can be broadly classified according to two criteria (a) characteristics of the effort, and (b) characteristics of application. Based on effort characteristics the strength can be either static or dynamic. Static strengths are also known as isometric strengths. Dynamic muscle strengths are further subdivided into: (i) isotonic muscle strengths, and (ii) isokinetic muscle strengths. Based on the characteristics of application, human strength can be classified under the following two categories: (1) static functional strength, and (2) dynamic functional strength. Static strengths can be: (i) simulated job static strengths, and (ii) continuous or repetitive static muscular strength. The dynamic strength on the basis of application can be divided into four categories: (i) isoinertial muscle strengths, (ii) psychophysical muscle strengths, (iii) simulated job dynamic strengths (SJDS), and (iv) repetitive dynamic strengths (RDS). Since additional application may emerge, other categories of strength may be added in future.

3.1. Static (isometric) muscle strengths

Static muscle strengths are the capacities of muscles to produce force or torque by a single maximal voluntary isometric exertion; the body segment involved remains stationary (displacement = 0). The internal muscular effort, amplified by the mechanical advantage of the body members involved, is measured as the external static effect in the form of force, torque, or torque across time. The length of the muscle though stated to remain constant, it does not. Since the external position of the body member remains unchanged the misconception of fixed muscle length arose. However, the external resistance force (static effect) varies in response to the magnitude of the muscular force.

3.1.1. Simulated job static strengths

These are static strengths that are measured while simulating job conditions in terms of the body posture (location of hands, feet, arm orientation, etc.).

3.1.2. Continuous static muscle strengths (endurance)

These are the static strength-duration of exertion relationships and depict how the strength declines with the duration of sustained exertion (also known as endurance time). In general, the decline in static strengths is very rapid in the first two minutes of the exertion; strength may decline by as much as 75% during this period. Static muscular exertions which are approximately 20% of the maximum can be sustained for several minutes.

3.1.3. Repetitive static muscle strengths

These are maximal static exertions applied at specified frequencies. The muscles may be allowed to relax between successive exertions. The muscular strength also declines with the duration of exertions but not quite as rapidly as in the case of continuous static muscular exertions. When using non-powered hand tools (wrenches and screwdrivers) and exerting at frequencies individuals feel most comfortable with, the torque declines nearly 30% after 2 min and almost 40% after 4 min.

3.2. Dynamic muscle strengths

In dynamic muscular exertions, body segments move and the muscle length changes significantly. The measured force resulting from muscle and body segment motion is referred to as dynamic strength. Even though several different dynamic strengths have been mentioned, from an industrial application standpoint, isotonic strengths are irrelevant.

3.2.1. Isotonic muscle strengths

Isotonic muscle exertion is the muscle action that involves production of a constant muscle force (tension). In practice, the term is also commonly used both when the joint movement is constant over a range of motion and when a constant load is being moved through a distance. However, it is important to understand that because the lever arm is changing throughout the motion, the force developed by the muscles in both these cases will actually be changing, rendering them non-isotonic. And it is this reason that, primarily, makes this kind of strengths irrelevant for industrial applications.

3.2.2. Isokinetic muscle strengths

Isokinetic muscular exertion is the muscle action in which the rate of shortening or lengthening of the muscle is constant. In practice, the term isokinetic strength applies either to a constant velocity of the force being applied or resisted or to a constant angular velocity of the joint and is the measure of a person's maximum voluntary muscle contraction when the body segments involved move at a constant speed.

3.3. Isoinertial muscle strengths

Isoinertial muscle exertions reflect a person's ability to overcome the initial static resistance (maximum weight a person can handle and move to a assigned point at freely chosen speed); the name is a misnomer as the inertia during the exertion changes due to the changing distribution of mass (external resistance plus muscle mass) changes.

3.4. Psychophysical muscle strengths

Psychophysical muscle strength is a person's measure of psychophysically determined maximum acceptable level of force application. This maximum force is considered a measure of a person's maximum dynamic strength in that category of activity.

3.5. Simulated job dynamic strengths (SJDS)

Simulated job dynamic strengths are dynamic (isokinetic or psychophysical) human muscle strengths measured while simulating the job in terms of body posture and speed of limb movement.

3.6. Repetitive dynamic strengths (RDS)

Repetitive dynamic strengths are isokinetic or psychophysical dynamic strengths that take into account the effect of frequency of exertion.

4. Measurement of human muscular strengths

Strength measurement procedures have become fairly standard. There are, however, procedural differences that must be accounted for in the assessment of either static or dynamic strengths.

4.1. Measurement of isometric strengths

The following factors are important in the assessment of isometric muscle strengths: (i) duration of exertion, (ii) strength measuring device, (iii) rest periods between repeated exertions, (iv) body position or posture, and (v) reporting of test conditions, subject demographic data, and strength data.

The duration of muscular exertion to reach the peak should be between 4 and 6 s. Individuals should maintain the peak exertion for a 4–6 s to permit recording of a 3 s mean value. A break of at least 30 s should be provided between successive exertions if only a few measurements are to be made. It is necessary to increase the rest duration to 2 min if about 15 measurements are to be made in one test session. This additional rest is necessary to recover from fatigue generated due to the isometric exertion. Longer rest breaks should be allowed if individuals being tested so desire.

The posture assumed by the individuals during isometric exertion influences strength. It is necessary, therefore, to specify and control the body position if the corresponding strength values of the individuals are to be compared. For measuring simulated job static strength, the body posture assumed by the individuals must duplicate the posture assumed by the individuals during the course of performing the task.

In general, the isometric strength measurement procedure requires individuals to build-up their muscular exertion slowly over a 4–6 s period, without jerking, and maintain the peak exertion for about 3 s. This peak exertion (3 s average) is the isometric strength of the individual. No external motivation should be provided. When measuring continuous static muscular exertion, the peak exertion is maintained throughout the duration of exertion and is recorded as a function of exertion duration. In case of repetitive static strengths, the duration between successive peak exertions is determined by the frequency of exertions. The frequency of exertions is either chosen by the individual or is specified.

A variety of isometric strength recording devices are available. For lifting strength one popular static strength measuring device (ISTU–Isometric Strength Testing Unit; cost can vary from $2000 to $25 000$^+$, depending upon the attachments) is marketed by Ergometrics, Inc. (Ann Arbor, MI). The device allows measurement of push/pull/lift and other types of forces through an adjustable frame. The frame permits adjustments in the horizontal and vertical distances. The device also comes with a personal computer and data analysis software. A similar device is available from Lafayette Instrument of Lafayette, IN (the Jackson Strength Evaluation System; approximate cost is $3000). It is less expensive, but limited in terms of the number of applications. In general, the strength measuring device should have the following characteristics: (i) it should be capable of recording peak, 3 s time averaged, and continuous exertions, (ii) it should not generate discomforting localized pressure, (iii) it should accommodate the population, and (iv) it should permit easy adjustments for recording different types of exertions (lifting, pulling, pushing, etc.).

Some of the widely used isometric strengths are: (i) arm strength, (ii) shoulder strength, (iii) composite (leg) strength, and (iv) back (torso) strength. These positions are briefly described below. In each case, the muscular exertion is upwards (vertical) and is applied by the hands.

The isometric arm strength measurement requires the forearms to be flexed at 90°, i.e., perpendicular to the individual's torso, and the upper arms are vertical, i.e., parallel and adjacent to the torso. The individual stands erect, with back and legs straight and feet flat. The exertion is upward, vertical and generated by only the arm muscles.

The isometric shoulder strength measurement requires the forearms to be vertical, i.e., parallel and adjacent to the torso and the upper arms horizontal, i.e., perpendicular to the torso. The position of the feet, back, etc., remains the same as in the case of arm strength.

The isometric composite strength measurement, or leg lifting strength measurement, requires the individual to take a semi-squat position, such that the hands are between the legs, at approximately 38 cm height. The individual's elbows must be extended and the contact between the upper and lower extremities is not permitted. The heads of the first metatarsals should be placed opposite one another and intersect the vertical plane of muscular exertion. The feet remain flat and the individual looks straight ahead.

The isometric back strength measurement, or torso lifting strength measurement, requires the hands to be located between the knees, at 75% of the knee height (tibial height) and approximately 38 cm in front of the ankle joint. The individual's feet are separated at shoulder width and both feet are kept at an equal distance from the line of force exertion. The individual flexes the torso in order to exert; the force is exerted by torso extension.

Grip strength is also of considerable significance in industrial setting, such as isometric muscular exertions (torques) using non-powered hand tools. Peak torques (maximum volitional torques) are exerted on a workpiece, using the specified hand tool. The measurement procedure is similar to the one described above. Peak muscular exertions are built-up slowly over a 4–6 s period, without jerking, and are maintained for about 3 s. The level at which force/torque is exerted (workpiece height level), the body posture, arm reach distance and orientation (angle), and the duration and frequency of exertion (in case of continuous and repetitive isometric strengths) are specified. There are several grip dynamometers on market for the measurement of grip strength. One such device is ComputAbility Hand Grip Dynamometer available for approximately $5 000 from ComputAbility Pine Brook, Corporation, NJ.

4.2. Measurement of dynamic strengths

In dynamic muscular exertions, body segments move and the muscle length changes significantly. The measured force resulting from muscle and body segment motion is referred to as dynamic strength. As stated earlier, there are several types of dynamic strengths and the measurement procedure varies with each kind of strength.

4.2.1. Measurement of isokinetic muscle strength

The prime requirement of isokinetic strength measurement is that the body segment velocity should remain constant during the maximum voluntary contraction. As with isometric strengths, individuals are required to exert maximal voluntary contraction by pulling a handle, attached to a rope, at a constant speed (no jerks). The magnitude of this exertion is measured by either a mechanical or hydraulic dynamometer.

Speed control is achieved by a mechanical or hydraulic device that does not allow the body segment to move faster than the pre-selected speed, the prime requirement of isokinetic strength measurement. The individual is asked to assume the required posture, including arm reach distance and arm orientation. Once the individual is ready, he/she is required to exert (pull) the handle to exert as much force as possible, without jerking. The speed of exertion is set such that it simulates the speed of the task. The maximum force of exertion is recorded. If other conditions, such as body posture, also simulate the task, the strength becomes the simulated job dynamic strength. However, it must be borne in mind that most of the so called "isokinetic" strengths are not isokinetic, because during the motion the angles of the muscles and body segments change. With this changing position the velocity of the body segment does not remain constant though the external object/interface travels at a constant velocity.

Rest breaks between successive exertions must be provided and the posture, speed of exertion, arm orientation, reach distance, and population must be reported with the data as in the case of isometric strength measurement.

A Super II Mini-Gym, build by Fitness Systems, Inc., located in Independence, MO, may be used for measuring isokinetic strengths. It costs approximately $300. Alternative equipment is available from CYBEX, Bay Shore, NY, at a substantially higher cost ($30 000$^+$).

In general, the following different kinds of isokinetic strengths have been used: (i) dynamic lift strength (the handle is pulled from the floor to chest height), (ii) dynamic back extension strength (handle is pulled from the floor to approximately knuckle height (approximately 80 cm)), and (iii) dynamic elbow flexion strength (pulling from knuckle height to chest height). In all cases, the movement of the rope is in the sagittal plane (plane dividing the body into left and right halves) and the measurement procedure is identical to the one described above.

In order to measure repetitive dynamic (isokinetic) strengths, the above procedure is modified slightly – the duration between successive exertions are determined by the frequency of exertions.

4.2.2. Measurement of isoinertial strengths

The isoinertial strength testing does not measure the peak muscular exertion. Instead, it is a method of measuring the maximum weight a person is willing to handle, at his/her own selected speed within a specified range of movements. The isoinertial strength really is a variation of the psychophysical muscle strength.

One equipment (among others) used to measure isoinertial strength is a modified Mach I "press station" with guide rails extended to 3 m. It has two handles, 46 cm apart, horizontal and parallel to each other, pointing forward. The strength measurement procedure requires the individual to grasp both handles, located 5 cm above the floor, and lift the weight to knuckle or overreach height and then lower the weight. The individual is started with an initial weight of 11.4 kg. If lifted successfully, an additional 11.4 kg weight is added. Increments of 11.4 kg weight are added until the cut-off limit is reached (77.3 kg in knuckle height tests and 45.5 kg in overreach height tests – to prevent the risk of overexertion injury) or an attempt fails. If an attempt fails, weight is reduced by 6.8 kg. If this weight is successfully lifted then 4.5 kg is added; otherwise 2.3 kg weight is subtracted.

4.2.3. Measurement of psychophysical strengths

Psychophysical strengths are generally defined as the maximum acceptable weight of lift. The measurement procedure requires the individual to adjust the weight of lift according to his/her perception of physical strain. All other factors, such as size of the container and range of lift, are controlled. Individuals are started randomly with either a very heavy load, that they cannot lift, or very light load, that they can easily lift, and allowed to make adjustments to it (subtract or add to the weight already in the container) until they arrive at a load they can sustain without strain or discomfort, and without becoming unusually tired, weakened or overheated, or out of breath. The final weight is the maximum acceptable weight of lift or the psychophysical strength of the individual.

When frequency is present and the container is lifted repeatedly at the specified frequency, the psychophysical strength becomes repetitive dynamic (psychophysical) strength. Usually in this case, the individual can either adjust the weight in the container or the frequency of lift. Sometimes, the frequency level is specified and the individual can only manipulate the load in order to arrive at the maximum acceptable weight of lift.

The equipment required to measure psychophysical strength is very simple and consists of a container, loose weights (lead shot, etc.), and, in case of repetitive psychophysical strengths, a metronome to pace the individual.

In the event, task conditions such as body posture, starting and ending points of lift, and frequency, are duplicated during the psychophysical strength measurement, the strength is called the simulated job dynamic (psychophysical) strength.

5. Data collection and analysis

Data collection involves measuring specified strength (static or dynamic) on a number of individuals, specifying their demographic data (age, gender, body height, body weight, etc.), and computing descriptive statistics (mean, standard deviation, and range of strength values). Once this information is compiled, it can be used in designing jobs, equipment and tools. Population percentiles may be computed from descriptive statistics and used to design for a specific proportion of the population. It should be noted that strengths are not completely normally distributed. The strength distributions are almost normal up to 70th percentile. Beyond the 70th percentile, the distribution is skewed. Correction factors given in Table 1 should be used to compute actual strength percentile values from the normal distribution percentile values.

Table 1
Correction factors to adjust normal distribution percentile values to actual (skewed) distribution percentile values

Percentile value[a]	Males	Females
Up to 70th	1.0000	1.0000
75th	1.0002	1.0004
80th	1.0189	1.0154
85th	1.0374	1.0304
90th	1.0561	1.0454
95th	1.0751	1.0603

[a] For intermediate values interpolation may be used.

6. Static versus dynamic strengths

Absence of body segment, object, and muscle movement during maximal voluntary contraction for assessing static strengths is the major distinction between static and dynamic strengths. Since there is no effective limb-object-muscle movement in case of static strengths, these strengths cannot account for the effect of inertial forces. This leads to underestimating musculoskeletal joint loading during the performance of a dynamic task. For this reason alone, the isometric strength exertion capability of individuals should not be used to assess their capability to perform dynamic tasks. Furthermore, since most industrial processes require a force application through a range of motion in a continuous activity, the design of tasks based on static strengths in a fixed posture has little relevance. A number of recent studies clearly indicate that dynamic strengths, not static strengths, are more appropriate measures of a person's work performance capabilities and, therefore, should be used in worker screening and form the basis for designing industrial jobs, work places, tools, equipment, and controls.

7. Factors affecting human strengths

A number personal and task factors influence human strengths. The factors that are particularly important are: (i) age, (ii) gender, (iii) posture, (iv) reach distance, (v) arm and wrist orientations, (vi) speed of exertion, and (vii) duration and frequency of exertions.

7.1. Effect of gender and age

The effect of age and gender on strength has been a subject of several investigations and while detailed information is not available, it is possible to summarize the net effect of these factors on strength. Table 2 expresses male and female strengths as a percentage of male strength at age 24. The relationships shown in Table 2 have been developed on the basis of whole-body isometric strength testing. Similar data for dynamic strength

Table 2
Decline in muscular strength capability with age

Age (yr)	Strength exertion capability (%)	
	Men	Women (% of men's capability)
24	100	62
30	97	59
35	94	56
40	91	52
45	89	50
50	83	46
55	80	43
60	75	40
65	71	38

Note: Published literature also indicates that isokinetic strengths of females is only 50% of the isokinetic strength of males. Isometric strength of females are approximately two-thirds of the isometric strengths of males.

are yet to be developed. Until that time, and in the absence of any other indicator, it is recommended that factors given in Table 2 be used to determine approximate whole body or segmental strengths of males and females of different ages.

7.2. Effect of posture

The muscle strength is significantly affected by posture. This is particularly true for isokinetic strengths, where, depending upon the posture assumed by the individual, the strength exertion capability of the individual can vary greatly – from almost no force to considerable force. Further, considerably more force is exerted (as much as 37% more) in the standing posture than in the sitting posture.

7.3. Effect of reach distance

Mechanical disadvantage increases the task difficulty. This causes reduction in strength output. Isometric strength increases with the reach distance up to a certain point if the body posture does not change significantly (note that significant stature differences for the same reach distance will cause a significant change in the posture). One-handed isokinetic pull strengths also increase, almost

Table 3
Relationship between torque (%) and duration of repeated exertions (min)

Hand tool	Regression equation
Pipe wrench	Torque = $1.05 - 0.301(Time) + 0.08(Time)^2$
Socket, crescent, and spanner wrenches	Torque = $1.056 - 0.304(Time) + 0.104(Time)^2$
Screwdrivers	Torque = $1.024 - 0.138(Time) + 0.04(Time)^2$

linearly, with the reach distance between 25 and 55 cm for the sitting posture and between 45 and 85 cm for the standing posture. If, however, the posture changes are significant, both static and dynamic strength exertion capabilities decline with the reach distance. For lifting type of tasks, people are able to generate greater forces closer to the body. As the distance of hand grip from the body increases, the strength exertion capability progressively decreases. An isokinetic effort at full reach in the sagittal plane is between 28% and 40% of peak isometric strength at half-reach.

7.4. Arm and wrist orientations

The orientation of the arm influences human isometric strength exertion capability. As the arm orientation changes the mechanical advantage also changes, resulting in weaker or stronger strength exertions. The horizontal isokinetic pull strength exertion is maximized when the arm is in the sagittal plane. Least strength is exerted when arm is in the frontal plane. The rates of decline in strength (Ng/degree) on either side of the sagittal plane are similar. One-handed isokinetic pull strength in the vertical plane is maximized when pulling straight up.

The wrist orientation is critical in generating isometric torques with non-powered hand tools. Approximately 70% more torque is exerted when wrenches are in the horizontal position than when they are in vertical positions (above or below the horizontal). For screwdrivers, maximum torque is exerted when the wrist is oriented 90° from the neutral wrist orientation (wrist is straight) in the counterclockwise direction. The torque in this wrist orientation is almost 11% greater than the torque exerted when the wrist is in neutral position.

7.5. Speed of exertion

Dynamic strength exertions can take place at variety of speeds. The dynamic strength peaks generally would be higher than static strength peaks if deceleration is avoided and momentum is allowed to contribute to the force output. The dynamic strength peaks would otherwise be significantly lower than static strength peaks. For one-handed isokinetic pull exertions in the vertical plane, the peak strength is exerted at the slowest speed and declines almost linearly by 25–30% as the speed of exertion increases from 0.3 to 0.75 m/s.

7.6. Duration and frequency of exertion

Prolonged exertions result in cumulative fatigue and reduced stress bearing capacity. The results of studies of repetitive dynamic (isokinetic) lifting strengths indicate that: (1) isokinetic strength declines exponentially with the duration of lifting, (2) the decline in strength is sharper for higher frequencies, and (3) repetitive dynamic strengths are better estimators of an individual's capability to perform tasks that require repeated heavy physical exertion.

In case of non-powered hand tools, the decline in torque (%) with the duration of exertion (minutes) can be determined by the equations given in Table 3.

For isometric exertions, strength declines very rapidly to almost 25% of the peak value after nearly 2 min of continuous exertions. After approximately 4 min of exertion, the strength declines to approximately 20% of the peak value and can be maintained at this level for several minutes.

8. Suggested strength guidelines

Numerous epidemiological studies have demonstrated a high association between job strength requirement on one hand and incidence and severity of injury on the other. This clearly interferes with productivity and increases the cost to industry as well as worker. In addition it may also result in turnover of the skilled workers increasing cost of replacement. For these reasons the following guidelines are being proposed.

8.1. Policy for use of the criterion

It is suggested that industry adopt the policy of use of strength as a criterion of job design as well as placement (where permissible). Having adopted such a policy various jobs could be classified on the basis of strength requirement of the body part concerned. Having gained this stamp of approval the criterion could be applied with legitimacy and discretion.

8.2. Problem identification (current and potential)

Having adopted the guideline 1 and developed a classification of jobs an assessment of current and future potential problem should be undertaken. In order to achieve this goal company's injury records should be examined for overexertion injuries. This will provide the extent of the current problem. A comparison of job strength requirement and the relevant population strength data should be undertaken. Such exercise would reveal all jobs with potential problems and means of assessment and selection of strategy to deal with the impending problem.

8.3. Selection of job-specific criteria

A thorough job analysis with a specific emphasis on strength requirement during different phases of the job in question should be carried out. This will enable one to separate specific stages in the job which may be responsible for causing or contributing towards the development of an overexertion injury.

8.4. Selection of appropriate predictive equations

Having determined the job specific criteria an appropriate database for the population should be searched. If available, it will indicate the mismatch or deficit. If such a database is not available then appropriate strengths available in the population may be predicted using appropriate regression equations (see Part II). These predicted values may then throw light on the extent of mismatch to select a strategy of choice.

8.5. Worker selection

If legally and socially permissible then for specially demanding jobs people who may be physically capable should be selected. These selected workers should be appropriately trained to handle the job with strategies to minimize any risk of injury.

8.6. Make an informed guess on cumulative effect

The design of job or selection of workers should be done such to provide optimum margin of safety. If such optimum is not possible for the criterion of strength, it should be compensated with rest provided. It has been shown through scientific studies that the cumulative load is a definite risk factor for injury causation.

8.7. Checklist

Use of the guidelines is suggested here as a checklist to ensure all steps are being considered until a better approach can be found.

9. Strength databases

A number of researchers have developed strength databases that can be used in design. While it is not possible to reproduce all these databases here, some prominent ones are provided (Tables 4–29). The details about the studies generating these databases can be obtained from Part II of this guide (Mital and Kumar, 1998).

Table 4
Isometric arm, shoulder, composite (leg), and back (torso) lifting strengths (N) of males and females

Strength	Males		Females	
	Mean	SD	Mean	SD
Arm lifting	382	127	200	78
Shoulder lifting	490	125	270	66
Composite lifting	942	340	416	198
Back lifting	545	243	266	138

Table 6
Isokinetic dynamic lift strength (DLS), back extension strength (DBES), and elbow flexion strength (DEFS) in Newtons

Gender	DLS		DBES		DEFS	
	Mean	SD	Mean	SD	Mean	SD
Men	601	100	540	101	324	46
Women	493	139	315	87	196	104

Table 5
Mean peak and average isometric strength during arm lift (N)

Strength	Plane	Gender	Stat.	0.5R	0.75R	1R
Peak	Sagittal	Male	x	590	295	184
			SD	140	67	34
		Female	x	261	182	118
			SD	87	63	33
	30° lateral	Male	x	483	274	165
			SD	148	67	37
		Female	x	249	160	105
			SD	100	45	45
	60° lateral	Male	x	385	206	130
			SD	101	55	32
		Female	x	204	130	88
			SD	75	40	23
Average	Sagittal	Male	x	465	237	151
			SD	109	51	26
		Female	x	210	147	92
			SD	64	50	26
	30° lateral	Male	x	385	219	135
			SD	105	50	29
		Female	x	205	128	83
			SD	78	36	26
	60° Lateral	Male	x	282	160	103
			SD	68	44	26
		Female	x	166	105	72
			SD	60	32	19

Note: R = Reach.

Table 7
Mean peak isometric stoop lifting strengths in Newtons(N) with two hip angles

Strength	Plane	Gender	Stat.	0.5R	0.75R	1R
Isometric hip = 60°	Sagittal	Male	x	647	342	206
			SD	177	64	89
		Female	x	319	216	139
			SD	33	46	25
	30° lateral	Male	x	508	313	202
			SD	147	97	21
		Female	x	277	218	141
			SD	39	42	41
	60° lateral	Male	x	440	227	151
			SD	108	59	25
		Female	x	230	161	106
			SD	65	28	28
Isometric hip = 90°	Sagittal	Male	x	697	370	208
			SD	238	105	47
		Female	x	323	254	155
			SD	51	48	11
	30° lateral	Male	x	580	318	194
			SD	185	110	71
		Female	x	321	222	155
			SD	64	45	25
	60° Lateral	Male	x	381	230	142
			SD	186	63	48
		Female	x	224	162	128
			SD	40	26	35

Note: R = Reach.

Table 8
Mean average isometric stoop lifting strength in Newtons (N) with two hip angles

Strength	Plane	Gender	Stat.	0.5R	0.75R	1R
Isometric hip = 60°	Sagittal	Male	x	517	272	168
			SD	191	85	40
		Female	x	264	181	114
			SD	34	40	19
	30° lateral	Male	x	396	253	158
			SD	154	89	57
		Female	x	231	181	113
			SD	38	37	31
	60° lateral	Male	x	331	180	121
			SD	139	52	38
		Female	x	190	131	84
			SD	54	26	17
Isometric hip = 90°	Sagittal	Male	x	560	287	164
			SD	176	70	35
		Female	x	273	206	126
			SD	46	39	11
	30° lateral	Male	x	460	255	149
			SD	151	91	46
		Female	x	266	185	125
			SD	52	34	14
	60° Lateral	Male	x	303	183	113
			SD	99	63	37
		Female	x	180	124	99
			SD	276	19	23

Table 9
Mean peak isometric squat lift strength in Newtons (N) with two knee angles

Strength	Plane	Gender	Stat.	0.5R	0.75R	1R
Isometric hip = 90°	Sagittal	Male	x	384	290	199
			SD	97	90	79
		Female	x	175	137	93
			SD	59	34	18
	30° lateral	Male	x	361	273	185
			SD	99	78	54
		Female	x	186	129	88
			SD	60	40	15
	60° lateral	Male	x	323	233	162
			SD	75	87	42
		Female	x	141	114	72
			SD	41	38	14
Isometric hip = 135°	Sagittal	Male	x	409	281	187
			SD	117	78	43
		Female	x	185	117	91
			SD	63	33	18
	30° lateral	Male	x	404	262	188
			SD	150	106	63
		Female	x	182	114	81
			SD	44	28	13
	60° Lateral	Male	x	323	233	162
			SD	75	87	42
		Female	x	141	114	72
			SD	41	38	14

Table 10
Mean average isometric squat lift strength in Newtons (N) with two knee angles

Strength	Plane	Gender	Stat.	0.5R	0.75R	1R
Knee = 90°	Sagittal	Male	x	303	236	167
			SD	86	77	76
		Female	x	142	114	78
			SD	48	26	17
	30° lateral	Male	x	286	218	150
			SD	82	65	47
		Female	x	152	103	70
			SD	49	27	11
	60° lateral	Male	x	252	183	127
			SD	64	75	39
		Female	x	110	89	56
			SD	36	31	12
Knee = 135°	Sagittal	Male	x	320	225	154
			SD	91	68	43
		Female	x	150	97	77
			SD	56	27	16
	30° lateral	Male	x	325	215	157
			SD	121	87	58
		Female	x	151	95	65
			SD	37	25	13
	60° Lateral	Male	x	252	183	127
			SD	64	75	39
		Female	x	110	89	56
			SD	36	31	12

Table 11
Mean peak and average isokinetic stoop lift strength in Newtons (N)

Strength	Plane	Gender	Stat.	0.5R	0.75R	1R
Peak	Sagittal	Male	x	542	294	233
			SD	215	93	47
		Female	x	296	250	202
			SD	121	104	78
	30° lateral	Male	x	505	299	207
			SD	184	118	72
		Female	x	272	237	162
			SD	122	84	56
	60° lateral	Male	x	334	233	168
			SD	130	82	47
		Female	x	243	174	135
			SD	103	70	39
Average	Sagittal	Male	x	359	192	142
			SD	136	43	36
		Female	x	189	166	134
			SD	84	79	45
	30° lateral	Male	x	328	190	139
			SD	131	60	22
		Female	x	183	159	108
			SD	94	67	35
	60° Lateral	Male	x	205	138	105
			SD	79	37	16
		Female	x	159	109	89
			SD	70	46	24

Table 12
Mean peak and average isokinetic squat lift strength in Newtons (N)

Strength	Plane	Gender	Stat.	0.5R	0.75R	1R
Peak	Sagittal	Male	x	441	320	215
			SD	202	74	33
		Female	x	232	184	155
			SD	91	69	31
	30° lateral	Male	x	416	269	215
			SD	193	77	37
		Female	x	225	176	155
			SD	59	49	23
	60° lateral	Male	x	296	236	182
			SD	76	90	45
		Female	x	186	143	111
			SD	36	43	42
Average	Sagittal	Male	x	273	228	153
			SD	152	54	30
		Female	x	148	125	107
			SD	68	54	34
	30° lateral	Male	x	279	187	154
			SD	161	61	32
		Female	x	145	116	108
			SD	52	39	24
	60° Lateral	Male	x	197	163	120
			SD	53	63	32
		Female	x	121	100	76
			SD	28	37	30

Table 13
One-handed isokinetic pull strengths (N) in the vertical plane

Variable	Males		Females	
	Mean	SD	Mean	SD
Speed of exertion (m/s)				
0.30	322	250	61	65
0.35	303	236	60	73
0.48	274	219	51	68
0.58	242	197	49	63
0.75	225	192	48	63
Angle of preferred arm (deg)				
− 30 (arm up and hyperextended; pull down)	173	159	24	6
0 (arm vertical, pull down)	400	265	59	47
30 (pull down)	269	200	43	23
60 (pull down)	220	186	36	21
90 (horizontal pull)	160	165	33	20
120 (pull up)	380	186	30	13
150 (pull up)	370	252	26	10
180 (arm vertical, pull up)	460	273	194	107
210 (pull up, arm hyperextended)	230	152	50	74
240 (pull up, arm hyperextended)	190	195	22	10

Table 14
Two handed pulling strength of males in isometric and isokinetic modes at low, medium, and high heights (strength in N)

Mode	Ht	Peak plane						Average plane					
		Sagittal		30° lateral		60° lateral		Sagittal		30° lateral		60° lateral	
		$\bar{x}$	SD	$\bar{x}$	SD	$\bar{x}$	SD	$\bar{x}$	SD	$\bar{x}$	SD	$\bar{x}$	SD
Isom.	Low	423	135	364	71	311	67	292	102	253	47	217	48
	Med.	537	133	432	64	338	65	387	94	300	50	237	50
	High	469	73	428	131	324	139	320	47	277	88	224	107
Isokin.	Low	337	92	326	73	266	31	106	18	93	14	76	10
	Med.	434	96	377	95	289	40	137	25	119	32	86	11
	High	390	88	316	65	235	46	127	30	98	16	75	13

Table 15
Two handed pulling strength of females in isometric and isokinetic modes at low, medium, and high heights (strength in N)

Mode	Ht	Peak plane						Average plane					
		Sagittal		30° lateral		60° lateral		Sagittal		30° lateral		60° lateral	
		$\bar{x}$	SD	$\bar{x}$	SD	$\bar{x}$	SD	$\bar{x}$	SD	$\bar{x}$	SD	$\bar{x}$	SD
Isom.	Low	306	80	303	82	247	67	219	63	220	61	176	56
	Med.	385	119	328	84	281	50	275	109	230	72	204	44
	High	368	72	306	92	281	107	267	59	221	75	197	78
Isokin.	Low	209	53	202	46	185	46	64	13	56	9	52	9
	Med.	292	59	268	46	230	42	91	16	78	13	67	13
	High	253	47	218	35	177	30	85	16	74	12	62	13

Table 16
Two-handed pushing strength of males in isometric and isokinetic modes at low, medium, and high heights (strength in N)

Mode	Ht	Peak plane						Average plane					
		Sagittal		30° lateral		60° lateral		Sagittal		30° lateral		60° lateral	
		$\bar{x}$	SD	$\bar{x}$	SD	$\bar{x}$	SD	$\bar{x}$	SD	$\bar{x}$	SD	$\bar{x}$	SD
Isom.	Low	363	92	335	74	281	59	258	73	233	55	199	47
	Med.	395	123	358	93	295	60	266	85	249	65	202	46
	High	320	44	274	68	229	62	216	74	191	48	156	39
Isokin.	Low	338	96	300	92	253	66	72	43	54	11	47	10
	Med.	339	85	306	76	281	60	60	11	56	9	52	8
	High	327	115	301	104	263	54	58	15	54	13	49	7

Table 17
Two handed pushing strength of females in isometric modes at low, medium, and high heights (strength in N)

Mode	Ht	Peak plane						Average plane					
		Sagittal		30° lateral		60° lateral		Sagittal		30° lateral		60° lateral	
		$\bar{x}$	SD	$\bar{x}$	SD	$\bar{x}$	SD	$\bar{x}$	SD	$\bar{x}$	SD	$\bar{x}$	SD
Isom.	Low	275	72	260	74	239	71	204	60	192	60	169	47
	Med.	288	72	271	47	227	37	207	52	189	41	158	27
	High	224	42	196	38	186	52	167	34	140	30	134	30
Isokin.	Low	171	40	197	40	160	41	44	20	42	7	36	7
	Med.	270	56	246	47	197	40	90	74	51	10	44	7
	High	220	28	200	32	191	27	49	8	45	8	44	7

Table 18
Mean peak isokinetic strength (kg) of males in various posture–reach distance–arm angle combinations

Posture	Reach distance (cm)		Arm angle					
			−90	−60	−30	0	+30	+60
Sitting	25	Mean	62	68	73	77	75	70
		SD	29	31	31	22	25	24
	40	Mean	79	93	93	96	94	87
		SD	28	29	29	25	28	29
	55	Mean	93	99	105	109	97	97
		SD	29	27	22	30	25	26
Standing	45	Mean	96	101	108	120	110	97
		SD	31	32	36	39	44	34
	65	Mean	113	120	128	136	121	124
		SD	32	27	28	35	32	39
	85	Mean	123	128	129	138	132	124
		SD	29	25	33	32	34	34

Note: Degree measured from mid-sagittal plane. Mid-sagittal plane–0 deg; − counterclockwise; + clockwise.

Table 19
Peak isokinetic strength (kg) profiles of men and women in various postures

Posture No.	Gender	Mean	Standard deviation	Range
1	Male	103	31	53–154
	Female	67	17	44–91
2	Male	56	19	27–119
	Female	10	17	0–50
3	Male	2	8	0–39
	Female	0	0	0
4	Male	113	26	78–169
	Female	61	15	34–75
5	Male	76	19	36–108
	Female	24	16	5–42
6	Male	76	20	49–128
	Female	29	7	19–39
7	Male	94	21	56–142
	Female	40	11	28–53
8	Male	86	23	54–151
	Female	38	12	20–52
9	Male	30	24	0–70
	Female	8	20	0–57
10	Male	11	19	0–67
	Female	0	0	0
11	Male	107	29	63–170
	Female	65	15	0–95
12	Male	130	25	76–189
	Female	67	21	21–94
13	Male	66	16	30–99
	Female	17	15	0–36
14	Male	59	24	18–110
	Female	17	9	3–30
15	Male	92	21	59–148
	Female	46	16	17–67

Note: Posture 1 – semi-squat using both hands at the knee height; Posture 2 – sitting using the preferred hand (upper arm vertical and forearm horizontal); Posture 3 – sitting using the preferred hand (upper and forearm horizontal); Posture 4 – semi-squat, torso twisted 60°, hands at the knee; Posture 5 – standing erect, forearms vertical, upper arms horizontal; Posture 6 – kneeling on both knees, forearms vertical and upper arms horizontal; Posture 7 – kneeling on one knee; forearms horizontal and upper arms vertical; Posture 8 – standing erect, torso twisted 60°, upper arms vertical and forearms horizontal; Posture 9 – sitting, preferred arm vertical and by the side of the body; Posture 10 – sitting, preferred hand, grip at the chin height in the mid-sagittal plane; Posture 11 – stooping, both arms fully stretched, torso twisted 60°; Posture 12 – stooping, both hands grip at height midway between the knee height and the floor; Posture 13 – sitting, forearms vertical, upper arms horizontal; Posture 14 – sitting, forearms vertical, upper arms horizontal, torso twisted 60°; Posture 15 – standing erect, forearms horizontal, upper arms vertical.

Table 20
Maximum volitional isometric torques (Nm) for various non-powered hand tool-posture combinations

Posture No.	Short screwdriver		Long screwdriver		Spanner wrench		Socket wrench	
	Mean	SD	Mean	SD	Mean	SD	Mean	SD
Males								
1	4.6	1.3	3.4	1.1	32.0	10.4	40.0	14.3
2	4.4	1.5	2.9	1.1	37.7	15.4	43.5	15.6
3,4	4.4	1.4	3.2	1.1	35.4	12.3	41.9	12.8
5	5.0	1.7	3.5	1.2	34.0	10.5	41.5	14.6
6,7	4.3	1.7	3.0	1.0	37.9	15.0	42.7	14.4
8	5.0	1.7	3.4	1.0	34.0	10.9	41.8	14.4
9,10,11	4.6	1.7	3.2	1.3	35.2	13.0	43.7	13.7
12	5.1	1.8	3.5	1.2	33.7	12.8	42.9	14.4
13,14,15	4.4	1.6	3.2	1.2	32.9	12.0	39.3	13.1
16	4.5	1.7	3.4	1.2	30.2	10.9	37.1	15.9
17	3.7	1.7	2.8	1.2	26.5	8.9	32.0	8.4
18,21	4.2	1.7	3.1	1.3	29.5	9.6	35.3	10.9
19	4.5	1.8	3.4	1.3	35.9	14.2	40.4	12.1
20	4.2	1.6	3.1	1.2	31.2	11.6	36.0	12.4
22	2.6	1.7	3.1	1.2	10.9	1.8	13.7	2.0
23	2.6	1.7	3.1	1.2	5.9	0.9	7.0	1.1
24	2.6	1.7	3.1	1.2	11.9	2.4	14.6	3.5
25	2.6	1.7	3.1	1.2	7.1	1.3	8.3	1.3
26	2.6	1.7	3.1	1.2	9.7	2.0	12.0	2.8
27[a]	3.4	1.6	3.2	1.4	27.4	10.2	32.5	11.9
28[b]	3.2	1.2	3.2	1.0	33.0	12.6	39.5	13.3
Females								
1	3.6	1.6	2.45	1.2	15.3	6.1	18.8	6.5
2	3.1	1.1	2.3	0.9	24.8	8.3	31.1	11.1
3	3.7	1.4	2.5	1.2	20.0	8.1	22.9	8.0
4	3.0	1.0	2.2	0.8	23.4	8.0	28.2	11.1
5	3.6	1.5	2.3	1.1	20.0	7.9	23.3	7.0
6	3.0	1.2	1.8	0.6	23.6	7.8	28.0	9.3
7	3.4	1.1	2.1	0.6	25.0	11.0	29.3	10.2
8,10	3.9	1.5	2.5	0.8	18.3	7.9	23.9	9.7
9,11	3.4	0.9	2.1	0.6	24.7	10.3	27.0	10.0
12	3.8	1.5	2.2	1.0	17.0	8.0	22.3	8.6
13	3.1	0.9	2.2	0.6	21.0	6.0	26.1	9.7
14	3.2	1.1	2.4	0.5	23.4	6.6	28.4	9.6
15	3.4	1.5	2.4	1.1	17.0	6.5	20.6	9.2
16	2.7	1.0	2.0	0.8	15.4	6.6	19.0	7.7
17	3.0	1.0	1.8	0.9	18.5	5.4	22.2	9.4
18,21	2.7	0.8	1.9	0.7	17.5	6.2	20.7	6.5
19	3.1	1.2	2.1	0.8	23.5	8.4	25.9	10.6
20	3.0	1.2	2.1	0.9	19.0	6.8	21.3	7.6
22	2.0	0.2	2.1	0.9	7.8	2.4	9.9	1.9
23	2.0	0.2	2.1	0.9	4.8	1.5	5.8	1.7
24	2.0	0.2	2.1	0.9	8.0	1.8	10.3	3.2
25	2.0	0.2	2.1	0.9	5.3	1.1	6.1	1.3
26	2.0	0.2	2.1	0.9	6.4	1.5	7.8	1.7
27[a]	3.6	0.9	2.8	0.9	16.6	5.6	19.7	6.0
28[b]	2.9	0.8	2.4	0.8	23.0	7.7	26.0	8.7

Table 20 (Footnotes)
Note: [a] – reach distance for the sitting posture: 46 cm; [b] – reach distance for the standing posture: 33 cm. Posture 1 – standing, tool axis vertical (upward); Posture 2 – standing, tool axis horizontal (outward); Posture 3 – standing, knees bent, back supported, tool axis vertical (upward); Posture 4 – standing, knees bent, back supported, tool axis horizontal (outward); Posture 5 – standing, bent at the waist, tool axis vertical (upward); Posture 6 – standing, bent at the waist, tool axis horizontal (outward); Posture 7 – standing, bent at the waist, tool axis horizontal (inward); Posture 8 – kneeling on one knee, tool axis vertical (upward); Posture 9 – kneeling on one knee, tool axis horizontal (outward); Posture 10 – kneeling on both knees, tool axis vertical (upward); Posture 11 – kneeling on both knees, tool axis horizontal (outward); Posture 12 – squat, tool axis vertical (upward); Posture 13 – squat, tool axis horizontal (outward) and between knees and shoulder; Posture 14 – squat, tool axis horizontal (outward) and between knees and ankles; Posture 15 – squat, tool axis vertical (downward), work surface-floor; Posture 16 – lying on the side, tool axis vertical (upward); Posture 17 – lying on the side, tool axis horizontal (outward); Posture 18 – lying on the back, tool axis vertical (upward); Posture 19 – lying on the back, tool axis horizontal (parallel to the body)(work piece remote from the body); Posture 20 – lying on the stomach, tool axis vertical (upward); Posture 21 – lying on the stomach, tool axis horizontal (outward); Posture 22 – leaning sideways from a ladder, tool axis horizontal (outward); Posture 23 – leaning sideways from a ladder, tool axis vertical (upward); Posture 24 – overhead, tool axis horizontal (outward); Posture 25 – overhead, tool axis vertical (upward); Posture 26 – squatting in a confined space, tool axis horizontal (outward); Posture 27 – sitting, torque exerted at the eye level, tool axis horizontal (outward); Posture 28 – standing, torque exerted at the eye level, tool axis horizontal (outward).

Table 21
Grip strength (kg) for combination of postures of shoulder, elbow and wrist for males

Shoulder flexion (deg)	Elbow flexion (deg)	Wrist flexion extent	Wrist ulnar deviation extent					
			None		1/3		2/3	
			Mean	SD	Mean	SD	Mean	SD
0	90	Neutral	38.9	9.9	30.2	9.7	28.9	10.8
		One-third	31.9	6.0	28.7	8.0	28.0	8.5
		Two-third	28.5	6.8	27.6	6.9	24.1	6.5
	45	Neutral	40.0	9.7	34.0	5.8	31.3	9.0
		One-third	32.4	5.3	29.7	6.7	29.7	6.8
		Two-third	31.2	6.0	27.6	7.2	27.6	6.3
	0	Neutral	38.1	8.1	31.2	7.1	27.8	7.5
		One-third	30.8	7.3	29.8	9.6	26.2	7.8
		Two-third	28.2	7.6	27.6	8.7	22.6	7.2
20	90	Neutral	36.4	11.2	30.2	7.8	29.9	10.7
		One-third	31.7	9.3	26.5	6.6	25.4	10.0
		Two-third	28.6	7.6	25.0	7.3	24.1	8.8
	45	Neutral	38.2	9.8	33.6	8.8	30.5	7.7
		One-third	32.9	8.2	29.0	6.2	27.6	7.2
		Two-third	29.8	7.0	28.2	7.1	27.8	7.2
	0	Neutral	36.2	9.3	32.1	8.9	26.8	7.0
		One-third	30.5	7.0	28.0	9.6	25.0	7.7
		Two-third	27.2	8.1	23.0	7.8	23.0	5.8

Table 22
Mean pistol grip strength (N) for tool handles with varying sizes of hands among university students

Hand Size	Age (yr)	Gender	Grip strength (N) Mean	SD
Small	22	F	236	47
	21	F	152	13
	20	F	220	22
	20	F	179	20
	20	F	122	28
	29	F	181	15
Medium	24	F	346	43
	26	F	195	21
	20	F	181	17
	20	F	115	24
	31	M	382	97
	24	F	273	46
Large	25	M	286	29
	29	M	308	26
	23	M	525	137
	26	M	433	58
	34	M	399	85
	22	M	492	86
Mean	24		279	
SD	4		133	

Table 23
Mean pistol grip strength (N) for tool handles with varying sizes of hands among industrial workers

Hand size	Age (yr)	Gender	Grip strength (N) Mean	SD
Medium	46	F	221	41
	41	M	411	36
	52	M	313	58
	46	F	259	17
	51	M	315	46
Large	51	M	328	71
	48	M	277	24
	49	M	391	120
	60	F	289	52
	43	M	384	83
	39	M	413	151
Mean	48		327	
SD	6		90	

Table 24
Mean and SD values of the total grip force and individual finger contribution to the total grip force for each of the object weight and diameter conditions in holding cylindrical objects with circular grip for males

Weight (kg)	Diameter (cm)	Total grip force Mean (N)	SD	Thumb Mean (%)	SD	Index finger Mean (%)	SD	Middle finger Mean (%)	SD	Ring finger Mean (%)	SD	Little finger Mean (%)	SD
0.5	7.5	47.2	17.8	40.0	4.5	10.2	3.6	11.9	4.5	17.8	4.1	20.1	5.2
1.0	7.5	106.5	31.0	42.1	4.1	9.4	2.8	12.3	4.2	18.7	3.5	17.6	4.2
2.0	7.5	233.1	68.2	43.5	4.1	7.9	2.3	13.4	4.6	19.0	4.1	16.2	4.3
1.0	5.0	118.4	34.2	38.6	2.9	10.8	3.4	14.8	5.1	18.0	4.2	17.7	4.2
1.0	10.0	137.8	28.9	40.2	2.8	9.6	3.5	10.6	2.8	17.4	3.3	22.2	3.3

Table 25
Mean and SD values of the total grip force and individual finger contribution to the total grip force for varied grip modes in holding cylindrical objects with circular grips for males

Grip mode	Total grip force		Thumb		Index finger		Middle finger		Ring finger	
	Mean (N)	SD	Mean (%)	SD	Mean (%)	SD	Mean (%)	SD	Mean (%)	SD
4-finger	107.5	30.5	45.5	2.9	11.6	3.8	17.9	4.9	25.0	4.6
3-finger	121.8	39.5	45.7	3.7	14.5	4.6	39.8	4.6		
2-finger	152.2	40.4	49.9	0.7	50.1	0.7				

Table 26
Peak pinch strengths (kgf) of 17 males at various pinch widths

Pinch width (cm)	Statistics	Chuck pinch	Lateral pinch	Pulp-2 pinch
2.0	Mean	16.9	14.2	11.6
	SD	7.0	5.6	3.4
3.2	Mean	17.7	14.2	11.7
	SD	5.3	4.8	3.5
4.4	Mean	18.4	13.5	11.7
	SD	6.8	4.2	3.6
5.6	Mean	17.2	12.8	11.0
	SD	7.0	4.0	3.2
6.8	Mean	14.9	10.1	11.1
	SD	4.7	4.0	3.5
8.0	Mean	14.8	9.7	10.4
	SD	5.8	5.1 ($n = 12$)	3.8
9.2	Mean	13.8	8.0	10.1
	SD	5.3	2.8 ($n = 7$)	3.7
10.4	Mean	11.2		8.6
	SD	4.9		3.7
11.6	Mean	10.5		7.1
	SD	4.6		3.6 ($n = 16$)
12.8	Mean	9.5		5.9
	SD	3.5 ($n = 13$)		3.3 ($n = 15$)
14.0	Mean	9.0		5.7
	SD	3.6 ($n = 7$)		2.8 ($n = 7$)

Table 27
Means of different pinch strengths at different wrist positions ($n = 30$) (in kgf) for males

	Wrist position									
	Netural		Radial deviation		Ulnar deviation		Dorsi-flexion		Palmar flexion	
	Mean	SE	Mean	SE	Mean	SE	Mean	SE	Mean	SE
Pinch										
Lateral	9.8	0.4	8.5	0.3	7.9	0.3	7.7	0.3	6.5	0.2
Chuck	9.6	0.3	7.4	0.2	6.9	0.2	6.8	0.3	5.5	0.2
Pulp-2	7.1	0.2	5.0	0.2	5.0	0.2	5.0	0.3	4.2	0.2
Pulp-3	6.9	0.2	5.0	0.2	4.7	0.2	4.6	0.2	3.9	0.1

Table 28
Peak pinch isometric strengths in kilograms for females with wrist in neutral, half and full wrist extension and wrist flexion

Pinch style	Neutral	Half extension	Half flexion	Full extension	Full flexion
Pulp-2	4.21 (0.68)	4.01 (0.77)	3.57 (0.62)	3.29 (0.66)	2.79 (0.63)
Pulp-3	3.97 (1.16)	3.83 (0.89)	3.46 (1.01)	3.29 (0.77)	3.00 (1.13)
Pulp-4	2.70 (0.82)	2.43 (0.69)	2.25 (0.66)	2.25 (0.46)	2.17 (0.52)
Pulp-5	1.85 (0.43)	1.73 (0.42)	1.65 (0.42)	1.63 (0.43)	1.54 (0.35)
Tip	4.78 (0.65)	4.74 (0.54)	4.42 (0.51)	3.77 (1.02)	3.67 (0.70)
Lateral	6.39 (0.65)	6.09 (0.82)	5.83 (1.12)	5.69 (1.01)	4.97 (0.93)
Chuck	6.35 (1.11)	6.27 (1.19)	5.32 (0.97)	4.97 (0.79)	3.99 (0.92)

Note: Pulp-2 = thumb and index finger; Pulp-3 = thumb and middle finger; Pulp-4 = thumb and ring finger; Pulp-5 = thumb and little finger; Lateral = thumb and lateral side of index finger; Chuck = thumb, index and middle finger.

Table 29
Finger pinch-pull isometric strengths (N) of males

Direction	Type of pull	Left hand	Right hand
Horizontal	LPP	92.0 (3.9)	94.8 (4.9)
	CPP	52.8 (2.1)	57.2 (2.9)
	PPP	36.5 (1.9)	37.5 (2.1)
Oblique	LPP	97.7 (4.9)	95.9 (4.2)
	CPP	53.2 (2.8)	57.7 (2.6)
	PPP	37.0 (1.8)	37.0 (1.9)

Note: Numbers in the parentheses are standard error of the mean; LPP – lateral pinch-pull strength exerted with the pad of the thumb and the radial lateral aspect of the index finger with other three fingers acting as a buttress to the index finger; CPP – chuck pinch-pull strength exerted with the pads of the index and middle fingers on one side and with the pad of the thumb on the other side; PPP – pulp pinch-pull strength exerted with the pads of the thumb and the index finger.

References

Mital, A., Kumar, S., 1998. Human muscle strength definitions, measurement and usage: Part II – The scientific basis (knowledge base) for the guide. International Journal of Industrial Ergonomics 22 (1–2), 123–144.

Human muscle strength definitions, measurement, and usage: Part II – The scientific basis (knowledge base) for the guide[1]

Anil Mital[a,*], Shrawan Kumar[b]

[a] *Ergonomics and Engineering Controls Research Laboratory, College of Engineering, University of Cincinnati, Cincinnati, OH 45221-0116, USA*
[b] *Department of Physical Therapy, University of Alberta, Edmonton, Alberta T6G 2G4, Canada*

1. Rationale for measuring and using human strengths

Many industrial activities are performed through human intervention. It is generally accepted that knowledge of what a person can or cannot do under specified circumstances is essential for efficient work design and injury prevention. In a variety of production processes, force application is an essential activity through which the productive phase is mediated. Besides industrial engineers/ergonomists, measurement of human strengths is of interest to many other professionals in many other disciplines, such as physiotherapy, sports medicine, physical education, rehabilitation, orthopaedics, etc.

Human strengths are a primary measure of an individual's physical capabilities, particularly those that permit a person to exert force or sustain external loading without inflicting personal injury. Researchers in the past 35 years have actively engaged in developing design databases and strength-based employee screening and job matching procedures (Chaffin et al., 1978; Keyserling et al., 1980; Mital and Das, 1987; Rühmann and Schmidtke, 1989; Kumar, 1991; Kumar and Garand, 1992; Mital et al., 1993a; Kumar et al., 1995c; Kumar, 1995). The recommendations for utilizing human strengths are particularly strong for manual materials handling activities (Chaffin and Park, 1973; Ayoub and McDaniel, 1974; Chaffin, 1974; Chaffin et al., 1978; Garg et al., 1980; Mital and Ayoub, 1980; Yates et al., 1980; Pytel and Kamon, 1981; Kamon et al., 1982; Kroemer, 1983, 1985; Aghazadeh and Ayoub, 1985; Mital and Karwowski, 1985; Ayoub et al., 1986; Dales et al., 1986; Mital et al., 1986a,b,c; Mital et al., 1987; Mital and Genaidy, 1989; Kumar and Garand, 1992; Kumar, 1995). Human strength recommendations for tool design, and work and workspace design have also been receiving considerable attention (Chaffin et al., 1983; Mital and

---

*Corresponding author.

[1] The recommendations provided in this guide are based on numerous published and unpublished scientific studies and are intended to enhance worker safety and productivity. These recommendations are neither intended to replace existing standards, if any, nor should be treated as standards. Furthermore, this document should not be construed to represent institutional policy.

The following individuals participated in the discussion of the earlier version of this guide. Their suggestions (written or verbal) were incorporated by the authors in this version: A. Aaras, Norway; J.E. Fernandez, U.S.A.; A. Freivalds, U.S.A.; T. Galleway, Ireland; M. Jager, Germany; S. Konz, U.S.A.; H. Krueger, Switzerland; K. Landau, Germany; A. Luttmann, Germany; J.D. Ramsey, USA; M-J. Wang, Taiwan.

Channaveeraiah, 1988; Mital and Genaidy, 1989; Mital and Faard, 1990; Kumar, 1991; Kumar and Garand, 1992; Kumar, 1996; Mital et al., 1995).

Lack of design guidelines and screening procedures can lead to overloading of the muscle-tendon–bone–joint system and, thereby, fatigue and possible consequent injury (Hettinger, 1961). As a matter of fact, Chaffin et al. (1978) have reported that the incidence rate of back injuries sustained on the job increased when the job strength requirements exceeded isometric strengths of the workers. Similarly, Keyserling et al. (1980) observed that the incidence rate among employees selected on the basis of isometric strengths was approximately one-third that of employees selected using traditional clinical criteria. The relationship between insufficient physical capability and injury is widely acknowledged not only for manual materials handling activities (NIOSH, 1981; Taber, 1982; National Safety Council, 1983; Asfour et al., 1983; Khalil et al., 1983) but also for tasks requiring hand tool usage (Mital and Aghazadeh, 1985; Aghazadeh and Mital, 1987). These two activities, materials handling and force/torque exertion with the aid of human-powered hand tools, account for approximately 45% of all industrial overexertion injuries in the United States. The total cost of these injuries in the US is estimated to be well over $150 billion per year (Aghazadeh and Mital, 1987; Mital et al., 1993b).

In addition to screening procedures and design information for selecting fastener devices (nuts, bolts, screws, etc.) which are tightened or loosened to restrain or unrestrain objects, a number of studies have focused on the development of work and workplace design data. Among some of the significant works are studies by Martin and Chaffin (1972), Ayoub and McDaniel (1974), Chaffin et al. (1978), Snook (1978), Davis and Stubbs (1980), Warwick et al. (1980), Chaffin et al., 1983, Mital (1984a,b), Mital et al. (1986a,b,c), Mital and Wang (1987), Kumar et al. (1988), Mital and Genaidy (1989), Mital and Faard (1990), Kumar (1991), Kumar and Garand (1992), Kumar et al. (1995c), and Mital et al. (1995).

Human strengths of individuals thus form the basis for many design data bases and screening procedures. The overall intent of these developments is to reduce injuries and, in the process, maximize industrial productivity.

2. Static versus dynamic strengths

Absence of body segment, object, and muscle movement during maximal voluntary contraction for assessing static strengths is the major distinction between static and dynamic strengths. Since there is no effective limb-object-muscle movement in case of static strengths, these strengths cannot account for the effect of inertial forces. This leads to underestimating musculoskeletal joint loading during the performance of a dynamic task (Leskinen et al., 1983; Hall, 1985; McGill and Norman, 1985; Bush-Joseph et al., 1988). Clearly, for this reason alone the static strength exertion capability of individuals should not be used to assess their capability to perform dynamic tasks. Furthermore, since most industrial processes require a force application through a range of motion in a continuous activity, the design of tasks based on static strengths in a fixed posture has lost its relevance.

A number of recent studies have shown that dynamic strengths of individuals are more highly correlated to their task performance capabilities than their static strengths (Kamon et al., 1982; Kroemer, 1983, 1985; Aghazadeh and Ayoub, 1985; Mital et al., 1986a,b, 1987; Dales et al., 1986; Smith and Ayoub, 1989). It has also been shown that even though two individuals might have the same static strength, their dynamic lifting capabilities could be substantially different (Mital et al., 1987; Kumar, 1991). Furthermore, static strengths tend to underestimate low-back joint loading estimated from psychophysical maximum acceptable weight of lift (Garg et al., 1982; Freivalds et al., 1984; Mital and Kromodihardjo, 1986; Kromodihardjo and Mital, 1987) and may underpredict or overpredict a person's capability to perform physical tasks (Garg et al., 1980; Kamon et al., 1982; Kroemer, 1983; Kumar et al., 1988; Mital et al., 1995).

Mital et al. (1995), in a study designed to determine isokinetic pull strengths of individuals in the vertical plane, found that the magnitude of isokinetic exertion was well below isometric exertions (Rohmert, 1966). The maximum isokinetic exertions for males and females, on the average, were approximately 58 and 36% of their respective body weight. For isometric strength in the vertical plane the peak exertion has been reported to be 120% of the body weight (Rohmert, 1966). It appears that the need to maintain stable posture during a dynamic exertion does not allow individuals to take advantage of the leverage that lower leg muscles provide during an isometric exertion.

The recent studies, thus, clearly indicate that dynamic strengths, not static strengths, are more appropriate measures of a person's work performance capabilities and, therefore, should be used in worker screening and form the basis for designing industrial jobs, work places, tools, equipment, and controls.

3. Task factors affecting dynamic strengths

Several task-related factors are known to influence dynamic strength exertions. In particular, the effects of the following factors have been investigated: (1) posture, (2) reach distance, (3) arm orientation, (4) speed at which exertions are recorded, and (5) frequency and duration of exertions. The reported effects of these factors on dynamic strength exertion capability of the individual are, however, occasionally confounded with other factors. Following is a brief review of recent investigations that have studied the effect of these factors on dynamic strength.

3.1. Posture

The dynamic strength has been shown to be significantly affected by posture (Pytel and Kamon, 1981; Kamon et al., 1982; Kroemer, 1983, 1985; Grieve, 1984; Kumar and Chaffin, 1985; Marras, 1985; Mital et al., 1986a, 1986b, 1986c, 1987; Kumar et al., 1988; Mital and Genaidy, 1989; Mital and Faard, 1990; Kumar, 1991; Kumar and Garand, 1992). Detailed studies of the postural effect on dynamic strength are, however, few (Mital and Genaidy, 1989; Mital and Faard, 1990).

Different postures result in different lever arms and muscle angles and involvement of different muscle groups. Mital and Genaidy observed that, depending upon the posture assumed by the individual, the isokinetic strength exertion capability of the individual can vary greatly – from almost no force to considerable force. Their study provides isokinetic strengths of males and females for infrequent vertical exertions in fifteen different working postures. The mean and standard deviations of force exertions in each posture are given in Table 19 of the Part I of this guide.

The average peak isokinetic pull-up strength of females varied from 0 kg, for the sitting posture with preferred arm fully extended (posture 3 – it should be noted that the recorded strength was 0 kg most likely due to the lack of precision of the recording instrument, not because the actual strength was 0 kg) and preferred arm grip at chin (posture 10), to approximately 67 kg, for the two-handed stoop posture (posture 12). For males, the average peak isokinetic pull-up strength varied from approximately 2 kg, for the sitting posture with the preferred arm fully extended, to approximately 130 kg, for the two-handed stoop posture. The data in Table 19 of the Part I of this guide also show that isokinetic strength exertions of females are approximately one-half of the male exertions, not 2/3rds as in the case of static strengths (Mital et al., 1986b).

The variation in strengths with posture are also not limited to just one particular kind of activity. Mital and Faard (1990) focussed their efforts on the sitting and standing posture and measured one-arm peak isokinetic pull strengths of males in the horizontal plane. They found that 37% more strength is exerted in the standing posture (119 kg) as compared to the sitting posture (87 kg). Such a large difference in strength exertion capability in different postures is expected to have a profound influence on work and workplace designs.

3.2. Reach distance

Mechanical disadvantage increases the task difficulty. This may cause reduction in strength output. Many previous studies have systematically explored the effect of reach distance on isometric strength and shown that isometric strength increases with the reach distance up to a certain point if the body posture does not change significantly (Clarke et al., 1950; Wakim et al., 1950; Martin and Chaffin, 1972; Mital et al., 1985b; Mital, 1986; Mital and Sanghavi, 1986; Mital and Channaveeraiah, 1988; Kumar, 1991; Kumar and Garand, 1992). If, however, the posture changes are significant, the strength exertion capability declines with the reach distance (Martin and Chaffin, 1972; Ayoub and McDaniel, 1974; Chaffin et al., 1983). The effects of reach distance on dynamic strength also appear to have similar effects.

Kumar (1987a,b, 1991) and Kumar and Garand (1992) found that for lifting type of tasks people are able to generate greater forces closer to the body. As the distance of hand grip from the body increased, the strength exertion capability progressively decreased. An isokinetic effort at full reach in the sagittal plane was between 28% and 40% of peak isometric strength at half-reach. Table 1 shows the mean peak and average isokinetic strengths of males and females during arm lift. The peak and average strengths of females as a percentage of males are shown in Table 2.

Mital and Faard (1990) investigated the effect of reach distance on one-handed peak isokinetic pull strength in the horizontal plane within a given posture (sitting or standing). The results indicated that the magnitude of isokinetic strength increases significantly and almost linearly with the reach distance between 25 and 55 cm for the sitting posture and between 45 and 85 cm for the standing posture. Fig. 1 shows the effect of reach distance on one-handed peak isokinetic pull strength in the horizontal plane.

The data given in Table 1 (mean and standard deviation) can be used to compute various population percentiles assuming a normal distribution. However, as demonstrated by Mital (1978), the distribution of strengths is not exactly normal, but slightly skewed beyond the 70th percentile. In order to compute exact percentile values, the normal percentile values need to be multiplied by a correction factor. These correction factors have been developed by Mital (1978) and are given in Table 1 of the Part I of this guide.

In order to correct the strength values for age, factors given in Table 2 of Part I should be used. These factors are based on the works of Hollmann and Hettinger (1980) and Grimby and Saltin (1983) on isometric strengths. In the absence of any dynamic strength study dealing with the effect of age, this is the best available information.

Table 1
Mean peak and average isokinetic strength during arm lift (N) (Kumar, 1991; Kumar and Garand, 1992)

Strength	Plane	Gender		0.5R	0.75R	1R
Peak	Sagittal	Male	X	311	225	162
			SD	87	52	28
		Female	X	184	139	101
			SD	39	33	26
	30° lateral	Male	X	284	214	156
			SD	69	58	31
		Female	X	177	145	98
			SD	44	29	27
	60° lateral	Male	X	258	171	129
			SD	69	32	29
		Female	X	160	122	81
			SD	40	31	22
Average	Sagittal	Male	X	236	180	137
			SD	61	40	24
		Female	X	141	109	81
			SD	25	29	22
	30° lateral	Male	X	219	174	130
			SD	59	45	28
		Female	X	137	114	80
			SD	28	22	24
	60° lateral	Male	X	200	140	110
			SD	51	27	30
		Female	X	126	98	62
			SD	28	26	19

Note: R - Reach.

Table 2
The peak and average isokinetic strengths of females (Table 1) as percentage of males (Kumar, 1991; Kumar and Garand, 1992)

Strength	Plane	0.5R	0.75R	1R
Peak	Sagittal	59	61	62
	30° lateral	62	67	63
	60° lateral	62	71	63
Average	Sagittal	60	60	59
	30° lateral	62	65	61
	60° lateral	63	70	56

Note: R - Reach.

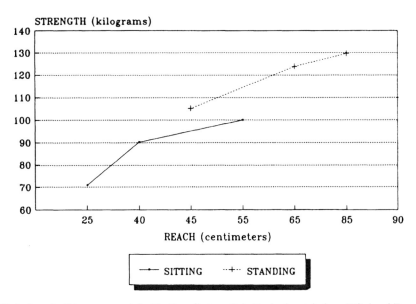

Fig. 1. Effect of reach distance on peak isokinetic pull strength in the horizontal plane (Mital and Faard, 1990).

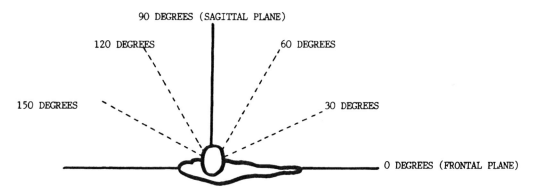

Fig. 2. Arm orientation with respect to the frontal plane (Mital and Faard, 1990).

3.3. Arm orientation

The orientation of the arm is also known to influence human isometric strength exertion capability (Caldwell, 1959; Rohmert, 1966; Rohmert and Jenik, 1972). As the arm orientation changes the mechanical advantage also changes, resulting in weaker or stronger strength exertions. To determine its influence on one-handed peak isokinetic pull strength in the horizontal plane, Mital and Faard (1990) studied the effect of angle of the stronger arm relative to the frontal plane. The following angles were included in the study: 0 (frontal plane), 30°, 60°, 90°, 120°, and 150° (Fig. 2). The results indicated that strength exertion was maximized when the arm was in the sagittal plane. Least strength was exerted in the frontal plane (0° angle) and 150° arm orientation. The rates of decline in strength (kg/deg) below and above 90° angle were similar. Fig. 3 shows the influence of arm orientation on one-handed peak isokinetic pull strength in the horizontal plane.

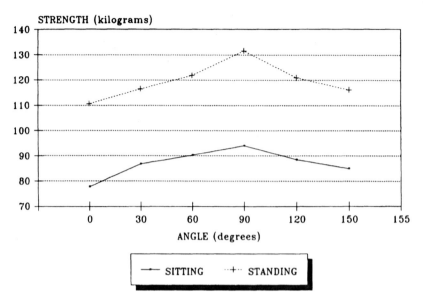

Fig. 3. Effect of preferred arm angle on peak isokinetic pull strength in the horizontal plane (Mital and Faard, 1990).

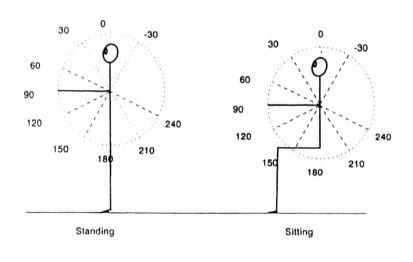

Fig. 4. Angular positions of the preferred arm in the vertical plane (Mital et al., 1995).

Kumar (1991) and Kumar and Garand (1992) also observed that isokinetic strengths in the sagittal plane are higher than isokinetic strengths in the lateral planes (Table 1).

Mital et al. (1995) studied the effect of preferred arm angle on one-handed isokinetic pull strength in the vertical plane. Using two postures, sitting and standing, the following angles of the preferred arm from the vertical were studied: $-30°$, $0°$, $30°$, $60°$, $90°$, $120°$, $150°$, $180°$, $210°$, and $240°$. These angles are shown in Fig. 4. The results of the study indicated that arm orientation also had a significant

Table 3
One-handed isokinetic pull strength (N) in the vertical plane (Mital et al., 1995)

Preferred arm angle (Degrees)	Males		Females	
	X	SD	X	SD
−30	173	159	24	6
0	400	265	59	47
30	269	200	43	23
60	220	186	36	21
90	160	165	33	20
120	380	186	30	13
150	370	252	26	10
180	460	273	194	107
210	230	152	50	74
240	190	195	22	10

influence on isokinetic strength in the vertical plane. Table 3 shows one-handed isokinetic pull strength of males and females in the vertical plane.

3.4. *Speed of exertion*

Measurement of dynamic strength, whether of a well-defined muscle group or on a whole-body basis, is much more complicated and difficult than static strength measurement. One of the factors that introduces a great deal of complexity is velocity of exertion. While in static strength measurements no motion is permitted, dynamic strength exertions can take place at variety of speeds. Body motions have several effects on muscular force capability. The motion of body segments may require significant muscle force simply to accelerate the body mass and overcome inertia. During the deceleration phase, momentum will assist the muscles in producing additional effective force output. The effective force output can vary considerably over time, especially when large body segments are accelerated and decelerated quickly (Kroemer, 1970; Miller and Nelson, 1976; Garg et al., 1988; Kumar et al., 1988; Mital et al., 1995). Figs. 5 and 6, from Garg et al. (1988), show force and velocity curves against time for typical male and female dynamic pulling strength. Furthermore, the peak strength for different speeds occurs at different time (Kumar et al., 1988, Table 4). The effect of speed of exertion

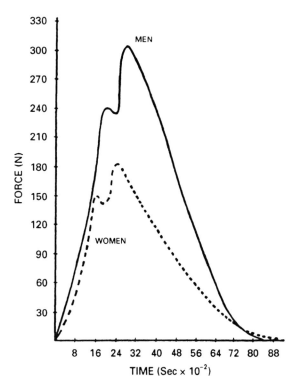

Fig. 5. Variation in dynamic pulling strength with time for typical male (___) and female (_ _ _ _) pulls (Garg et al., 1988).

on one-handed pulling strength in the vertical plane is shown in Table 5 (Mital et al., 1995).

Isometric strengths are greater than dynamic strengths (Kumar et al., 1988; Kumar, 1991; Kumar and Garand, 1992). The study by Mital et al. (1995) further reinforced this conclusion. The peak strength in their study was exerted at the slowest speed (0.30 m/s^{-1}) and declined almost linearly by almost 30% for males and 25% for females, as the speed of exertion increased to 0.75 m/s^{-1} (Table 5). Decline in strength with velocity was also observed by Imrhan and Ayoub (1990).

The dynamic strength exertion is also affected by the force–velocity property of the muscle and moment arms of the muscles acting about a joint. The moment arms vary with the joint angle and as a muscle is rapidly required to shorten and produce force, the maximum force is reduced as a function of the velocity of muscle shortening. However, the relationship between static and dynamic strength

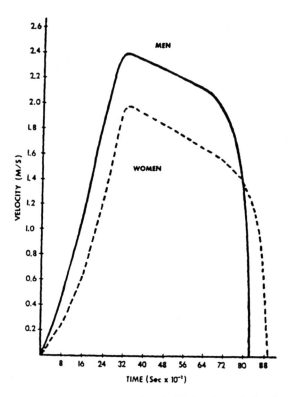

Fig. 6. Variation in velocity of pull with time for typical male (___) and female (_ _ _ _) pulls (Garg et al., 1988).

magnitudes depends upon how these strengths are measured. The dynamic strength peaks generally would be higher than static strength peaks if static strengths closely simulate actual job conditions or if deceleration is avoided and momentum is allowed to contribute to the force output (Garg et al., 1980; Kroemer, 1983, 1985; Jiang, 1984; Mital et al., 1986a,b,c; Mital and Genaidy, 1989; Mital and Faard, 1990). The dynamic strength peaks would otherwise be significantly lower than static strength peaks (Asmussen et al., 1965; Carlson, 1970; Aghazadeh and Ayoub, 1985; Marras, 1985; Kumar et al., 1988; Garg et al., 1988; Kumar, 1991; Kumar and Garand, 1992; Mital et al., 1995).

3.5. Frequency and duration of exertions

All biological tissues are viscoelastic in nature. Prolonged exertions result in cumulative fatigue and reduced stress bearing capacity. The use of strengths in gauging physical exertion and overexertion is of strategic significance, particularly in activities which are known to cause injuries, for example manual materials handling. The obvious question is where exertion becomes overexertion? Can exertion and overexertion be gauged by the magnitude of singular dynamic strength exertion or the concept of repetition and cumulative fatigue must be introduced? Historically, singular static or dynamic strength exertions have been used to gauge physical exertion and overexertion (Ayoub and Mital, 1989; Chaffin and Anderson, 1991; Mital et al., 1993b). Yet, there continues to be significant and recurring epidemiological evidence that majority of injuries and industrial accidents are being caused by overexertion (NIOSH, 1981; Holbrook et al., 1984; Statistics Canada, 1988; Kumar, 1989; Kumar, 1990).

In the context of isometric strength, Mital and Channaveeraiah (1988) investigated the effect of the duration of repeated exertions, applied at self-determined frequency, on volitional forces/torques exerted with common non-powered hand tools. They concluded that the decline in strength is very rapid during the initial period of exertion. Table 3 in Part I of this guide provides equations developed by them that describe the relationship between torque and the duration of exertion for various hand tools.

However, in order to get a better handle on gauging exertion/overexertion, the effect of repetition on dynamic strength needs to be established. With this in mind, Mital et al. (1986b, 1987) investigated the effect of repetition on isokinetic lifting strengths. Figs. 7 and 8b show the relationship between repetitive dynamic strength (RDS) and the duration of repetition for lifting height and frequency of lifting. The results of these studies indicated that: (1) isokinetic strength declines exponentially with the duration of lifting, (2) the decline in strength is sharper for higher frequencies, and (3) repetitive dynamic strengths are better estimators of an individual's capability to perform tasks that require repeated heavy physical exertion.

While the ability of RDS to gauge physical exertion may have been established, its ability to gauge overexertion is yet to be demonstrated.

Table 4
Mean peak isokinetic strengths (N), their corresponding vertical hand location (CM), activity duration (S), and the percent of activity (Kumar et al., 1988).

	Peak	Strength speed	Peak strength	Vertical strength	Activity of duration	Time of occurrence
Males	Back	Stationary	726	5.0	4.5	45
		Slow	672	19.3	4.0	18
		Medium	639	33.9	1.3	44
		Fast	597	42.4	0.8	51
	Arm	Stationary	521	82.7	4.5	58
		Slow	399	95.1	3.2	17
		Medium	332	116.1	0.9	49
		Fast	275	114.0	0.7	49
Females	Back	Stationary	503	5.0	4.5	41
		Slow	487	15.0	4.2	19
		Medium	432	35.6	1.7	48
		Fast	436	42.2	1.4	63
	Arm	Stationary	296	76.2	4.5	60
		Slow	266	89.1	3.4	17
		Medium	221	95.3	1.4	34
		Fast	192	103.7	1.2	53

Table 5
Effect of pulling speed on one-handed isokinetic pulling strength (N) in the vertical plane (Mital et al., 1995).

Speed of exertion (m s^{-1})	Males		Females	
	X	SD	X	SD
0.30	322	250	61	65
0.35	303	236	60	73
0.48	274	219	51	68
0.58	242	197	49	63
0.75	225	192	48	63

4. Dynamic strength – requirements and measurement

While our argument rationalizes the use of human dynamic strengths in industrial applications, several questions, such as "what kind of dynamic strength should be used?", "should we use peak

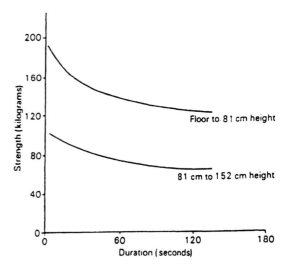

Fig. 7. Relationship between repetitive dynamic strength (RDS), maximum dynamic strength (MDS) and duration of frequent exertions for two heights of lifting (Mital et al., 1986b).

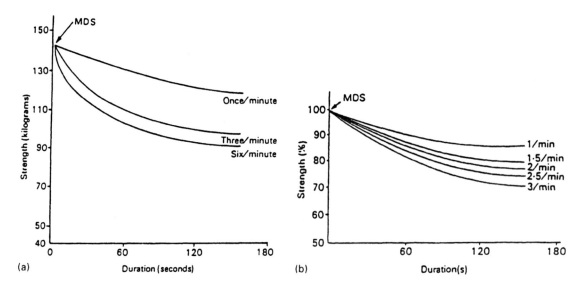

Fig. 8. (a) Relationship between repetitive dynamic strength (RDS), maximum dynamic strength (MDS) and duration of frequent exertions for various lifting frequencies (Mital et al., 1986b). (b) Relationship between repetitive dynamic strength (RDS), maximum dynamic strength (MDS) and duration of frequent exertions for various lifting frequencies (Mital et al., 1987).

strength or average strength?", "how should we measure dynamic strength?", "what kind of device should be used to measure dynamic strength?", etc., remain unanswered.

4.1. Job simulated strength

Logic suggests that dynamic strengths that simulates the job should be used. It is critical that the measured strength be job specific due to the large variability in the measured strength of an individual or within a given population. For example, many factors influence dynamic strength exertion capability of the individual (see Section 3) and these factors vary from job to job. In addition, since muscular strength is significantly affected by motivation and instructions given to subjects to perform (Kroemer and Howard, 1970; Caldwell et al., 1974; Miller and Nelson, 1976; Kroemer and Marras, 1981), standardization of dynamic strength measurement procedures is essential (Kroemer et al., 1990).

4.2. Peak versus average strength

In spite of a reasonable clarity and general agreement among the experts regarding the definition of various terms, ambiguity exists regarding the aspects or derivations of such characteristics to be employed in design or safety consideration. The most obvious is "peak or average" dilemma. Given any set of conditions for exerting maximal voluntary effort, we always obtain one reading for peak and another for average. So far considerable emphasis has been placed on the peak value. Most of the studies have emphasized this variable and most industrial applications have used it as a design criterion. While it is reasonable to use the peak value as a design ceiling, its use to design jobs that must be performed for a prolonged period of time and require sustained strength arguably is inappropriate. As shown in Section 3.5, strength is influenced by the frequency and duration of exertion. In such cases, it is appropriate to use repetitive strengths. At the very least, consideration should be given to using average strengths.

The use of peak strength in gauging exertion/overexertion is, however, difficult to judge. Whether peak strengths or average strengths are appropriate in such situations is yet to be established. While it appears that strengths that take into consideration the effect of repetition (lower than peak strength values) correlate better with a person's capability to perform heavy manual materials handling tasks repeatedly, their ability to determine when an exertion becomes overexertion and, thus, potentially hazardous is yet to be demonstrated.

4.3. Dynamic strength measuring devices and their applications

Several different devices have been developed to measure dynamic strengths. Some of these devices include LIFTEST equipment (Kroemer, 1983, 1985), B200 Isolation (a triaxial dynamometer from Isotechnologies, Inc. - Parnianpour et al., 1988), a servo-controlled motorized dynamic strength tester (Kumar et al., 1988), a modified mini-gym (Pytel and Kamon, 1981; Kamon et al., 1982; Mital et al., 1986b; Karwowski and Pongpatanasuegsa, 1988; Mital and Genaidy, 1989; Mital and Faard, 1990; Mital et al., 1995), a Cybex II isokinetic dynamometer (Marras et al., 1984, 1985; Langrana and Lee, 1984; Smith et al., 1985), and a Biokinetic ergometer from Isokinetic Inc. (Garg et al., 1988). Some applications of dynamic strength include measurement of whole body lifting strength (Pytel and Kamon, 1981; Kamon et al., 1982; Kroemer, 1970, Kroemer, 1983, 1985; Mital et al., 1986b, 1987); back and arm lifting strengths (Kumar et al., 1988); trunk flexion, extension, and rotation strengths (Smith et al., 1985); lifting strength in teamwork (Karwowski and Mital, 1986; Karwowski and Pongpatanasuegsa, 1988); one-handed pull-up and pull strengths (Mital and Genaidy, 1989; Mital and Faard, 1990; Mital et al., 1995). Marras et al. (1984) Marras et al. (1985) have used isokinetic devices to study the effects of trunk angle and speed on the torque production capability of the back and the relationship between the onset of intra-abdominal pressure and torque produced by the back. Several studies have demonstrated a correlation between actual maximum lifting capability and isokinetic strengths (Pytel and Kamon, 1981; Kamon et al., 1982; Mital et al., 1986c) and maximum acceptable weight of lift for repetitive tasks and repetitive isokinetic strengths (Mital et al., 1986b, 1987). Langrana and Lee (1984) attempted to identify those individuals who are at risk because of weak trunk muscles. Attempts have also been made to distinguish between submaximal and maximal efforts using Trunk Extension/Flexion and Lift ask machines (Lumex, Inc. - Hazard et al., 1988). The B200 isolation system has been used to study motor output and movement patterns (Parnianpour et al., 1988). Garg et al. (1988) used one-handed pulling strength to simulate starting of lawn mower engines.

Dynamic strength measurements can be either force produced at some point (either on the body or in spatial reach) or moment or torque produced about a given body joint. Both of these measures are equally important depending upon the particular application. Force measures are generally used for whole body/segment exertions (such as load lifting or pulling chords). Moment or torque measures are generally used to study muscle moment capability of a specific body joint in developing biomechanical human strength models and to describe specific body action such as turning a screwdriver or a knob. Since most industrial jobs involve different motion trajectories and ranges, durations, velocities, and accelerations, the dynamic strength measuring device should permit accurate simulation of job conditions. Also, since the force output varies with the velocity of exertion (Kumar et al., 1988; Garg et al., 1988), which may change with time for a given exertion or may differ from exertion to exertion, the measurement device should be sensitive to fluctuations in the force output. At present no such device is available.

4.4. Limitations of dynamic strength measurement

Given the difficulties described above, researchers have adopted some simplifications and used isoinertial and isokinetic strengths for industrial applications. Both of these, however, have some serious limitations.

The use of isoinertial strength has been limited to only one type of industrial activity. That is, predicting a person's capability to lift loads infrequently. While this method tends to simulate the velocities and accelerations involved, it makes no provision for the trajectory of motion, duration of the activity, technique, coupling or other important job parameters. However, it does appear to provide a better worker screening alternative than static strengths (Kroemer, 1983, 1985; Dales et al., 1986; Smith and Ayoub, 1989).

The isokinetic strength is more flexible and has been used to assess workers' capacity to do a variety of jobs (Mital et al., 1986b, 1987; Mital and Genaidy, 1989; Kumar, 1991; Mital et al., 1995; and others). While it can simulate job velocity, frequency, posture and range of motion, it also has some serious flaws. The most serious limitation is the failure of isokinetic strengths to account for motion acceleration and deceleration. The other occasional limitation could be the inability of isokinetic strengths to simulate the trajectory of motion. This means that the effective force output, which varies throughout the motion, is not recorded.

In spite of these difficulties and limitations, major progress has been made in the use of dynamic strengths for industrial applications. Some notable accomplishments are: (i) development of screening procedures for frequent and infrequent manual lifting activities (Kamon et al., 1982; Kroemer, 1983, 1985; Aghazadeh and Ayoub, 1985; Mital et al., 1986a, 1986b, 1986c, 1987; Dales et al., 1986; Kumar et al., 1988; Kumar, 1990; Smith and Ayoub, 1989), (ii) development of design databases for a variety of maintenance and work place activities (Mital and Genaidy, 1989; Mital and Faard, 1990; Kumar, 1991; Kumar and Garand, 1992; Mital et al., 1995), and (iii) development of criteria for work and equipment design (Karwowski and Mital, 1986; Karwowski and Pongpatanasuegsa, 1988; Kumar, 1987a,b; Garg et al., 1988, 1989). However, much remains to be done.

5. Strength data bases

Strength data bases, describing the capabilities of the population, are necessary for design purposes. A number of investigations have focused around the development of such databases. Among the well known works are works of Martin and Chaffin (1972), Ayoub and McDaniel (1974), Chaffin et al. (1978), Snook (1978), Davis and Stubbs (1980), Warwick et al. (1980), Chaffin et al. (1983), Mital (1984a,b), Mital et al. (1986a,b,c), Mital and Wang (1987), Mital and Genaidy (1989), Mital and Faard (1990), Kumar (1991), Kumar and Garand (1992), Mital et al. (1995), and Kumar et al. (1995a,b).

While much of this guide has focused on dynamic strengths, dynamic strength databases are few and small. In some cases, dynamic data even does not exist. For instance, in using non-powered hand tools. In such cases, one has to rely on isometric strength data. Historical isometric data are also helpful in characterizing worker populations and comparing results from different studies.

Even though there are only a limited number of studies that have developed strength data bases, it is not possible to cover all of them in detail. Some basic data from these studies are provided in Part I of this guide. Table 4 of Part I of this guide provides mean and standard deviations of isometric arm, shoulder, composite, and back lifting strengths. These data are summarized from the works of Chaffin et al. (1977) and Ayoub et al. (1978).

Pytel and Kamon (1981) and Kamon et al. (1982) pioneered isokinetic strength measurements. The data from their studies are summarized in Table 6 of the Part I. It contains means and standard deviations for lift strength, back extension strength, and elbow flexion strength.

Table 6 and 7 summarize one-handed isokinetic pull strength data. Table 6 is reproduced from the work by Mital et al. (1995) and Table 7 is based on the work of Mital and Faard (1990). In this context, the reader should also refer to Tables 1 and 2 of this part of the guide.

Table 19 of Part I of the guide contains two-handed isokinetic lift strength data for a variety of postures encountered in routine maintenance activities. These data are reproduced from Mital and Genaidy (1989). The maintenance activity postures are described at the end of the table in a footnote.

Table 20 of Part I of the guide provides maximum isometric torques that can be exerted voluntarily using a variety of non-powered hand tools in a variety of postures. The postures are described at the bottom of the table.

Table 27 of Part I of the guide provides isometric finger pinch-pull strengths (Imrhan and Sundararajan, 1992). Three different kinds of pinch grips are defined: lateral pinch-pull (LPP) which is pinching strength with the pad of the thumb and the radial lateral aspect of the index finger with other three fingers acting as a buttress to the index finger; chuck pinch-pull (CPP) which is pinching strength with the pads of the index and middle fingers on one side and with the pad of the thumb on the other side; pulp pinch-pull (PPP) which is pinching strength with the pads of the thumb and the index finger.

There are many situations that require more than one person to complete the task. In order to design such tasks, we must either know team strengths or how strengths of two or more people can be added. Karwowski and Mital (1986) and Karwowski and Pongpatanasuegsa (1988) tested the additivity assumption of isokinetic lifting and back extension strengths (the additivity assumption states that strength of a team is equal to the sum of individual members' strengths). They found that, on the average, strength of two people teams is about 68% of the sum of the strengths of its members. For three member teams, male and female teams generate only 58% and 68.4% of the sum of the strengths of its members, respectively. For both genders, the isokinetic team strengths were much lower than static team strengths. Table 6 shows isokinetic strengths of males and females in teamwork.

Table 6
Isokinetic strengths (N) of males and females in teamwork (Karwowski and Mital, 1986; Karwowski and Pongpatanasuegsa, 1988)

Variable	Mean	Std. Dev.
Isokinetic lifting strength		
Individual Strength		
Males	861.3	96.1
Females	543.9	123.3
Team of two people		
Males	1146.3	103.3
% Diff.(sum-actual)	33.4	5.2
Females	752.8	69.9
% Diff.(sum-actual)	28.2	10.2
Team of three people		
Males	1558.6	133.4
% Diff.(sum-actual)	39.7	3.5
Females	1169.0	128.4
% Diff.(sum-actual)	26.6	9.3
Isokinetic back-extension strength		
Individual strength		
Males	683.7	76.5
Females	473.0	119.6
Team of two people		
Males	929.0	74.5
% Diff.(sum-actual)	31.8	5.5
Females	599.5	111.8
% Diff.(sum-actual)	35.7	11.7
Team of three people		
Males	1136.7	81.4
% Diff.(sum-actual)	44.4	3.5
Females	888.2	110.1
% Diff.(sum-actual)	36.6	10.1

6. Prediction of human strengths

Efforts to predict human strengths go back at least 30 years. The prediction models that have been developed can be broadly classified in two categories: (1) prediction models that utilize different strengths to predict maximum acceptable weight of lift (psychophysical lifting strength) and (2) prediction models to predict different kinds of dynamic strengths. Models to predict isometric strengths are not covered in this guide.

6.1. Lifting capability prediction models

Manual lifting of materials is the major cause of industrial overexertion injuries. One of the approaches to protect workers from the strenuous demands of lifting materials manually, and consequent overexertion injuries, has been to match the physical demands of the job with their strength capabilities. Serious efforts to utilize human strengths in the development of worker screening procedures for manual lifting activities date as far back as 1971 (Poulsen and Jorgensen, 1971). As recently as the beginning of the last decade, static strengths were frequently being used in models to predict the lifting capabilities of individuals. The low correlations between static strengths and material lifting capability of individuals, however, did not permit the development of reliable and accurate lifting capability prediction models. The failure of static strengths to account for inertial forces resulting from the movement of the object and body segment is believed to be the main reason. This belief has been confirmed by many researchers who have attempted to use dynamic strengths to predict maximum acceptable lifting capabilities of individuals. In all cases, compared to static strengths, relatively higher correlations between dynamic strengths and acceptable lifting capabilities have been reported and simpler and more reliable lifting capability prediction models have resulted (Pytel and Kamon, 1981; Kamon et al., 1982; Kroemer, 1983, 1985; Aghazadeh and Ayoub, 1985; Mital et al., 1986a,b,c; Mital et al., 1987; Smith and Ayoub, 1989). In fact Smith and Ayoub (1989) report that LIFTEST based on isoinertial strength has emerged as the best predictor of materials handling capacity of workers for infrequent tasks. The general conclusion of all these studies is that the use of static strength based lifting capability prediction models should be avoided. Some of the recently proposed worker screening models are described in the following subsections.

6.1.1. Isokinetic strength based lifting capacity prediction models

Several different models have been developed to predict maximum acceptable weight of lift for occasional and frequent manual lifting. These models are based on data reported by Mital et al. (1986b) and Aghazadeh and Ayoub (1985) and are given below:

Maximum acceptable weight of lift (kg) for floor to knuckle height

$$= 485.47 - 12.955\, X_1 + 0.085\,(X_1)^2 \quad (1)$$
$$+ 1.133\, X_2 + 0.761\, \text{LOG}(X_3)$$
$$R_2 = 88.7\%,$$

where

X_1 = abdominal circumference (cm),

X_2 = grip strength (kg),

X_3 = floor to knuckle (F–K) isokinetic lifting strength (kg).

Maximum acceptable weight of lift (kg) for knuckle to shoulder height

$$= 62.20 - 2.135\, X_1 + 0.16\,(X_1)^2 \quad (2)$$
$$+ 0.009\,(X_2)^2 + 5.347\, \text{LOG}\,(X_3),$$
$$R^2 = 85.6\%,$$

where

X_1 = body weight (kg),

X_2 = grip strength (kg),

X_3 = knuckle to shoulder (K–S), isokinetic lifting strength (kg).

Maximum acceptable weight of lift (kg) for F–K and K–S Heights

$$= 16.88 - 0.004\, X_1 - 1.14\, X_2 + 0.11\, X_3, \quad (3)$$
$$R_2 = 77.5\%$$

where

X_1 = lifting height (= 1 for F–K height; 2 for K–S height),

X_2 = frequency of lift/minute,

X_3 = Isokinetic lifting strength (kg) for F–K or K–S height.

6.1.2. Isoinertial strength based lifting capacity prediction models

Smith and Ayoub (1989) have developed several general models for predicting maximum acceptable weight of lift based on 6 ft incremental lift (isoinertial) test. All subjects participating in the study were instructed that the test was to be a maximum lift test to 6 ft with the weight of the final lift to be attained by the addition of supplementary weights until subjects could lift no additional weight under the specified conditions. The weight machine was marked with a white tape to indicate the 6 ft height. The subject assumed an overhand grip position with one hand on each handle of the test apparatus (Kroemer, 1983, 1985) and proceed to lift the initial weight of 40 lb to the 6 ft height. After reaching the 6 ft height the subject lowered the load and the experimenter added 10 lb to the previous weight. The lift was repeated with the addition of 10 lb until the subject could no longer reach the 6 ft mark with the current weight. The last weight lifted to the 6 ft level was recorded as the score for the incremental 6 ft lift test.

The various prediction models developed are:

$$L6 = 1.18 X_1 + 11.22; \quad R^2 = 73\%, \quad (4)$$

$$L7 = 0.99 X_1 + 16.73; \quad R^2 = 64\%, \quad (5)$$

$$L8 = 0.52 X_1 + 15.11; \quad R^2 = 72\%, \quad (6)$$

$$L9 = 0.45 X_1 + 8.24; \quad R^2 = 67\%, \quad (7)$$

$$K\text{-}S = 0.27 X_1 + 8.79; \quad R^2 = 67\%, \quad (8)$$

$$F\text{-}S = 0.20 X_1 + 10.01; \quad R^2 = 62\%, \quad (9)$$

$$K\text{-}30 = 0.78 X_1 + 23.60; \quad R^2 = 77\%. \quad (10)$$

where

- X1 = maximum weight (lb) lifted to 6 ft on incremental weight machine,
- L6 = capacity (lb) to lift a tool box from floor to 30 in in one hand,
- L7 = capacity (lb) to lift handled box from floor to knuckle height,
- L8 = capacity (lb) to lift handled box from floor to shoulder height,
- L9 = capacity (lb) to lift handled box from floor to 6 ft height,
- K-S = capacity (lb) to lift box without handles from knuckle to shoulder height 6 times per minute,
- F-S = capacity (lb) to lift box without handles from floor to shoulder height at 6 times per minute,
- K-30 = capacity (lb) to lift a wooden box with handles from knuckle height to 30 in height.

6.2. Prediction of dynamic strengths

Prediction equations have been developed to predict isokinetic and isoinertial strengths. Mital et al. (1985a) developed models to predict isokinetic strength for the knuckle to shoulder height and floor to shoulder height. These models are provided below:

Isokinetic strength for the knuckle to shoulder height $= -0.00008 + 1.002\, T_4$

$$R_2 = 0.789; \quad MSE = 208.43$$

where

$$T_4 = -0.411 + 1.029\, T_3,$$

$$T_3 = 13.749 + 0.013\, T_1 T_2,$$

$$T_2 = 6.633 + 0.064(\text{biceps circumference})^2$$
$$\quad - 0.621(\text{gender})(\text{biceps circumference}),$$

$$T_1 = 21.422 \times 0.064(\text{chest width})^2$$
$$\quad - 0.633(\text{gender})(\text{chest width}),$$

Gender = 1 for males; 2 for females,

Isokinetic strength for the floor to shoulder height
$= 31.98 + 0.007\, T_1 T_2$

where

$$T_1 = 129.392 - 19.509(\text{gender})^2,$$

$$T_2 = -83.048 + 0.038(\text{acromial height})(\text{biceps circumference}),$$

Gender = 1 for males; 2 for females.

Kroemer (1983, 1985) developed regression equations to predict isoinertial strengths for the knuckle and overhead reach heights. The final models developed were:

Isoinertial strength for the overhead reach height
$= 63.225 - 16.11(S) + 0.309(H) - 0.322(W)$
$\quad - 0.0507(A) - 0.0038(L) + 0.077(B);$

$$R^2 = 0.90,$$

Isoinertial strength for the knuckle height
$= 72.35 + 1.055(W) - 0.62(H) + 0.507(A)$
$\quad - 0.131(L) + 0.151(B);$

$$R^2 = 0.61$$

where

$S = 0$ formales; 1 for females,

W = body weight in kg,

H = stature in cm,

A = average isometric arm strength in kg,

L = average isometric leg or composite strength in kg,

B = average isometric back or torso strength in kg.

The above equations clearly indicate that isometric arm, back, and composite strength must be known in order to predict overreach and knuckle height isoinertial strengths. If these isometric strengths cannot be measured, they can be predicted from the works of Mital and Manivasagan (1982, 1984), utilizing only anthropometric measures. These equations are, however, quite cumbersome to be reproduced here.

Imrhan and Ayoub (1988) have developed equations to predict linear isokinetic pull strength of the upper extremity. These equations, however, require rotary isokinetic strengths to predict isokinetic pull strength.

Kumar (1995) has developed an extensive set of 108 regression equations to predict peak and average isokinetic lifting strengths in the stoop and squat postures. These equations require input of other isokinetic and/or isometric strengths. By using an appropriate combination of reported equations strength can be predicted for any activity in three-dimensional space. Since these deal with empirical relationships on various activities, these may be population neutral though developed with data generated from North American sample on the assumption that populations may not be different in their internal pattern and relationship. However, this assumption has not been validated yet. In absence of extensive data bases and specific prediction equations for other populations these equations may be used with caution.

In general, prediction of strengths is possible but not very accurate. Further, from a design standpoint it is desirable to have strength databases. It should also be realized that in some countries, for instance the United States, recent legislation (The American With Disabilities Act (ADA) 1990) does not permit worker screening. The use of these prediction equations, therefore, is of little practical significance.

7. Grip strength

High and/or repetitive force application through hands has been associated with sore and injured hands (Silverstein et al., 1986). Working with tools frequently requires application of significant force.

Forceful exertions with inadequate recovery (Welch, 1972; Armstrong, 1986a,b; Hertzberg, 1955), direct mechanical pressure, hand-arm vibration (Armstrong et al., 1989) all have been reported to be associated with the musculoskeletal disorders of the upper extremity. Thus the majority of the risk factors cause mechanical stress on muscles and tissues of the upper extremities. Since upper extremity has many joints with limited to extensive range of motion and joint position has significant effect on muscle mechanical advantage, measurement of grips strength in a variety of postures has become an essential first step to design of both, tools and jobs. Grip strength is used in carrying out industrial tasks in a variety of ways. Each configuration has its own limitation in terms of human capability and advantage in performing a job efficiently. Kattel et al. (1996) investigated the effect of upper extremity posture on grip strength of the dominant hand among 15 male subjects. Using an adjustable Jamar hydraulic hand dynamometer (Lafayette Instruments Co., Inc.), they found significant effect of shoulder, elbow and wrist angles on the magnitude of the grip strength (Table 21, Part I).

The maximum grip strength was recorded in neutral shoulder and wrist position while the elbow was flexed by 45°. In the 56 postures in which Kattel et al. measured the grip strength the value ranged between 58% and 103% of neutral shoulder, 90° elbow flexion and neutral wrist posture (no flexion and no deviation). Thus a 40% variation recorded in a sample of 15 male university students is an indication for appropriate tool and task design for enhanced safety and efficiency. With respect to the wrist maximal strength has been reported in neutral posture (Putz-Anderson, 1988). In 45° extension of wrist the forces drop by 18% to 25% and in 45° flexion they drop by 28% to 40% (Hallbeck and McMullin, 1993; McMullin and Hallbeck, 1991; Putz-Anderson, 1988).

Non-dominant hand has been reported to produce a grip strength lower than the dominant hand by many authors. The values reported range between 90% (O'Driscoll et al., 1992), 93% (Hunter et al., 1978), 94% (McMullin and Hallbeck, 1991), 95% (Hallbeck and McMullin, 1993), and 97% (McMullin and Hallbeck, 1992) of that of the dominant hand. On the contrary Harkonen et al. (1993) reported no significant difference between dominant and non-dominant hand. The effect of grip span has also been reported to have a significant effect on the grip strength. A span of 6 cm has been reported to provide maximal strength (Wang, 1992; Harkonen et al., 1993). Ramakrishnan et al. (1994) investigated the mixed effects of grip span, gender, wrist posture and both hands. They reported significant effect of gender, wrist posture, hand dominance and grip span with significant interaction between grip span and gender, wrist posture and hand, wrist posture and grip span, and hand and grip span. Thus grip span is an extremely important variable in generating grip strength.

Fransson and Winkel (1991) investigated and reported the effect of grip span and grip type on total grip force and the forces shared by each of the digits in a cross tool. In their study both resultant force and finger forces varied with the grip span. Both in traditional and reverse grips of the tool, the authors found maximal force was generated with a span of 5 to 6 cm among females and 5.5 to 6.5 cm among males. They found that an increase of every cm in the width of separation there was a 10% drop in the grip strength. The force producing capability of the individual digits did not depend only on its own span but the spans of other fingers. They found that 35% of differences in the grip strength between the gender was explained by the size of the hand. Grip span is also critically involved in the design of pistol trigger type tools where several factors become important. Pistol grip strength for university students and industrial workers were investigated by Oh and Radwin (1993) and are reported in Tables 22 and 23 respectively (Part I).

They found that as the span increased from 4 to 7 cm average peak finger force increased 24%, peak palmar force increased 22% and average finger and palmar tool-holding forces increased 20%. The authors reported, the peak finger force decreased 9%, peak palmar force decreased 8%, finger tool holding force decreased 65%, and palmar tool holding force decreased 48%.

Ohtsuki (1981) reported maximal voluntary isometric finger flexion strength of individual digits in unidigital and multidigital contraction. Each finger produced highest force in unidigital efforts and progressive decrement occurred as the number of fingers contributing to the contraction increased. He found that the ratio of this decrease relative to the strength under unidigital condition were higher in index and little fingers (30%) than in the middle and ring fingers (15–25%). He proposed an efferent synergistic inhibition to explain this pattern of strength decline in the fingers.

8. Pinch strength

In manufacturing and assembly jobs grasping and manipulating components or objects is the most frequent and important. To do this, muscular forces are applied in pinching and gripping. The parts and products may be of varying size and weight. It is, therefore, very relevant to know and understand the pinch characteristic in terms of the effect of separation and wrist posture. These data will furnish strategic information for the design of parts, tools and jobs. Database for three common types of pinch are presented in Table 26 of Part I of this guide (Imrhan and Rehman, 1995). In a sample of 17 right-handed males they reported that the pinch strength was not significantly different between 2.0 and 5.6 cm for chuck and lateral pinches, nor between 2.0 and 9.2 cm for pulp-2 pinch but decreased significantly at greater widths. Imrhan and Rehman found that the strength differences between different types of pinches depended on the width of the pinch. The authors also found that the chuck pinch was greater than lateral or pulp-2 pinch at all widths, and lateral pinch was greater than pulp-2 pinch from 2.0 to 5.6 cm width but weaker at greater widths.

Hazelton et al. (1975) studied the effect of wrist position on the force produced by the finger flexors and reported that the percentage distribution of the total force produced by the finger flexors to each individual finger bore a constant relationship regardless of the wrist position. However, the magnitude of the total force did vary with the wrist position. In palmar flexion of the wrist the middle and distal phalannx produced least amount of force whereas maximum force was generated in ulnar deviation.

Imrhan (1991) and Fernandez et al. (1992) reported the effect of the wrist position on several pinch strength. In a study with 30 males Imrhan (1991) found that the wrist deviation resulted in decreased pinch strengths. The decline was greatest in wrist flexion and the least in radial deviation. The decline in strength ranged from 14% to 43% depending on the wrist position and the type of the pinch. The lateral pinch was least affected. Among females, Fernandez et al. (1992) also found that the wrist deviation was associated with decline in the pinch strengths. The percentage decline ranged from 1% to 10% for mid-extension, 8% to 17% for mid-flexion, 11% to 22% for full extension, and 20% to 37% for full flexion. The wrist flexion was associated with the largest decline in pinch strength. Furthermore, the authors reported that the female pinch strength was between 60% and 91% of their male counterparts. Thus, these capability comparisons among the same sample may be reflective of comparative risk for musculoskeletal disorder of hands. Such information, therefore, must be borne in mind while designing the tasks and components thereof to minimize the risk and optimize the safety.

When exerting submaximal forces in a five finger static pinch prehension, Radwin et al. (1992) studied the contribution of individual fingers with 10%, 20%, and 30% of their maximal voluntary prehensile strength using spans of 4.5–6.5 cm. They also recorded total pinch force and individual finger forces while grasping a dynamometer supporting fixed weights of 1.0, 1.5 and 2.0 kg loads. They found that the average proportion of forces developed by the index, middle, ring, and little fingers were 33%, 33%, 17%, and 15% respectively. With increase in exertion level from 10% to 30% the contribution of the middle finger was not constant increasing from 25% to 38%. Such mechanical behaviours of different digits in the background of anatomical and mechanical information may hold a strategic clue to minimize these musculoskeletal afflictions.

9. Need for further research

Some of the discussion in Section 4 pointed out to future research needs. It has been stated repeatedly that dynamic strengths are more suitable for industrial job, equipment, and tool design, and, if needed, for worker screening than isometric strengths. However, before dynamic strengths can be effectively used in industrial applications, at least the following must be accomplished:

- development of a measuring device that can simulate all aspects of the job and its motion characteristics,
- development of a standardized dynamic strength testing procedure which can lead to safe, repeatable, and simple tests,
- development of dynamic strength tests, for applications in industry, which can correlate with injury risk potential,
- development of design databases for a variety of industrial applications, and
- development of generalized physical capability models and dynamic strength thresholds (job severity indices) for screening workers for physically demanding tasks.

It is imperative that dynamic strength measurements be related to job physical demands. As our review and discussion have shown, a number of factors influence dynamic strength exertion capability of a person. These factors include velocity, acceleration, type of motion (smooth versus jerk), motion trajectory, distance of movement, duration of exertion, horizontal and vertical locations of the object, body posture, arm orientation, reach distance, etc. Only some of these variables have been systematically studied. Those yet to be investigated need to be related to dynamic strength exertion capability and job physical demands.

References

Aghazadeh, F., Ayoub, M.M., 1985. A comparison of dynamic and static strength models for prediction of lifting capacity. Ergonomics 28, 1409–1417.

Aghazadeh, F., Mital, A., 1987. Injuries due to hand tools: results of a questionnaire. Applied Ergonomics 18, 273–278.

Armstrong, T.J., 1986a. Ergonomics and cumulative trauma disorders. Hand Clinics 2, 553–566.

Armstrong, T.J., 1986b. Upper-extremity posture: Definition, measurement and control. In: Corlett, N. (Ed.), Ergonomics of Working Posture. Taylor and Francis, Philadelphia, PA, pp. 59–73.

Armstrong, T.J., Ulin, S., Ways, C., 1989. Hand tools and control of cumulative trauma disorders of the upper limb. In: Haslegrave, C.M., Wilson, J.R., Corlett, E.N. (Eds.), Proceedings of the 3rd International Occupational Ergonomics Symposium, Yugoslavia. Taylor and Francis, Philadelphia, PA, pp. 43–50.

Asfour, S.S., Khalil, T.M., Moty, E.A., Steele, R., Rosomoff, H.L., 1983. Back pain: a challenge to productivity. In: Raouf, A., Basic, M., Armad, S.I. (Eds.), Proceedings of the VIIth International Conference on Production Research. University of Windsor, Windsor, Ontario, Canada, pp. 813–818.

Asmussen, E., Hansen, O., Lammert, O., 1965. The relation between isometric and dynamic muscle strength in man. Communications of the Testing and Observation Institute, No. 20, Danish National Association for Infantile Paralysis, Hellerup, Denmark.

Ayoub, M.M., Bethea, N.J., Deivanayagam, S., Asfour, S.S., Bakken, G.M., Liles, D., Mital, A., Sherif M., 1978. Determination and modeling of lifting capacity. National Institute of Occupational Safety and Health, Final Report, Grant No. 5R01-OH-00545-02, Cincinnati, OH.

Ayoub, M.M., McDaniel, J.W., 1974. Effects of operator stance on pushing and pulling tasks. Transactions of the American Institute of Industrial Engineers 6, 185–195.

Ayoub, M.M., Mital, A., 1989. Manual Materials Handling. Taylor and Francis, London.

Ayoub, M.M., Selan, J.L., Chen, H.C., 1986. Human strength as a predictor of lifting capacity. In: Proceedings of the Human Factors Society Annual Meeting, Human Factors Society, Santa Monica, CA, pp. 960–963.

Bush-Joseph, C., Schipplein, O., Andersson, G.B.J., Andriacchi, T.P., 1988. Influence of dynamic factors on the lumbar spine moment in lifting. Ergonomics 31, 211–216.

Caldwell, L.S., 1959. The effect of the special position of a control on the strength of six linear hand movements. U.S. Army Medical Research Laboratory, Report No. 411, Fort Knox, Kentucky.

Caldwell, L.S., Chaffin, D.B., Dukes-Dobos, F.N., Kroemer, K.H.E., Laubach, L.L., Snook, S.H., Wasserman, D.E., 1974. A proposed standard procedure for static muscle strength testing. American Industrial Hygiene Association Journal 35, 201–206.

Carlson, B.R., 1970. Relationship between isometric and isotonic strength. Archives of Physical Medicine and Rehabilitation 51, 176–179.

Chaffin, D.B., 1974. Human strength capability and low back pain. Journal of Occupational Medicine 16, 248–254.

Chaffin, D.B., Andres, R.O., Garg, A., 1983. Volitional postures during maximal push/pull exertions in the sagittal plane. Human Factors 25, 541–550.

Chaffin, D.B., Andersson, G.B.J., 1991. Occupational Biomechanics. 2nd ed., Wiley, New York.

Chaffin, D.B., Herrin, G.D., Keyserling, W.M., 1978. Preemployment strength testing: an updated position. Journal of Occupational Medicine 20, 403–408.

Chaffin, D.B., Herrin, G.D., Keyserling, W.M., Foulke, J.A., 1977. Pre-employment strength testing in selecting workers for materials handling jobs. National Institute of Occupational Safety and Health, Publication No. CDC-99-74-62, Cincinnati, OH.

Chaffin, D.B., Park, K.S., 1973. A longitudinal study of low back pain as associated with occupational weight lifting factor. American Industrial Hygiene Association Journal 34, 513–525.

Clarke, H.H., Elkins, E.C., Martin, G.M., Wakim, K.G., 1950. Relationship between body position and the application of muscle power to movements of joints. Archives of Physical Medicine 31, 81–89.

Dales, L.J., MacDonald, E.B., Anderson, J.A.D., 1986. The "LIFTEST" strength test-an accurate method of dynamic strength assessment?. Clinical Biomechanics 1, 11–13.

Davis, P.R., Stubbs, D.A., 1980. Force Limits in Manual Work. IPC Science and Technology Press, Guildford, UK.

Fernandez, J.E., Dahalan, J.B., Halpern, C.A., Fredericks, T.K., 1992. The effect of deviated wrist posture on pinch strength for females. In: Advances in Industrial Ergonomics and Safety, vol. IV, Taylor and Francis, London, pp. 693–700.

Fransson, C., Winkel, J., 1991. Hand strength: the influence of grip span and grip type. Ergonomics 34, 881–892.

Freivalds, A., Chaffin, D.B., Garg, A., Lee, K., 1984. A dynamic biomechanical evaluation of lifting maximum acceptable loads. Journal of Biomechanics 17, 251–262.

Garg, A., Chaffin, D.B., Freivalds, A., 1982. Biomechanical stresses from manual load lifting: a static vs. dynamic evaluation. Transactions of the Institute of Industrial Engineers 14, 272–281.

Garg, A., Funke, S., Janisch, D., 1988. One-handed dynamic pulling strength with special application to lawn mowers. Ergonomics 31, 1139–1153.

Garg, A., Mital, A., Asfour, S.S., 1980. A comparison of isometric strength and dynamic lifting capability. Ergonomics 23, 13–27.

Grieve, D.W., 1984. The influence of posture on power output generated in a single pulling movements. Applied Ergonomics 15, 115–117.

Grimby, G., Saltin, B., 1983. The aging muscle. Clinical Physiology 3, 209.

Hall, S.J., 1985. Effect of attempted lifting speed on forces and torque exerted on the lumbar spine. Medicine and Science in Sports and Exercise 17, 440–444.

Hallbeck, M.S., McMullin, D.L., 1993. Maximal power grasp and three-jaw chuck pinch force as a function of wrist position, age, and glove type. International Journal of Industrial Ergonomics 11, 195–206.

Hazard, R.G., Reid, S., Fenwick, J., Reeves, V., 1988. Isokinetic trunk and lifting strength measurements: variability as an indicator of effort. Spine 13, 54–57.

Hazelton, F.T., Smidt, G.L., Flatt, A.E., Stephens, R.I., 1975. The influence of wrist position on the force produced by the finger flexors. Journal of Biomechanics 8, 301–306.

Hertzberg, H.T.E., 1955. Some contributions of applied physical anthropology to human engineering. Annals of the New York Academy of Science 63, 616–629.

Hettinger, T., 1961. Physiology of Strength. Charles C. Thomas, Springfield.

Holbrook, T.L., Grazier, K., Kelsey, J., Stauffer, R.N., 1984. The frequency of occurrence, impact and cost of selected musculoskeletal conditions in the united states. Paper Presented at the Annual Conference of the American Academy of Orthopaedic Surgeons.

Hollmann, W., Hettinger, T., 1980. Sports Medizin-Arbeits- und Trainings-Grundlagen. F.K. Schattauer, Stuttgart, Germany.

Imrhan, S.N., 1991. The influence of wrist position on different types of pinch strength. Applied Ergonomics 22, 379–384.

Imrhan, S.N., Ayoub, M.M., 1990. Predictive models of upper extremity rotary and linear pull strength. Human Factors 30, 83–94.

Imrhan, S.N., Ayoub, M.M., 1990. The arm configuration at the point of peak dynamic pull strength. International Journal of Industrial Ergonomics 6, 9–15.

Imrhan, S.N., Rehman, R., 1995. The effects of pinch width on pinch strengths of adult males using realistic pinch-handle coupling. International Journal of Industrial Ergonomics 16, 123–134.

Imrhan, S.N., Sundararajan, K., 1992. An investigation of finger pull strengths. Ergonomics 35, 289–299.

Jiang, B.C., 1984. Psychophysical capacity modeling of individual and combined manual materials handling activities. Ph.D. Dissertation, Texas Tech University, Lubbock, TX.

Kamon, E., Kiser, D., Pytel, L.J., 1982. Dynamic and static lifting capacity and muscular strength of steelmill workers. American Industrial Hygiene Association Journal 43, 853–857.

Karwowski, W., Mital, A., 1986. Isometric and isokinetic testing of lifting strength of males in teamwork. Ergonomics 29, 869–878.

Karwowski, W., Pongpatanasuegsa, N., 1988. Testing of isometric and isokinetic lifting strengths of untrained females in teamwork. Ergonomics 31, 291–301.

Kattel, B.P., Fredericks, T.K., Fernandez, J.E., Lee, D.C., 1996. The effect of upper-extremity posture on maximum grip strength. International Journal of Industrial Ergonomics 18, 423–429.

Keyserling, W.M., Herrin, G.D., Chaffin, D.B., 1980. Isometric strength testing as a means of controlling medical incidents on strenuous jobs. Journal of Occupational Medicine 22, 332–336.

Khalil, T.M., Asfour, S.S., Moty, E.A., Rosomoff, H.L., Steele, R., 1983. The management of low back pain: a comprehensive approach. In: Proceedings of the Annual International Industrial Engineering Conference. Institute of Industrial Engineers, Norcross, GA, pp. 199–204.

Kroemer, K.H.E., 1970. Human strength: terminology, measurement, and interpretation of data. Human Factors 12, 297–313.

Kroemer, K.H.E., 1983. An isoinertial technique to assess individual lifting capability. Human Factors 25, 493–506.

Kroemer, K.H.E., 1985. Testing individual capability to lift material: repeatability of a dynamic test compared with static testing. Journal of Safety Research 16, 1–7.

Kroemer, K.H.E., Howard, J.M., 1970. Towards the standardization of muscle strength testing. Medicine and Science in Sports 2, 224–230.

Kroemer, K.H.E., Marras, W.S., 1981. Evaluation of maximal and submaximal static muscle exertions. Human Factors 23, 643–653.

Kroemer, K.H.E., Marras, W.S., McGlothlin, J.D., McIntyre, D.R., Nordin, M., 1990. On the measurement of human strength. International Journal of Industrial Ergonomics 6, 199–210.

Kromodihardjo, S., Mital, A., 1987. biomechanical analysis of manual lifting tasks. Journal of Biomechanical Engineering 109, 132–138.

Kumar, S., 1987a. Arm-lift strength variation due to task parameters. In: Buckle, P. (Ed.), Musculoskeletal Disorders at Work. Taylor and Francis, London, pp. 37–42.

Kumar, S., 1987b. Arm-Lift strength at different reach distances. In: Asfour, S.S. (Ed.), Trends in Ergonomics/Human Factors, vol. IV, North-Holland, Amsterdam, pp. 623–630.

Kumar, S., 1989. Load history and backache among institutional aides. In: Mital, A. (Ed.), Advances in Ergonomics, vol. I, Taylor and Francis, London, pp. 757–765.

Kumar, S., 1990. Cumulative load as a risk factor for low back pain. Spine 15, 1311–1316.

Kumar, S., 1991. Arm lift strength in work space. Applied Ergonomics 22, 317–328.

Kumar, S., 1995. Development of predictive equations for lifting strengths. Applied Ergonomics 26, 327–341.

Kumar, S., Chaffin, D.B., 1985. Static and dynamic lifting strengths of young males. Paper Presented at the American Society of Biomechanics.

Kumar, S., Chaffin, D.B., Redfern, M., 1988. Isometric and isokinetic back and arm lifting strengths: devices and measurement. Journal of Biomechanics 21, 35–44.

Kumar, S., Dufresne, R., Van Schoor, T., 1995a. Human trunk strength profile in flexion and extension. Spine 20, 160–168.

Kumar, S., Dufresne, R., Van Schoor, T., 1995b. A pilot study in human trunk strength profile in lateral flexion and axial rotation. Spine 20, 169–177.

Kumar, S., Garand, D., 1992. Static and dynamic lifting strength at different reach distances in symmetrical and asymmetrical planes. Ergonomics 35, 861–880.

Kumar, S., Narayan, Y., Bacchus, C., 1995c. Symmetric and asymmetric two-handed pull-push strength of young adults. Human Factors 37, 854–865.

Langrana, N.A., Lee, C.K., 1984. Isokinetic evaluation of trunk muscles. Spine 9, 171–175.

Leskinen, T.P.J., Stalhammer, H.R., Kuorinka, I.A.A., Troup, J.D.G., 1983. The effect of inertial factors on spinal stress when lifting. Engineering in Medicine 12, 87–89.

Martin, J.B., Chaffin, D.B., 1972. Biomechanical computerized simulation of human strength in sagittal plane activities. Transactions of the American Institute of Industrial Engineers 4, 19–28.

Marras, W.S., 1985. Isometric vs isokinetic back lifting performance. National institute of Occupational Safety and Health, Final Report, No. R03-OH-01755, Cincinnati, OH.

Marras, W.S., King, A.I., Joynt, R.L., 1984. Measurements of loads on the lumbar spine under isometric and isokinetic conditions. Spine 9, 176–187.

Marras, W.S., Joynt, R.L., King, A.I., 1985. The force-velocity relation and intra-abdominal pressure during lifting activities. Ergonomics 28, 603–613.

McGill, S.M., Norman, R.W., 1985. Dynamically and statically determined low back moments during lifting. Journal of Biomechanics 18, 877–885.

McMullin, D.L., Hallbeck, M.S., 1991. Maximal power grasp force as a function of wrist position, age, and glove type. A pilot study. In: Proceedings of the Human Factors Society 35th Annual Meeting. Human Factors Society, Santa Monica, CA, pp. 733–737.

McMullin, D.L., Hallbeck, M.S., 1992. Comparison of power grasp and three-jaw chuck pinch static strength and endurance between individual workers and college students: a pilot study. In: Proceedings of the Human Factors Society 36th Annual Meeting. Human Factors Society, Santa Monica, CA, pp. 770–774.

Miller, D.I., Nelson, R.C., 1976. Biomechanics of Sports. Lea and Febiger, Philadelphia.

Mital, A., 1978. Strength and lifting capacity: data norms and prediction models. Technical Report, Department of Industrial Engineering, Texas Tech University, Lubbock, TX.

Mital, A., 1984a. Maximum weights of lift acceptable to male and female industrial workers for extended work shifts. Ergonomics 27, 1115–1126.

Mital, A., 1984b. Comprehensive maximum acceptable weight of lift data base for regular 8-hour work shifts. Ergonomics 27, 1127–1138.

Mital, A., 1986. Effect of body posture and common hand tools on peak torque exertion capabilities. Applied Ergonomics 17, 87–96.

Mital, A., 1987. Patterns of difference between the maximum weights of lift acceptable to experienced and inexperienced materials handlers. Ergonomics 30, 1137–1147.

Mital, A., Aghazadeh, F., 1985. A review of hand tool injuries. In: Brown, I.D., Goldsmith, R., Coombs, K., Sinclair, M.A. (Eds.), Ergonomics International 85. Taylor and Francis, London, pp. 85–87.

Mital, A., Aghazadeh, F., Karwowski, W., 1986a. relative importance of isometric and isokinetic lifting strengths in estimating maximum lifting capabilities. Journal of Safety Research 17, 65–71.

Mital, A., Aghazadeh, F., Ramanan, S., 1985a. Use of GMDH to predict dynamic strengths. Computers and Industrial Engineering 9, 371–377.

Mital, A., Ayoub, M.M., 1980. Modelling of isometric strength and lifting capacity. Human Factors 22, 285–290.

Mital, A., Channaveeraiah, C., 1988. Peak volitional torques for wrenches and screwdrivers. International Journal of Industrial Ergonomics 3, 41–64.

Mital, A., Channaveeraiah, C., Faard, F., Khaledi, H., 1986b. Reliability of repetitive dynamic strengths as a screening tool for manual tasks. Clinical Biomechanics 1, 125–129.

Mital, A., Das, B., 1987. Human strengths and occupational safety. Clinical Biomechanics 2, 97–106.

Mital, A., Faard, H.F., 1990. Effects of sitting and standing, reach distance, and arm orientation on isokinetic pull strengths in the horizontal plane. International Journal of Industrial Ergonomics 6, 241–248.

Mital, A., Garg, A., Karwowski, W., Kumar, S., Smith, J.L., Ayoub, M.M., 1993a. Status in human strength research and application. IIE Transactions 25, 57–69.

Mital, A., Genaidy, A.M., 1989. Isokinetic pull-up strength profiles of men and women in different working postures. Clinical Biomechanics 4, 168–172.

Mital, A., Karwowski, W., 1985. Use of simulated job dynamic strengths (SJDS) in screening workers for manual lifting tasks. In: Proceedings of the Human Factors Society Annual Meeting, Human Factors Society, Santa Monica, CA, pp. 513–516.

Mital, A., Karwowski, W., Mazouz, A.K., Orsarh, E., 1986c. Prediction of maximum acceptable weight of lift in the horizontal and vertical planes using simulated job dynamic strengths. American Industrial Hygiene Association Journal 47, 288–292.

Mital, A., Kopardekar, P., Motorwala, A., 1995. Isokinetic pull strengths in the vertical plane: effects of speed and arm angle. Clinical Biomechanics 10, 110–112.

Mital, A., Kromodihardjo, S., 1986. Kinetic analysis of manual lifting activities: Part II-biomechanical analysis of task variables. International Journal of Industrial Ergonomics 1, 91–101.

Mital, A., Manivasagan, I., 1982. Application of a heuristic technique in polynomial identification. In: Proceedings of the IEEE Systems, Man, and Cybernetics Society International Conference, Institute of Electrical and Electronics Engineers, New York, pp. 347–353.

Mital, A., Manivasagan, I., 1984. Development of non-linear polynomials in identifying human isometric strength behavior. Computer and Industrial Engineering 8, 1–9.

Mital, A., Nicholson, A.S., Ayoub, M.M., 1993b. A Guide to Manual Materials Handling. Taylor and Francis, London, UK.

Mital, A., Sanghavi, N., 1986. Comparison of maximum volitional torque exertion capabilities of males and females with common handtools. Human Factors 28, 283–294.

Mital, A., Sanghavi, N., Huston, T., 1985b. A study of factors defining the 'operator-hand tool system' at the workplace. International Journal of Production Research 23, 297–314.

Mital, A., Wang, L.W., Faard, H.F., 1987. Boundary line between the strength and endurance regions in manual lifting. Clinical Biomechanics 2, 220–222.

National Institute for Occupational Safety and Health, 1981. Work practices guide for manual lifting. NIOSH Report No. 81-122, Cincinnati, OH.

National Safety Council, 1983. Back injury prevention through ergonomics (Videotape). Chicago, ILL.

Oh, S., Radwin, R.G., 1993. Pistol grip power tool handle and trigger size effects on grip exertions and operator preference. Human Factors 35, 551–569.

Ohtsuki, T., 1981. Inhibition of individual fingers during grip strength exertion. Ergonomics 24 (1), 21–36.

Parnianpour, M., Nordin, M., Kahannovitz, N., Frankel, V., 1988. The triaxial coupling of torque generation of trunk muscles during isometric exertions and the effect of fatiguing isoinertial movements on the motor output and movement patterns. Spine 13, 982–992.

Poulsen, E., Jorgensen, K., 1971. Back muscle strength, lifting and stooped working postures. Applied Ergonomics 2, 133–137.

Putz-Anderson, V. (Ed)., 1988. Cumulative Trauma Disorders: a Manual for Musculoskeletal Diseases of the Upper Limbs. Taylor and Francis, London.

Pytel, L.J., Kamon, E., 1981. Dynamic strength as a predictor for maximum acceptable lifting. Ergonomics 24, 663–672.

Radwin, R.G., Oh, S., 1992. External finger forces in submaximal five finger static pinch prehension. Ergonomics 35, 275–288.

Ramakrishnan, B., Bronkema, L.A., Hallbeck, M.S., 1994. Effects of grip span, wrist position, hand, and gender on grip strength. In: Proceedings of the Human Factors and the Ergonomics Society-38th Annual meeting, Human Factors Society, Santa Monica, CA, pp. 554–558.

Rohmert, W., 1966. Maximalkrafte von Mannern im Bewegungsraum der Arme und Beine. Forschungsberichte des Landes Nordrhein-Westfalen no. 1616. West Deutscher, Cologne, Germany.

Rohmert, W., Jenik, P., 1972. Maximalkrafte von Frauen im Bewegungsraum der Arme und Beine. Arbeitswissenschaft und Praxis. Beuth-Vertrieb, Berlin, Germany.

Ruhmann, H., Schmidtke, H., 1989. Human strength: measurement of maximum isometric forces in industry. Ergonomics 32, 865–879.

Silverstein, B.A., Fine, L.J., Armstrong, T.J., 1986. Hand wrist cumulative trauma disorders in industry. British Journal of Industrial Medicine 43, 779–784.

Smith, J.L., Ayoub, M.M., 1989. Isoinertial strength-best predictor of infrequent manual materials handling capacities. Technical Report, Department of Industrial Engineering, Texas Tech University, Lubbock, TX.

Snook, S.H., 1978. The Design of Manual Handling Tasks. Ergonomics 21, 963–985.

Statistics Canada, 1988. Work Injuries. Ottawa, Canada.

Taber, M., 1982. Reconstructing the scene: back injury. Occupational Health and Safety 51, 16–22.

Yates, J.W., Kamon, E., Rodgers, S.H., Champney, P.C., 1980. Static lifting strength and maximal isometric voluntary contractions of the back, arm, and shoulder muscles. Ergonomics 23, 37–47.

Wakim, K.G., Gersten, J.W., Elkins, E.C., Martin, G.M., 1950. Objective recording of muscle strength. Archives of Physical Medicine 31, 90–99.

Wang, M., 1982. A study of grip strength from static efforts and anthropometric measurement. Unpublished Masters Thesis, University of Nebraska, Lincoln, NE.

Warwick, D., Novak, G., Schultz, A., Berkson, M., 1980. Maximum voluntary strengths of male adults in some lifting, pushing, and pulling activities. Ergonomics 23, 49–54.

Welch, R., 1972. The causes of tenosynovitis in industry. Industrial Medicine 41, 16–19.

Repetitive work of the upper extremity: Part I–Guidelines for the practitioner *

Åsa Kilbom

National Institute of Occupational Health, S-17184 Solna, Sweden

1. Who is the practitioner?

The practitioner is anyone involved in evaluating, designing or redesigning work places or work systems. The occupational categories include production engineers, industrial designers, supervisors, occupational health and safety professionals, ergonomists and labor inspectors.

---

* The recommendations provided in this guide are based on numerous published and unpublished scientific studies and are intended to enhance worker safety and productivity. These recommendations are neither intended to replace existing standards, if any, nor should be treated as standards. Furthermore, this document should not be construed to represent institutional policy.

The following individuals participated in the discussion of the earlier version of this guide. Their suggestions (written or verbal) were incorporated by the authors in this version: Arne Aaras, *Norway*; Fred Aghazadeh, *USA*; Roland Andersson, *Sweden*; Jan Dul, *The Netherlands*; Jeffrey Fernandez, *USA*; Ingvar Holmér, *Sweden*; Matthias Jäger, *Germany*; Anders Kjellberg, *Sweden*; Olli Korhonen, *Finland*; Helmut Krueger, *Switzerland*; Shrawan Kumar, *Canada*; Ulf Landström, *Sweden*; Tom Leamon, *USA*; Anil Mital, *USA*; Ruth Nielsen, *Denmark*; Jerry Ramsey, *USA*; Murray Sinclair, *UK*; Rolf Westgaard, *Norway*; Ann Williamson, *Australia*; Jørgen Winkel, *Sweden*; Pia Zätterström, *Sweden*. The guide was also reviewed in depth by several anonymous reviewers.

2. When and where should this guideline be used?

The aim of this guideline is to provide assistance in primary and secondary prevention of work-related musculoskeletal disorders of the upper extremities, associated with repetitive work. Repetitive work occurs in industry, in offices and in many service jobs like e.g. cleaning and catering. The guideline applies to design of new work stations and work organizations, as well as to work redesign when repetitive work occurs and when certain musculoskeletal disorders have arisen.

3. Problem identification

Repetitive work of the upper extremity is defined as the performance of similar work cycles, again and again. No consensus exists concerning the exact definition of "similar". A tentative definition for the purpose of this guideline is that the output of the work is similar from one work cycle to the next (e.g. assembling circuit boards, packing boxes on an assembly line). Each work cycle should also closely resemble the next with regard to the time sequence, force exertion pattern and spatial characteristics of the movements. Repetitive work of the upper extremities implies the performance of movements and muscle contractions of the shoulder, arm or hand. For the purpose of this guideline work is considered *repetitive* when the duration of the work cycle is less than 30 seconds, or when one fundamental work cycle constitutes more than 50% of the total cycle, independent of its length. It must be emphasized, however, that this time limit is arbitrary and that a potential risk of musculoskeletal disorders may also exist for longer work cycles. The physiological and biomechanical characteristics of repetitive work can be categorized as either intermittent static, i.e. external movements are small or negligible, or dynamic, i.e. movements around joints are easily distinguishable. The reader is referred to Part II of this guideline for further presentation and discussion of these definitions.

There is no consensus concerning minimum duration of repetitive work. For this guideline, it is assumed that the repetitive work must be performed continuously for a minimum of 60 minutes in order to be considered repetitive. Symptoms and signs of musculoskeletal disorders have been observed in physically demanding repetitive work exceeding one hour. Usually, musculoskeletal disorders associated with repetitive work of the upper extremity have been observed when the repetitive work was performed for nearly full work days.

Disorders related to repetitive jobs are commonly encountered in tendons, muscles and nerves of the shoulder, forearm, wrist and hand. The clinical diagnoses include tendinitis, peritendinitis, tenosynovitis, myalgias and distal nerve entrapment, e.g. carpal tunnel syndrome. In these disorders, repetitiveness of work is one of several risk factors. Others include exertion of external force, static work load, posture and probably also speed of movements. The duration of the exposure – per day and per number of years – is yet another risk factor. Thus, although the focus of this guideline is on repetitiveness of work, its interaction with other risk factors will be emphasized.

A potential problem of repetitive work is at hand, when:
(a) a repetitive work is performed for more than one hour during the work day, and/or
(b) a case of musculoskeletal disorder of the upper extremity (see above), with no evident cause external to the work, is encountered at the work place.

4. Data collection

4.1. Rationale for task analysis

4.1.1. Repetitive work and tendon disorders

There is epidemiological evidence of an association between frequency and duration of repetitive work and disorders of tendons and adjacent tissues, including nerves. The association is further strengthened by experimental studies demonstrating mechanisms likely to be involved. These include friction and viscous strain of tendons, which suggests that both the rate of repetition and the duration of recovery between repetitions are important time factors. Epidemiologic evidence suggests that work cycles shorter than 30 seconds, or one fundamental work cycle constituting more than 50% of the total cycle, is strongly related to disorders of the forearm and wrist. Knowledge concerning the quantitative exposure–effect relationship is however still lacking. Moreover, there is no support for considering work cycles longer than 30 seconds as safe. For further information see Part II of this guideline.

The pathophysiological mechanism of tendon disorders appears to be linked to the frequency of movements. Long cycle times do not necessarily imply a low rate of movements, since one work cycle may contain several fundamental cycles and each fundamental cycle may contain several work elements (see Part II, section 3.1 and Fig. 1 of this guideline for an example and definitions of work cycles, fundamental work cycles and work elements.) Therefore a risk assessment should rely on movement frequencies rather than on cycle times. So far the epidemiological evidence for an increase in risk above certain rates of movements is incomplete, but some tentative recommendations can be given (see also Table 1). Epidemiological data suggest that fingers may tolerate a higher frequency of movements than the wrist, and that the wrist may tolerate a higher frequency of movements than the elbow and shoulder.

4.1.2. Repetitive work and muscle disorder

Muscle disorders also appear to be associated with repetitive work, but the epidemiological support is weaker. The disorders observed in many studies may, at least in part, be associated with static loads rather than with an intermittent work pattern. Experimental studies of intermittent static contractions suggest that an optimum work/rest regime exists, where muscle fatigue can be avoided. Whether adherence to such a regime in repetitive jobs will prevent disorders remains to be demonstrated. The pathophysiological mechanisms of muscle disorders appear to be linked to muscle fatigue and lack of recovery, which suggests that short contraction periods followed by sufficient periods for recovery are important factors. There is no support for the notion that the risk of muscle disorders is lower when the cycle time is long (as is the case of tendon disorders). For further information and discussion see Part II of this guideline.

4.1.3. Interaction between repetitive work and other factors

A number of factors interact with the repetitiveness and duration of the work cycles, increasing the risk of disorder and fatigue. The main factor is the exertion of external force, which has been well documented both epidemiologically and experimentally. Force and repetitiveness interact so that high force and high repetitiveness increase the risk in a multiplicative way. Moreover, the exertion of a high force may be acceptable at a given low rate, but not at higher rates, and vice versa. The quantitative relationship between force exertion and disorders in repetitive work is still insufficiently documented, whereas the relationship between force exertion and fatigue after intermittent static exercise is better known from experimental studies.

Repetitive work is sometimes superimposed onto a static load especially in the shoulder. This is likely to increase the risk of disorders further, but no quantitative data are available. Repetitive work is often performed by distal parts of the upper extremity, while the proximal part (shoulder) stabilizes the arm and thereby performs static work. This is likely to be a common cause for shoulder disorders, and reduction of the repetitivity of the distal work is then likely to have a preventive effect on shoulder disorders.

Both epidemiological and experimental data support posture as a modifying factor. Thus extreme postures increase the risk of disorders, although the quantitative relationships are unclear.

Similarly, a high speed of motion (and possibly acceleration) appears to increase the risk of disorders according to epidemiological studies. No quantitative data are known.

In similarity with other work-related musculoskeletal disorders, it is likely that the duration of exposure - in minutes per day, and number of years - influences the risk of disorders in repetitive work. Cases of disorders have been reported even after a few full days of unaccustomed repetitive work, which indicates that the duration of exposure per work day is important but that factors like skill and training for the work tasks also influence the level of risk. No epidemiological data is available today in support of statements concerning acceptable duration of repetitive work.

Psychosocial factors, like lack of control over work tasks, perceived monotony and/or time pressure are factors likely to reduce the tolerance to repetitive work. A temporary increase in demands on work output is likely to potentiate the effect of repetitive work. The mechanisms are poorly understood, and the quantitative data is still scarce, and therefore these factors are not reviewed in Part II.

4.2. Task analysis

If a repetitive work task has been identified or if cases of musculoskeletal disorder (see section 3) occur, the task should be analysed with regard to its time, force, posture and speed characteristics. This analysis should be performed for shoulder, elbow, wrist and fingers. It is common that large movements are only observed around one joint (e.g. wrist), however discreet movements may simultaneously occur around others (e.g. shoulders).

Time parameters include:
frequency of specific movements or contractions, duration of work cycle, duration of fundamental work cycle, duration of work elements (MTM), duty cycle (when intermittent static contractions occur), exposure time (min/day).

If resources for MTM (method-time measurement) analysis are available, the duration of all work elements constituting the work cycle should be recorded, including duration of rest periods. The rate of specific movements like elevation of arm, flexion/extension and ulnar/radial deviation of wrist and flexion/extension of fingers should be assessed. If the work cycles include large elements of static workloads, their duration and the duration of subsequent relaxations/alternative movements should be measured and the duty cycle calculated.

The *static load and the external force parameter* can be dichotomized, using subjective assessment, into high or low demands on the active joint or muscle group in relation to a given reference level. No quantitative definitions exist; thus the identification of high demands is subjective.

The *posture parameter* should be dichotomized into neutral/small deviation or moderate/extreme deviation. The extreme is defined as close to the extreme range of motion for the active joint, but no quantitative recommendations can be given.

The *speed parameter* should be dichotomized into static/slow or fast movements. This definition too is non-quantitative, and the identification of fast movements is subjective.

The above recommendations for task data collection are discussed and commented in Part II of this guideline (section 3 and Table 2).

The identification of psychosocial and training/skill risk factors can only be done by observations of the work performed, in combination with interviews with employers and employees. It is not possible without an extensive interview with several employees to form a distinct opinion concerning the occurrence of these risk factors. For the purpose of this guideline, only a very crude evaluation can be made.

4.3. Gathering data on disorder rates

When a case of disorder (see above) is encountered, an analysis of the tasks performed and a survey of occurrence of disorders in the group of workers performing similar work tasks should be made. If resources for statistical analysis are available and if the group is sufficiently large, the results can be compared with disorder rates among groups of workers performing other, more variable work tasks. Data can be gathered by questionnaire, worker compensation claims, or records from Occupational Health Service. Clinical examination of the cases is advisable, both as a basis for treatment and for a subdivision of cases into tendon vs. muscle disorders. Disorders should always be classified by upper extremity segment, i.e. shoulder, upper arm, forearm, hand.

5. Data analysis

Table 1 can be used for an assessment of risk when a repetitive task has been identified. It must be stressed, however, that the epidemiological support for these risk levels is relatively weak and that cases may occur even at lower exposure levels.

In repetitive work with strong elements of intermittent static work. (e.g. handgrip exercise) muscular disorders are more likely to occur than tendon disorders. For an assessment of upper arm and forearm intermittent static exercise the recommendations presented by Byström and Dul can be used (see Part II). They give detailed recommendations for calculation of acceptable work/rest regimes.

Intermittent static work performed by the shoulders constitutes a special risk assessment problem because of the difficulty of analyzing the work pattern of shoulder muscles without EGM and/or posture recordings. It is recommended that, if simultaneous repetitive work of the upper arm, forearm and hand can be identified, then risks should be assessed by those body regions. If intermittent shoulder static exercise occurs (or is believed to occur) in isolation, then it should be assessed according to the Guideline presented by

Winkel and Westgaard (see Part II). This guideline subdivides exposure into three levels, by the quality of work station design, posture and exertion of external forces. It does not, however, address the work/rest characteristics of intermittent static shoulder exertions.

The recommendation concerning rate of finger movement concerns number of keystrokes per minute and is only tentative (see Part II).

The survey of disorder rates and clinical examinations should be performed for screening purposes, so that suspected cases of disorders can be referred for treatment. If resources for statistical analysis are available, the magnitude of the problem should be assessed by comparison with a reference group not exposed to repetitive work. This will provide more background information as regards the urgency of redesign of work. For such an analysis basic principles of epidemiological studies should be followed.

6. Solutions

Repetitive work of the upper extremity – as defined above – should be avoided due to high risks of musculoskeletal disorders of shoulder, forearm, wrist, hand and fingers. Primary and secondary prevention implies that repetitiveness, as well as other risk factors like static load/force/posture/speed, psychosocial factors and skill should be reduced when they appear. No scientific studies have been performed where the effect of intervention against one factor has been compared to that of intervention against another. Thus there is no scientific support for giving priority to changing one certain risk factor, because it has been demonstrated to be superior in terms of prevention. However, since the factors repetitiveness and force exertion are best documented scientifically, prevention should primarily aim at those two risk factors.

The risk of tendon/nerve disorders increases as the rate of movements increases or duration of work cycles shortens. However, no "safe" levels have been identified. Therefore, it is unlikely that prevention can be achieved by a moderate increase of work cycle duration. Moreover, a lengthening of the work cycle often implies that one fundamental work cycle is performed during a larger proportion of the total cycle, and the task will thereby retain its characteristics of repetitive-

Table 1
Recommendations for risk assessment in repetitive work

Body region	Type of exercise	Frequency of movement or contraction	Risk assessment	Risk modification – very high risk
Shoulder	Dynamic	> 2.5/min	high	One of the following high external force, high speed, and high static load, extreme posture, lack of training, high demands on output, monotony, lack of control, long duration of repetitive work
	Interm. static		see Winkel Westgaard	
Upper arm, elbow	Dynamic	> 10/min	high	
	Interm. static		see Dul et al.	
Forearm, wrist	Dynamic	> 10/min	high	
	Interm. static		see Byström	
Finger	Dynamic	> 200/min?	high	

ness. It appears more likely that a reduction of the rate of movements or contractions, and of the duration per day of repetitive work would reduce the risks. Reducing the rate of movements or contractions is also likely to increase the time for recovery. As yet, no quantitative recommendations can be given concerning maximal acceptable duration of repetitive work per day, or acceptable rate of movements or contractions per time unit.

The following tentative list of priorities can be given for intervention against tendon/nerve disorders associated with repetitive work:

(1) Is the work repetitive (i.e. cycle time < 30s, fundamental cycle > 50% of total cycle, repetitive task performed > 1 hr per day)?
(2) If yes, does it induce high or a very high risk (see Table 1)?
(2.1) Work induces high risk: Reduce rate of movements/contractions and total exposure time.
(2.2) Work induces very high risk: Reduce rate of movements/contractions and total exposure time considerably. Simultaneously reduce external force demands if high.
(2.2.1) Intervene against other risk factors in the following order:
  – extreme postures and static work
  – lack of control
  – all other risk factors – high speed, lack of skill, high demands on output, monotony.

As previously mentioned, muscular disorders associated with repetitive work occur mainly in conjunction with intermittent static exercise. The Guideline (see Part II) by Winkel and Westgaard provides advice with regard to shoulder disorders and can be followed when no repetitive work of upper arm and forearm can be identified. The recommendations by Byström and by Dul et al. (see Part II) provide advice with regard to optimum work–rest regimes for different contraction intensities in intermittent static work. They focus on time and force parameters of intermittent static exercise. When other risk factors – extreme postures, lack of control, lack of skill, high demands on output and speed, and monotony – occur simultaneously, they should also be intervened against.

High demands on force can be reduced by work station or tool design – see ergonomic textbooks.

Extreme postures can be influenced by work station and tool design and by improved lighting – see ergonomic textbooks.

High demands on speed of movements can be influenced by work station and tool design, by training and by work reorganization.

Lack of control can be influenced by work reorganization. The reader is referred to textbooks of occupational psychology. The most important measure is to provide opportunities for the worker to control when the repetitive work should be performed and when breaks should be taken.

Lack of skill can be ameliorated by appropriate training and by a gradual onset of repetitive work.

Reference

Kilbom, Å., Repetitive work of the upper extremity: Part II – The scientific basis (knowledge base) for the guide. International Journal of Industrial Ergonomics, Vol. 14, Nos. 1–2: 59–86 (this issue).

Repetitive work of the upper extremity:
Part II – The scientific basis (knowledge base) for the guide

Åsa Kilbom
National Institute of Occupational Health, S-17184 Solna, Sweden

1. Introduction

The concept "repetitive work" refers to similar work tasks performed again and again. By necessity, repetitive work of the upper extremity implies a motor component, which can be defined in terms of time and force. Repetitive work of the upper extremity is considered one of several physical work load factors, associated with symptoms and injuries of the musculoskeletal system. Other important factors are static loads, postures, and exertion of external forces. Scientific literature does not yet provide sufficient support for a separate assessment of risk for each one of these factors. They occur simultaneously or during alternating tasks within the same occupational work, and their effects concur and interact. Usually, harmful effects on the musculoskeletal system cannot be identified separately for each factor. Thus neither epidemiological, clinical or physiological/biomechanical studies support the view that repetitive work should be considered the prime exposure factor, without simultaneous consideration of forces and postures. Physical exposures should therefore be described and quantified with composite measures that consider all these factors (Drury, 1987; Putz-Anderson, 1988; Wells et al., 1989). In such exposure measures duration of exposure (hours, days, years) should also be included. For a complete risk assessment the characteristics of the individual and psychological and social conditions at work should also be included. No such quantitative risk assessment models have so far been developed, and the methodological problems of developing and validating them must be considered extremely large.

There are several reasons to summarize present knowledge about the relationships between repetitive work and musculoskeletal disorders, while simultaneously considering the demands on force, posture and duration of exposure. The concept repetitive work is used frequently both in scientific and popular science publications and assumed to be a causative factor in musculoskeletal disorders, without further definition and without quantifying the magnitude of the exposure. Repetitive work is considered a risk factor for some types of musculoskeletal disorders and less so for others. Thus there is a need to define repetitive work more stringently and to describe the musculoskeletal injuries typically associated with it. Such a review might assist in the understanding of the mechanisms of injury, and also help to direct prevention towards the most effective measures.

Thus, the aim of this review is to
- define repetitive work of the upper extremity and the parameters by which it should be described;
- review scientific literature with regard to the exposure–effect relationship between repetitive work of the upper extremity and short- and long-term effects, especially musculoskeletal disorders;

– evaluate if there is sufficient support for recommendations that will reduce the risk of such injuries.

Sometimes the concept "repetitive work" is used synonymously with "monotonous work". This is unfortunate, since these concepts appear to be understood differently by different readers. Both imply that the demands of each work task are very similar to the previous and/or subsequent work tasks, and that these tasks are performed again and again. The concept "monotonous work" is however wider and less well defined. The demands in monotonous work are not necessarily physical, and may in many cases be predominantly cognitive or sensory. Monotonous work also implies certain unwanted psychological responses like mental fatigue and stress, which may, or may not, be part of the response to repetitive work (Weber et al., 1980; Lundberg et al., 1989). For example, repetitive work of the upper extremity, like e.g. knitting, or of the lower extremity, like e.g. walking, can be rewarding and does not necessarily lead to mental fatigue and stress. Repetitive work is frequently combined with monotony, but monotonous work is not necessarily repetitive.

This review is based on a biomechanical/physiological model of mechanisms for musculoskeletal injuries, because the scientific documentation in this area appears to have a high degree of relevance and is more comprehensive than in other areas (Kuorinka, 1981). Another reason is that biomechanical/physiological parameters can be quantified and are therefore more easily accessible for ergonomic guidelines. Other scientific models may also have a high degree of relevance, e.g. those focusing on psychological and social factors (Karasek and Theorell, 1990), or skill and learning. For a review of the role of psychological and social factors in musculoskeletal disorders see Bongers et al. (1993). The mechanisms of interaction between physiological/biomechanical and psychological effects of repetitive work, and subsequent musculoskeletal injury are only beginning to be recognized. This review therefore concentrates on the physiological/biomechanical factors, while it is acknowledged that the factors mentioned above can modify the response. In a final model of repetitive work psychological and social factors, as well as factors related to skill and learning should be incorporated. For this review it must be kept in mind that in repetitive work, where the influence over the work process is limited, where the work is performed under time pressure, or when the worker is untrained for the task, the tolerance to repetitive work can be further reduced.

2. Methods

This review is based upon scientific literature in the areas of biomechanics, ergonomics, occupational medicine, orthopedics, physiology and rheumatology. The basis is scientific articles collected over a long period of time, but further documentation has also been compiled through literature searches in the databases CISILO, Medline 1983–92 and NIOSHTIC.

The literature can be subdivided into two main areas. The first concerns exposure circumstances, pathophysiology and clinical features of occupational musculoskeletal disorders. The effects thus are clinical, often of a chronic character, and the studies have mostly been performed with epidemiological and ergonomic methods.

The other main area concerns the association between repetitive work and various signs of fatigue and/or lack of recovery. The results have been obtained in experimental studies, above all in biomechanics and work physiology, and are thus concerned with acute and short-lasting responses to repetitive work.

The literature search in the first area yielded many hundreds of references. The abstracts were sorted with regard to the quality of exposure assessment. Only those studies where the abstract suggested that the physical nature of the work had been assessed were included. Thus merely the information that the work was repetitive did not qualify for selection into this review. In those cases where the abstract suggested that exposure might have been assessed, the article was studied. Most of them were case studies of occupational groups or patients, some presented newly developed methods with applications, and several were reviews. Only those studies presented in Table 3 fulfilled even low demands on exposure assessment. Usually these assessments included quantitative information on number of work cycles per time unit. Only very few studies with more complete information on work/rest schedules, postures and exertion of external force have been published. In conclusion, the literature search yielded relatively meagre results.

3. Defining and quantifying repetitive work

In the introduction repetitive work was defined as similar physical work tasks performed repetitively. In order to be applicable in practical work this definition needs to be specified with regard to similarity of work cycles and repetitiveness, i.e. the duration/frequency of work cycles. For applications concerned with prevention of musculoskeletal disorders, it is desirable that repetitive work is defined and quantified in such a way that harmful work patterns can be identified. A description of the work tasks therefore needs to include (where applicable) the parameters static loads, external force and posture (see introduction). These parameters should be operationalized into measures that can be applied in occupational studies. Moreover the engaged body region and the duration of exposure should be specified. The definition of repetitive work has also been discussed by Moore and Wells (1992).

Attempts have been made to assess the degree of repetitive work using questionnaires. Experience from the Stockholm MUSIC 1 study indicates that the reliability of questions concerning occurrence of repetitive hand movements is relatively high, and for repetitive finger movements high (Wiktorin et al., 1991). However, no attempts were made to quantify exposures in terms of duration of repetitive work, or simultaneous exertion of forces. Thus repetitive work should be quantified using observations and/or measurements.

The physiological/biomechanical characteristics of repetitive work can be categorized as either intermittent static, i.e. external movements are small or negligible, or dynamic, i.e. movements around a joint are easily distinguished. In either case there are breaks (or micropauses) in muscle activation between contractions. Intermittent static must be distinguished from static (or sustained) contractions, where the contraction is maintained for longer periods without breaks. No consensus regarding the minimum duration of static contractions is available. Most authors appear to use the term "static" for contractions exceeding 30–60s.

3.1. Work cycles, fundamental work cycles and work elements

Repetitive work cycles are often subdivided into fundamental work cycles (Konz, 1990; Silverstein et al., 1986, 1987a,b). For example, if the work task is to pack fruit into boxes, the duration of one work cycle, i.e. packing one box, may be a few minutes. The packing of each fruit is then a fundamental work cycle, which may only take a few seconds. From the biomechanical/physiological viewpoint these fundamental work cycles can be subdivided into several work elements, each one of which puts different demands on the worker. In the above example, these include: stretch arm, grasp fruit, lift fruit, wrap it into paper (which includes several repeated elements

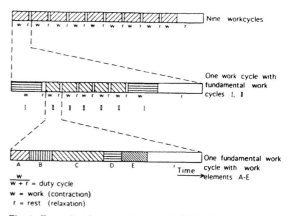

Fig. 1. Example of one work cycle subdivided into fundamental work cycles and work elements.

of rotation), lift the fruit into the box. An example of work cycles, fundamental work cycles and work elements is given in Fig. 1. In a similar way an assembly work or the work of a check-out cashier can be subdivided into fundamental work cycles, and these, in turn, into work elements. Standard times for such work elements have been presented from industrial time studies, e.g. MTM (Kanawaty, 1992) or MOST (Zandin, 1990).

In industrial studies it is usually not obvious which are the harmful exposures, since many different work cycles, fundamental work cycles and work elements may occur during the work day. Therefore experimental studies have been conducted, where specific work elements have been simulated in the laboratory. Repetitive tasks like flexion/extension, rotation or holding of objects have been performed while varying the work/rest schedule, posture, speed of motion and force (see section 6).

In experimental studies the aim is usually to quantify the tolerance for well-defined work elements with regard to the risk of fatigue. The underlying hypothesis is that fatigue precedes musculoskeletal injuries.

3.2. Repetitiveness

Ergonomic textbooks give some help in defining repetitiveness. Konz (1990) states that a work is repetitive if the duration of the fundamental work cycle is below 30 seconds. The duration of the total work cycle may be longer. Laurig (in Luczak, 1983) defines repetitive work as work elements that take place more than 15 times per minute and engage less than 1/7 of the muscle mass. Rodgers (1986) defines repetitiveness as work cycles shorter than 2 minutes repeated for one whole work shift, while highly repetitive work is defined as cycle times below 30 seconds. An indirect definition of repetitive work can be deduced from "Introduction to work study" presented by ILO (Kanawaty, 1992). Work cycles shorter than 0.17 minutes are considered to induce muscular fatigue and therefore require a certain rest allowance, which increases in duration as the duration of the work cycle diminishes.

In an epidemiological study of hand-wrist disorders in industry Silverstein et al. used an operational definition of highly repetitive work as follows: highly repetitive is work cycles less than 30

Table 1
Definitions of repetitive work in literature

	Cycle time	Additional classification	Comments
Huppes, 1992	< 30 s	–	"Very short cycled"
Konz, 1990	< 30 s	–	"Fundamental work cycle"
Kuorinka and Koskinen, 1979	2–9 s 7–26 s	–	Short cycled Long-cycled *Operational definition for epidem. study*
Laurig, in Luczak, 1983	< 4 s	< 1/7 of muscle mass active	"Einseitig dynamische Muskelarbeit". Criteria: steady state in HR and EMG
Rodgers, 1986	< 30 s < 2 min	≥ one work shift	Highly repetitive Repetitive
Silverstein et al., 1986	< 30 s or > 50% of work cycle fundamental	simultaneous class. of high/low force	*Operational definitions for epidem. study*

seconds, or the same fundamental work cycle performed during more than 50% of the total cycle time (Silverstein et al., 1986; 1987a,b). In Table 1 some definitions of repetitive work have been summarized. The definitions of cycle times in repetitive work vary from 4 to 120 seconds, with a majority of authors using 30 seconds.

3.3. Similarity of work cycles

None of the definitions presented in Table 1 include statements about the similarity of work cycles. Similarities between repeated motor actions can be defined with regard to
- time, i.e. the time sequence of one work cycle is similar to the other,
- space, i.e. each cycle or movement starts and ends with the same posture and follows the same spatial pattern, and
- force, i.e. each work cycle or motor action is similiar to the other with regard to the exertion of force over time.

It can be assumed that in studies of occupational jobs, similarity between work cycles has been evaluated with regard to similarity of work output rather than similarity of movements. Thus it is usually assumed, that as long as the task performed is the same from one cycle to the other, then there is also similarity of time, space and force parameters from one cycle to another. Whether this is the case remains to be demonstrated. Certainly there can be some variations in cycle time, e.g. with onset of fatigue, and in force, e.g. when the demands on precision are low. However, there is neurophysiological support for the assumption that once a motor task has been thoroughly learnt, the degree of variation in its performance is very low. At this stage no exact quantitative definition of similarity of work tasks can be given. Small variations in task characteristics, e.g. in shape and weight of materials and tools, working height, etc., can probably be accepted.

3.4. Quantifying static loads, external force, posture, duration of exposure and body region

Static loads frequently occur simultaneously with repetitive work, performed either by the same muscle group or by another. For example, repetitive work by the forearms/hands is often combined with static, or intermittent static exertions of the shoulder muscles when stabilizing the arms. High demands on force and precision of the repetitive work performed by the distal part of the arms is known to increase the static load on proximal muscles (Milerad and Ericson, 1994). Whether simultaneous static work by one group of muscles potentiates the effect of repetitive work is not known. When repetitive actions are superimposed upon static exertions of the same muscle group it is usually not possible to ascribe the local responses to one of the two activities. This is often the case for the shoulder; i.e. upper arm movements may be clearly distinguishable but simultaneous sustained static shoulder muscle activity can often by identified by EMG recordings. Thus repetitive shoulder/upper arm activities are difficult to evaluate without extensive recordings of both movements and EMG. Interpretation of the results is hampered by the still incomplete knowledge of the mechanisms whereby muscle injuries develop (see section 4.1). For tendon disorders of the shoulder the pathophysiological mechanisms are better known and the interpretation of exposure pattern therefore more clearcut (see section 4.2). Static workloads are usually quantified in percent of maximal voluntary contraction (MVC) or as force levels in Newtons. Considerable uncertainty exists concerning optimal measurement and calibration procedures.

Repetitive work is frequently performed in combination with exertion of *external forces*, i.e. manual lifting of objects or tools, and exertion of forces when pushing, pulling and holding objects. Both long-term clinical effects and short-term physiological/biomechanical effects are profoundly influenced by the magnitude of the forces exerted (see sections 5 and 6). Quantitative assessment of force exertion can be done by direct measurement using force transducers or via calibrated EMG recordings; however, no consensus exists on how to standardize measurements or evaluate the results. Silverstein et al. (1986, 1987a,b) dichotomized the exertion of external hand forces into either high (> 4 kg) or low (< 1 kg). External force generation was subjectively assessed at the workplace, and EMG recordings were made at selected workplaces. Only in five out of 34 jobs studied, the assessments of force had to be changed after the EMG studies. Thus it appears that external force can be assessed subjectively with reasonable accuracy, at least if the assessment is dichotomized. The definitions of force exertion introduced by Silverstein et al. have subsequently been used in several other epidemiological studies.

Posture during repetitive work appears to exert a strong effect on the risk of musculoskeletal disorders. Two likely mechanisms have been identified. In certain postures an increased pressure and mechanical wear of exposed tissues (especially tendons and nerves) occur. This mechanism is discussed further in section 4.2. Another likely mechanism is the effect of muscle length on force exertion capacity. A somewhat extended muscle is able to exert a larger maximal force, whereas shortened or overextended muscles have a lower maximal capacity. This also affects fatigue at a given submaximal level. Thus the muscle force–length relationship has consequences for repetitive wrist motions and postures, and will increase the strain during flexed and hyperextended, and ulnar or radial deviated wrist postures. Tichauer and Gage (1977) showed that the repetitive use of tongs that forced the wrist into ulnar deviation, produced a high prevalence of

Table 2
Suggested parameters for quantifying repetitive work. Factors with an asterisk are optional and can be used to supplement other measures when resources are available

Parameter	Laboratory studies	Field studies
Time	duration of work cycle duration fundamental work cycle* duration work element* duration contraction duration relaxation duty cycle duration experiment duration recovery endurance/max. endurance*	duration of work cycle duration of fundamental work cycle duration of work element (MTM)* frequency of specific movements or contractions duty cycle* duration of exposure (min/day, years*)
External force	force over time* mean force of work cycle (N and % MVC)	max. during cycle mean during cycle* (dichotomized > ref level or < ref level *or* ordinal scale)
Static work of adjacent regions	mean of work cycle N and % MVC	mean of work cycle (dichotomized into low or high level)
Posture of engaged joint(s)	posture over time* start of work cycle extreme of work cycle mean during work cycle (degrees in relation to neutral)	start of work cycle extreme of work cycle (dichotomized into neutral/mild or moderate/extreme deviation *or* ordinal scale)
Region	engaged muscle groups, tendons joint(s)	same as lab studies
Speed	speed over time* max during work cycle* mean during work cycle*	extreme of work cycle* (dichotomized *or* ordinal scale)
Acceleration	acceleration over time*	

tenosynovites, epicondylites and carpal tunnel syndromes in building apprentices. In this case the likely mechanism is friction and pressure increase in the carpal canal, in addition to fatigue of the wrist radial extensors. The American College of Orthopaedic Surgery (1965) has reached a consensus agreement about the definition of postures for different joints. Deviations from neutral position are quantified in degrees, in two dimensions. This however requires measument with goniometers during work, which has so far only been performed in research projects. Posture sometimes deviates significantly from neutral already at the start of a repetitive work cycle (e.g. ulnar deviation of the wrist in typing). Therefore it should be recorded both at the beginning and at the extreme of the work cycle.

In parallell with findings in epidemiological studies of work-related shoulder-neck complaints (Winkel and Westgaard, 1992) the *duration of exposure* (expressed in minutes per time unit and in months or years) is likely to influence the risk of disorders in repetitive work. In Table 1 only the definition by Rodgers includes criteria for the duration of the repetitive work.

Neither is there consensus on how to identify the *body region* involved in repetitive work. Usually it is classified by joint(s) and segment (shoulder, upper arm, forearm, hand).

3.5. New methodological development in the workplace assessment of repetitive work

In recent years several studies have been reported where more extensive instrumentation was used at the work place. They include the use of goniometers to record speed and acceleration of movements (Marras and Schoenmarklin, 1993), and use of spectral analysis of goniometer recordings to assess repetitiveness and postures (Radwin and Lin, 1993). The EMG method (EVA) presented by Mathiassen and Winkel (1991) is based on simultaneous assessment of both duration and intensity of work, which permits a quantitative assessment of both force/repetitivity characteristics and similarity of repeated work tasks. The above measurements can only be used by experts using sophisticated instrumentation, and so far they serve more to study mechanisms of injury than to assess the ergonomic conditions. They may later develop into easily available standard methods for assessing exposure.

3.6. Recommended parameters for quantifying repetitive work in field studies

The parameters selected for assessment of repetitive work in occupational work are those which can be quantified by means of observation, time studies and simple measurements at the work place (Table 2). Such measures can be used both for assessment of acceptability and to quantify exposure for epidemiological studies.

The *time parameters* include the duration of all work cycles and fundamental work cycles occurring during the work day, and the duration of exposure in minutes per day and years. If competence for time study is available, the work study should also include time studies of the work elements using MTM or MOST. Evaluation of duty cycles (i.e. the proportion of the work cycle spent in physical work) may be impossible to perform in fast repetitive work, and in field studies it usually requires a prior MTM analysis. From these data the frequency of work cycles and fundamental work cycles can be calculated by hour or day. In those few epidemiological/ ergonomic field studies which have been performed, the frequency of work cycles or fundamental work cycles has usually been recorded (Table 3). The frequency of specific movements (e.g. wrist flexion/extension, handgrips, elbow flexion, arm elevation) should either be obtained from time studies or registered separately.

Static work and exertion of external force cannot be quantified unless more advanced measurement techniques are available. Subjectively assessed, dichotomized evaluations (higher or lower than reference level) can probably be obtained at the workplace. The possibilities of using several levels on an ordinal or interval scale should be investigated.

Body posture should be quantified using angular displacements from neutral position, for the start of the work cycle and for the extreme postures reached during the cycle. The measures should be dichotomized into neutral/mild deviation and moderate/extreme deviation. The possible use of ordinal or interval scales should be investigated.

Apart from the parameters time, external force, static work and posture the joints and muscle groups involved in repetitive work should be reported. This can usually be achieved using basic knowledge of functional anatomy and requires EMG recordings only occasionally.

3.7. Recommended parameters for quantifying repetitive work in experimental studies

Those work elements that form a part of repetitive work tasks in industry can be analysed in experimental models at the laboratory. The effect measure is usually fatigue, operationalized as non-steady state in blood flow, blood pressure, heart rate, EMG frequency, electrolyte balance, or subjectively perceived fatigue. Other indices of fatigue include insufficient recovery after work (muscle strength or response to test contractions) and reduced productivity (Kilbom et al., 1993). In Table 2 recommendations are given for a minimum number of measures that should be used in order to quantify the experimental situation, i.e. the exposure. The validity of experimental studies and the comparability between experimental and field studies partly depend on selection of suitable exposure measures. This list, as well as the response measures, can naturally be expanded depending on the scientific aim of the study.

Different work elements usually engage several tissues, anatomically located around several joints. The simple task "stretch out arm – pick up object – lift it for inspection – turn it around for inspection – put it back" engages joints, tendons, muscles and nerves from neck-shoulder down to fingers. If such a task is to be simulated the choice is either between studying only one of the elements, considered the most "critical" with regard to fatigue and injury, or simulating the entire series of elements.

In conclusion, an experiment simulating repetitive work must be preceded by an analysis of the tissues engaged in the activity, selection of the most critical tissues with regard to fatigue and injury, and the selection of effect measures to be used to quantify fatigue. When selecting methods the results should as far as possible be expressed in a way that permits comparisons with field studies. Some recommendations are given in Table 2.

The *time parameters* are duration of contractions and of relaxations (i.e. work–rest), duty cycle, cycle time, and frequency of contractions or movements. Frequencies may seem unnecessary to report since they can easily be calculated. However, frequencies are commonly reported in ergonomic field studies and should therefore also be reported from experimental studies, for the sake of comparisons. The duration of each work element should also be reported when applicable, together with the total exercise period, in relation to maximal endurance (when applicable) and the time to recovery of different parameters.

4. Mechanisms of injury

Recently Armstrong et al. (1993) proposed a conceptual model for work-related neck and upper-limb disorders. The model emphasizes the interaction between exposure (external), dose (internal), capacity and response, and highlights the cascading events within a tissue that will eventually lead to an injury. This approach facilitates the understanding of the pathophysiological mechanisms and can also be applied to the following review of mechanisms in repetitive work.

4.1. Muscle

Findings indicating muscle injuries have been made in clinical studies, during explorative operations and at muscle biopsies of patients exposed to repetitive work tasks. In the earliest clinical studies of patients with repetitive work these observations included not only tendon disorders, but also in many cases tenderness on palpation of forearm muscles. Howard (1937, 1938) and Ranney et al. (1992) have pointed out that tenderness on palpation of the forearm muscles is common especially in electronics assembly, metal and cash register work, i.e. jobs characterized by a high degree of repetitiveness. These findings have partly been verified at operation. Thus Howard observed already in 1937 that especially patients with peritendinitis crepitans in the forearm, but also patients with tendinitis in the shoulder and upper arm, had inflammatory and degenerative changes in the muscles. Among his patients the majority had repetitive work tasks, but static work loads may also have occurred. In many cases the onset was preceded by a temporary increase in workload, and in several cases external trauma to muscles had occurred. Both these factors may have contributed to an increase of pressure within the muscle fascia. The results of microscopical examination indicate close similarity to findings in compartment syndromes in athletes (Styv and Körner, 1987; Styv et al., 1987). Repetitive exercise in endurance athletes is known to be associated with edema and increase in intramuscular pressure of muscles surrounded by a tight fascia, e.g. the calf extensor muscles.

In 1988 Dennett and Fry published a muscle biopsy study from the hand muscles (m interosseus 1) of patients with repetitive strain injury (RSI). The majority of patients worked on keyboards and thus had repetitive work tasks of the fingers. Among the results were an increased proportion of type 1 fibres, a reduced proportion of type 2 fibres and certain abnormalities of the mitochondria. The changes were more advanced in those patients who had severe symptoms, and there were obvious differences between the right and left side in those patients who had unilateral symptoms.

Several mechanisms have been proposed to explain the occurrence of muscle disorders and injuries associated with repetitive work. The mechanism via increased intramuscular pressure has received strong scientific support (Järvholm et al., 1988). According to this mechanism the increase in pressure arises due to mechanical deformation of the muscle within the tight fascia, and intramuscular edema. Electrolyte disturbances, membrane damage and insufficient pauses for recovery between contractions all contribute to formation of edema. The pressure increase leads to reduced inflow of oxygen and nutrients, and subsequently, through mechanisms yet incompletely known, to degenerative and inflammatory changes in the muscles (Armstrong et al., 1993).

According to another hypothesis, repeated periods of partial ischemia and subsequent reperfusion would lead to formation of free oxygen radicals, which exert a toxic effect on muscle tissue, especially its membranes. McCord (1985) and Corbucci et al. (1984) found some support for this hypothesis among marathon runners. However, in an experimental study of repetitive static thigh exercise (10 s work, 10 s rest at 30% MVC) Sahlin et al. (1992) did not find support for formation of free oxygen radicals.

Neurophysiological mechanisms were recently suggested to explain muscular disorders in repetitive work (Johansson, 1991). According to this hypothesis local muscle fatigue leads to further muscular contractions via activation by the gamma-motor system. This prevents muscles from relaxing between the contractions in repetitive tasks, and in turn leads to formation of more metabolites, further activation during the short time periods between each work task, etc. This hypothesis has been supported by empirical studies by Elert, who observed a partial inability to relax during the relaxation phase of repetitive arm elevations in patients with shoulder myalgia (Elert, 1991).

According to a hypothesis recently put forward, intermittent static contractions may lead to a prolongation of exposure, and this would eventually lead to an increase in risk (Byström et al., 1991). A conceivable mechanism is that since perception of fatigue is delayed and less pronounced during intermittent contractions (compared to the same amount of work performed continuously), this may lead to a prolongation of total work time. However, no empirical support for the hypothesis has as yet been given.

The mechanism of muscle injury via increased intramuscular pressure has several characteristics in common with the mechanisms responsible for injuries in static contractions. Thus electrolyte disturbances, membrane damage and edema formation occur in static contractions too (Armstrong et al., 1993).

As a mechanism for muscle injuries following

static contractions, the so-called Cinderella hypothesis has been presented (Hägg, 1991). In low-level static contractions, the same low-threshold motor units are active repeatedly and for prolonged periods, and this may lead to injury of these units even though the total work load is very low. This hypothesis was recently supported by a longitudinal study by Veiersted et al. (1993), who investigated the number and duration of pauses in the EMG recording from shoulder muscles. Among subjects performing machine-paced repetitive packing work, those with symptoms had fewer pauses (0.9 vs. 8.4 per minute) and a tendency of shorter total duration of pauses in their trapezius-EMG. The subjects did not change their EMG activity pattern when changing from healthy to "patient" status, which rules out the neurophysiological mechanism mentioned above. There was also a significantly higher static work load (1.9 vs. 1.3% MVC) in those subjects who later developed symptoms. This study, as well as one by Jensen et al. (1993) highlights the importance of micropauses for recovery. It is also noteworthy that the work tasks in those two studies were described as repetitive. In these cases it is likely that forearms/hands performed the repetitive dynamic work, while the shoulder muscle load should be characterized as intermittent static, with micropauses.

Several of the mechanisms mentioned above build on the hypothesis that accumulation of fatigue and lack of recovery leads to permanent muscle injury. The empirical and experimental support for this hypothesis is not yet sufficient. However, the scientific literature supports the basic notion that muscle injuries can arise from repetitive work. At present the most likely mechanism appears to be increase of intramuscular pressure with concomitant metabolic disturbances, possibly in combination with inability to relax during short pauses due to a feedback loop in the gamma-motor system.

4.2. Tendons

Disorders localized to tendons are probably the most common musculoskeletal manifestation of repetitive work. The body regions most often affected are the hand, wrist, forearm and shoulder. The association between repetitive workloads and disorders of the hand and wrist tendons is relatively well established (Armstrong et al. (1987, 1993). Pathological conditions include inflammations of tendons (tendinitis), of tendon sheaths (tenosynovitis) and of peritendinous tissues (peritendinitis) (Kurppa et al., 1979). As a combined effect of aging and tendinitis degenerative changes occur, which may lead to tendon ruptures especially in the shoulder (Herberts and Kadefors, 1976; Herberts et al., 1984). Moreover, chronic thickening of tendons and surrounding connective tissue, with subsequent functional impairment (e.g. trigger finger) and adhesions is observed (Lampier et al., 1965; Phalen, 1966). These conditions probably correspond to conditions in athletes like peritendinitis in the achilles, and tendinitis in the patellar and shoulder tendons (Kennedy et al., 1978).

The pathophysiological mechanisms have been discussed by Armstrong et al. (1993). During muscle contractions the tendon is exposed to mechanical load both along the tendon and from surrounding tissues. This tension leads to mechanical viscoelastic deformation (strain). The viscous deformation requires a certain time period for recovery, and during the period of increased tension and insufficient recovery blood supply and metabolism of the tendon is affected. Ischemic conditions in the tendon together with friction against surrounding tissues may produce an inflammatory reaction that increases the pressure. This increase in pressure can exert secondary effects on surrounding tissues. In the carpal tunnel, for example, it may cause injury to the median nerve. Similar mechanisms have been demonstrated in the shoulder, where the tendon of m supraspinatus is squeezed when the arm is elevated. Repeated arm elevations increase the friction and thereby the risk of tendinitis. In a case-control study of industrial workers with acute shoulder-neck disorders Bjelle et al. (1981) demonstrated that the cases had a more highly repetitive work and that they worked with elevated arms during a longer part of the work day than the controls. Studies of building workers (Herberts et al., 1984) demonstrate similar findings.

Friction against surrounding tissues is thus a common denominator for many tendon disorders. In experimental studies Wells et al. (1989) modelled high force/high repetition and low force/low repetition tasks. The increase in internal tendon frictional work and the changes in forearm EMG amplitude pattern (APDF) parallelled the increase in risk associated with high force/high repetition jobs, as observed by Silverstein et al. in epidemiological studies.

Recovery after viscous strain is also an important pathophysiological factor. In an experimental postmortem study, Goldstein et al. (1987) demonstrated that the work/rest regime influenced the rate of recovery. Thus, further studies of tendons using varying work/rest regimes might be used as a basis for recommendations concerning repetitive work.

4.3. Nerves

Nerves react with a functional impairment when exposed to increased pressure. In those anatomical regions where nerves pass through canals with limited space the risk of impaired function is high. On the forearm, in the wrist and in the hand the anatomical prerequisites for nerve compression and functional impairment exist (Armstrong et al., 1993). The most common form of occupational peripheral nerve injury is carpal tunnel syndrome, where repetitive work tasks with simultaneous exertion of high forces is an important factor (Barnhart et al., 1991; Feldman et al., 1987; Silverstein et al., 1987a), with a 5–15 times increase of risk compared to non repetitive, low force jobs. Functional impairment of the ulnar nerve has also been reported in industrial populations (Stetson et al., 1993). The impairment was most marked in symptomatic workers, but unsymptomatic workers too had smaller amplitudes over the sensory nerves and longer motor and sensory latencies, both in the medial and ulnar nerve. The impairments seemed to be associated with forceful hand exertions rather than with repetitivity of work, as asssessed by a checklist. No certain associations with repetitive work tasks have been documented in other forms of occupational nerve injury either (sensory lesions of the fingers, compression at Frohse's arcade). External pressure, vibration and traumata are more likely mechanisms in these cases.

4.4. Cartilage and bone

Scientific reports highlighting repetitive work versus cartilage and bone disorders are scarce. Ebara et al. (1989, 1990) observed that cervical spondylosis occurred already at a young age among patients with cerebral palsy and fast, repetitive head movements. In order to study the mechanisms they performed an animal experiment where rabbits were exposed to electrically provoked flexion/extension movements of the neck. The animals most heavily exposed (200,000 movements in 60 days) had more advanced changes in annulus fibrosus than the control animals, and moreover they had developed early signs of osteophytes in the lower cervical spine. Whether these findings are relevant to disorders of the cervical spine in healthy individuals is uncertain, since both patients and animals were highly exposed.

5. Associations between repetitive work and musculoskeletal disorders in epidemiological studies

5.1. Outcome

In Table 3 epidemiological studies of repetitive work and musculoskeletal disorders of the upper extremity have been listed. Only studies with a quantitative evaluation of the character of repetitive work have been included. In most studies disorders of forearms, wrists and hands have been studied. The most commonly described injuries affect nerves and tendons with adjacent structures, like in carpal tunnel syndrome (CTS), tendinitis, tenosynovitis, especially de Quervain's disease, and peritendinitis. In the US these diagnoses are commonly lumped under the umbrella term Cumulative Trauma Disorders (CTD) (Silverstein et al., 1986, 1987b; Marras and Schoenmarklin, 1993).

Muscular disorders in the shoulder and arm have been described by Ferguson (1971), Jensen

Table 3
Summary of some epidemiological studies where exposure to repetitive work has been quantified. CTD = cumulative trauma disorder, CTS = carpal tunnel syndrome, OCD = occupational cervicobrachial disorder, HiR = highly repetitive, HiF = high force, LoR = low repetitive, LoF = low force, OR = odds ratio

	Design	n	Occ group	Exposure	Effects	Comments
Amano et al., 1988	Cross-sect.	102 102	shoe manuf. assembly line controls (age, sex matched)	Observation and film of 29 shoe workers. 3400 arm actions/day (grasp, extend, bend, hold)	Reduced pain and vibration sensib., increased prevalence finger flexor tenosynov. and muscle tenderness in shoe workers	Predominantly left-sided disorders due to work organization
Bjelle et al., 1981	Case-control	20+ +26	Industrial jobs	Film + biomech. analysis	Cases reported to OHS, acute shoulder-neck complaints. No sign diff. in cycle time (9 vs. 12 min). Cases longer duration and higher frequency of shoulder abd., flexion (2.5/min)	Age, anthropometics, strength not ass. with disorders
Chiang et al., 1993	Cross-sect.	207 61 118 28	Fish processing LoR, LoF HiR or HiF HiR and HiF	Observation and EMG of typical jobs. HiR cycle time < 30 s or <50% of cycle fundamental, HiF > 3 kg	Self-reported and clin. exam: prevalence HiR+HiF > HiR or HiF > LoR, LoF (shoulder girdle pain, CTS). OR (log. regr. shoulder pain) HiR 1.6 HiF 1.8. OR (log. regr. CTS) HiR 1.1 HiF 1.8	Job seniority not risk factor. Load on shoulders may have been static although forearm worked repetitively. Gender (f) riskfactor for CTS
Feuerstein and Fitzgerald, 1992	Case-control	16 13	Sign interpreters	Videoanalysis of wrist postures/movements, HR, subj. ratings during standard interpretation	Cases: work with shoulder wrist-pain. Controls: work without. Hand-wrist deviations/min 10.2/7.5, rest breaks/min 0.8/1.7, work envelope excursions 2.7/1.0	Wrist movements faster among cases
Ferguson, 1971	Cross-sect.	516	Morse and keyboard telegraphists	Observation of typical jobs Morse telegraphy 515 muscle contractions/min Keyboard 204/min	Interviews: Occ cramp (previous or present) 14%, myalgia 5%	Cramp strongly associated with neurosis
Jensen et al., 1993	Cross-sect.	39 32	chocolate workers office workers	EMG: static load level 2.1 vs. 1.2% of max EMG. No. of short EMG gaps 5.8 vs. 9.0/min. No. of long EMG gaps 1.3 vs. 2.9/min	Symptom score similar between groups. Weak relationship Emg – symptoms	Current techniques (EMG) for measuring shoulder load inadequate for prediction

Reference	Study type	N	Population	Method	Findings	Comments
Kuorinka and Koskinen, 1979	Cross-sect.	93 143	Scissor manuf. shop assistants	Observation + video of typ. jobs: Short cycles 2–9.5 s Longer cycles 7.3–26 s	Tension-neck 61.3% Muscle-tendon forearm 18.3%. No diff. short-longer cycles. No. of objects handled/year related to disorders.	No effect of age. Short-cycled jobs faster? Skill modifies?
Luopajärvi et al., 1979	Cross-sect.	163 143	Food packers Shop assistants (controls)	Observation of typ. jobs. 60 finger/hand movements/min, max 25000/day. Also extreme wrist postures	Upper extremity tendon disorders, epicondylit. RR 3.6	Static loads common. Food packers' job causative, uncertain which exposure factor
Marras and Schoenmarklin, 1991	Case-control	20+ +20	Industrial jobs	High = low risk jobs ∼ 60 wrist movements/min. Wrist angular position, velocity and acceleration monitored in radio/ulnar, flexion/ext, pronation/supination	"Cases" (without disorders) from CTD high-risk jobs. "Controls" from CTD low risk jobs. Velocity and acceleration (exp flex/ext) sign ass. with high-risk jobs, OR 2–6	Wrist position not ass. with high risk jobs
Nathan et al., 1988	Cross-sect.	471	Industrial workers	Observations: job classified by force and repetitiveness (ordinal scales)	Sensory conduction of median nerve (CTS). Slowing in 39%. No diff. between exposure groups	No relationship with job seniority
Obolenskaja, 1927	1-year study	660	Teapacking	Job 1: 7600–12000 hand movements/day, job 2: 25000/day	Forearm-wrist tendon disorders, 1-year incidence 41%	No control group. Disorders more common in "new" workers. Chronic disorders common. Also exper. studies!
Ohara et al., 1976	Cross-sect.	117 379	packing machine op. Cash register op.	Observation (2 operators) Packing ∼ 3 articles/min	Occ cerv. brach dis (OCD) Grade II and III ∼ 17% (packing) Grade II and III ∼ 22% (cashing)	Severity of OCD increased with job seniority
Onishi et al., 1976	Cross-sect.	47 95 109 74	VDT op A lamp assembly B film rolling C office work D (controls)	Observations: A: 15–20000 strokes/day B: Cycle time 3.5–4.5 s C: Cycle time 2.5–5 s D: non-repetitive	Interview, PPT (tenderness) Shoulder stiffness: RR 1.60, 1.90, 1.70 (A, B and C vs. D) Tenderness: RR 1.48, 1.77 (B and C vs. D)	Lower muscle strength in subjects with stiffness/tenderness. EMG disclosed static shoulder loads
Silverstein et al., 1986	Cross-sect.	574	34 industrial jobs > 20 workers each	Obs. of indiv. jobs. Video and EMG of typical jobs. HiR cycle time < 30 s or > 50% of cycle fundamental. HiF > 4 kg, LoF < 1 kg	Cumulative trauma disorders (CTD) OR (log. regr) HiF, HiR 29.1, HiF, LoR 5.2, LoF, HiR 3.3 (vs. LoF, LoR)	Force stronger risk factor than repetitiveness

[Table 3 continued overleaf]

Table 3 (continued)

	Design	n	Occ group	Exposure	Effects	Comments
Silverstein et al., 1987a	Cross-sect.	652	39 ind. jobs > 20 workers each	Obs. of indiv. jobs. Video and EMG of typical jobs. HiR cycle time < 30 s or > 50% of cycle fundamental. HiF > 4kg LoF < 1 kg	Carpal tunnel syndrom (CTS) OR (log. regr.): HiF, HiR 15.5, HiF, LoR 1.8, LoF, HiR 2.7 (vs. LoF, LoR)	Repetitiveness stronger risk factor than force. Rep. and force multiplicative. Slight (ns) age effects, no gender effects. Vibrations likely risk factor
Silverstein et al., 1987b	3-year follow-up	136	Industrial jobs	Obs. of indiv. jobs. Video and EMG of typical jobs. HiR cycle time < 30 s or > 50% of cycle fundamental. HiF > 4 kg, LoF < 1 kg No change in job class in 3 years	Cumulative Trauma Disorders (CTD) at study start (9,5%) and end (16,8). Increase in all exp cat. except HiF, HiR (transfer from these jobs)	Hand, wrist posture recorded. Job satisfaction was high and not related to CTD. CTD more common in female. Job seniority negative ass. with CTD (survivor effect)
Vihma et al., 1982	Cross-sect.	40 20	sewing machine op. seamstresses (controls)	Observation: Machine op. cycle time 30–60 s. Seamstresses longer	Neck-shoulder complaints RR 1.60	Machine operators more static postures

et al. (1993) Kuorinka and Koskinen (1979), Ranney et al. (1992) and Vihma et al. (1982). In many cases of muscular disorders of the shoulder it is not possible to conclude whether the disorder was predominantly caused by repetitive dynamic, intermittent static or static exercise (see also section 4.1).

Shoulder tendon disorders in repetitive jobs have been studied less (Bjelle et al., 1979). It is likely, however, that several of the occupational groups studied by Herberts et al. were also exposed to repetitive work tasks, although the repetitiveness of the exposures have not been quantified (Herberts and Kadefors, 1976; Herberts et al., 1981, 1984).

Usually clinical examinations with well-defined criteria have been used. In some cases (Ferguson 1971, Ohara et al., 1976; Onishi et al., 1976, Marras and Schoenmarklin, 1993) the diagnoses have been given on the basis of interviews, which makes the evaluation of associations with exposures more uncertain. Nathan et al. (1988) and Stetson et al. (1993) used measurements of sensory conduction velocity of the median nerve and/or ulnar nerve as indicators of nerve impairment or CTS, but no clinical examinations were performed.

5.2. Exposure – repetitiveness

One aim of this review is to evaluate if a quantitative exposure–effect relationship exists between repetitive work and musculoskeletal disorders of the upper extremity.

The studies listed in Table 3 have used different exposure and outcome measures, and are therefore difficult to compare for such a purpose. Exposures have sometimes been recorded as cycle times, and sometimes as rate of movements, which has been assessed for fingers, hands/fingers and wrists respectively. Studies using cycle times and rates of movements cannot readily be compared, since a certain cycle time may imply several movements of fingers, hands and wrists. Cycle times for shoulder movements have only been recorded by Bjelle et al. (1981).

Rates of movements vary over a large range. As can be expected finger movements are recorded at a substantially higher rate than wrist movements. Ferguson observed a very high rate of 515 finger movements per minute in morse telegraphy, which he estimated to be twice that in typing (Ferguson, 1971). This was associated with a prevalence (previous or present) of occupational cramp (14%) and myalgia (5%). These figures are apparently above normal, at least for occupational cramp, but since the prevalence in the control group is not given, the magnitude of this increase cannot be estimated. In a recent study by NIOSH no relationships between forearm/wrist disorders and number of keystrokes per day were observed (NIOSH, 1992). However, the relatively low number of keystrokes (average 40/min) and the relatively narrow range observed "limits the ability to generalize the results to other data entry employees who may perform up to 80,000 keystrokes per day".

515 finger movements per minute can be compared with rates of hand/finger movements, estimated by Obolenskaja and Goljanitzki (1927) and Luopajärvi et al. (1979) to be around 25–60 and 60 movements per minute respectively. These exposure levels yielded a one-year incidence of forearm-wrist tendon disorders of 41% in Obolenskaja's study, and a relative risk of 3.6 (compared to shop assistants) in Luopajärvi's study. Contributing to the elevated risk in Luopajärvi's study was probably that the work also induced static loads and extreme wrist postures.

After supplementary animal experiments, Obolenskaja concluded that in untrained workers movement rates of 25–33 movements per minute should not be exceeded if tendon disorders were to be prevented. However, she also concluded that if the force requirements are high, this figure must be lowered.

Specifically observing wrist deviations, Feuerstein and Fitzgerald (1992) recorded a higher movement rate among sign interpreters with pain, than in those without pain, i.e. 10 vs. 7.5 excursions per minute. These frequencies are considerably lower than those cited above, and it is especially noteworthy that the movements were not combined with exertion of external forces.

In several studies *cycle times* in repetitive work have been recorded. The shortest work cycles reported are 2.5-5 s, observed in lamp assembly and film rolling work (Onishi et al., 1976). Compared with office work, these short cycled tasks yielded shoulder stiffness and tenderness with relative risks of 1.9 and 1.6 respectively. Kuorinka and Koskinen (1979) compared short cycles – 2-9.5 s – with longer cycles – 7.3-26 s – in scissor manufacturing. On the group level there was an exposure-effect relationship between the number of objects handled per year and the number of muscle-tendon symptoms of the forearm-wrist, but there was probably not sufficient contrast between the two cycle time categories to demonstrate a difference. The prevalence of disorders was elevated in comparison with shop assistants, but lower than in factory workers studied with the same outcome criteria by Luopajärvi et al. (1979).

The most extensive studies of repetitive jobs in industry have been performed by Silverstein et al. (1986; 1987a,b). They have systematically contrasted highly repetitive jobs, defined as work cycles < 30 s or fundamental work cycles constituting more than 50% of the work cycle, with low repetitive jobs, for a large range of industrial tasks. Four exposure categories have been created by combining repetitiveness (high or low) with force exertion (high force > 4 kg, low force < 1 kg). A combination of high repetitiveness and high force exertions resulted in odds ratios of 29 and 15.5 for CTD and CTS respectively, in multiple logistic regression analysis. High repetitiveness alone also created elevated risks, with odds ratios of 2.8 and 5.5 respectively.

The above studies all indicate an association between the repetitiveness of work and forearm-wrist muscle-tendon disorders. The only "negative" study is the one performed by Nathan et al. (1988), who could not demonstrate a relationship between repetitiveness/force exertion and measurements of sensory conduction velocity of the median nerve (as an indicator of CTS). However, no quantitative data on forcefulness and repetitiveness, nor on symptoms of the study groups are reported.

The situation is more complex when shoulder disorders are considered. The study by Bjelle et al. (1981) indicates that the risk of shoulder tendon disorders increases with the rate of shoulder movements. Muscle disorders of the shoulder do not, however, seem to follow the same straightforward relationship. The study by Jensen et al. (1993) indicates that short work cycles imply a lower risk than long cycles, which agrees with the micropause-mechanism of muscle disorders presented in section 4.1.

5.3. Exposure – External force, static load, posture, speed and acceleration

Only in a few of the studies listed in Table 3 risk factors other than repetitiveness have been quantified.

Exertion of external force was well established as a risk factor, interacting with repetitiveness in a multiplicative way, by Silverstein et al. (1986; 1987a,b). Exertion of external forces does not, however, appear to be a prerequisite for pain according to Feuerstein and Fitzgerald (1992).

Only Jensen et al. (1993) have quantified the static load levels using EMG of the trapezius muscle. Industrial workers had somewhat higher static load levels and fewer EMG gaps (period with no appreciable EMG activity) than office workers. However, since the association between EMG parameters and symptoms were weak, they conclude that present methods for EMG recording seem inadequate for a prediction of muscle pain symptoms.

The possible interaction between postures and repetitiveness is less well established. Tichauer and Gage (1977) observed ulnar deviation of the wrist in repetitive building work, associated with a high prevalence of CTD. Silverstein et al. (1987a) investigated wrist posture but found no association with CTS. Feuerstein and Fitzgerald (1992), Luopajärvi et al. (1979) and Silverstein et al. (1987b) have observed a high proportion of extreme wrist postures in their studies, but no quantitative risk assessments have been made.

By means of recently developed goniometers, it has become possible to measure the speed and acceleration of movements around joints in field studies. Such measurements have been applied to study wrist movements by Marras and Schoenmarklin (1993). In a case-control ("cases" were non-symptomatic and selected from high risk jobs) study, they demonstrated that speed and acceleration of movements, especially in extension/flexion, was significantly higher in high than in low risk jobs. Workers in both high and low risk jobs performed about 60 wrist movements per minute, and there was no difference in wrist position (flexion/extension, ulnar/radial deviation, pronation/supination) between the two groups.

Feuerstein and Fitzgerald (1992) video-recorded hand movements in sign interpreters, and observed faster hand movements among cases with CTD. In their cross-sectional study Kuorinka and Koskinen (1979) observed that the short-cycled work operations seemed to take place with faster movements. The above findings indicate that frequency, force and position parameters may not be sufficient to estimate risks of musculoskeletal disorders in repetitive jobs, and that speed and acceleration of movements may also be risk factors.

5.4. Duration of exposure

The reviewed studies do not generally support the notion that job seniority in repetitive work is a factor that further increases the risk of tendon, muscle or nerve disorders of the upper extremity. The only exception is the study by Ohara et al. (1976). Obolenskaja and Goljanitzki (1927), Nathan et al. (1988) and Silverstein et al. (1987a) observed a negative relationship. Since all these studies were cross-sectional, the possibility of survivor effects cannot be ruled out. In no study was the duration of repetitive work per day reported. However, it is noteworthy that Hagberg provoked symptoms and signs of tendinitis in the descending part of the trapezius and biceps muscles in experimentally performed arm elevations (rate 15/min) after only one hour of work (see 6.2) (Hagberg, 1981).

5.5. Conclusions

All studies of hand, wrist and forearm disorders reviewed above, except the one by Nathan et al. (1988) indicate that repetitive jobs are associated with an elevated occurrence of tendon disorders of the shoulder, hand, forearm and wrist. The exposure-effect relationship is still unclear, as disorders have been observed at a large range of cycle times and movement frequencies in different studies. The reason for this large variation is probably that in most studies the exertion of force has not been considered or controlled. Thus the risk is further elevated in combination with exertion of high external forces and probably also in extreme postures, in combination with static loads, and with a high speed and acceleration of movements. There is no basis for conclusions concerning minimum rates of repetitiveness (or maximum cycle times) that will abolish the risk. Cycle times up to 30 s, especially in combination with exertion of high forces, are clearly associated with an increased risk of forearm-wrist tendon disorders and CTS. Wrist movements at frequencies as low as 10 per minute and shoulder movements at frequencies of 2,5 movements/min are associated with tendon disorders, whereas the tolerance for repetitive finger movements appears higher.

Concerning disorders of muscles there is some indication of an association between repetitive work and shoulder muscular disorders (Ferguson, 1971; Jensen et al. 1993; Kuorinka and Koskinen, 1979; Ohara et al., 1976; Onishi et al., 1976). Exposures have mainly been evaluated as frequency of hand manipulation, and therefore it is uncertain whether the observed effects have arisen secondary to repetitive or static activity of the shoulder muscles. When work cycles have been studied using EMG, it appears that short cycles, usually in the form of static work interrupted by micropauses, i.e. intermittent static exercise, implies a lower risk than long cycles. The risk of musculoskeletal disorders of the shoulders in relation to ergonomic factors at work has recently been reviewed by Winkel and Westgaard (1992).

6. Experimental studies

Both intermittent static and repetitive dynamic exercise have been used experimentally to study the effects of repetitive work. Intermittent static exercise has mainly been used to model fatigue and risk of injuries to muscles, using physiological and subjective criteria. Repetitive dynamic exercise has been used to a more limited extent, mainly to model the risk of tendon fatigue and disorders, using either subjective fatigue/ endurance or biomechanical criteria. Since both muscular and tendon disorders are main endpoints of repetitive work (see 4.1 and 4.2) both these approaches are relevant. Risks of nerve injuries in repetitive work seem to be strongly associated with the mechanisms of tendon disorders, as nerve lesions usually appear anatomically close to tendons under strain.

6.1. Validity of experimental models

For ethical reasons, it is not possible to pursue experimental studies in human subjects to the endpoint of clinically identifiable disorders. The relevance of experimental studies on animals is questionable, since it is not possible to simulate voluntary activity. Several of the experiments and models presented in this review therefore rest on the hypothesis that acute fatigue or discomfort indicates an increased risk of musculoskeletal disease. However, it must be remembered that the mechanisms leading from tissue fatigue to injury are still incompletely known, and that alternative mechanisms may be responsible too. Such alternative mechanisms may for example include extreme vulnerability in certain individuals and the effects of single trauma. Neither do these models take into consideration additional strain due to e.g. vibration and psychosocial factors.

The validity of experimental studies, with regard to musculoskeletal disorders in repetitive work, therefore needs further study. This could be achieved by comparing the work/rest regimes and the rates of repetition recommended through modelling, to the rate of disorders and work characteristics of repetitive work performed in industry. Such comparisons have been performed to a very limited extent. Moore and Garg (1993) developed a strain index for upper extremity disorders, based on estimations of intensity, posture and speed of work, and measurements of duty cycle, frequency of efforts and duration of task per day. This index was developed for the range of tasks occurring at a pork processing plant, and showed a reasonably good agreement with rates of elbow, forearm and hand disorders reported in the preceding time period at the same workplace. The validity of the index should be tested in other workplaces, using larger study groups and additional outcome criteria.

Another sign of validity is that the selected exposure measures create sufficient contrast between exposed groups and that this contrast is reflected in the risk ratio of exposed versus unexposed groups. Wells et al. (1989) made biomechanical calculations of frictional force of wrist tendons and measured forearm extensor EMG, contrasting high repetitive/high force against low repetitive/low force tasks in the laboratory. They found a good agreement between the contrasts between these exposure measures, and the increased risk of CTD in high repetitive/high force vs. low repetitive/low force tasks observed by Silverstein et al. in epidemiological studies. The tasks studied by Wells were set up to simulate an industrial task. Recently, Loslever and Ranaivosoa (1993) presented extensive studies of some biomechanical factors relevant for carpal tunnel syndrome, and compared those with the prevalence of CTS in an epidemiological study. They found a close correspondence between relative time spent in flexion and in exerting high forces, and the prevalence of CTS. Their biomechanical measurements were performed in industry, which supports the view that with more sophisticated biomechanical monitoring it will become possible to quantify more precisely the association between exposure and musculoskeletal disorders at the workplace.

The ecologic validity, i.e. to which extent the experimental situation resembles an industrial task, is another important point to consider in the design of experiments. The experimenter must decide what is the first priority: a detailed study of one work element (see section 3.6) or a more realistic simulation of a series of work elements constituting one work cycle. The first alternative offers certain advantages as it is easier to vary and control several parameters (force, posture, duration of work/rest). Moreover, the response can with certainty be ascribed to one specific work element. The first alternative has been chosen in the majority of experimental studies (Björkstén and Jonsson, 1977; Bishu et al., 1990; Byström, 1991; Dahalan and Fernandez, 1993; Dul et al. 1991; Mathiassen, 1993 (paper D); Milner, 1985; Müller, 1935; Pottier et al., 1969; Rohmert, 1973a; Serratos-Perez and Haslegrave, 1992; Sheeley et al., 1991; Snook et al., 1992). The second alternative, i.e. simulating an entire work cycle, was chosen by Sundelin (1992) and Mathiassen (1993). They simulated word processing and an industrial assembly task, respectively, and assessed the physiological and psychological responses at different MTM rates and work/rest regimes. Similarly, Kim and Fernandez (1993) and Marley and Fernandez (1991) performed a psychophysical experiment of simulated drilling. This experimental design provides a more realistic simulation of an occupational work, but the interpretation of responses is more difficult as a whole sequence of work elements is performed. A commonly used experimental design is to simulate repetitive handgrip contractions (Byström, 1991; Dahalan and Fernandez, 1993) which have the advantage of being common in industry as well as containing only one work element.

In conclusion, none of the experimental studies, except those performed by Wells and by Loslever and Ranaivosoa have conclusively proven their validity with regard to study design and selection of exposure measures. This needs to be done especially for the models developed by Byström (1991) and Dul et al. (1991), which are intended to prevent musculoskeletal fatigue, and supposedly disorders.

6.2. Criteria of acceptability

In Table 4 experimental studies of human subjects performing repetitive or intermittent static exercise have been listed. In addition, some studies are included where remaining endurance or recovery after single static contractions was measured (Milner, 1985; Rohmert, 1973a; Serratos-Perez and Haslegrave, 1992).

From a theoretical viewpoint the extrapolation of results obtained at single contractions to predictions of the performance at repeated contractions can be questioned, mainly because it may not be possible to identify slight fatigue or lack of recovery after only one contraction. Milner studied forward stoop of the trunk, and measured the remaining endurance after contractions and rest periods of various durations. Based on his results a model for predicting remaining endurance was developed. This model has subsequently been compared with experimental data on endurance of shoulder abduction, but was found to overestimate endurance for this muscle group. Milner's model was also modified to predict remaining endurance after repeated static postures with variable durations of contractions and relaxations. The absolute durations (in seconds) are not reported, but since no external loads were held and the back extensors have a high endurance, it is likely that even the shortest submaximal postures were held for at least a minute. The applicability of this model for repetitive exertion of less than one minute can therefore be questioned. Dul later expanded the model to encompass several muscle groups (Dul et al., 1991). Preliminary reports indicate a relatively good agreement between experimentally obtained results and the model predictions.

Some of the studies in Table 4 have been conducted explicitly to give guidance for recommendations concerning prevention of musculoskeletal disorders associated with repetitive work in industry (Dul et al., 1991; Byström, 1991; Bishu et al., 1990; Sheeley et al., 1991; Dahalan and Fernandez, 1993; Snook et al., 1992; Kim and Fernandez, 1993; Marley and Fernandez, 1993). Others (Rohmert, 1973a; Müller, 1935; Björkstén and Jonsson, 1977; Pottier et al., 1969; Sundelin 1992) aim mainly at reducing fatigue and/or increasing productivity in industry. Ahlthough their aim is not to reduce injuries, they are frequently used for such recommendations. Several other studies of intermittent exercise (not reviewed here) have been performed, aiming at understanding the physiological responses in intermittent exercise (for review see Mathiassen and Winkel, 1991). Many of these studies have been influenced by sports physiology, which means that high forces have been studied rather than low

Table 4
Some experimental studies of intermittent static and repetitive dynamic exercise

	Type of exercise, body region	Independent variables	Dependent variables	Main result
Björkstén and Jonsson, 1977	Sustained and intermittent static elbow flexion 90° until exhaustion or up to 60 min	5.5–60% MVC duty cycle 0.3–0.7 contractions 3–7 s	time to exhaustion	endurance time by mean contraction intensity. mean accept. contr. int <14% MVC
Bishu et al., 1990	repetitive dynamic wrist flexion - extension	weight 3, 5, 11 lb frequency 60/min no glove/glove	subjective exhaustion	max. no. of repetitions by force
Byström, 1991	sustained and intermittent handgrip	10–40% MVC duty cycle 0.25–1.0 contractions 5, 10 s relaxations 2–20 s duration exercise 3–30 min tension time constant	during exercise: EMG freq., subj. fatigue, local blood flow. After exercise: low frequency fatigue, MVC, K+, EMG freq.	acceptability of either: contraction intensity, duty cycle, contraction and relax. duration, exercise duration mean accept contr int. <17% MVC
Dahalan and Fernandez, 1993	repetitive handgrip	force 20, 30, 50, 70% MVC duration of grip 1.5, 3.5, 7 s	psychophysical assessment of frequency. Also HR, BP, EMG, RPE	maximal acceptable frequency reduced as force and duration increased. Mean accept contr. intensity <14% MVC
Dul et al., 1991	intermittent static several muscle groups (exper. test on arm elevation)	model input: duty cycle, duration of contraction, and relaxation	subjective fatigue	model output: endurance, remaining endurance, "muscle fitness"
Kim and Fernandez, 1993	simulated drilling task, arms	force 2.7–10.9 kg, wrist flexion 0, 10, 20°	psychophysical assessment of frequency. Also HR, BP, EMG, RPE	maximal acceptable frequency reduced as force and wrist flexion increased
Marley and Fernandez, 1991	simulated drilling task, arms	weight, posture, task constant	psychophys, assessment of frequency. Also HR, BP, EMG, RPE	max no. of acceptable tasks
Mathiassen (paper D) 1993	sustained and intermittent static arm elevation until exhaustion or up to 60 min	1–3× arm gravitational torque duty cycle 0.33–1.0 contractions 20–300 s relaxations 0–60s mean load = 1× gravit. torque	during and after exercise: EMG, BP, HR, perceived fatigue. After exercise: K+, La+, NH3, MVC pressure pain threshold (PPT)	cycle time and duty cycle influenced physiological response and endurance
Mathiassen (paper E), 1993	simulated assembly work, shoulder+arm	100 and 120 MTM duration work 2, 4, 6 hrs breaks (active or passive) 20 min/2 hrs	during and after work: EMG, HR, perceived exertion, PPT after work: MVC, proprioceptive performance	steady state not maintained during work, less fatigue at 120 MTM. No diff in recovery between protocols

Reference	Task	Measurement	Results
Milner, 1985	single and intermittent static trunk flexions	duty cycle 0.11–0.67 contractions 25–100% of max holding times (MHT) relaxations 25–100% MHT	remaining endurance	model output: remaining endurance after single or intermittent contr.
Müller, 1935	sustained or intermittent static elbow flexion (90°)	3–20 kg duty cycle 0.13–1.0 contractions 8 s–8 min relaxations 4 s–6 min tension time 20, 40, 60 kg×min	remaining endurance	No. of pauses does not influence remaining endurance! Optimum duty cycle 0.25
Pottier et al., 1969	sustained and intermittent static contractions	duty cycle 0.1–1.0		mean acceptable contraction intensity <17% MVC
Rohmert, 1973a	single static contractions, several muscle groups	15–100% MVC	heart rate recovery	model output: required rest allowance after each contraction, by intensity and duration of contraction
Serralos-Perez and Haslegrave, 1992	single static shoulder abductions (60°)	duty cycle 0.5–0.67 contractions 25–100% MHT relaxation 12.5–100% MHT	remaining endurance	remaining endurance lower than Milner's model
Sheeley et al., 1991	repetitive dynamic wist flexion–extension	frequency 20, 60/min weight 4, 6, 10 lb	subjective exhaustion, RPE	time to fatigue by pace and force
Snook et al., 1992	repetitive dynamic wrist flexion (pinch and powergrip) and wrist extension (powergrip)	frequency 2.5, 10, 15, 20/min	psychophysical assessment of acceptable torque. Max wrist strength, tactile sensitivity, symptoms	maximum acceptable torque by repetition rate, movement and grip
Sundelin, 1992 paper I and II	word processing	I: 3–5 hrs with or without pauses II: 3×30 min without pauses, with passive and active pauses every 6th min resp.	EMG trapezius, levator scapulae, discomfort ratings	Neg. corr. between spontaneous pauses and right trapezius static load (I) Introduced pauses did not influence EMG responses. Subjects preferred active pauses (II)
Sundelin, 1992 paper IV and V	repetitive arm elevations + grasp	IV: MTM – 110 = 41 cycles/min, V: 1st hr 41 cycles/min, 2nd hr 49 cycles/min, 1 min pause every 6 min	EMG trapezius, infraspinatus, RPE	IV: EMG signs of fatigue in most muscles and subjects V: EMG signs of fatigue less marked with pauses, but subj. ratings and discomfort more marked

ones and that the emphasis has been more on complete exhaustion than on early signs of fatigue. For example, intermittent contractions superimposed on static low level intensities have hardly been studied at all.

As discussed above, the validity of experimental studies for prevention of muskuloskeletal injuries depends on the validity of the experimental design and on the selection of exposure measures. In addition, the validity depends on the criteria used for estimating acceptability. In several studies *endurance* measurements have been used, either by measuring the maximal duration of a single or intermittent task (Björkstén and Jonsson, 1977; Mathiassen, 1993 (paper D); Bishu et al., 1990; Sheeley et al., 1991) or by measuring the remaining endurance (in per cent of the maximal) after a repetitive task has been performed for a certain time period (Dul et al., 1991; Milner, 1985; Müller, 1935; Serratos-Perez and Haslegrave, 1992). The use of experimental endurance data has been questioned (Byström, 1991) since endurance can be influenced by subject motivation. Especially at low work intensities, maximal endurance for a given task may vary in the same individual due to the experimental circumstances. Recently the use of subjective measures of fatigue and exhaustion has been questioned by Mathiassen, since ratings of perceived exertion failed to pick up important physiological events (Mathiassen, 1993). Thus subjective exhaustion or fatigue ratings at low exercise intensities are not well-defined concepts.

Psychophysical criteria have been used extensively to develop acceptability limits for manual materials handling. The method has a high degree of test–retest reliability. Recently it has also been used to model acceptability of repetitive work. Snook et al. (1992) used the method to calculate acceptable torque in wrist flexion while varying frequencies, while Marley and Fernandez (1991), Kim and Fernandez (1993) and Dalahan and Fernandez (1993) used it to assess acceptable frequencies in drilling and handgrips, while varying external force, duration and wrist posture.

Physiological criteria have been used to a relatively limited extent. Rohmert (1973a) used recovery of heart rate after single static contractions as the criterion for acceptability. Based on his results, predictions can be made concerning acceptable work/rest periods in repeated static contractions. Marley and Fernandez (1991) used physiological criteria in addition to psychophysical ones. In a series of experiments, Byström (1991) used a large range of physiological measures of muscular fatigue to develop a model for acceptability in intermittent or sustained static handgrip contractions. For each physiological parameter (and also for ratings of perceived exertion) a decision was made as to the magnitude of change, from resting or steady state conditions, that was considered acceptable. Applying these limits to the experiments, it was found that different physiological criteria frequently indicated the same level of acceptability. This enabled Byström to conclude which work/rest regimes should be considered acceptable and which should not. The experiments were usually not conducted until subjective exhaustion, and in all experiments the total amount of "work" (i.e. tension-time) was kept constant, while the duty cycle, duration of contraction and relaxation, and contraction intensity were varied. Based on these results Byström could develop a model for acceptability of handgrip contractions, where either of the parameters contraction intensity, duty cycle, duration of contraction/relaxation, or duration of exercise can be estimated provided the others are known. The model is only valid at or below the tension-time studied experimentally, and will probably have to be modified for higher tension-times.

In the studies recently reported by Mathiassen (1993) physiological and psychological indices of fatigue of the shoulders were monitored in intermittent static contractions and in simulated assembly work, respectively. Since shoulder loads are usually sustained occupationally for a longer duration than forearm loads, the duration of intermittent contractions was varied between 10 and 360 seconds. The physiological responses were related both to cycle time and duty cycle. However, static shoulder exercise and assembly work differed considerably in responses, maybe due to the possibility of using alternating motor unit recruitment in assembly work. Mathiassen therefore concludes that intermittent static shoulder contractions may not be a valid model for industrial assembly tasks. Sundelin (1992) used EMG recordings and psychophysical ratings to study possible effects of introduced pauses. In general, an association between pauses and EMG pattern was observed, whereas subjective responses were more variable.

Clinical criteria were used by Hagberg in his study of repetitive arm elevations (Hagberg, 1981). After one hour of work at a rate of 15/min the subjects developed symptoms and signs of shoulder tendinitis.

Energy requirements have also been used as acceptability criteria in several studies (Price, 1990; Åberg, 1968; Freivalds and Goldberg, 1988; Rohmert, 1973b; Kanawaty, 1992). Their relevance for repetitive work of the upper extremity is less obvious, as they usually model longer and more variable work tasks with exertion of high force, which is not common in repetitive work.

6.3. Response measures in experimental studies

It is evident from Table 4, that the dependent variables (responses) of the experimental studies vary. In studies of intermittent static contractions, the end results have been expressed as e.g. endurance of exercise, remaining endurance after exercise, required rest allowances, and mean acceptable contraction intensity. It is interesting to note that several authors, although using different criteria and experimental design, reach the same conclusion concerning mean acceptable contraction intensity (cf. discussion by Byström, 1991). Byström's studies as well as those by Björkstén and Jonsson (1977) and Bellemare and Grassino (1982) indicate that mean levels of contractions below around 15% MVC are acceptable in intermittent static exercise, while levels above 10% MVC should not be exceeded in sustained contractions. It is noteworthy that Dahalan and Fernandez (1993), using psychophysical criteria, identify mean handgrip contraction intensities below 14% MVC as acceptable. Thus there seems to be some agreement between psychophysically and physiologically based limits.

Only the models developed by Dul et al. (1991), Byström (1991), Rohmert (1973a) and Dahalan and Fernandez (1993) allow conclusions concerning acceptable duty cycles and duration of contraction/relaxation. Rohmert's and Byström's results show good agreement at high contraction intensities, while Byström's model requires considerably longer relaxation periods between low intensity contractions. The model developed by Dul mainly uses subjective responses as dependent variables and as basis for recommendations.

In experimental studies of repetitive dynamic (or dynamic + static) work, the effect measures have been maximal frequency, maximal acceptable frequency, or the predicted time to fatigue at a given work pace and mean percentage of MVC.

6.4. Conclusions

As yet there is no consensus on which criteria to use when assessing the acceptability of experimentally performed repetitive work. Recommendations based on endurance data are questionable and those based on psychophysical studies need more experimental support. Physiological data are partly convincing especially when several parameters indicate a similar level of acceptability. It is likely, however, that physiological parameters relevant for one type of intermittent exercise may be irrelevant for another.

The validity of experimentally determined recommendations for work/rest regimes and rate of repetitions needs more study, both with regard to ecological validity, and with regard to the underlying assumption of a cause–effect relationship between fatigue and injury. The models developed should be applied in industry and evaluated against the rate of injuries reported in jobs identified as not acceptable. For intermittent static exercise, the models developed by Byström and Dul are the most versatile, and these models should especially be evaluated. For repetitive dynamic work (especially wrist extension/flexion) available models still need more experimental support. Thus there is at present only limited support in experimental studies for guidelines on frequencies/forces of repetitive industrial jobs.

7. Previous guidelines

In 1989, Tanaka and McGlothlin presented a conceptual model to assess musculoskeletal stress, as a basis for prevention of hand and wrist CTD (Tanaka and McGlothlin, 1989). They suggest that the three factors "work done by hand", repetition rate and wrist angles, in certain combinations described mathematically, will increase the risk of CTDs above an action limit (AD), and further when a maximal permissible limit (MPL) is exceeded. The model cannot yet be used quantitatively, as the coefficients and constants of the formula need to be defined through epidemiological research. Moreover, it does not consider the potential risk factor duration of exposure.

As mentioned above, the models presented by Byström (1991) and Dul et al. (1991) have been developed to provide guidelines for intermittent static work. The models are based on experimental data but have not been validated in industry, and therefore it is not possible at this stage to conclude that their application will prevent musculoskeletal disorders associated with repetitive work. One main objection to the theoretical framework of these models is that they predict fatigue of one repeated work element only. Repetitive work in industry often contains several work elements, and their effects may interact to produce a fatigue effect that exceeds the effect predicted by these simple models. Moreover, they predict musculoskeletal disorders on the basis of acute responses, and not on the long-term effects as measured by disease rates.

As pointed out above, the recommendations of Byström are based on muscular fatigue criteria, which may not be applicable in cases of tendon disorders. It may be argued that since muscle and tendon form one functional unit, muscle fatigue, since it can be measured more easily, is a suitable proxy also for tendon strain. This however needs to be proven. Psychophysical criteria may be applicable both for muscle and tendon strain, but this also needs to be demonstrated. Recommendations based on the tendon strain concept were recently published (Fisher et al., 1993). They also include considerations of how to optimize productivity while maintaining a low risk of tendon injury. Using elaborate mathematical modelling the authors give recommendations concerning work rates and rest brakes. The model still needs both experimental and epidemiological support, however. The same comment can be made concerning the model (strain index) developed by Moore and Garg (1993).

Guidelines more easily applied in industrial work have been presented by Williams (1973) and by the International Labor Organisation (Kanawaty, 1992). Both acknowledge the fatiguing effect of repetitive work and provide exact data on the need for fatigue allowances when cycle time is below a certain level. However, the maximum cycle times requiring fatigue allowances are very short; in Williams' recommendations the maximal cycle time is 6 seconds, and in the ILO recommendations the maximum is 10 seconds if additional rest is to be provided. This is obviously much shorter than the 30 s cycle time demonstrated by Silverstein et al. (1986, 1987a, b) to be associated with musculoskeletal disorders. Neither do these recommendations take into consideration the force and/or speed exerted at work. Thus the application of these recommendations is unlikely to prevent disorders, neither were they developed with such an aim.

The guideline presented by Winkel and Westgaard (1992) can be used for shoulder-neck complaints, especially when inadvertent postures and exertion of large force occurs. It does not specifically address work/rest characteristics, cycle times, or frequency of movements.

8. Need for further research

Neither epidemiological nor experimental studies performed to date have been able to fully clarify the relationship between repetitive work of the upper extremity and musculoskeletal disorders. Thus more research is needed both on epidemiological exposure–effect relationships and pathophysiological mechanisms. The following list of urgent research topics can serve as a starting point, but will probably need to be further expanded.

Epidemiological studies
– Exposure–effect relationships, using a larger range of cycle time categories than previously studied, for tendon and muscle disorders separately.

- Use of more detailed exposure measurements, especially frequency of movements of shoulder, wrist and fingers, and force and speed of movements, in order to study interaction between risk factors.
- Incorporation of exposure duration (per day and number of years) in exposure measurements.
- Application of experimentally deduced models for repetitive static contractions in epidemiological studies.
- Development of more valid measures for assessment of outcome.

Experimental studies
- Pathophysiological mechanisms of tendon, muscle and nerve disorders in repetitive exercise.
- Identification of optimal work-rest regimes for tendons.
- Modelling of experimental studies by simple industrial tasks.
- Use of a wider range of criteria for fatigue and acceptability.

References

American College of Orthopaedic Surgeons, 1965. Joint Motion. Churchill Livingstone, Edinburgh.

Amano, M., Umeda, G., Nakajima, H. and Yatsuki, K., 1988. Characteristics of work actions of shoe manufacturing assembly line workers and a cross-sectional factor-control study on occupational cervicobrachial disorders. Japan Journal of Industrial Health, 30: 3–12.

Armstrong, T., Buckle, P., Fine, L., Hagberg, M., Jonsson, B., Kilbom, Å., Kuorinka, I., Silverstein, B., Sjøgaard, G. and Viikari-Juntura, E., 1993. A conceptual model for work-related neck and upper-limb musculoskeletal disorders. Scandinavian Journal of Work, Environment and Health, 19: 73–84.

Armstrong, T., Fine, L.J., Goldstein, S.A., Lifshitz, Y.R. and Silverstein, B.A., 1987. Ergonomics considerations in hand and wrist tendinitis. The Journal of Hand Surgery, 12A(5/2): 830–837.

Barnhart, S., Demers, P., Miller, M., Longstreth, W. and Rosenstock, L., 1991. Carpal tunnel syndrome among ski manufacturing workers. Scandinavian Journal of Work, Environment and Health, 17: 46–52.

Bellemare, F. and Grassino, A., 1982. Effect of pressure and timing of contraction on human diaphragm fatigue. Journal of Applied Physiology, 53: 1190–1195.

Bishu, R.R., Manjunath, S. and Hallbeck, M., 1990. A fatigue mechanics approach to cumulative trauma disorders. In: B. Das (Ed.), Proceedings of the Advances in Industrial Ergonomics and Safety II, Taylor & Francis, London, pp. 215–222.

Bjelle, A., Hagberg, M. and Michaelsson, G., 1979. Clinical and ergonomic factors in prolonged shoulder pain among industrial workers. Scandinavian Journal of Work, Environment and Health, 5: 205–210.

Bjelle, A., Hagberg, M. and Michaelson, G., 1981. Occupational and individual factors in acute shoulder-neck disorders among industrial workers. British Journal of Industrial Medicine, 38: 356–363.

Björkstén, M. and Jonsson, B., 1977. Endurance limit of force in long-term intermittent static contraction. Scandinavian Journal of Work, Environment and Health, 3: 23–37.

Bongers, P.M., de Winter, C.R., Kompier, M.A.J. and Hildebrandt, V.H., 1993. Psychosocial factors at work and musculoskeletal disease. Scandinavian Journal of Work, Environment and Health, 19: 297–312.

Byström, S., 1991. Physiological response and acceptability of isometric intermittent handgrip contractions. Thesis, Arbete och Hälsa, no. 38, National Institute of Occupational Health, Solna.

Byström, S., Mathiassen, S. and Fransson-Hall, C., 1991. Physiological effects of micropauses in isometric handgrip exercise. European Journal of Applied Physiology, 63: 405–411.

Chiang, H.-C., Ko, Y.-C., Chen, S.-S., Yu, H.-S., Wu, T.-N. and Chang, P.-Y., 1993. Prevalence of shoulder and upper-limb disorders among workers in the fish-processing industry. Scandinavian Journal of Work, Environment and Health, 19: 126–131.

Corbucci, C., Montanari, G., Cooper, M., Jones, D. and Edwards, R., 1984. The effect of exertion on mitochondrial oxidative capacity and on some antioxidant mechanisms in muscle from marathon runners. International Journal of Sports Medicine, 5: 135–140.

Dahalan, J.B. and Fernandez, J.E., 1993. Psychophysical frequency for a gripping task. International Journal of Industrial Ergonomics, 12: 219–230.

Dennett, X. and Fry, H., 1988. Overuse syndrome: A muscle biopsy study. The Lancet, April 23:905–908.

Drury, C., 1987. A biomechanical evaluation of the repetitive motion injury potential of industrial jobs. Seminars in Occupational Medicine, 2: 41–49.

Dul, J., Douwes, M. and Smitt, P., 1991. A Work-Rest-Model for Static Postures. In: Y. Quéinnec and F. Daniellou (Eds.), Proceedings of the The 11th Congress of the IEA. Taylor & Francis, London, pp. 93–95.

Ebara, S., Harada, T. and Yamazaki, Y., 1989. Unstable cervical spine in athetoid cerebral palsy. Spine, 14: 1154–1189.

Ebara, S., Yamazaki, Y. and Harada, T., 1990. Motion analysis of the cervical spine in athetoid cerebral palsy. Spine, 15: 1097–1103.

Elert, J., 1991. The pattern of activation and relaxation during

fatiguing isokinetic contractions in subjects with and without muscle pain. Thesis, University of Umeå.
Feldman, R., Hyland Travers, P., Chirico-Post, J. and Keyserling, W.M., 1987. Risk assessment in electronic assembly workers: Carpal tunnel syndrome. Journal of Hand Surgery, 12 A: 849-855.
Ferguson, D., 1971. An Australian study of telegraphists' cramp. British Journal of Industrial Medicine, 28: 280-285.
Feuerstein, M. and Fitzgerald, T., 1992. Biomechanical factors affecting upper extremity cumulative trauma disorders in sign language interpreters. Journal of Occupational Medicine, 34: 257-264.
Fisher, D.L., Andres, R.O., Airth, D. and Smith, S.S., 1993. Repetitive motion disorders: The design of optimal rate-rest profiles. Human Factors, 35: 283-304.
Freivalds, A. and Goldberg, J.H., 1988. A methodology for assigning variable relaxation allowances: Manual work and environmental conditions. In: F. Aghazadeh (Ed.), Trends in Ergonomics/Human Factors V. Elsevier Science, pp. 457-464.
Goldstein, S., Armstrong, T., Chaffin, D. and Matthews, L., 1987. Analysis of cumulative strain in tendons and tendon sheaths. Journal of Biomechanics, 20: 1-6.
Hagberg, M., 1981. Work load and fatigue in repetitive arm elevations. Ergonomics, 24 (7): 543-555.
Herberts, P. and Kadefors, R., 1976. A study of painful shoulders in welders. Acta Orthopaedica Scandinavica, 47: 381-387.
Herberts, P., Kadefors, R., Andersson, G. and Petersén, I., 1981. Shoulder pain in industry: An epidemiological study on welders. Acta Orthopaedica Scandinavica, 52: 299-306.
Herberts, P., Kadefors, R., Högfors, C. and Sigholm, G., 1984. Shoulder pain and heavy manual labor. Clinical Orthopaedics and Related Research, 191: 166-178.
Howard, N., 1937. Peritendinitis crepitans. Journal of Bone and Joint Surgery, 19: 447-459.
Howard, N., 1938. A new concept of tenosynovitis and the pathology of physiologic effort. American Journal of Surgery, 42: 723-730.
Hägg, G., 1991. Static work loads and occupational myalgia – A new explanation model. In: P. Anderson, D. Hobart and J. Danoff (Eds.), Electromyographical Kinesiology. Elsevier Science, New York, pp. 141-144.
Jensen, C., Hansen, K. and Westgaard, R., 1993. Trapezius muscle load as a risk indicator for occupational shoulder-neck complaints. International Archives for Environmental Health, 64: 415-423.
Johansson, H. and Sojka, P., 1991. Pathophysiological mechanisms involved in genesis and spread of muscular tension in occupational muscle pain and in chronic musculoskeletal pain syndromes: A hypothesis. Medical Hypothesis, 35: 196-203.
Järvholm, U., Styf, J., Suurkula, M. and Herberts, P., 1988. Intramuscular pressure and muscle blood flow in supraspinatus. European Journal of Applied Physiology, 58: 219-224.

Kanawaty, G., 1992. Introduction to work study. International Labour Organisation, Geneva.
Karasek, R. and Theorell, T., 1990. Healthy Work. Stress, Productivity and the Reconstruction of Working Life. Basic Books, New York.
Kennedy, C., Hawkins, R. and Krossoff, W., 1978. Orthopaedic manifestations of swimming. American Journal of Sports Medicine, 6: 309-322.
Kilbom, Å., Mäkäräinen, M., Sperling, L., Kadefors, R. and Liedberg, L., 1993. Tool design, user characteristics and performance: A case study on plate-shears. Applied Ergonomics, 24(3): 221-230.
Kim, C.-H. and Fernandez, J.E., 1993. Psychophysical frequency for a drilling task. International Journal of Industrial Ergonomics, 12: 209-218.
Konz, S., 1990. Work Design: Industrial Ergonomics. Publishing Horizons, Worthington.
Kuorinka, I., 1981. Work movements in semi-paced tasks. In: G. Salvendy and T. Smith (Eds.), Machine pacing and occupational stress. Taylor & Francis, London, pp. 143-149.
Kuorinka, I. and Koskinen, P., 1979. Occupational rheumatic diseases and upper limb strain in manual jobs in a light mechanical industry. Scandinavian Journal of Work, Environment and Health, 5 (suppl. 3): 39-47.
Kurppa, K., Waris, P. and Rokkanen, P., 1979. Peritendinitis and tenosynovitis. Scandinavian Journal of Work, Environment and Health, 5 (suppl. 3): 19-24.
Lampier, T., Crooker, C. and Crooker, J., 1965. De Quervain's disease. Ind. Medicine Surgery, 34: 847-856.
Loslever, P. and Ranaivosoa, A., 1993. Biomechanical and epidemiological investigation of carpal tunnel syndrome at workplaces with high risk factors. Ergonomics, 36: 537-554.
Luczak, H., 1983. Ermüdung. In: W. Rohmert and J. Rutenfranz (Eds.), Praktische Arbeitsphysiologie. Georg Thieme, Stuttgart, pp. 71-85.
Lundberg, U., Granqvist, M., Hansson, T., Magnusson, M. and Wallin, L., 1989. Psychological and physiological stress responses during repetitive work at an assembly line. Work & Stress, 3: 143-153.
Luopajärvi, T., Kuorinka, I., Virolainen, M. and Holmberg, M., 1979. Prevalence of tenosynovitis and other injuries of the upper extremities in repetitive work. Scandinavian Journal of Work, Environment and Health, 5 (suppl. 3): 48-55.
Marley, R. and Fernandez, J., 1991. A psychophysical approach to establish maximum acceptable frequency for hand/wrist work. In: W. Karwowski and J. Yates (Eds.), Proceedings of the International safety and industrial ergonomics conference. Taylor & Francis, London, pp. 75-82.
Marras, W. and Schoenmarklin, R., 1993. Wrist motions in industry. Ergonomics, 36: 341-351.
Mathiassen, S.E., 1993. Variations in shoulder-neck activity. Thesis, Arbete och Hälsa, no. 7, National Institute of Occupational Health, Solna.

Mathiassen, S.E. and Winkel, J., 1991. Quantifying variation in physical load using exposure-vs.-time data. Ergonomics, 34(12): 1455–1468.
McCord, J., 1985. Oxygen-derived free radicals in postischemic tissue injury. New England Journal of Medicine, 312: 159–163.
Milerad, E. and Ericson, M., 1994. Effects of precision and force demands, grip diameter, and arm support during manual work: An electromyographic study. Ergonomics, 37: 255–264.
Milner, N., 1985. Modelling fatigue and recovery in static postural exercise. Thesis, University of Nottingham.
Moore, A.E. and Wells, R., 1992. Towards a definition of repetitiveness in manual work. In: M. Mattila and W. Karwowski (Eds.), Computer Applications in Ergonomics, Occupational Safety and Health. Elsevier Science B.V., Amsterdam, pp. 401–408.
Moore, J.S. and Garg, A., 1993. A job analysis method for predicting risk of upper extremity disorders at work: preliminary results. In: R. Nielsen and K. Jörgensen (Eds.), Advances in Industrial Ergonomics and Safety V. Taylor & Francis, London, pp. 163–169.
Müller, E.A., 1935. Der Einfluß von Arbeitsgröße, Pausenlänge und Pausenverteilung auf die Ermüdung bei statischer Haltearbeit. Arbeitsphysiologie, 8: 435–445.
Nathan, P., Meadows, K. and Doyle, L., 1988. Occupation as a risk factor for impaired sensory conduction of the median nerve at the carpal tunnel. Journal of Hand Surgery, 13: 167–170.
NIOSH (National Institute for Occupational Safety and Health), 1992. HETA 89-299-2230, US West Communications Phoenix, Aroizona, Minneapolis Minnesota, Denver Colorado, Health Hazard Evaluation report. US Department of Health and Human Services, Public Health Service.
Obolenskaja, A. and Goljanitzki, J., 1927. Die seröse Tendovaginitis in der Klinik und im Experiment. Deutsche Zeitschrift Chirurgie, 201: 388–399.
Ohara, H., Nakagiri, S., Itani, T., Wake, K. and Aoyama, H., 1976. Occupational health hazards resulting from elevated work rate situations. J. Human Ergology, 5: 173–182.
Onishi, N., Nomura, H., Sakai, K., Yamamoto, T., Hirayama, K. and Itani, T., 1976. Shoulder muscle tenderness and physical features of female industrial workers. Journal of Human Ergology, 5: 87–102.
Phalen, G., 1966. The carpal tunnel syndrome, seventeen years' experience in diagnosis and treatment of six hundred fifty-four hands. Journal of Bone and Joint Surgery of America, 48: 211–228.
Pottier, M., Lille, F., Phuon, M. and Monod, H., 1969. Étude de la contraction statique intermittente. Travail Humain, 32: 271–284.
Price, A.D.F., 1990. Calculating relaxation allowances for construction operatives – Part 2: Local Muscle Fatigue. Applied Ergonomics, 21 (4): 318–324.
Putz-Anderson, V., 1988. Cumulative Trauma Disorders: A Manual for Musculoskeletal Diseases of the Upper Limbs. Taylor & Francis, London. New York, Philadelphia.
Radwin, R.G. and Lin, M.L., 1993. An analytical method for characterizing repetitive motion and postural stress using spectral analysis. Ergonomics, 36: 379–389.
Ranney, D., Wells, R. and Moore, A., 1992. Forearm muscle strains: The forgotten work-related musculoskeletal disorders. In: M. Hagberg and Å. Kilbom (Eds.), Proceedings of PREMUS 92. National Institute of Occupational Health, Stockholm, pp. 240–241.
Rodgers, S., 1986. Repetitive work. In: S. Rodgers (Ed.), Ergonomic Design for People at Work. Van Nostrand Reinhold, New York, pp. 246–258.
Rohmert, W., 1973a. Problems in determining rest allowances, Part 1: Use of modern methods to evaluate stress and strain in static muscular work. Applied Ergonomics, 4(2): 91–95.
Rohmert, W., 1973b. Problems of determining rest allowances. Part 2: Determining rest allowances in different human tasks. Applied Ergonomics, 4: 158–162.
Sahlin, K., Cizinsky, S., Warholm, M. and Högberg, J., 1992. Repetitive static muscle contractions in humans – A trigger of metabolic and oxidative stress? European Journal of Applied Physiology, 64: 228–236.
Serratos-Perez, J. and Haslegrave, C., 1992. Modelling fatigue and recovery in working postures. In: E. Lovesey (Ed.), Contemporary Ergonomics. Taylor & Francis, London, pp. 66–71.
Sheeley, G.A., Hallbeck, M.S., Bishu, R.R., 1991. Wrist fatigue in flexion and extension. In: W. Karwowski and J.W. Yates (Eds.): Proceedings of the International Safety and Industrial Ergonomics Conference. Taylor & Francis, London, pp. 55–60.
Silverstein, B.A., Fine, L.J., and Armstrong, T.J., 1986. Hand wrist cumulative trauma disorders in industry. British Journal of Industrial Medicine, 43: 779–784.
Silverstein, B., Fine, L.J. and Armstrong, T.J., 1987a. Occupational Factors and Carpal Tunnel Syndrome. American Journal of Industrial Medicine, 11: 343–358.
Silverstein, B., Fine, L. and Stetson, D., 1987b. Hand-wrist disorders among investment casting plant workers. The Journal of Hand Surgery, 12A (5/2): 838–844.
Snook, S.H., Vaillancourt, D.R., Ciriello, V.M., Webster, B.S., 1992. Psychophysical studies of repetitive wrist motion. Report from Liberty Mutual Insurance Company, Hopkinton, Massachusetts.
Stetson, D.S., Silverstein, B.A., Keyserling, W.M., Wolfe, R.A., and Albers, J.W., 1993. Median sensory distal amplitude and latency: Comparisons between nonexposed managerial/professional employees and industrial workers. American Journal of Industrial Medicine, 24: 175–189.
Styv, J.R. and Körner, L.M., 1987. Diagnosis of chronic anterior compartment syndrome in the lower leg. Acta Ortophaedica Scandinavica, 58: 139–144.
Styv, J., Körner, L. and Suurküla, M., 1987. Intra-muscular pressure and muscle blood flow during exercise in chronic

compartment syndrome. Bone and Joint Surgery (Br), 69-B(2): 301–305.

Sundelin, G., 1992. Electromyography of shoulder muscles – The effect of drafts and repetitive work cycles. Thesis, Arbete och Hälsa, no. 16, National Institute of Occupational Health, Umeå.

Tanaka, S. and McGlothlin, J.D., 1989. A conceptual model to assess musculoskeletal stress of manual work for establishment of quantitative guidelines to prevent hand and wrist cumulative trauma disorders (CTDs). In: A. Mital (Ed.), Advances in Industrial Ergonomics and Safety. Taylor & Francis, London, pp. 419–426.

Tichauer, E.R. and Gage, H., 1977. Ergonomics principles basic to hand tool design. American Industrial Hygiene Association Journal, 38: 622–634.

Veiersted, B., Westgaard, R.H. and Andersen, P., 1993. Electromyographic evaluation of muscular load pattern as a predictor of trapezius myalgia. Scandinavian Journal of Work, Environment and Health, 19: 284–290.

Vihma, T., Nurminen, M. and Mutanen, P., 1982. Sewing-machine operators' work and musculo-skeletal complaints. Ergonomics, 25(4): 295–298.

Weber, A., Fussler, C., O'Hanlon, J., Gierer, R. and Grandjean, E., 1980. Psychophysiological effects of repetitive tasks. Ergonomics, 23: 1033–1046.

Wells, R., Moore, A. and Ranney, D., 1989. Evaluation of hand intensive tasks using biomechanical internal load factors. In: N. Corlett, J. Wilson and I. Manenica (Eds.), Proceedings of the The 3rd Occupational Ergonomics Symposium; Work Design in Practice. Taylor & Francis, London, pp. 67–73.

Wiktorin, C., Karlqvist, L., Nygård, C.-H., Winkel, J. and the Stockholm-MUSIC 1 Study Group, 1991. Design and reliability of a questionnaire for estimation of physical load in epidemiologic studies. In: Y. Quéinnec and F. Daniellou (Eds.), Proceedings of the International Ergonomics Association Conference. Taylor & Francis, pp. 230–232.

Williams, H., 1973. Developing a table of relaxation allowances. Industrial Engineering, 5: 18–22.

Winkel, J. and Westgaard, R., 1992. Occupational and individual risk factors for shoulder-neck complaints: Part I – Guidelines for the practitioner, Part II – The scientific basis (literature review) for the guide. International Journal of Industrial Ergonomics, 10: 79–104.

Zandin, K.B., 1990. MOST work measurement systems. Marcel Dekker, New York.

Åberg, U., Elgstrand, K., Magnus, P., Lindholm, A., 1968. Analysis of the components and prediction of energy expenditure in manual tasks. International Journal of Production Research, 6: 189–196.

The reduction of slip and fall injuries: Part I – Guidelines for the practitioner *

Tom B. Leamon

Liberty Mutual Research Center, 71 Frankland Road, Hopkinton, MA 01748, USA

A. The practitioner

Slipping, and the injuries which result are so ubiquitous that almost everyone responsible for building premises might be regarded as a practitioner in reducing slipping accidents. This ranges from householders (slipping accidents are a major cause of injury in the home), but it is particularly significant for those in businesses which involve inviting the public onto the premises and those responsible for maintaining public buildings and spaces. The phenomenon of slipping is very complex and typically only part of the environment is under the control of the practitioner. For example, it is not possible to require all supermarket customers to wear a certain type of footwear. Consequently, the guidelines are meant to address those issues which are controllable by the practitioner and which are likely to affect the incidence of slips.

B. Context

These guidelines may be used in any area where people walk. As indicated in the body of the report, certain activities such as pushing, pulling or load carrying while walking, are subsets of this activity. However, the approach followed by the paper emphasizes the problems caused at the heel during foot touchdown. It is believed that following the guidelines based upon this consideration will not conflict with improving the safety of other specific tasks. That is, improvements to the walking/slipping environment are likely to benefit the pushing and pulling environments.

C. Identify the problem

As with many approaches to accident prevention, determining the scope of the problem will facilitate the allocation of the appropriate level of resources to tackling it.

D. Data collection

Accident reporting forms should highlight the possible, contributory role of a slip or a trip in other accidents. For example, accidents recorded

---

Correspondence to: T.B. Leamon, Liberty Mutual Research Center, 71 Frankland Road, Hopkinton, MA 01748, USA.

* The recommendations provided in this guide are based on numerous published and unpublished scientific studies and are intended to enhance worker safety and productivity. These recommendations are neither intended to replace existing standards, if any, nor should be treated as standards. Furthermore, this document should not be construed to represent institutional policy.

The following individuals participated in the discussion of the earlier version of this guide. Their suggestions (written or verbal) were incorporated by the authors in this version: Roland Andersson, *Sweden*; Alvah Bittner, *USA*; Peter Buckle, *UK*; Jan Dul, *The Netherlands*; Bahador Ghahramani, *USA*; Juhani Ilmarinen, *Finland*; Sheik Imrhan, *USA*; Asa Kilbom, *Sweden*; Stephan Konz, *USA*; Shrawan Kumar, *Canada*; Mark Lehto, *USA*; William Marras, *USA*; Barbara McPhee, *Australia*; James Miller, *USA*; Anil Mital, *USA*; Don Morelli, *USA*; Maurice Oxenburgh, *Australia*; Jerry Purswell, *USA*; Jorma Saari, *Finland*; W. (Tom) Singleton, *UK*; Juhani Smolander, *Finland*; Terry Stobbe, *USA*; Rolf Westgaard, *Norway*; Jørgen Winkel, *Sweden*. The guide was also reviewed in depth by several anonymous reviewers.

as manual handling, moving equipment and 'struck by' type of accidents may all have resulted from, or have been exacerbated, by a slip. An injury received from contacting a hard sharp surface, such as a shelf, a projection or a moving component, may be more serious than the floor-impact injury involved in a fall. In Sweden, 37% of non-falling accidents involved a slip (Andersson and Lagerlof, 1983). It is not surprising that an accident report of an incident involving amputation of a hand in a moving blade may not record the spilt sand in front of the machine which resulted in the original slip and the loss of balance. It is necessary that specific reporting requirements be included in every accident report to determine the possible role of slipping and falling. Factors would include light levels on the floor, footwear, load carrying, turning, change of level, etc. A traditional managerial response to such a proposition is that should this level of detail be required for every factor, then the recording systems would collapse. The significance of slipping in the total universe of accidental deaths and injuries is the most powerful factor for discounting such an argument. Simply put, the number of slips warrant special attention during any accident reporting system.

Of increasing concern is slipping and falling injuries of delivery drivers, who are often involved in highly unstructured situations, particularly in the unloading operation, and in variable weather and other environmental factors. The difficulties of determining the true reasons for such incidents were identified by Nicholson and David (1985). It is likely that a pro-active approach, based upon a critical incidence procedure, is the only solution to this problem. The approach involved would be the regular completion of questionnaires, or structured interviews, by significant numbers of drivers during various days of work.

F. Solutions

The United States Code of Federal Regulations (1989) addresses innumerable aspects of occupational exposure to hazards but does not include slipping. The National Safety Council Safety Manual addresses slips and trips without establishing standards or guidelines of good practice. Similarly, building codes throughout the nation are largely silent on antislip surfaces and, where synonymous terms are used, there is usually no attempt to define what is meant by antislip. In the legal area, a measurement, using the ASTM (1975) method, on the James machine greater than 0.5 is accepted as a non-slip surface: the significance of this to slipping and falling is unclear.

(i) Walkthroughs. Housekeeping walkthroughs are a traditional form of industrial safety practice. Visual observation of contaminating spills, dust, such as flour dust which produces highly slippery conditions, loose components, etc. should also be routinely recorded. Projecting components, discarded tools, etc. pose an obvious tripping hazard and should be routinely evaluated by supervision. It is unfortunate, however, that frequently such attention will be discounted by managers as a form of nit-picking or mere justification of the safety officers' responsibilities and are frequently down played in relationship to 'important' considerations such as electrical hazards, poisonings, and 'caught between' type of injuries. The inappropriateness of such concerns with what might be called 'apparent risk', as opposed to real risk, are revealed by the annual statistics. In all industries, 17% of injuries involved falls, which is almost 50% greater than the sum of 'caught between', 'contact with radiation', 'caustics, etc.', 'rubbed or abraded', and 'temperature extremes'. Similarly, in industrial fatalities, falls account for 12.5% and this is greater than the total of 'electrical current', 'fires', 'burns', and 'poisons of all types'. Consequently, such audits should be increased and the significance of slips and falls raised for production management. In some cases, a specific, floor care focus would have considerable benefits; both in preventing accidents and avoiding litigation. Water spills, for example, by drinking fountains, refrigerator cabinets and especially from rain water tracked in on feet, can create a major hazard by either physical interaction with the floor polish, causing softening, or by modifying the coefficients of friction of the shoe material.

(ii) Janitorial audit. Auditing the janitorial process of floor finishing is equally significant and should be introduced. The finish and polish in-

dustry have spent considerable time in developing sophisticated polishes which, by and large, are known to reduce, rather than increase, the coefficient of friction on a wide range of rigid surfaces. These sophisticated surface finishes are applied and maintained by a grade of labor which is frequently among the least educated and supervised of any within an organization. The practices in use should be one subject of such an audit. In particular, many floor polishes require a significant time to harden prior to pedestrian traffic in order to maintain true slip resistant potential. A janitorial staff member, attempting to remove the obstruction of floor cleaning procedures to enhance pedestrian flow, creates a severe risk. Conversely, janitors who attempt to disrupt busy traffic flows in order to allow for full drying time will be subject to considerable pressure, especially when the floor may look dry. It is clearly an organizational responsibility to support janitors in the correct procedures for using these finishes, and the scheduling of floor polishing operations should reflect this. A further janitorial practice, which should be monitored, is that of using oil-treated mops to remove dust. Excessive use of such oils can result in softening of many floor surfaces with a concomitant increase in slipperiness. Polish material selection involves the consideration of solvents in certain polishes, which may have undesirable effects. In particular, solvent based polishes are likely to attack and soften bituminous tile floors.

Wax build-up has been identified in several legal actions as a cause of slips but there is little evidence of this as a significant factor in falling. The very low film thicknesses involved, usually less than half of one thousandth of an inch (0.0125 mm) indicate the 'build-up', even if visible, is not significant. Furthermore, it is believed that the surface finish of the final coat has the same surface properties as previously applied coats. There is some ambiguity whether the high speed buffing of polished surfaces destroys the slip resistant properties of those surfaces. The chemical manufacturers do not appear to think so, although the empirical position established by litigation has identified this as a cause of negligence (Steinle, 1961).

(iii) Job analysis. As part of the prevention of slips and falls, any task analysis carried out should involve consideration of slips and falls. Many of the remarks in this paper have addressed normal, pedestrian locomotion, for example, in identifying the significance of the heel in slides. However, in certain slipping and falling accidents, such as those involved in pushing loads, the design of the sole material, the sole treads and floor slip treatments may become significant features. This task analysis should reveal when excessive frictional requirements at the feet occur, and hence when solutions should be sought. For example, by providing a hip bar, which would allow an operator to safely reach over and grasp a load, in order to pull it forward, without relying on the foot:floor reaction. Similarly, task analysis should identify where strongly mounted hand rails and poles will avoid reliance upon foot:floor friction to provide the reaction for the task.

(iv) Visual environment. The significance of visually recognizing slippery surfaces has been established and this certainly requires appropriate light levels. In particular, the analysis should concentrate on changes of illumination, including changes of floor coloring, which might conceal a change in coefficients of friction between adjacent surfaces. Possibly, one of the most dangerous situations is an unanticipated change in friction which would prevent anticipatory changes in gait. This same rationale applies to the audit of floor care programs, waxed and polished areas should extend throughout the area wherever possible, and at the very least, should extend to natural breaks, thus avoiding 'hidden' discontinuities in the frictional properties of the walking surfaces.

(v) Floor surface specification. Surface material specification will continue to be required and it will probably be wise to maintain the UL criterion of a static coefficient of friction measured, on the James machine, greater than 0.5. However, this should be seen to be part of the overall strategy as already outlined, for there is conflicting evidence on the level of protection provided by such a measurement in normal locomotion and activities. Its main advantage might be as a legal defense, rather than as a particularly useful tool in producing safer work places. It is sobering for practitioners, who are seeking to improve their performance with regard to slipping and falling accidents, to find that the 1989 publication of the

Chemical Specialties Manufacturers, Inc. on 'Waxes, Polishes, and Floor Finishes Test Methods' recognizes, in a statement quoted by Steinle (1961): "it is now generally accepted by those engaged in this study that machine measurements of the coefficient of friction cannot correlate in all cases with foot tests on the floor or with safety in use". Furthermore, ... "there are presently no standards of floor safety that can be expressed in terms of accident frequency, coefficient of friction, or subjective foot tests in the field".

This is the challenge in 1992 to both the laboratory researcher and to the practitioner.

Reference

Leamon, T.B., 1992. The reduction of slip and fall injuries: Part II – The scientific basis (knowledge base) for the guide. International Journal of Industrial Ergonomics, 10: 29–34 (this issue).

The reduction of slip and fall injuries: Part II – The scientific basis (knowledge base) for the guide

Tom B. Leamon
Liberty Mutual Research Center, 71 Frankland Road, Hopkinton, MA 01748, USA

1. The problem

Falling on the same level may result from slipping or tripping. Tripping occurs where the foot, usually the leading foot, is arrested by an obstruction, preventing the center of gravity moving smoothly past the expected foot location point. Slipping arises when the coefficient of friction between the shoe material and walkway surface provides insufficient resistance to counteract the forward, resultant forces at the point of contact. Slips can occur at either push-off (the rear foot) as the force necessary to propel the center of gravity forward is generated, or at touch-down (the leading foot) as the vertical component of the weight is transferred. Of these two types of slip, the rearward slip of the push-off foot is usually the least serious. Such a slip would normally be expected at a point in the walking cycle at which the angle from the hip to the toe of the push-off leg is a maximum and the vertical force a minimum. Thus, the weight of the subject has already begun to be transferred to the leading foot during the touchdown cycle and, even if the rear foot does break away, the leading foot is already chiefly responsible for maintaining a stable posture. However, a slip of the leading foot at touchdown can be extremely hazardous, as this will normally occur when the hip:heel angle subtended at the floor is small: if the heel accelerates to a velocity larger than the velocity of the center of gravity, then the angle and the horizontal component will continue to increase, leading to a loss of balance.

In such forward slips, it is useful to describe three categories based upon slip length. A *microslip* is a slip shorter than approximately 2 cms, a *slip* is normally as long as 8–10 cms, and a *slide* was coined to describe the uncontrolled movement of the heel, which would typically arise if the slip length exceeds about 10 cms (Leamon and Son, 1989). These categories are significant to the study of human locomotion, a *microslip* is not normally perceived by the human operator and, moreover, it has been shown that these occur very frequently, perhaps as often as half the steps involve a microslip. A *slip* would normally be perceived by the human subject who will generate postural responses to the sensation at the heel, which may involve jerking the upper body, lunging forward, moving the arms, etc. A *slide* will undoubtedly be recognized by the human subject, as these involve a total loss of control and, in most cases, will lead to a loss of balance and impact with the floor.

Surprisingly, considering the very complex psychomotor activity involved in locomotion and the visual control required to accommodate the walking environment, falls are extremely rare events. Their significance arises from the fact that despite these very low probabilities, the exposure is enormous: for example, as defined by the number of steps taken by inhabitants of the United States in a 24-hour period.

Correspondence to: T.B. Leamon, Liberty Mutual Research Center, 71 Frankland Road, Hopkinton, MA 01748, USA.

2. Scope and significance of the problem

Falling on the same level is a largely ignored national and international disaster. Brungraber (1976) estimated that 8,000,000 falls occur in the home each year. The National Safety Council identified the death rate from falls in the United States as 11,700 during 1989. This figure excludes falls in or while boarding or alighting transport vehicles. Staggeringly, 4,400 people died from falling over in a public place. The lack of a public outcry contrasts with the hypothetical situation expected if one jumbo jet should crash in the continental USA each month for a year. Public accidents in air transport, water transport, railroad transport and other transport amounted to approximately half that of the falling total.

The largest single cause of deaths in the home is falling, which involves over 6,500 deaths. Occupational exposures result in approximately 1,500 deaths and approximately one-third of a million injuries. Comparable figures have been reported in the United Kingdom (Portitt, 1985; Buck and Coleman, 1985) and in Scandinavia. However, slips are frequently not recorded by current accident recording procedures. For example, analysis in Scandinavia of low back injuries, which are normally classified as overexertion injuries in many recording systems, identified a slip or a trip as a contributing factor while load carrying in over 30% of all such cases. Clearly, by analogy slips may also be involved in statistics on scalds, burns, cuts, and injuries from moving machinery.

The significance of slipping and falling is second only to automobile accidents in both social and economic significance. However, in stark contrast with the enormous efforts to improve highway traffic safety by federal, state, private sector and public organizations, slipping is truly perceived to be a pedestrian activity, worthy of little attention.

One significant exception to this is in the legal field of torts, where significant activity does exist and empirical criteria have been, and are being, established by juries with various degrees of relevance to the engineering parameters of slipping and falling. Such concerns have led to continuing activities of bodies, such as the Chemical Specialties Manufacturers Association, Inc. (which is concerned with waxes and floor polishes), Underwriters Laboratories, and many businesses serving the public, such as retail and fast food services.

Unfortunately, the empiricism of the legal system makes its mark here and many cases are settled out of court (as a cost control measure), which significantly closets the problem and inhibits an approach which might ultimately lead to a reduction in the appalling statistics.

3. Background

The coefficient of friction (COF) between the shoe material and the walking surface is clearly of significance to slipping accidents. Friction is classically measured in one of two ways. Firstly, so-called static friction which is the ratio of the horizontal force to the vertical force acting at a point, which is just sufficient to cause the point to move. Dynamic friction is the same ratio measured with a horizontal force large enough to maintain the motion. It is usually expected that coefficient of dynamic friction is lower than that of static friction.

Classical friction considerations are based on centuries of experimentation, from Leonardo onwards, and specify that
(a) frictional force is proportional to the normal force,
(b) it is not affected by the contact area or
(c) by the rate of movement.

Traditionally, much research has been done on the assumption that the heel is stationary and in contact with the floor immediately prior to a slip. This assumption would indicate that the static coefficient-of-friction would be a good index of floor slipperiness and many devices were developed with this assumption (over 60 coefficient of friction measuring devices have been developed, which typically show low correlations with each other) (Strandberg, 1983).

Three basic approaches have been used in the past to measure friction coefficients (Brungraber, 1976). First, pulling an instrumented load across the surface; second, allowing a pendulum to strike the surface under test, and finally, displacing an articulated strut. The drag meter consists of a force measuring device attached to a sample of shoe material under a known vertical load. In principle, this device can be used to measure the peak force which will be normally assumed to be the static friction, or the steady state force required to keep the load moving. Pendulum devices are meant to measure dynamic friction only and consist of a pendulum carrying a shoe material patch and designed to sweep across the floor material, allowing the contact between the two surfaces with a predetermined time dependency. The loss of energy is intended to reflect dynamic friction. The articulated strut involves applying a known vertical force through a rigid member to the shoe material in contact with the floor. The vertical angle is increased until a slip takes place and the tangent of this angle of displacement, giving an indication of static friction. A variation of the latter approach involves dragging the strut across the floor (Hunter machine) until uncontrolled slipping occurs, which was seen as a *disadvantage by* Brungraber in that it measures the dynamic friction. Five currently used devices, illustrative of the various techniques, have been evaluated by Andres and Chaffin (1985) and an example of the need for portable, field measurements and the tendency to develop yet another approach is illustrated by Ballance et al. (1985).

The technique used most widely in the United States is based on the original James machine ASTM (1975). The main purpose of such a machine has been to establish the static COF of various walking surfaces and the test utilizes specially prepared leather and specially prepared floor surfaces in order to achieve repeatability. This is currently the method used by the Underwriter Laboratories, Inc. which uses the criterion that a COF greater than 0.5 indicates a non-slip material. The question of the significance of using clean leather on carefully prepared surfaces to predict slipping and falling has rarely been considered. Indeed, the approach incorporated in the James machine has been justified by regarding the human leg as a simple strut, without attempting to validate the significance of such an assumption (Ekkebus and Killey, 1971, in CSMA, 1989).

More modern research has established that in normal locomotion a slide begins before the heel stops moving (Perkins and Wilson, 1983; and Leamon, 1989). Additionally, synthetic materials do not always behave classically, for example, the vertical force may indeed alter the coefficient of friction (Tisserand, 1985). The predictive value of dry leather testing for slipperiness of contaminated synthetic shoe materials may be very low. The results of tribological evaluations indicate that certain synthetic materials have extremely low slipping rates which blur the difference between the static and dynamic states (Bartenev and Lavrentev, 1961; Tisserand, 1985, and Robinowicz, 1956). The practice of reinforcing leather heels with synthetic materials, at the heel contact point, raises doubts of the validity of static testing as currently performed. Further evidence (Leamon and Son, 1989) indicates that slides usually occur at the heel and typically the subject falls *before* the sole of the shoe comes in contact with the floor material, suggesting that heel design and material must be very carefully considered, and that a three-inch square sample of shoe material may not be representative of human foot/floor contact.

Furthermore, dynamic friction coefficients for synthetic materials have been shown to vary with the load applied and with the slipping velocity (Perkins and Wilson, 1983). Consequently, if such measures are to be predictive of slipping and falling behavior, the measurement parameters must be related to those occurring in human locomotion. If a large number of slips occur at the microslip level, but stop, before progressing to a slide, perhaps the question to be addressed should be how does the human operator stop a microslip from becoming a slip and a slip becoming a slide. This again suggests that studying an appropriate dynamic coefficient of friction would be most useful in determining whether a combination of shoe material, flooring and surface contamination is hazardous or not.

Beyond the tribology of the shoe:floor combination, it is clear that human locomotion is far more complicated than a consideration of, say the foot, with its anatomical structure of 26 bones, 107 ligaments and 19 muscles. There is a number of characteristics of human gait which may be very significant to slipping and falling. For example, load carrying, which has at least two consequences; it results in a shift in the center of gravity, which must be accounted for during locomotion and may alter the frictional demands on the floor, and it increases the vertical load. The question of an increase of the vertical load at the foot is significant. Several researchers have identified the significance of a peak in the ratio F_h/F_v (when F_v is the vertical force, F_h is the horizontal force) measured during the slipping cycle, a few milliseconds following touchdown as indicating the maximum slip potential. This has been termed 'friction use' by Strandberg (1983), and several investigators have followed his original assumption that this is the point at which a slip is most likely to occur. However, this is not necessarily so, if the amount of frictionally based resistance varies with load and velocity during this critical part of the cycle: a more appropriate description might be frictional demand. In other words, the ratio between the available frictional force and the frictional demand would be the actual predictor of slipping potential. To determine this ratio, namely frictional demand over friction available, a more sophisticated measuring technique will be required, capable of determining frictional changes within the half second period of a slip. The characteristics of such a measuring system are likely to be concomitant with the requirements for slip predictive meters and both laboratory and field devices are under development which have the overall aim of closely following the parameters of force, velocity, and timing of the heel during locomotion (Gronqvist et al., 1989, Ertas et al., 1991). An interesting corollary of the Strandberg position would be the prediction that an increase in the vertical force should decrease the chance of a slip as the ratio is decreased. This may appear counterintuitive and at Texas Tech, carrying a load closely strapped to the chest or to the back in a way to minimize restrictions on arm swing, has been shown to increase significantly the length of a slip while walking on a low friction surface (Leamon and Li, 1991). Secondly, a load may interfere with the normal technique of balancing, which involves rotation around the hips and arm swings which are used to maintain the position of the center of gravity over the appropriate foot. Such factors have not been systematically investigated, much less related to frictional requirements between the floor and the foot.

Experiments have indicated that people manipulate gait when walking on slippery surfaces. It is common experience that people can, and do, walk on ice and other very slippery surfaces, but in doing so, typically they reduce the stride length and use a more flat-footed posture in transferring the weight to the leading foot.

Proprioceptive recognition of the slipperiness of the surface is probably automatically monitored during walking to allow these adjustments to be made. Discontinuities in surface slipperiness have been implicated as a significant factor in many slips and these occur when the subject, walking on an accustomed surface unexpectedly encounters a patch with lower frictional characteristics, or (theoretically at least) of higher frictional characteristics. This is the ultimate basis for innumerable law claims typified by customers stepping in a pool of water in the middle of an otherwise non-slippery supermarket floor. It appears likely that such a slide occurs on the initial contact, thus precluding any chance of proprioceptive monitoring and postural adjustment. Hence any possibility of modifications to posture must be dependent on anticipatory, visual evaluation of the surfaces. In common with the performance of all visual skills, the lighting environment, peripheral and foveal vision are clearly potential factors in slipping and falling. Indeed, the safest possible surface might be a surface with a high coefficient of friction but an appearance of slipperiness.

4. Future research

The major thrust of slipping and falling research in the past five years has identified a new direction to that of earlier studies, which chiefly concern laboratory measurement of static coefficients of friction. While many avenues are being taken by researchers throughout the world, a general approach may be discerned. That is a progress from the consideration of classical coefficients of friction to the 'real world' evaluation of slipperiness. One interesting approach involves the selection of parameters for standardization which approximate human locomotion, for example, Bunterngchit (1990). Other researchers are approaching the tribological questions of significance to the specification of floors, such as surface roughness, stiction and adhesion, and film penetration (Jones et al., 1991; Grönquist, 1991). Specific activities generate specific slip potential, for example, load carrying (Bloswick and Love, 1991; Leamon and Li, 1991) and the design of decking on motor vehicles (Miller et al., 1987).

Of clear significance is the need to determine an agreed measuring technique likely to have predictive value for the occurrence of slips and falls. Thus, research findings, based on first principles, need to be evaluated using the incidence of slips and falls. Such a device, or approach, will then unlock research opportunities into many interventions involving materials, tread design, and walking surfaces – in the current situation valid measurements are almost impossible to make.

An extension of such a device should be capable of accommodating actual shoes, in order to establish a forensic analysis capability, which is likely to further validate the intervention research effort. Finally, in order to be truly effective, such devices should be field portable, although it is clear that any such device is unlikely to be as convenient as the current static coefficient devices.

In addition to the foot:floor coupling question, there are three other areas: the built environment, special populations, and falling outcomes. For example, the role of the design of visual environments on slipping and falling in public places must be investigated. The evaluation of assistive devices in special environments such as nursing homes requires specific investigation and the identification of the personal characteristics of 'fallers', if they exist, could provide means to reduce slips and falls. Finally, the extension of current 3D measurement systems widely used in slipping and falling research would allow the development of predictive computer models of a slip and fall and lead to the study of the physics of the impact energy exchange between the faller and the environment.

References

Andersson, R. and Lagerlof, E., 1983. Accident data in the new Swedish information system on occupational injuries. Ergonomics, 26 (1): 33–42.
Andres, R.O. and Chaffin, D.B., 1985. Ergonomic analysis of slip-resistance measurement devices. Ergonomics, 28(7): 1065–1079.
ASTM (American Society for Testing and Materials), 1975. Designation D 2047-75, Standard method of test for static coefficient of friction of polish-waxed floor surfaces as measured by the James machine. American Society for Testing and Materials, Philadelphia, PA, U.S.A.
Ballance, P.E., Morgan, J. and Senior, D., 1985. Operational experience with a portable, friction testing device in university buildings. Ergonomics, 28(7) 1043–1054.
Bartenev, G.M. and Lavrentev, V.V., 1961. The nature of static friction in elastomers. Rubber Chemistry and Technology, 34: 461–465.
Bloswick, D.S. and Love, A.C., 1991. Slip potential during load carrying activities. Slip, Trip, Fall. University of Surrey, Guildford, United Kingdom.
Brungraber, R.J., 1977. A new, portable tester for the evaluation of slip-resistance of walkway surfaces. U.S. Department of Commerce.
Brungraber, R.J., 1976. An overview of floor slip-resistance research with an annotated bibliography. U.S. Department of Commerce, NBS Technical Note 895, January 1976.
Buck, P.C. and Coleman, V.P., 1985. Slipping, tripping and falling accidents: A national picture. Ergonomics, 28(7): 949–958.
Bunterngchit, Y., 1990. Measurement of dynamic friction available between shoes and floors, appropriate to friction demands in walking. Ph.D. Thesis, Center for Safety Science, Faculty of Engineering, University of New South Wales, Sydney, Australia.
Ekkebus, C.F. and Killey, W., 1971. Validity of 0.5 static coefficient of friction (James machine) as a measure of safe walkway surfaces. SMA 67th Mid-Year Meeting, May 1971. Reproduced in: Chemical Specialties Manufacturing Association Waxes, Polishes and Floor Finishes, Test Methods and General Information Manual, 5th Edition; 1989. Washington, DC.
Ertas, A., Mustafa, G. and Leamon, T.B., 1991. Design of an automated, simulated, stepping dynamic coefficient of friction measurement device. Mechatronics (in press).
Grönquist, R., Roine, J., Jarvinen, E. and Korhonen, E., 1989. An apparatus and a method for determining the slip resistance of shoes and floors by simulation of human foot motions. Ergonomics, 32(8): 979–995.
Grönquist, R., 1991. Mechanisms of friction and the assessment of slip resistance of new and used footwear soles on contaminated floors. Slip, Trip, Fall. University of Surrey, Guildford, United Kingdom.
Jones, C., Manning, D.P. and Bruce, M., 1991. Detecting and eliminating slippery footware. Slip, Trip, Fall. University of Surrey, Guildford, United Kingdom.
Leamon, T.B. and Son, D.H., 1990. The natural history of a microslip. In: A. Mital (Ed.), Adv. in Industrial Ergonomics and Safety I. Taylor and Francis, pp. 633–638.
Leamon, T.B. and Li, K.W., 1991. Load carrying and slip length. Human Factors Society, Proceedings of the 35th Annual Meeting, San Francisco, CA.
Miller, J.M., Rhoades, T.P. and Lehto, M.R., 1987. Slip resistance predictions for various metal step materials, shoe soles and contaminant conditions, SAE Technical Paper Series 872288, Truck and Bus Meeting, Dearborn, MI, Nov. 16–19.
Nicholson, A.S. and David, G.C., 1985. Slipping, tripping and

falling accidents to delivery drivers, Ergonomics. 28(7): 977-93.

Perkins, P.J. and Wilson, M.P., 1983, Slip resistance rating of shoes - New development. Ergonomics, 26(1): 73-82.

Portitt, 1985. Slipping, tripping and falling: Familiarity breeds contempt. Ergonomics, 28(7): 947-948.

Robinowicz, E., 1956, Stick and slip. Scientific American, 194: 109-118.

Steinle, J.V., 1961. The technical aspects of alleged negligence from the use of floor wax. CSMA's 47th Mid-Year Meeting, May.

Strandberg, L., 1983. Ergonomics applied to slipping accidents. In: T.O. Kvalseth (Ed.), Ergonomics of Work Station Design. London: Butterworth.

Strandberg, L., 1983. On accident analysis and slip resistance measurement. Ergonomics, 26(1): 11-32.

Tisserand, M., 1985. Progress in the prevention of falls caused by slipping. Ergonomics, 28(7): 1027-1942.

U.S. Governments, Code of Federal Regulations, 29 Port 1910, revised July 1, 1989. U.S. Office of the Federal Register, National Archives and Records Administration.

Job design for the aged with regard to decline in their maximal aerobic capacity: Part I – Guidelines for the practitioner *

Juhani Ilmarinen

Institute of Occupational Health, Laajaniityntie 1, SF-01620 Vantaa, Finland

A. Who is the practitioner?

The audience for these guidelines include the following professional groups: (1) occupational health and safety personnel, (2) foremen and supervisors, and (3) the senior staff in the company, whose duties include the design and development of jobs.

Why this audience? The changes in human physical, mental and social capabilities due to the aging process are an extremely complicated issue because the true biological changes during aging have interactions with incidence of diseases, life style factors and with factors of work. Therefore, basic knowledge of normal and premature aging related to work is very important. The professional group having this basic knowledge is the occupational health service personnel, which includes occupational health physicians, nurses and physiotherapists.

The understanding of the aging process should be combined with knowledge of work systems, work content, and work demands. The specialists in this field can be found among the industrial engineers, occupational safety personnel, and also among the supervisors and foremen in the companies. The design and redesign of jobs cannot be done without their expertise and support. This audience is probably the key audience on factory floor level when age-related redesign activities are going to be undertaken.

The senior workers in the companies represent the third audience, which has a large potential for planning age-related changes in single jobs. Their expertise is based on a long experience in the given job. Their task in job design is to focus on the question how to utilize their personal experience for job design.

Job design for the aged is a cooperation process of the three groups mentioned above. It is possible, however, that the operational group is strengthened by other experts or decision makers. So, the final collaborative group depends on the company. If large investments are needed the role of company leadership is naturally emphasized.

Correspondence to: Juhani Ilmarinen, Institute of Occupational Health, Laajaniityntie 1, SF-01620 Vantaa, Finland

* The recommendations provided in this guide are based on numerous published and unpublished scientific studies and are intended to enhance worker safety and productivity. These recommendations are neither intended to replace existing standards, if any, nor should be treated as standards. Furthermore, this document should not be construed to represent institutional policy.

The following individuals participated in the discussion of the earlier version of this guide. Their suggestions (written or verbal) were incorporated by the authors in this version: Roland Andersson, *Sweden*; Alvah Bittner, *USA*; Peter Buckle, *UK*; Jan Dul, *The Netherlands*; Bahador Ghahramani, *USA*; Sheik Imrhan, *USA*; Asa Kilbom, *Sweden*; Stephan Konz, *USA*; Shrawan Kumar, *Canada*; Tom Leamon, *USA*; Mark Lehto, *USA*; William Marras, *USA*; Barbara McPhee, *Australia*; James Miller, *USA*; Anil Mital, *USA*; Don Morelli, *USA*; Maurice Oxenburgh, *Australia*; Jerry Purswell, *USA*; Jorma Saari, *Finland*; W. (Tom) Singleton, *UK*; Juhani Smolander, *Finland*; Terry Stobbe, *USA*; Rolf Westgaard, *Norway*; Jørgen Winkel, *Sweden*. The guide was also reviewed in depth by several anonymous reviewers.

B. Aims of the guidelines

The guidelines have the following aims:
(1) to prevent premature aging
(2) to improve work satisfaction
(3) to improve productivity
(4) to reduce early retirement of the aging work population.

C. Context in which the guidelines should be used

In general, the guidelines for the aged workforce should be used as a part of occupational health and safety strategy of the company. Several circumstances can be indicated where these guidelines are urgently needed and should be implemented in daily working: (1) changes in demographic data of the work force, (2) changes in retirement policy, (3) implementation of new technology, (4) changes in the cost/benefit analysis, (5) changes in well-being and in company image.

C.1. Changes in demographic data of the work force

In many companies the work force are on average getting older than ever before due to several reasons (see Part II). The distribution of the employees in different age groups over time can be fairly well predicted. At least three trends can be identified in different companies: (1) the mean age is increasing, but the distribution is fairly normal, (2) the distribution is skewed, so that the number of employees over 45 years of age are the major part of manpower, and (3) the distribution is twofold; the two largest age groups are the 'youngsters' (under 30 years of age) and the 'elders' (over 50 years of age). It is obvious that the changes in age structure can also be strongly job-related. Each trend mentioned above carries different problems which should be solved with specific actions.

C.2. Changes in retirement policy

Several OECD-countries are using early retirement procedures as a part of the work disability pension. Also part-time working, not only before but also after old age pension, has received serious attention. The existing possibilities for early retirement can markedly lower the actual retirement age. On the other hand, available part-time working has made work possible in older age in spite of the decreased work ability. The model used in retirement policy in the company also seriously influences work motivation. Any changes in the retirement system can have an impact on productivity and well-being in the company.

C.3. Implementation of new technology

Changes in the production system on the factory floor or computerizing the office environment produce problems which are different among the older and younger workers. The technical development has improved the content of several traditionally heavy physical jobs. However, in many occupations the physical load is still hazardous due to its changed nature (see Part II). It is obvious that for learning new skills and techniques the older worker needs special support.

C.4. Changes in cost/benefit analysis

The increase in absenteeism, turnover rate, industrial injury and in health insurance premiums decreases the productivity of the company. Investments in job design and, on the other hand, employment fitness and life-style programs are potential ways to solve several economic problems. The changes in the 'health status' of the company is mainly related to the health, work ability and well-being of the older employees and workers in the company. Increasing health cost can be compensated by better support of the aging workers.

C.5. Company image

Worker satisfaction is the basis for improving the company image. Among the actions for improving the company image the aging worker should be specially taken into consideration. A good image includes acknowledgement and appreciation of the older worker. The company can show that the experience gained over time by the older worker is still valid and valuable for the company. It is also obvious, that the collaboration between younger and older people is a sign of fruitful leadership.

The guidelines for the aged in this article can be used in different contexts. However, some restrictions have been made. Instead of covering both physical, mental and social aspects for the aged, these guidelines focus on the aspects of physical capabilities. The main reason for this restriction is that the physical capacity declines much earlier than the other capabilities. After 40–45 years of age the first signs of the aging process can be seen (see Part II). Therefore, the age of 45 years has been taken here as the reference for the aging worker. The physical capabilities can be further divided into cardiorespiratory and musculoskeletal capabilities. The present guidelines focus on the cardiorespiratory problems.

D. Problem identification

The identification of the problem with the aged is based on data collection and analysis. Data is needed from the following sources: (1) age structure, (2) health risks of the work, (3) assessment of work ability and functional capacity, (4) assessment of job demands, (5) identification of the imbalance between work demands and functional capabilities.

D.1. Age structure

The changes in age structure over the last 20 years and including the next 10–15 years should be identified first. The company or branch statistics are needed. The main interest is to find out how the proportion of 45–65-year-old active work population of the entire work population (15–65 years) has changed and is going to change in the near future.

From the statistics the following information should be analysed:
– Average trends of demographic changes in the age structure of the active work population (see Section C)
– Trends of the aging of the active work population by sex and job

– Trends of retirement statistics. Information is needed concerning rates of early retirement, part-time working, old age pension and, work disability pension.

D.2. Health risks of the work

The next step is to identify the risk factors of work which have a negative influence both on the health and on the work ability of the aging worker. Such a list of risk factors has recently been identified (see Part II) and it can be summarized as follows:

(a) Physical demands that are too high:
– static muscular work and use of muscular strength
– lifting and carrying
– sudden peak loads
– repetitive movements
– simultaneously bent and twisted work postures

(b) Stressful and dangerous work environment:
– dirty and wet workplace
– high risk of work accidents
– hot workplaces
– cold workplaces
– changes in temperature during workday

(c) Poorly organized work:
– role conflicts
– supervision and tackling of work
– fear of failure and mistake
– lack of freedom of choice
– time pressure
– lack of influence on own work
– lack of professional development
– lack of acknowledgment and appreciation

The three sources (*a*, *b*, *c*) of risk factors can be identified by using standardised questionnaires or interviews. It is notable that the list given above is actually the key list of factors which should be in the focus of improvements needed for the aging work population. It should be emphasized here that action is needed in all sources of risk factors before effective improvements can be expected. The present guidelines, however, concentrates mostly on the first group of the risk factors, namely 'Physical demands that are too high'.

D.3. Assessment of work ability and functional capacity

(a) Assessment of work ability

Work ability can be assessed by questionnaire and interview techniques. The main interest is to evaluate the self-experienced ability to work. The following items should be included for the assessment of work ability:
(1) Present work ability compared with the lifetime's best
(2) Work ability in relation to physical work demands
(3) Work ability in relation to mental work demands
(4) Number of diagnosed, chronic diseases
(5) Work impairment due to disease
(6) Sickness absence during past year
(7) Prognosis of work ability after 2 years
(8) Psychological resources

Based on these items a work ability index can be constructed. Standard scales and calculation of the index value are available (see Part II).

(b) Assessment of functional capacity

The assessment of physical work capacity should be carried out using standardised methods, which include the bicycle ergometer or treadmill test protocols for measuring the maximal cardiorespiratory capacity. Because the prevalence rates of chronic diseases of 45–65-year-old work population can be higher than expected, special attention should be given to the medical screening and safety aspects of the tests. Submaximal tests are recommended instead of maximal, all-out aerobic tests. The aim of these tests is to predict the maximal oxygen consumption (VO_2max) of the aging worker. It should be noted here, that although the VO_2max is the measure of aerobic work capacity it can be used also as a measure of overall physical work capacity and health. The VO_2max can be expressed as an absolute value in liters of oxygen that can be consumed per minute (VO_2max l/min) or in milliliter of oxygen per kg body weight that can be consumed per minute (VO_2max ml/kg/min).

The assessment of physical work capacity can be extended to the assessment of musculoskeletal capacity. Only the list of strength and endurance measures needed for aging workers is given here:
– Hand grip
– Knee extension
– Elbow flexion
– Trunk extension
– Trunk flexion

Beside using single test values an overall index of musculoskeletal capacity can be constructed.

The aim of the assessment of physical capabilities is to create a many-sided profile of the individual physical resources.

D.4. Assessment of job demands

The aim of the job analysis is to identify the actual demands of the daily routines, including physical, mental and social aspects. A large variety of methods is available. The practitioner can start the assessment of the physical demands of work by (1) assessment of physical activities. For further analysis, like (2) work load measurements based on oxygen consumption, the practitioner can be referred to specialists.

D.4.1. Assessment of physical activities

The physical activities during work can be assessment by using observation technique. It is based either on minute-to-minute analyses over the entire shift or covering the main work tasks. The activities are classified into seven different levels according to the multiples of basal metabolic rate (Edholm scale). The activity classes are the following:
– Lying asleep
– Lying
– Sitting without special movements
– Standing with light activity
– Activity with low intensity
– Activity with moderate intensity
– Activity with high intensity

The aim of the activity assessment is to find out, (1) how long an older person works in different activity levels, using absolute and relative number of minutes of the shift, and (2) how the activity levels are distributed in real time analysis during the shift. It is recommended that the activity analysis be combined with task analysis and with recordings of strain variables (e.g. heart rate recordings). Several occupational tasks have been classified according to the Edholm scale, and this information can be used before the real time analysis at the work place is started (see Part II).

D.4.2. Work load measurements

The practitioner can refer to specialists for measuring the actual work load. Physical work load can be directly measured in dynamic, aerobic activities. The oxygen consumption (VO_2) is a measure of aerobic demands of work. Ambulatory methods are today available (e.g. Morgan Oxylog). These methods are complicated to use and they need a special training. Because of their irreplaceable information, however, the measurement of VO_2 is highly recommended whenever the work includes heavy physical tasks. It should be noted here that a VO_2 level from 1.0 to 1.5 liter per minute at work can already be a relatively heavy physical load for an older person.

It is recommended to measure oxygen consumption during work at least under the following conditions:
- work that includes much use of muscle strength;
- work that includes much lifting and carrying;
- work that includes peak loads lasting several minutes.

Measurements of oxygen consumption should be performed during 5 to 60 minutes continuously, and should be analysed minute by minute. Data should be given as (1) an average, and (2) task specific, and (3) in real time. The aim is to identify the heaviest tasks of job, the time needed to carry out the heaviest tasks, as well as to find out how much the VO_2 is reduced during the breaks available.

D.5. Assessment of strain during work

(a) Heart rate measurements

Heart rate is a valuable measure of physiological strain. The level of strain measured by heart rate per minute is strongly related to the physical work load and to the physical work capacity of the subject. Heart rate measurements can be combined with the activity studies using Edholm scale or with the VO_2 measurements. Several reliable techniques over 8 to 12 hours continuous recordings are available (e.g. Sport tester). The heart rate data should be analysed to identify the following aspects:
- work tasks producing the highest strain;
- number of minutes over 50% of relative aerobic strain (RAS);
- recovery during breaks;
- trends during the 8-hour work shift.

The heart rate can be analysed as a % of maximum heart rate (%HRmax) or % of heart range (%HRR). If HR and VO_2 have been measured simultaneously during work, the HR/VO_2 relationship can be used for predicting the VO_2 for other similar tasks, where the VO_2 measurements have not been carried out.

(b) Relative aerobic strain (RAS)

The practitioner can refer to the specialist for measuring the relative aerobic strain (RAS).

The relative aerobic strain is a measure of oxygen consumption during work related to the maximal oxygen consumption of the same subject measured in the laboratory. For example, if the VO_2 at work was on average 1.0 liter per minute and the VO_2max of the same person was 3.0 liter per minute the RAS at work is 33%. If the VO_2 during the heaviest task was 1.5 liter per minute, the RAS during the heaviest task is 50%. Recommendations have been given concerning the acceptable level of relative aerobic strain. However, new tables are necessary (see Section E).

D.6. Identification of the imbalance between physical work demands and physical work capacity

Based on the information collected by several methods and using only five steps (age structure, health risk, work ability and functional capacity, job demands, strain) the practitioner is in the position to evaluate
- whether there is a balance between work demands and individual capacity, and if not
- what should be done?

In the next section tables needed to solve the problems of physical load for aging individuals are presented.

E. Problem solving

E.1. Principles

In principle, physical work demands should decrease with increasing age because of the natu-

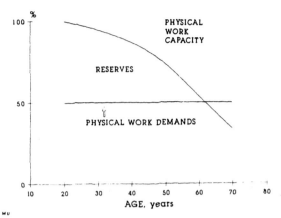

Fig. 1. The relationship between physical work capacity and physical work demands during aging. A normal decline of capacity is not compensated by a respective decrease of work demands. After the age of 50 the reserves available for the recovery from work get smaller if the work demands are unchanged. This model is pushing older people out of working life.

ral decline of the physical work capacity. The amount of decrease of work demands should follow the natural decline of work capacity (figures 1 and 2). Secondly, the premature decline of physical work capacity should be prevented by regular, moderate physical exercise for the cardiovascular and musculoskeletal capacity. Thirdly, among those involved in physical work at an older age (45–65 years) the work should include possi-

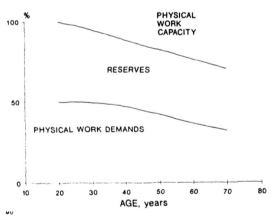

Fig. 2. The optimal relation between physical work capacity and physical work demands with age. The physical work demands decline with age and, the physical work capacity has been maintained in a sufficient level during aging by regular cardiovascular and musculoskeletal exercises. The reserves available for the recovery from work stress remain the same over time.

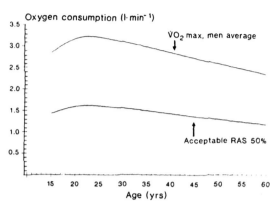

Fig. 3. Age and acceptable oxygen consumption during work for men. The figure indicates the decline of maximal oxygen consumption (VO_2max, liter/min) with age and the 50% RAS curve acceptable for mixed dynamic/static work, or for dynamic work with rest pauses (see text).

bilities to maintain the sufficient capacity for work. Finally, the cardiovascular and musculoskeletal demands of work should be at least 20% less for older women than for older men. The critical question related to these principles is, how the decreased work demands should be compensated.

E.2. *Age and acceptable physical work load*

In the following, the list of figures and tables needed are introduced. Research is still needed to complete the tables. Therefore, the values given are in most cases illustrative in nature.

(a) *Age and acceptable oxygen consumption during work for men*

Figure 3 indicates the normal decline of maximal oxygen consumption (VO_2max) with age. For prolonged work over 8 hours two recommendations can be given:
(1) The RAS should not exceed 50% (RAS 50%), on the presumption that rest pauses are available, and
(2) The RAS should not exceed 33%, on the presumption that rest pauses are not available

On the basis of these presumptions the corresponding curves can be drawn (figure 3). From the figure the highest acceptable oxygen consumption during work can be obtained for any age or age group. It should be noted that the figure is based on average values and there are marked individual differences in VO$_2$max at every age. It is also obvious that the variability of VO$_2$max increases with age. The figure, however, gives the general magnitude of how much the aerobic work demands should decrease with age. On the other hand the figure clearly illustrates the importance of maintaining sufficient VO$_2$max during aging.

The practitioner can use this figure in the following way. First, it should be checked if the VO$_2$max of the worker is situated on the average level according to his/her age. Secondly, the upper limit of acceptable level of VO$_2$ at work can be taken from the figure. Thirdly, the data collected from the work site should be compared to the acceptable level. If VO$_2$ data is available, the comparison is simple; it is recommended that the VO$_2$ of different work tasks be checked. It is possible that the average VO$_2$ at work is at an acceptable level, but the heaviest tasks are not. It should be noted, that the choice of 50% or 33% level of RAS is essential. Therefore the information of the breaks available should be analysed using the Edholm scale (see Section D.4.1, Assessment of physical activities) before the upper acceptable level can be fixed.

(b) Age and acceptable oxygen consumption during work for women

The highest acceptable VO$_2$ during work is systematically lower for women than for men, because of the systematic differences in VO$_2$max between men and women over the whole range of ages (figure 4). The practitioner can use this figure as indicated in (a) above for men.

(c) Age and acceptable heart rate during the work for men and women

Because the collection of the VO$_2$ data during work is a much larger effort than the recording of heart rate, it is necessary to construct tables were the highest acceptable heart rate levels can be related to age, gender and VO$_2$max. Table 1 is one example of the tables needed.

Table 1 illustrates that the same work load (VO$_2$ liter/min) produces different levels of heart rates depending on the fitness (VO$_2$max) of the subjects. For a 50-year-old person, with normal

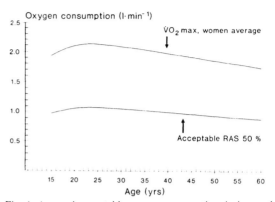

Fig. 4. Age and acceptable oxygen consumption during work for women. The figure indicates the decline of maximal oxygen consumption (VO$_2$max, liter/min) with age and the 50% RAS curve acceptable for mixed dynamic/static, or dynamic work with rest pauses (see text for figure 3).

VO$_2$max decline with age, working at a moderate level of 1.0 liter oxygen/minute will induce a heart rate of 100 beats/minute. An unfit subject of the same age will have a heart rate of 120 beats/minute in the same job. When the work load is higher the differences in heart rate response between fit and unfit subjects will increase. Table 1 indicates that working on the level of 1.5 and 2.0 liter oxygen/minute is not acceptable for unfit persons because of the too high average level of circulatory strain. Working on the level of 2.0 liter oxygen per minute is also too heavy for an average fit person (table 1). It should be noted that the heart rate values given in this table are based on experience from field

Table 1

Acceptable and not acceptable mean heart rates (beats/min) [a] for men in dynamic work related to work load (VO$_2$ l/min) and fitness (VO$_2$max).

VO$_2$max (l/min)	Work load (VO$_2$ l/min)				
	1.00	1.25	1.50	1.75	2.00
high > average	80		100		120
average < average	100		120		(150) [b]
low	120 [c]		(150) [c]		(175) [b]

[a] illustrative values
[b] Not acceptable at all
[c] Not acceptable without regular breaks

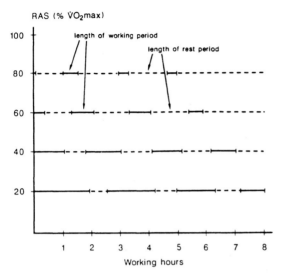

Fig. 5. Work and rest schedules related to the level of relative aerobic strain (RAS) during work. The figure illustrates that the periods of work are shorter and the rest periods are longer with increasing RAS level. Also the length of working periods becomes shorter and the rest periods longer during the working hours. The optimal model of work and rest prevents physiological tiredness due to the work. The periods presented in this figure are arbitrary. Research and practical experiences are needed to construct the optimal work rest schedules for physical work. Adjustments can be needed for age and gender and for the type of physical work.

studies. The values can be modified when new data of field studies are available.

The practitioner has two choices when the upper limit of heart rate is exceeded during work either in average or in specific work tasks:
- the work load should be decreased by cutting the peak loads or by planning more effective rest schedules, and/or
- the fitness of the worker should be improved by aerobic physical exercise

There is an urgent need to produce valid tables for men and women which give the highest acceptable heart rates. These tables should consider the age, gender, work load and fitness (VO_2max) of the subjects.

(d) Work and rest schedules related to relative aerobic strain during work

Figure 5 illustrates how the relative work load and rest schedules should be integrated into the work shift. This schematic example gives some guidelines on how the number and length of the rest pauses should increase with increasing aerobic strain. At a low level of RAS, e.g. 20%, the working periods can be long but the duration of the breaks should increase during the shift. At a level of 40% RAS, which is still within the acceptable level, the work periods should be short and the rest periods long enough to guarantee full recovery from work load.

In practice, the work includes tasks on different levels of RAS. Therefore a combination of different rest periods is needed. The practitioner can use the information of figure 5 to construct an appropriate model for each job.

(e) Work and rest schedules related to the heart rate during work

The needs of rest pauses based on the heart rate recordings should also be worked out as a complement to the RAS in figure 5. It should be noted here that heart rate is the measure of physiological strain which is sensitive also to other stress sources than the physical load. Therefore the heart rate figures and tables are less specific than the VO_2 and RAS guidelines. However, due to the high feasibility of heart rate recordings in all types of occupational tasks it is necessary to construct guidelines for interpretation of the heart rate data.

(f) Adjustments needed for the acceptable physical work load

These guidelines have focused on physiological aspects of physical work carried out by the aging worker (45 +). The information available is mainly based on dynamic muscular work. Therefore adjustments are needed for other types of physical work, which includes:
- static work,
- leg work, and
- arm work.

It is obvious that the health of the aging worker plays an important role when the acceptability of a certain work load is considered. These guidelines presume that the functional capacity is an indirect indicator both of the health status and of the work ability with advanced age. In some cases, however, the tables should be adjusted also to the health status. For example, the acceptable level in dynamic work is different for healthy 55-year-old subject than that for a subject in the same age who has cardiovascular disease, like hypertension. The same can be true when musculoskeletal diseases are present. The reason why health should be taken into consideration is that after the age of 50–55 years more than 50% of the people have at least one diagnosed chronic disease, mostly musculoskeletal or cardiovascular disease. So, much research is still needed before the guidelines for the aged are ready for the everyday praxis.

Reference

Ilmarinen, J., 1992. Job design for the aged with regard to decline in their maximal aerobic capacity: Part II – The scientific basis for the guide. International Journal of Industrial Ergonomics, 10: 65–77 (this issue).

Job design for the aged with regard to decline in their maximal aerobic capacity: Part II – The scientific basis for the guide

Juhani Ilmarinen

Institute of Occupational Health, Laajaniityntie 1, SF-01620 Vantaa, Finland

1. Problem description

Physical work capacity decreases naturally during aging, but the physical demands of work do not (see Part I, figure 1). Consequently, the reserves needed for the recovery, during work or in leisure time, decline with age. After the age of 55, the decreased reserves are not sufficient for recovery from work load during the rest pauses if the physical demands of work are left unchanged. The insufficient possibilities for recovery lead to accumulation of fatique as well as to other symptoms. In the worst scenario, life style habits during leisure time do not support the needs for recovery or the needs for improving the functional capacities.

Aging is also combined with increasing prevalence and incidence rates of chronic diseases. Therefore the load which is acceptable for the young healthy man is not acceptable for a 55-year-old man with musculoskeletal disease in the low back. The interactions of true biological aging, disease, life style and work are strong (figure 1). The role of work in the various interactions is least known. The factors mentioned in figure 1 increase the individual variations with age. Therefore, the problem to be solved is how we can design the job in such a way that enough flexibility and individual aspects can be taken into consideration in daily work.

2. Scope of the problem

Nearly 1/3 of the workers worldwide are over 45 years (Ilmarinen, 1991). Their number today is over 500 million and their annual increase is about 8 million workers. The rate of increase will double during the next decade when the post world war II baby boom reaches the age of 45. The work forces in many OECD countries is also becoming older than ever before. On the average, in 1980 32.0% of the working population (15-64 years) were older than 45. By the year 2000, the proportion of 45+ in the work force will be 35.5% and in 2025 41.3% (OECD 1984).

At the same time, while the proportion of 45+ workers is increasing there is the trend that the economically active population between 55-64 years is decreasing. From 1960 to 1990 the activity rate of the male age group of 55-64 decreased

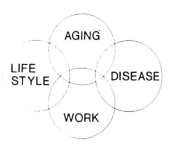

Fig. 1. Interactions of true biological aging, disease, life style and work (see text).

Correspondence to: Juhani Ilmarinen, Institute of Occupational Health, Laajaniityntie 1, SF-01620 Vantaa, Finland

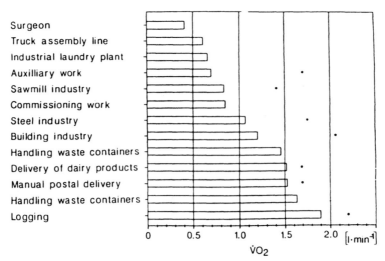

Fig. 2. Mean oxygen consumption (VO$_2$ liter/min) measured in different jobs. The dots indicate the peak value measured during single work tasks.

from 87.2 to 70.6% in the USA, from 81.0 to 61.7% in the FRG and in several countries it was about 50% in 1990 (France, Finland, Austria, Italy). Only in Japan the rate is about the same between 1960 and 1990, 88.6 and 87.2 respectively (ILO, 1986). This trend illustrates that in most countries early retirement is available and that it has been used increasingly.

The physically demanding occupations still cover about 20-25% of the working population in industrialized countries (Rutenfranz et al., 1990). The proportion of the work force doing physical work in non-industrialized countries is presumably much higher. Exact figures are not available because the definition of physical work is not uniform worldwide.

The physical demands during work are in many occupations still moderate or high (figure 2). Studies where oxygen consumption has been measured during work in England, Germany and

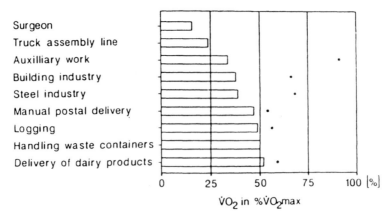

Fig. 3. The mean relative aerobic strain (VO$_2$ at work in % of VO$_2$max) measured in different jobs. The dots are the peak values measured during single work tasks.

Finland indicate that the average level of oxygen consumption (VO$_2$ liter/min) exceeded 1.0 liter/min in the steel and building industry; the level of 1.5 liter/min was exceeded in handling waste containers, in delivery of dairy products, in manual postal delivery and in logging (Ilmarinen, 1989). The peak loads during work often exceeded 1.5 liter oxygen per minute and in building industry and in logging over 2.0 liter oxygen per minute were measured.

It was notable that the relative aerobic strain (RAS) was near 50% RAS in several jobs. The peak loads, however, exceed 60% in building and steel industry. In auxiliary jobs the highest RAS was found: the 90% RAS was explained by the old age and low VO$_2$max of the subject (figure 3).

The third aspect introducing the large scope of the problem is that the work load does not have similar training effects on individual physical work capacity as aerobic physical exercise. Therefore the physical work capacity declines with age in physical occupations as much, or even more, as in mentally demanding occupations (Ilmarinen et al., 1991; Nygård et al., 1991). The decline of VO$_2$max, of muscle strength and endurance is a process with which everyone of us comes into contact. On the other hand, a large individual potential exists to prevent the premature decline of physical work capacity. Preventive measures like physical exercise, integrated both in working life and in life style, offer several possibilities to solve the problem of aging workers.

3. Background

3.1. Decline of physical work capacity with age

In general, the decline of maximal oxygen consumption (VO$_2$max) is about 1–2% per year after the age of 20–25, when the peak level has been reached (Åstrand and Rodahl, 1970; Seliger and

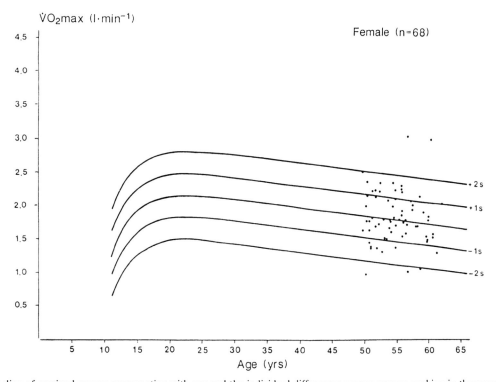

Fig. 4. Decline of maximal oxygen consumption with age and the individual differences among women working in the same branch. It is notable that the difference in VO$_2$max in liter/min can be 2–3-fold in the same age.

Bartunek, 1976; Saltin, 1990). The curve of decline is rather linear both among men and women. The individual variation is large but it seems to get slightly smaller among the oldest part of the population. It is notable, however, that after the age of 50 the differences in VO$_2$max can vary from 1.0 to 3.0 liter/min among working women or from 1.5 to 4.0 liter/min among working men in the same branch (figures 4 and 5, data from Ilmarinen et al., 1991).

At the individual level the decline of VO$_2$max can be much larger than 1–2% per year. If the life-style does not include any physical training, the decline has been shown to be 20–25% during a 4-year follow-up (Ilmarinen et al., 1991). Both among the women and among men in the age range of 45 to 60 the individual decline of VO$_2$max in liter/min was over 20% in 4 years. The subjects were all economically active and were working in the same, municipal branch during the follow up (figures 6 and 7, data from Ilmarinen et al., 1991).

There are at least two significant reasons why the cardiovascular function declines with age: occult disease and physical inactivity. The findings of autopsy data, based on the Baltimore Longitudinal Study On Aging, have indicated that among men aged 50 the prevalence rate of hearts with severe stenosis was about 35%; among women the rate was about 25% at the age of 60 (Lakatta, 1987). Also the average daily caloric expenditure for physical activity declined in a rather linear way from age 30 to age 60 (Lakatta, 1987).

The musculoskeletal capacity also declines with age. The differences of hand grip, knee extension, elbow flexion, trunk extension and in trunk flexion between the 31–35-year-old and the 51–55-year-old cohort was about 20% (Viitasalo et al., 1985). A four-year follow-up between the age of 51 and 55 has indicated that the decline of trunk flexion strength is about 20% among men. Among women, the musculoskeletal disability, based on 11 different tests, decreased more than 20% in four years after the age of 51. Interest-

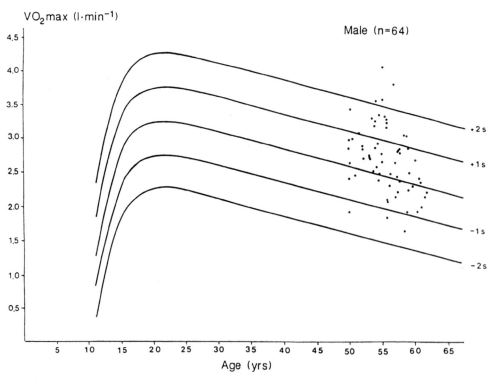

Fig. 5. Decline of maximal oxygen consumption with age and the individual differences among men working in the same branch. The 2-fold differences in VO$_2$max liter/min can be seen at the same age.

ingly, the decline was independent of the occupational demands. So, those involved with physical work were as strongly affected as those doing mental work (Nygård et al., 1991).

The decline of muscle strength with age has often been attributed to declining muscle mass in older subjects. Grip strength, however, is more strongly correlated with age than muscle mass. Other yet undetermined factors beyond declining muscle mass remain to explain some of the loss of strength seen with aging (Kallman et al., 1990).

3.2. Gender differences in physical work capacity

In working life the female population is also involved with physical work load. The proportion of women in auxiliary work, with jobs like cleaner, hospital assistant, kitchen helper, construction worker, street sweeper, park worker, or in home

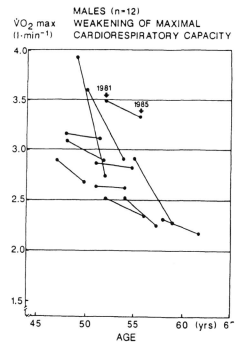

Fig. 7. The individual decline of maximal oxygen consumption of working men during a four-year follow-up period. This sample of men was not doing any aerobic training during the follow-up. Individuals with remarkable decline can be identified. After the follow-up all men in this sample were working in the same job as before.

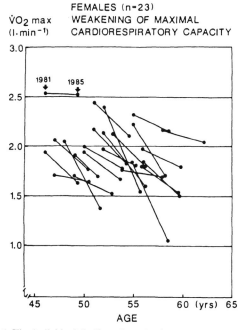

Fig. 6. The individual decline of maximal oxygen consumption of working women during a follow-up period of four years. The sample in this figure was not doing any regular aerobic training during the follow-up. It is notable that the VO_2max can decline 20–25% in four years after age 45 and independent of the base line level. After the follow-up all the women were still working in the same job as before.

care work, such as bathing and domestic help, is dominating. In agriculture women are often doing the same tasks as the men. In physical work capacity, however, systematic differences between men and women exist. The VO_2max of women at age 51 and 55 was about 69% and 64% of that of the men. The women's VO_2max in relation to body weight was about 81% at age 51 and about 73% at age 55 of that of the men. The gender differences were only slightly related to the work content. In the physically demanding occupations the differences between the men and the women were practically unchanged from age 51 to 55 (Ilmarinen et al., 1991).

In musculoskeletal strength similar differences exist between aging men and women. The trunk flexion and extension strength as well as the trunk endurance of women after the age of 50 were about 60% of that of the men. The difference was the same among women doing physical and mental work but in those women doing mixed physical and mental work (e.g. nurses) the trunk strength was about 80% of that of the men (Nygård et al., 1991).

Age and gender probably influence muscular endurance. Strength and absolute endurance, isometric and isokinetic, decline at the same rate with age. Absolute endurance in men is greater because of strength differences, whereas relative endurance is similar in men and women (Skinner et al., 1990). Age- and gender-specific norms for strength and endurance are needed for specific populations, especially at varying speeds of contraction.

3.3. Age, work and diseases

3.3.1. Mortality

Public health aspects of population aging have been recently illustrated (WHO, 1989). The assessment of health status is limited because the comprehensive data of mortality and especially of morbidity is lacking. The WHO reports that "roughly 50% of all deaths at ages 65–74 years, in the industrialized countries at least, are attributable to cardiovascular diseases. Cancer accounts for one-quarter of deaths and about 7% of deaths are due to respiratory diseases. There is considerable heterogeneity in mortality levels, even among the industrialized countries.

Age-standardised mortality rates for men and women, aged 30–69 years, indicate that an increasing trend in mortality (all causes) during the last ten-year period was observed among the men in Hungary, Czechoslovakia, Poland, Romania, Bulgaria and Denmark. Among women an increasing trend was seen in Hungary, Poland and Denmark. The mortality rates of all cardiovascular diseases tended to increase among men and women during the last ten years in East-European countries but also among men in Greece (Pisa and Uemura, 1989).

3.3.2. Morbidity

Comprehensive morbidity statistics are lacking and it is therefore difficult to see rising or falling trends in morbidity of the population. A recent study from the USA indicates that morbidity from leading causes of ill-health (cancer, heart disease, diabetes, hypertension, and arteriosclerosis) has increased, while a decline in morbidity among the elderly was found in skin diseases, visual and hearing-related problems and multiple orthopeadic problems (WHO, 1989).

A recent study of Finnish aging workers indicated that musculoskeletal diseases are most prevalent among men and women aged 45–64 years (Tuomi et al., 1991). At the mean age of 50 years the prevalence of musculoskeletal diseases was 37% and at the age of 54 years 53% among the women. The respective rates for men were 33% and 46%. The prevalence rates for cardiovascular diseases were much lower.

It is obvious that prevalence rates of cardiovascular and musculoskeletal diseases after the age of 50 are somewhat different from country to country and the health status of occupational groups may vary. The age-standardised mortality rates indicate, however, that the rates among men and women aged 30 to 69 are in several countries so high (Pisa and Uemura, 1989) that health should be seriously taken into consideration when designing jobs for aging workers.

3.4. Working and retirement

An increasing number of people is leaving the work force before old age pension at the age of 65 years. The ILO statistics projecting the year 1990 indicate that the economically active population of the male age group of 55–64 has decreased markedly in 30 years in all countries except Japan (ILO, 1986). Marked differences can be found between the countries: the activity rate of Japan is 87.2, USA 70.6, Germany 61.7 but in France, Finland, Austria and Italy only about 50% of the 55–64-year-old male population are economically active (table 1). The figure is different for female population. The activity rate has not decreased markedly in the 1980s although the trend is similar to that of the men. The highest activity rate of women aged 55–64 in 1990 is estimated in Sweden (51.0%) and the lowest in Italy (11.7%) (see table 2).

Table 1

Activity rate of the male age group 55–64 per country. (Rank order as of the year 1985).

	1960	1970	1980	1985	1990
1. JAP	88.6	90.3	88.6	88.4	87.2
2. SWIT	92.7	91.4	88.9	88.0	87.7
3. GDR	89.4	87.6	88.0	86.9	85.8
4. UK	92.0	87.7	84.3	83.1	83.1
5. SWED	85.6	82.4	79.5	78.3	78.2
6. CAN	87.2	83.6	75.8	75.0	74.4
7. DEN	90.0	88.2	76.2	74.8	74.7
8. USA	87.2	81.9	72.0	71.1	70.6
9. AUSTRL	88.6	83.1	71.5	70.5	69.5
10. NL	82.0	74.9	68.3	67.3	67.2
11. FRG	81.0	78.8	65.6	63.2	61.7
12. BELG	75.0	67.9	64.8	61.3	61.2
13. USSR	69.8	57.5	59.2	61.3	52.1
14. FRA	76.3	68.0	59.2	54.2	53.7
15. FIN	83.1	68.2	53.5	52.4	51.0
16. AUST	77.8	64.1	56.8	52.1	50.1
17. ITAL	74.0	62.7	52.6	49.0	48.2

Source: Figures based on: ILO: Economically active population, 1986. The figures through the year 1980 are estimates; the following years are projections.

Table 2

Activity rate of the female age group 55–64 per country. (Rank order as of the year 1985).

	1960	1970	1980	1985	1990
1. SWED	26.8	37.3	54.3	52.0	51.0
2. GDR	43.3	46.3	53.4	47.8	46.2
3. JAP	44.7	49.0	45.4	43.9	42.5
4. USA	37.3	42.3	41.6	40.4	39.6
5. DEN	30.7	34.3	41.5	39.5	38.5
6. UK	27.1	33.6	41.5	39.3	39.2
7. FIN	45.2	43.7	41.9	39.0	36.7
8. CAN	22.9	30.7	35.5	34.3	33.4
9. SWIT	29.6	32.6	34.3	32.9	32.1
10. FRA	37.6	35.4	33.6	29.9	29.0
11. AUSTRL	19.0	23.2	25.9	25.0	24.0
12. FRG	26.6	26.7	27.7	24.8	24.7
13. AUST	30.4	24.4	23.5	19.7	19.5
14. USSR	41.3	18.5	18.5	17.7	14.9
15. NL	12.0	14.8	16.1	15.4	15.1
16. BELG	14.7	13.6	14.2	12.6	12.5
17. ITAL	15.5	14.4	13.4	12.1	11.7

Source: Figures based on: ILO: Economically active population, 1950–2025, volumes IV and V, Geneva, 1986. The figures through the year 1980 are estimates; the following years are projections.

Several factors predict early retirement. Among men the strongest predictors of early retirement and of age of retirement include both structural factors and subjective factors, such as self-rated health and attitudes. When retirement is defined by amount of employment, job characteristics are more important predictors than all others combined (Palmore et al., 1982). Age, general medical work capacity, use of drugs, attitude to work environment, work performance and systolic blood pressure have been reported as independent variables of early retirement (Åstrand et al., 1988). On the other hand, experienced work ability and work characteristics seem to play important roles when working up to the 65 years. Experienced work ability indicates already at the age of 51 marked differences between the jobs (Tuomi et al., 1991). More than 20% of women in auxiliary and home care work seemed to have a poor work ability at the age of 55. Among men the highest prevalence rates of poor work ability were found among installation, transport and auxiliary jobs (figures 8 and 9, data from Tuomi et al., 1991).

The experienced work ability at the age of 51 predicted well the outcome during the follow-up. Of those women and men who, at the age of 51 years, had a poor work ability, about one-third became disabled for work during the 4-year follow-up period (table 3).

POOR WORK ABILITY

FEMALES, AGE 55 yrs

WORK	N	%
DENTIST	77	3.9
ADMINISTRATION	374	4.6
OFFICE	301	6.3
TEACHING	198	7.1
DOCTOR	22	9.1
SUPERVISION	120	10.0
NURSING	849	10.7
HOME CARE	411	20.0
AUXILLIARY	608	21.0
ALL	2960	12.4

Fig. 8. The prevalence rate (%) of poor work ability of women in different jobs at age 55. The work ability was assessed by questionnaire.

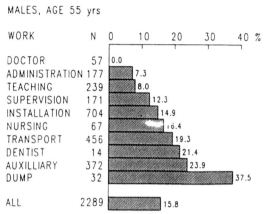

Fig. 9. The prevalence rate (%) of poor work ability of men in different jobs at age 55. The work ability was assessed by questionnaire.

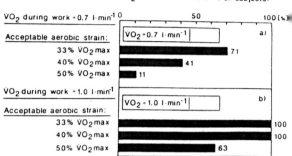

Fig. 10. Physical work demands and the prevalence rate (%) of not-sufficient work capacity among women at age 55. With dynamic work at the level of 0.7 liter/min including rest pauses (50% RAS acceptable), 11% of women did not have a sufficient aerobic capacity. At a work level of 1.0 liter/min, 63% of women did not have a sufficient capacity. For all women, working without rest pauses (33% RAS acceptable) was out of the question at the work level of 1.0 liter oxygen/min.

The findings of several studies emphasize that the work content and work environment need special attention if the work career aims to continue up to the normal old age pension.

3.5. Physical work demands and sufficient capacity

The cardiorespiratory load of work has been described earlier in this article (see section 2, Scope of the problem, figures 2 and 3). On the other hand, the musculoskeletal load has been ranked the highest in installation work, followed by auxiliary, home care and transport work (Nygård et al., 1987).

Table 3

Experienced work ability of 51-year-old workers and the work disability pension rate during the follow-up period from 51 to 55 years of age by gender.

Work ability at the age of 51 years	Work disability pension between 51–55 years of age			
	Men		Women	
	Number	Percent	Number	Percent
Poor	119	37.8	109	33.3
Average	120	8.4	85	4.4
Good	39	0.8	7	1.5
All	242	11.5	201	7.5

The cardiorespiratory load of 1.0 or even 0.7 liter oxygen per minute can become critical for older, unfit people. The large individual variation in each age group has important consequences. The individual data in figure 4 for women and in figure 5 for men leads to the following results. The VO_2max is not sufficient for 11% of the women when the VO_2 at work is 0.7 liter/min and rest pauses are available (RAS 50% acceptable). If rest regimens are not available (RAS 33% acceptable) about 71% of the women aged 55 years do not have a sufficient aerobic capacity (see figure 10). At working level of 1.0 liter oxygen/minute already 63% of older women do not have a sufficient capacity. Working without breaks and keeping the RAS below 33% was not possible for women working at the level of 1.0 liter oxygen/minute.

For men working at a level of 0.7 liter/min is not critical. Only work without rest regimen means restrictions for 37% of men aged 55 years (see figure 11). At working level 1.0 liter/min oxygen all men had a sufficient capacity but the restrictions arose when the RAS 33% should be used: 75% of men did not have a sufficient capacity for continuous activities at the level of 1.0 liter oxygen/minute.

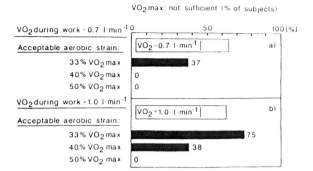

Fig. 11. Physical work demands and insufficient work capacity among men at age 55. Restrictions can be found only when working without rest pauses (33% RAS acceptable).

These examples indicate first, how critical a low aerobic capacity is already at a relatively low level of aerobic work load. Secondly, it also emphasizes the importance of effective work–rest regimens for older people in physically demanding occupations.

One crucial question is why 50% RAS is acceptable with breaks and 33% RAS without breaks during work shift. The main reason for setting acceptable limits for occupational induced strain is to avoid anaerobic metabolism during work. If work includes anaerobic phases the consequence is an accumulation of lactid acid in the muscle group used, and tiring of the muscles. The 50% RAS is a physiological level, which in normal conditions does not activate the anaerobic metabolism. However, in daily physical work the dynamic muscle work is often combined with static muscle contractions and the optimal, dynamic muscular rhythm (like in cycling or in walking) is disturbed. Therefore, breaks are needed at the 50% RAS level to avoid the tiring effects of possible anaerobic metabolism. According to the experiences in laboratory and field studies, the 33% RAS is not critical in respect to anaerobic metabolism. More information on this issue is given in the next section.

3.6. Previous guidelines

The question of acceptable work load has interested scientists during the last 40 years. One of the oldest attempts was made by Bink (1962). He used the data of Robinson (1939) concerning the maximal aerobic capacity per min/kg body weight of a normal man of 35 years old (15.06 kcal/min), the data of Lehmann (1953) for the allowable gross energy expenditure for 8 hours (5.2 kcal/work min) and, the data of 'Food tables in the Netherlands' for the mean food intake of a man of 35 years old during 24 hours (2.85 kcal/min). The physical work capacity at working times of 4 minutes, 480 minutes and 1440 minutes were calculated. The linear relation between physical working capacity and the logarithm of the working time was expressed by the formula:

$$\text{PWC} = \frac{\log 5700 - \log T}{3.1} \times A \, \text{kcal/min}, \quad (1)$$

where PWC = physical working capacity in kcal/min, T = working time in min, and A = aerobic capacity in kcal/min.

The relation of aerobic capacity and age was taken from the data of 81 men (Robinson, 1939) and from 35 women (Åstrand, 1960). The Binks formula suggests in general that 1/3 of the maximal aerobic capacity is acceptable for energy expenditure for 8 hours work. Bonjer (1962) used the Binks formula in mild delivery in apartment houses and the cleaning department. He was interested in calculating the degree of loading for shorter or longer parts of the day. In general, his calculations suggested that there is no objection against peak loads if there is a guarantee for sufficient recovery during the periods of moderate exercise or rest.

In order to be able to put "the right man in the right place" in manual labor Åstrand (1964) emphasized that absolute load of work should be determined with indirect calorimetry and the actual oxygen intake must be judged with regard to the capacity of the working groups of muscles. To avoid tiring effects in work with large muscular groups the oxygen intake should not exceed about 50% of the aerobic capacity. Åstrand mentioned also that there are no physiological objections to an application of the '50 percent load' to female labor.

Other levels than 50% RAS have also been suggested. For eight hours sustained work a reasonable RAS of 35% was introduced by Saha et al. (1979). Army males have been reported to work at 30–40% RAS for 6.5 hr/day for six days (Myles and Saunders, 1979). Evans et al. (1980)

suggested that for sustained physical work of 1 to 2 hours duration the free-paced strain level is the same for males and females and is adjusted to the level of 45% RAS. In manual mail delivery the 40% RAS was used in order to stay on the safe side of possible overstrain among older men and women (Ilmarinen et al., 1984). The maximal allowable delivery times for different delivery modes were calculated for age groups lower and higher than 50 years. Maximal allowable working times based on 40% RAS for different transport methods and weights while skiing and hauling a sledge or carrying a backpack have also been reported (Ilmarinen et al., 1986).

For mixed physical work, including handling operations, an upper general tolerance limit over an 8-hour work day, an acceptable level of 30–35% RAS has been suggested (Jörgensen, 1985). The reason for this low acceptable level was, that VO_2max is reduced about 20–40% in lifting exercise compared with dynamic exercise on a bicycle ergometer. Using psychophysical methods Legg and Myles (1985) found that soldiers can lift on 21% RAS level repetitively for an 8-hour workday without metabolic, cardiovascular or subjective evidence of fatigue. For extended work shifts Mital (1984) found that males selected weights that, on average, resulted in 23% RAS for 12 hours lifting. The maximum acceptable weight of lift was not influenced by the age of the subject. It was suggested that the experience on the job compensates for any physiological changes due to aging. It is obvious that the static component of the work strongly reduces the acceptable RAS level in practical work.

In general, it is not desirable to assign working periods that require the worker to continue until exhausted. Optimal intermittent work schedule to prevent fatigue has been introduced. One example is given by Kamon (1982). Workloads demanding more than the 50% of VO_2max call for rest allowances because of the exponential increase in the lactid acid (LA) production. Kamon used LA production as a criterion for rest. The time required to reduce the blood lactid acid to about twice the level found during rest is

$$T_r = 8.8 Ln(f VO_2 max - 0.5) + 24.6 \qquad (2)$$

where T_r is resting period in minutes (Kamon, 1982).

In the literature the levels from 33 to 50% RAS for physical work have been repeatedly given as overall acceptable limit avoiding anaerobic metabolism during work. Although this criterion can still be valid in general, more specific attempts are urgently needed. Recently Rutenfranz et al. (1990) suggested that the upper limit of energy expenditure at work should be:
- 30% RAS for dynamic work without special rest pauses, or
- 50% RAS for mixed dynamic/static work, or for dynamic work with rest pauses.

Matching the job demand to a person's physical characteristics is widely studied in manual materials-handling tasks. Maximal aerobic capacity and age is also a limiting factor in manual materials handling. Maximum acceptable weights and forces for lifting, lowering, pushing, pulling and carrying have been published recently by Snook and Ciriello (1991). The practitioner can find more information on maximum acceptable loads from reports of Jiang and Ayoub (1987), Mital (1987), and Karwowski (1991).

Need for further research

Several problems have to be solved before an optimal guideline for the aged can be constructed. The research data needed for acceptable heart rates and rest periods have been described in the table and figures in Part I (Problem solving). There are, however, several other problems to be solved, where research is also needed.

4.1. Not only physical demands are critical for the aged

The list of risk factors of work leading to health problems and deteriorating the work ability also covers the physical work environment and the organizational aspects of work. It should be emphasized that, e.g., working in the heat or cold are critical for the aged because of their decreased tolerance against environmental temperatures. Aging can also increase the risk of falling and although the work accident rates are not higher among older workers than among younger workers, the accidents seem to be more serious and the time for rehabilitation longer among the older workers than among the younger workers.

The organisational aspects of work to be improved are the third key reason why guidelines are urgently needed. The list has been given earlier in this guide, but the aspects of working time need a special attention. The working time is probably the most potential topic for improvements. The question is not only the lower tolerance to shiftwork of the older worker but also the length of daily working and the implementation of rest pauses. To prevent the cumulative effects of fatigue due to the work, effective rest regimens are needed.

The physical, environmental and organizational aspects of work introduce the consideration of combined effects to be included in the future guidelines. It is obvious, that e.g. a combination of heavy lifting, cold and working time or e.g. static work, heat and time pressure needs further research before special guidelines can be given.

4.2. How to consider health and disease?

It is clear that the prevalence and incidence rates of impairments, disabilities, and handicaps increase with age. The guidelines for healthy or for 'less' healthy older people at work should be different. The individual differences in physical, mental and social capabilities after the age of 50 years are large and factors explaining the differences are coming from several sources (disease, life style, work, genetics). A comprehensive model is needed before the complexity 'Aging and Work' can be transferred into a practical guide.

4.3. The dimensions of physical work capacity

Assessment of physical work capacity includes measures of flexibility, body composition, muscle strength and endurance, anaerobic abilities and aerobic abilities (Skinner et al., 1990). The present guidelines focus primarily on the aerobic ability and secondarily on the muscle strength and endurance. The reason for this restriction is that information is widely available for these aerobic dimensions. Some aerobic and muscular aspects, however, need still more intensive attention. The first question is the relation between the aerobic capacity and muscular capacity during aging. This knowledge is needed, because the general aspects of the cardiovascular system are often used to explain also the local muscle abilities.

The correlation of muscle strength and work ability index has been higher than the correlation between VO_2max and work ability index, suggesting that other laboratory tests than the VO_2max should be improved if the aim is to predict the status of work ability (Nygård et al., 1991b). It is obvious that the laboratory tests should be as job-specific as possible. If the VO_2max in the specific work task is known, the acceptable level of work load and the RAS can be identified in a much more valid way. Depending on the work tasks and the demands of moving one's own body weight also the VO_2max in ml per kg body weight should be taken into consideration.

During aging the blood pressure and prevalence of hypertension increase. Therefore the load of the heart can increase more than expected due to the RAS. Research is needed to find out the total load of heart and circulation during dynamic and static occupational work. The effects of drugs against hypertension on the cardiovascular strain during work are not clear.

Because the experienced abilities and perceived strain measures include prognostic power in the continuity of the work career, such measures should be taken more seriously into consideration beside the traditionally physiological methods. Both the subjective work ability and the ratings of perceived exertion (RPE) are valuable methods for job design among aging workers.

References

Aromaa, A., Heliövaara, M., Impivaara, O. et al., 1989. Health, functional limitations and need for care in Finland. Basic results from the Mini-Finland Health Survey. English summary. Helsinki and Turku: Publications of the Social Insurance Institution, Finland, AL: 32, 793 pp.

Bink, B., 1962. The physical working capacity in relation to working time and age. Ergonomics, 5: 25–28.

Bonjer, F.H., 1962. Actual energy expenditure in relation to the physical work capacity. Ergonomics, 5: 29–31.

Evans, W.J., Winsmann, F.R., Pandolf, K.B. and Goldman, R.F., 1980. Self-based hard work comparing men and women. Ergonomics, 7: 613–621.

Ilmarinen, J., Knauth, P., Klimmer, F. and Rutenfranz, J.,

1979. The applicability of the Edholm scale for activity studies in industry. Ergonomics, 22(4): 369–376.
Ilmarinen, J., Louhevaara, V. and Oja, P., 1984. Oxygen consumption and heart rate in different modes of manual postal delivery. Ergonomics, 27(3): 331–339.
Ilmarinen, J., Sarviharju, P. and Aarnisalo, T., 1986. Strain while skiing and hauling a sledge or carrying a backpack. Eur. J. Appl. Physiol., 55: 597–603.
Ilmarinen, J., 1989. Die Bedeutung der kardiopulmonalen Leistungsfähigkeit für praktische Tätigkeiten. In: W. Rohmert and H.G. Wenzel (Eds.), Aspekte der Leistungsfähigkeit (Different aspects of performance. Studies in industrial and organizational psychology). Verlag Peter Lang, Frankfurt am Main, Bern, New York, Paris, Volume 8: 210–219.
Ilmarinen, J., Louhevaara, V., Korhonen, O., Nygård, C-H, Hakola, T. and Suvanto, S., 1991. Changes in maximal cardiorespiratory capacity among aging municipal employees. Scand. J. Work. Environ. Health, 17 (suppl 1): 99–109.
Ilmarinen, J., Tuomi, K., Eskelinen, L., Nygård, C-H., Huuhtanen, P. and Klockars, M., 1991. Summary and recommendations of a project involving cross-sectional and follow-up studies on the aging worker in Finnish municipal occupations (1981–1985). Scand. J. Work Environ. Health, 17 (suppl 1): 135–41.
ILO, 1986. Economically active population, 1950–2025, volumes IV and V, Geneva.
Jiang, B.C. and Ayoub, M.M., 1987. Modelling of maximum acceptable load of lifting by physical factors. Ergonomics, 30(3): 529–538.
Jörgensen, K., 1985. Permissible loads based on energy expenditure measurements. Ergonomics, 28(1): 365–369.
Kallman, D.A., Plato, C.C. and Tobin, J.D., 1990. The role of muscle loss in the age-related decline of grip strength: Cross-sectional and longitudinal perspectives. Journal of Gerontology: Medical Sciences, 45(3): M82–88.
Kamon, E., 1982. In: G. Salvendy (Ed.), Handbook of Industrial Engineering, Chapter 6.4, Wiley, New York.
Karwowski, W., 1991. Psychophysical acceptability and perception of load heaviness by females. Ergonomics, 34(4): 487–496.
Lakatta, E.G., 1987. Why cardiovascular function may decline with age. Geriatrics, 42(6): 84–94.
Legg, S.J. and Myles, W.S., 1985. Metabolic and cardiovascular cost, and perceived effort over an 8 hour day when lifting loads selected by the psychophysical method. Ergonomics, 28(1): 337–343.
Lehmann, G., 1953. Praktische Arbeitsphysiologie. Thieme, Stuttgart.
Mital, A., 1984. Maximum weights of lift acceptable to male and female industrial workers for extended shifts. Ergonomics, 27(11): 1115–1126.
Mital, A., 1987. Patterns of differences between the maximum weights of lift acceptable to experienced and inexperienced material handlers. Ergonomics, 30(8): 1137–1147.
Myles, W.S. and Saunders, P.L., 1979. The physiological cost of carrying light and heavy loads. Europ. J. Appl. Physiol. and Occup. Physiol., 42: 125–131.
Nygård, C-H., Suurnäkki, T., Landau, K. and Ilmarinen, J., 1987. Musculoskeletal load of municipal employees aged 44 to 58 years in different occupational groups. Int. Arch. Occup. Environ. Health, 59: 251–261.
Nygård, C-H., Luopajärvi, T. and Ilmarinen, J., 1991a. Musculoskeletal capacity of elderly men and women in different types of work. Scand. J. Work Environ. Health, 17 (suppl 1): 110–117.
Nygård, C-H., Eskelinen, L., Suvanto, S., Tuomi, K. and Ilmarinen, J., 1991b. Associations between functional capacity and work ability among elderly men and women. Scand. J. Work Environ. Health, 17 (suppl 1): 122–127.
OECD, 1984. Working party and social policy 29–31.10.1984. The social and labour market implications of aging populations. The changing population age structure.
Palmore, E.B., George, L.K. and Fillenbaum, G.G., 1982. Predictors of retirement. Journal of Gerontology, 37(6): 733–742.
Pisa, Z. and Uemura, K., 1989. International differences in developing improvements in cardiovascular health. Ann. Med., 21(3): 193–197.
Robinson, S., 1939. Experimental studies of physical fitness in relation to age. Arbeitsphysiologie, 10: 251–324.
Rutenfranz, J., Ilmarinen, J., Klimmer, F. and Kylian, H., 1990. Work load and demanded physical performance capacity under different industrial working conditions. In: M. Kaneko (Ed.), Fitness for the aged, disabled, and industrial worker. International Series on Sport Sciences, volume 20. Human Kinetics Books, Champaign, Illinois, pp. 217–238.
Saha, P.N., Datta, S.R., Banerjee, P.K. and Narayane, G.G., 1979. An acceptable work load for Indian workers. Ergonomics, 22: 1059–1071.
Saltin, B., 1990. Cardiovascular and pulmonary adaptation to physical activity. In: C. Bouchard, R.J. Shephard, T. Stephens, J.R. Sutton and B.M. McPherson (Eds.), Exercise, Fitness and Health. A Consensus of Current Knowledge. Human Kinetics Books, Champaign, Illinois. Part III: Human Adaptation to Physical Activity, Chapter 18, pp. 187–203.
Seliger, V. and Bartunek, Z. (Eds.), 1976. Mean values of various indices of physical fitness in the investigation of Czechoslovak population aged 12–55 years. International Biological Programme Results of Investigations 1968–1974. CSTV, Praha (CSSR): Czechoslovak Association of Physical Culture, 117 pp.
Skinner, J.S., Baldini, F.D. and Gardner, A.M., 1990. Assessment of fitness. In: C. Bouchard, R.J. Shephard, T. Stephens, J.R. Sutton and B.D. McPherson (Eds.), Exercise, Fitness, and Health. A Consensus of Current Knowledge. Human Kinetics Books, Champaign, Illinois. Part II, Chapter 9, pp. 109–119.
Snook, S.H. and Ciriello, V.M., 1991. The design of manual handling tasks: Revised tables of maximum acceptable weights and forces. Ergonomics, 34(9): 1197–1213.
Tuomi, K., Ilmarinen, J., Eskelinen, L., Järvinen, E., Toikkanen, J. and Klockars, M., 1991. Prevalence and incidence rates of diseases and work ability in different work categories of municipal occupations. Scand. J. Work Environ. Health, 17 (suppl 1): 67–74.
Viitasalo, J.T., Era, P., Leskinen, A.-L. and Heikkinen, E., 1985. Muscular strength profiles and anthropometry in

random samples of men aged 31-35, 51-55 and 71-75 years. Ergonomics, 28(11): 1563-1574.

Åstrand, I., 1960. Aerobic work capacity in men and women with special reference to age. Acta Physiol. Scand., 49, suppl. 169.

Åstrand, N-E., Isacsson, S-O. and Olhagen, G.O., 1988. Prediction of early retirement on the basis of a health examination. Scand. J. Work Environ. Health, 14: 110-117.

Åstrand, P-O., 1964. Human physical fitness with special reference to sex and age. In: E. Jokl and E. Simon (Eds.), International Research in Sport and Physical Education. Charles C. Thomas Publ., Springfield, Illinois, pp. 517-558.

Åstrand, P-O. and Rodahl, K., 1990. Textbook of work physiology. McGraw-Hill Book Comp., New York.

Design, selection and use of hand tools to alleviate trauma of the upper extremities: Part I – Guidelines for the practitioner *

Anil Mital [a] and Asa Kilbom [b]

[a] *Industrial Engineering, University of Cincinnati, Cincinnati, OH 45221-0116, USA*
[b] *National Institute of Occupational Health, S-171 84 Solna, Sweden*

A. Audience

This guide is intended for practitioners. The practitioner in the context of this guide is defined as the individual who is responsible for designing hand tools. This person may be the design engineer, occupational safety and health professional, ergonomist, product designer, industrial or production engineer or a professional entrusted with the task of designing hand tools.

The practitioners should explicitly understand that this guide is intended for the purpose of designing and evaluating only hand tools. Proper design of hand tools alone may not be sufficient to correct the 'ergonomic' problem. The working situation, as a whole, may also have to be evaluated and corrected to alleviate the problem. For instance, a properly designed hand tool, used in an awkward posture, most likely, will not solve the problem. The guidelines provided here may be modified as a result of future research and experience.

B. Problem identification

The problem can be identified from occurrence of accidents to the hands involving tools, and cases of cumulative trauma disorders of the hand, wrist, forearm, shoulder, and neck. These cases should be investigated by collecting appropriate data pertaining to injuries, the task, and the tool(s). These data should be analyzed and results should be evaluated objectively.

C. Data collection

In a broad sense, this involves gathering data pertaining to occurrence of cumulative trauma disorders from the workplace under investigation, and preferably also a referent workplace for comparison purposes.

Correspondence to: A. Mital, Industrial Engineering, University of Cincinnati, Cincinnati, OH 45221-0116, USA.

* The recommendations provided in this guide are based on numerous published and unpublished scientific studies and are intended to enhance worker safety and productivity. These recommendations are neither intended to replace existing standards, if any, nor should be treated as standards. Furthermore, this document should not be construed to represent institutional policy.

The following individuals participated in the discussion of the earlier version of this guide. Their suggestions (written or verbal) were incorporated by the authors in this version: Roland Andersson, *Sweden*; Alvah Bittner, *USA*; Peter Buckle, *UK*; Jan Dul, *The Netherlands*; Bahador Ghahramani, *USA*; Juhani Ilmarinen, *Finland*; Sheik Imrhan, *USA*; Stephan Konz, *USA*; Shrawan Kumar, *Canada*; Tom Leamon, *USA*; Mark Lehto, *USA*; William Marras, *USA*; Barbara McPhee, *Australia*; James Miller, *USA*; Don Morelli, *USA*; Maurice Oxenburgh, *Australia*; Jerry Purswell, *USA*; Jorma Saari, *Finland*; W. (Tom) Singleton, *UK*; Juhani Smolander, *Finland*; Terry Stobbe, *USA*; Rolf Westgaard, *Norway*, Jörgen Winkel, *Sweden*. The guide was also reviewed in depth by several anonymous reviewers.

C.1. Pertinent data gathering techniques

C.1.1. Statistical information

Gather statistical information from occupational health and safety records, official statistics, questionnaires and/or company accident and injury reports. The aim of these statistics is to provide information on which workplaces constitute high risk, as well as to evaluate the need of medical assistance (e.g., rehabilitation). If no company records are maintained, steps should be taken immediately for gathering such data. As a minimum, the records should include the following: part of the body injured, medical diagnosis, duration of sick leave, workplace, age, gender, and experience of the worker. A questionnaire on discomfort/disorders (e.g., of the forearms/hands) will provide more specific information on problems possibly related to hand tools, but is not sufficient for a medical diagnosis. The number of workers at each plant/workplace must be recorded.

C.1.2. Ergonomic work analysis

This includes duration and frequency of work tasks, listing of tools being used, observation of work postures and handgrips, measurement of tool vibration and force exertions. The following question – What are the requirements on precision and force exertions? – should be asked and answered completely. Work analysis at high risk workplaces includes scrutiny of the work tasks, motion studies, observation of grips and work postures, and listing of tools. When power grips are used repetitively, hand anthropometry and grip strength of workers should also be measured, especially in case of women and older workers. The analysis should also include interviews with workers and supervisors, concerning the above factors, about the steps taken to date to alleviate the problems, and about procedures for purchasing and maintaining tools.

C.1.3. Ergonomic analysis of the tools

This includes effective weight (weight of the tool supported by the worker), shape, and use of hand tools. This step should include weighing the tool, determining how much of the tool weight is supported by the worker, and measuring its characteristics, such as vibration signature, analyzing its various functions in all work tasks, and in some cases biomechanical and physiological measurements on the worker. The latter is applicable only if resources are available to expand the investigation into a research project.

C.2. Equipment requirements

Normally, no sophisticated equipment is necessary. Video recordings are helpful for work analysis, and in some cases measurements of worker handgrip strength and hand size are advisable. For handgrip strength, a strain gauge handgrip (hand grip dynamometer) with adjustable grip should be used, and it should be calibrated regularly against known weights. For hand size measurements, an anthropometer to measure palmer hand length and metacarpal hand width are sufficient. For measuring vibration and workpiece strain, devices such as accelerometers and strainmeters are needed.

D. Data analysis

D.1. Data analysis procedures

D.1.1. Statistical analysis of injuries

The population under investigation is usually too small for this analysis to be conclusive, and the results of the statistical analysis will, therefore, provide *indications* of an association between the workplace/task and a disorder, rather than proof. Data analysis depends on the kind of data available and on the size of the population. A reference population – either in an office area, in another plant or another production unit – is desirable but not always available. Prevalence rates of injuries at the studied production unit, compared to those of the reference population, should be calculated. The results of the statistical analysis should also be compared to similar analyses performed during the time period after the tool, or task, redesign. Note that for small production units these data may not be conclusive. Even if the case(s) is not sufficiently serious, investigation should proceed to the next step.

D.1.2. Analysis of workplace data

This involves calculation of frequency and duration of work tasks, use of tools, arm-, wrist- and finger-postures and type of grip. Data gathered at the workplace should be analyzed for frequency and duration of each work task and the use of each tool, for different grips and arm, wrist and finger postures.

D.1.3. Ergonomic tool analysis

Tools' design should be compared to recommendations given in the following sections. Specifically, the following kinds of questions should be asked – Effective weight, balance, surface, grip design? Does the tool have only one specific function or several (i.e., what are the demands on versatility)? Risks of pinching or squeezing of hand or fingers? Does the tool give feedback of proper function? Can it be easily maintained? Does it vibrate? Require excessive force?

D.2. Results of data analysis

D.2.1. Occurrence of injuries

As mentioned above, the statistical analysis will often provide indications rather than proof of an association between work tasks and injuries. However, it is not possible to specify stages beyond which a further investigation (items D.1.2 and D.1.3) must be undertaken.

D.2.2. Results of workplace analysis

The difficulty in evaluating the workplace data lies in the interaction between the time factor and the physical characteristics of the tool/workstation. Obviously, it is not harmful now and then to exert large handgrip forces, hold the wrist in extreme postures or have pressure concentration points in the palm. A tool used infrequently can cause such exposures without causing any harm. Thus, it is mainly in case of hand tools that are used for prolonged periods and/or repeatedly that the design of hand tool and workstation becomes critical. If extreme wrist postures, forceful handgrips or pressure concentrations in the palm occur repeatedly, this can be rectified either by reducing the exposure time or redesigning the hand tool and workplace.

One way to divide demands on time is as follows:

low – less than one hour distributed over the entire day, or less than 10 minutes continuously or repetitively.

moderate – one to four hours distributed over the entire day, or 10–30 minutes continuously or repetitively.

high – more than 4 hours distributed over the entire day, or more than 30 minutes continuously or repetitively.

High demands on time, in combination with high force or precision requirements, should be considered unacceptable. Combinations of moderate demands on time and high force or precision requirements should also be analyzed.

The choice of action-reduction of exposure time and/or redesign of tool/workstation depends on the feasibility of different solutions. In general, there is so much evidence of harmful effects of repetitive work, both in terms of injuries and work satisfaction, that an attempt should always be made to reduce the time of exposure for the individual worker. This can be done by job enlargement or job rotation.

D.2.3. Results of ergonomic tool analysis

If the tools analyzed do not meet the recommendations specified in section E.3, they should be replaced or redesigned.

E. Solutions

E.1. Work redesign or not?

No 'acceptability' limit for the statistical risk measures determined can be provided. Decisions on whether to proceed with work redesign depend on the validity of the data gathered in each individual case.

E.2. Reduction of exposure time?

Tasks performed at a high frequency or for prolonged periods of time should be redesigned in order to reduce the exposure. Cycle times < 30 s, including forceful grips, should not be allowed if performed > 4 hours distributed over the entire working day or > 1 hour continuously. Even if these limits are not reached, it is advisable to reduce the exposure. The following tentative advice can be given: Repetitive work, including forceful grips or high demands on precision, should not be accepted if performed for more than 4 hours.

E.3. Reduction of exposure by tool redesign?

If the tools used do not conform to the recommendations provided below, they should be redesigned.

Design of grip:
For precision – grip between thumb and fingers.
For power/force – grip with the entire hand (four fingers forming one jaw, thumb the other).
Orientation of handle – for exertion of large forces, a pistol grip should be provided (at an angle of 80° from the long axis of the tool).
Grip thickness for precision – between 8 and 13 mm.
Grip thickness for power/force – between 50–60 mm.
Grip length for precision – minimum 100 mm.
Grip length for power/force – minimum 120 mm.
Grip length for use with gloves – minimum 125 mm.
Grip guard – minimum 16 mm.
Grip surface – smooth, slightly compressible, and non-conductive.
Grip shape – non-cylindrical, preferably triangular (periphery 110 mm).
Grip force for power grip – maximum 100 N.
Grip and handle bent for hammers, etc. – 10°.

Effective weight (weight of the tool supported by the worker) of the tool:
For precision tools – as low as possible but not more than 1.75 kg.
For power tools – preferably about 1.12 kg, but no more than 2.3 kg.

Design of trigger:
For tasks requiring precision for prolonged duration – trigger should be designed for use by distal phalanges of the finger(s) of either hand and should have a locking mechanism (e.g., latch) which can be deactivated quickly in case of an emergency.
For tasks requiring force exertions for prolonged duration – trigger should be designed for activation by thumb muscles and should have a locking mechanism.
Pressure to activate the trigger – maximum 10 N.

Vibration characteristics:
Avoid vibration in the range of 2 and 200 Hz for tasks performed repeatedly.
Eliminate vibrating tools if possible (segmental vibration above 1000 Hz is relatively less critical than segmental vibration below 1000 Hz).
Control exposure if elimination not possible.
Use damping materials to absorb vibration.

User characteristics:
Design tools for use by either hand.
Avoid extreme variations in posture.
Keep the wrist straight (handshake orientation).

General considerations:
Consider mechanical power source.
Avoid tool noise in excess of 85 dBA for a full-day exposure.
Tool weight should be balanced about the grip axis.
Use special purpose tool if the usage is continuous and/or prolonged.
Support the tool for reaction force if the torque exceeds 6 Nm for straight tool (good for downward action), 12 Nm for pistol type tools (good horizontal action), and 50 Nm for angled tools (good for both downward and upward action).
Design for safety – eliminate the risks of pinching or squeezing of hands and fingers.

Reference

Mital, A. and Kilbom, Å., 1992. Design, selection and use of hand tools to alleviate trauma of the upper extremities: Part II – the scientific basis (knowledge base) for the guide. International Journal of Industrial Ergonomics, 10: 7–21 (this issue).

Design, selection and use of hand tools to alleviate trauma of the upper extremities: Part II – The scientific basis (knowledge base) for the guide

Anil Mital [a] and Asa Kilbom [b]

[a] *Industrial Engineering, University of Cincinnati, Cincinnati, OH 45221-0116, USA*
[b] *National Institute of Occupational Health, S-171 84 Solna, Sweden*

1. Problem description

Poor design and excessive use of hand tools are associated with increased incidence of acute trauma and subacute/chronic disorders of the hand, wrist, and forearm (Armstrong, 1983; Swedish Information System on Occupational Injuries – ISA, 1986; Aghazadeh and Mital, 1987).

Acute trauma includes burns, cuts, lacerations, abrasions, fractures, sprains, strains, dislocations and amputations caused by hands or fingers being caught, cut or burnt by a tool.

Subacute/chronic disorders are so-called cumulative trauma disorders (CTD), also named repetition strain injury (RSI), occupational cervicobrachial disorders (OCD) or work-related neck and upper limb disorders. These range clinically from poorly defined musculoskeletal discomfort/pain syndromes, to well defined clinical entities like carpal tunnel syndrome. The force, posture and frequency of movement of the hand, wrist, forearm and shoulder has been associated with such disorders. Since poor design and excessive use of hand tools lead to exertion of large grip forces, extreme wrist and finger deviations, high local pressures in the hand, unsuitable shoulder/neck postures and repetitive movements (Tichauer and Gage, 1977; Armstrong, 1983; Viikari-Juntura, 1987), hand tools are implicated in a large proportion of upper extremity musculoskeletal disorders.

2. Scope of the problem

According to the Swedish statistics on Occupational Injuries, 32% of all reported accidents involve fingers, hands or wrists. Some occupational groups, characterized by frequent use of cutting and striking tools, have a substantially higher proportion (food manufacturing – 58%, wood work – 44%). On the average, these accidents lead to 20 days of sick leave per occurrence, and 15% of them lead to permanent disability. Wood-working equipment has been reported to produce approximately 720,000 upper extremity injuries per year in the USA (Kelsey et al., 1980). The Massachusetts Medical Society reported that in 1982 more than 1 million occupational hand and finger injuries were treated in hospital emergency rooms, constituting 35% of the total injuries (Massachusetts Medical Society, 1982).

Overexertion diseases of the hand/wrist have been reported in the Swedish official statistics to occur at an increased rate among male slaughterhouse workers and meat cutters (standardized incidence ratio – 12.8). Armstrong (1983) reported an incidence rate of cumulative trauma of 25.6 in an automobile upholstery plant and 35.8 in an athletic products plant. Increased preva-

Correspondence to: A. Mital, Industrial Engineering, University of Cincinnati, Cincinnati, OH 45221-0116, USA.

lence of tendon-related disorders have been observed in jobs that require repetitive handgrips and exertion of large grip forces, for example in packaging work (Luopajarvi et al., 1979), and among meat cutters (Roto and Kivi, 1984). Silverstein et al. (1987a, b) observed an odds ratio of more than 15 for carpal tunnel syndromes when workers using high grip forces and repetitive handgrips were compared with those using low grip forces/low repetitive handgrips.

In most studies of overexertion diseases the role played by hand tools has not been specifically investigated. However, Tichauer (1966a) showed that unsuitable design of pliers precipitated the development of symptoms from the hand and forearm. Furthermore, poor design and excessive usage of tools was related to carpal tunnel syndrome in a study of electronic workers (Feldman et al., 1987).

Short-term fatigue and discomfort have been considered risk factors for musculoskeletal syndromes, and have therefore been used as criteria in ergonomic guidelines and standards. Fatigue and discomfort have been related to handle angle and work orientation in hammering (Schoenmarklin and Marras, 1989), and to tool shape and work height in work with screwdrivers (Ulin et al., 1990; Ulin and Armstrong, 1991). Also, it is well established that poor design of the grip of a tool leads to exertion of higher grip forces (Cochran and Riley, 1986; Kilbom et al., 1991) and to extreme wrist deviations (e.g., work with pliers – Tichauer and Gage, 1977) and, therefore, to more fatigue.

Thus, experimental and epidemiological studies support the view that poor design and excessive use of hand tools can lead to accidents, fatigue and musculoskeletal disorders.

In 1984, the Ergonomics and Engineering Controls Research Laboratory at the University of Cincinnati mailed a questionnaire to various federal and state agencies in the United States to determine the frequency, severity and annual cost of hand tool-related injuries in industry and to identify problem areas with regard to tool type, accident type, nature of injury, parts of body affected, type of industry and characteristics of the injured worker. The findings of this survey were reported in detail by Aghazadeh and Mital (1987). Response data from the questionnaire were tabulated and grouped and then analyzed. The results of the questionnaire showed that in the United States hand tool injuries account for 9% of all industrial injuries and cost approximately $10 billion annually. Following is a brief summary of the survey results:

Frequency. Hand tool injuries comprise about 9% of all work-related compensable injuries (approximately 265,570 injuries per year).

Severity. Approximately 3.9% of all amputations are caused by non-powered hand tools; 5.1% of all amputations are caused by powered tools. The most severe of the injuries are caused by knives among non-powered hand tools and saws among powered hand tools.

Costs. The annual cost (indemnity cost plus cost of work day losses, wage losses, insurance and other direct and indirect costs) was estimated to be $10 billion.

Tool type. The majority of injuries (80%) are inflicted by non-powered hand tools due to the widespread use of non-powered hand tools. The four highest injury-causing non-powered hand tools are knife, hammer, wrench, and shovel. Among the powered hand tools, saws, drills, grinders, and hammers are involved in more injuries than any other hand tool.

Type of accident/exposure. In the case of non-powered hand tools, 71.2% of all injuries are caused by 'striking by' or 'striking against' the tool. Overexertion is the second leading cause. 'Caught in' or 'betweens' and 'falls' contribute the least. The same types of accidents are responsible for most injuries in the case of powered hand tools. 'Struck by' and 'struck against' types of accidents are caused by saws, drills, hammers, and grinders more than any other powered hand tool.

Nature of injury. Cuts and lacerations, sprains and strains are the most frequent injuries for both powered and non-powered hand tools.

Part of body affected. The parts of the body most injured are the upper extremities. Thirty percent of all body parts affected are fingers. Head, neck, and the eyes are the least injured parts.

Industry type. Hand tool related injuries are numerous in manufacturing, mining and construction, and wholesale and retail groups. The service group (building maintenance workers, automotive and other repair shops, garden and health care services) also contribute significantly as the hand tools are also used frequently in these areas.

Occupation of the injured worker. Laborers, craftsmen and kindred workers, and operatives are the three groups of workers who are most frequently injured.

Age and hand tool injuries. Workers between ages of 18 and 35 years suffer more from hand tool injuries than any other age group.

3. Background

The investigations summarized in the previous section clearly demonstrate the existence of a strong relationship between the occurrence of upper extremity trauma and excessive use of poorly designed hand tools. The need for developing clear and concise ergonomic guidelines for designing and using hand tools is also obvious. The intended users of such guidelines are tool designers and manufacturers, buyers and users and occupational health and safety experts.

Even though the basic ergonomic principles of hand tool design have been known for several decades (Herig, 1964; Tichauer and Gage, 1977; Greenberg and Chaffin, 1977; Fraser, 1980; Konz, 1990; Mital, 1991), limited attention has been paid to these principles in designing widely used industrial tools. There may be several reasons for a lack of success in applying ergonomic principles to hand tool design:
(a) Guidelines have been general, lacking specificity, and frequently conflicting.
(b) Manufacturers generally do not perceive ergonomics an important consideration in hand tool design.
(c) Some guidelines may be unrealistic and have undergone insufficient practical evaluation.
(d) A wide information gap frequently exists between the management, designers, managers and ergonomists that does not permit implementation of scientifically developed design principles.

Many of these limitations, at least in part, are continually being overcome. Particularly, in the recent years, the following have helped:
(1) research to fill existing gaps in knowledge;

(2) increased awareness, on the part of the management, of direct and indirect costs of occupational injuries and illnesses;
(3) increased emphasis on thorough testing of new designs;
(4) availability of new materials and production methods;
(5) increased willingness on the part of ergonomists to use a team approach for tool and workstation design and work reorganization.

The factors associated with trauma are:

acute trauma:
fingers, hands or arms caught, cut or burnt by tool.

subacute / chronic trauma:
high forces
extreme postures of fingers, wrist and shoulders
repetitive movements
high localized pressure
segmental vibration

For all subacute/chronic trauma exposure factors, the duration of exposure is of crucial importance. Corresponding to these exposure factors are the following hand tool and task design features and hand tool user considerations:
– design of grip
– weight
– design of trigger
– special purpose
– human versus external power
– vibration characteristics
– duration and frequency of use
– gloves
– hand tool user considerations

3.1. Design of grip

The dexterity of the hand permits various kinds of grips. The two most basic grips have been defined by Napier (1956) as the power grip and the precision grip. In a power grip, tool axis is perpendicular to the forearm axis and the hand makes a fist with four fingers on one side and the thumb on the other side. This grip is used when large forces are to be exerted. Depending upon the direction of the line of force, three subcategories of the power grip are defined: (1) force parallel to the forearm (e.g., saw), (2) force at an angle to the forearm (e.g., hammer), and (3) torque about the forearm (e.g., corkscrew).

In a precision grip, the tool is pinched between the thumb and fingers. This grip is primarily used for precision work which requires control rather than exertion of large forces. The precision grip is classified into two categories: internal precision grip and external precision grip. In an internal precision grip, the shaft of the tool is internal to the hand (e.g., knife). In the external precision grip, the shaft of the tool is external to the hand (e.g., pencil).

There are some situations that require a combination of precision and power grip (e.g., serving in tennis). In industrial work, such combination grips are not recommended because the use of control muscles, which are smaller, to produce power can accelerate fatigue (Greenberg and Chaffin, 1977). In addition, a hook grip is also sometimes defined. This grip is used in situations when loads are held by the fingers and the fingers form a hook (e.g., holding a suitcase). Such grips should be avoided if the hand is to be used for precision tasks in the immediate future (within hours – Patkin, 1969).

Precision / power: Precision grips, on the average, provide only 20% strength of a power grip (Swanson et al., 1970). As discussed above, this is primarily due to the fact that a precision grip involves small muscles whereas a power grip involves larger muscle groups. Tools designed primarily for exertion of force, such as a hammer, should use a power grip; tools designed for manipulation, such as a surgical knife, should use a precision grip.

Angles of the forearm, grip and tool: Wrist deviations lead to various illnesses as well as loss in productivity (Tichauer, 1976) and grip strength (by as much as 14–27% – Terrell and Purswell, 1976). To avoid chronic illnesses and loss in productivity, the wrist should be kept straight (handshake orientation). If any bending is required, the tool, rather than the wrist should be bent (Tichauer and Gage, 1977). By bending the handles and increasing the length of the upper handle, it is possible to keep the wrist straight and avoid nerve, tissue, and blood vessel compression. Yoder et al. (1973) also used bent-handle pliers and successfully reduced complaints of wrist fatigue and stress. Similar bends have also been included in the designs of soldering irons (Tichauer, 1966a) and hammers (Konz, 1986; Schoenmarklin and Marras, 1989a, b).

When large forces are exerted on the handle(s) of a straight tool, the wrist is deviated in the direction of the ulna. This causes reduced strength and endurance and leads to fatigue. In addition, the carpal tunnel is probably compressed and there is a risk of carpal tunnel syndrome. To avoid these situations, Armstrong (1983) recommends that large forces, when exerted by hand/arm on a workpiece through a hand tool, should be parallel to the long axis (length axis) of the forearm. The tool handle axis should be oriented at 80° from the long axis of the tool (pistol grip) whenever large forces are to be exerted on a workpiece (Fraser, 1980).

For hammers, Konz (1986) concluded that a 10° bend is preferable to subjects. Both, Konz (1986) and Schoenmarklin and Marras (1989a), concluded that the angle of bend does not influence performance. Vertical hammering, however, was reported to be more difficult, less accurate, and more stressful and fatiguing (Schoenmarklin and Marras, 1989b).

Grip thickness: A large range of grip thicknesses has been recommended in the published literature. For *precision grips* these recommendations are slightly contradictory. While Hunt (1934) prescribed an 8 mm screwdriver handle diameter rather than a 16 mm diameter because of slower work with the thicker handle, Kao (1976) recommended 13 mm diameter for pens. However, Sperling (1986) observed less fatigue, lower recorded strain and larger maximal force exertions with a pen diameter of 30 mm, compared to 10 mm diameter. For *power grips*, the recommended diameters are influenced by the hand size (Greenberg and Chaffin, 1977; Kilbom et al., 1991). According to a Swedish expert group (Jonsson et al., 1977), power grips around a cylindrical object should surround more than half the circumference of the cylinder, but the fingers and the thumb should not meet. Up to a certain grip diameter, grip strength increases with grip diameter, but beyond a certain point the grip strength starts decreasing as the grip diameter increases. The optimum grip diameter is 65 mm according to Hertzberg (1973), and 51 mm according to Ayoub and LoPresti (1971). Greenberg and Chaffin (1977) observed an optimum between 50–85 mm and recommended 50 mm. Fransson and Winkel (1991), on the other hand, recorded maximum grip force between 50 and 60 mm diameter for females and between 55 and 65 mm for males.

Several additional recommendations for optimum grip diameter have also been published, but in most cases the recommended grip diameter falls between 50 and 60 mm. For practical purposes, it is important to keep in mind that people with small hands should not have to perform repetitively with power grips that have grip diameter larger than 60 mm. This recommendation is further reinforced by a positive relationship between hand size and grip strength. In *cross-action tools*, where the shafts/grips of the tool move (e.g., plate shears, scissors), the maximum span should be 100mm; the minimum should be 50 mm (Greenberg and Chaffin, 1977). Since the muscles closing the hand are stronger than the muscles opening the hand (Radonjic and Long, 1971), a spring should be used to open the handles (Kilbom et al., 1991). The spring will relieve the extensor muscles during the opening of the tool.

Grip length: Hand width ranges from approximately 79 mm for a 1st percentile female (Garrett, 1969a; 1971) to approximately 99 mm for a 99th percentile male (Garrett, 1969b; 1971). Using these dimensions as the basis, Konz (1990) recommended a minimum grip length of 100 mm; 125 mm grip length is more comfortable. The Eastman Kodak Company (1983) recommended a grip length of 120 mm. If gloves are to be worn or the grip is enclosed (e.g., hand saw), the minimum grip length should be 125 mm (Konz, 1990). Lindstrom (1973), on the other hand, recommended a handle length of 110 mm for men and 100 mm for women. In reality, it is impractical to provide different handle length tools for the same work done by males and females. In general, the grip length should be such that it would not limit the tool head opening and at the same time avoid excessive compressive forces or pressure concentration on the tender parts of the palm.

For external precision grip, the tool shaft must be at least 100 mm in length and must be long enough to be supported at the base of the first finger or thumb. For internal precision grip, the tool should extend past the tender palm (Konz, 1990). It certainly must not end close to the central part of the palm (Fransson and Kilbom, 1991).

Magill and Konz (1986) reported that torque exerted by a screwdriver was proportional to the handle grip length. The stem length of screwdrivers, however, do not influence human torque exertion capability (Mital, 1991).

Grip force: The average grip strength of randomly selected Swedish male and female subjects, using individually optimized grip diameters were 525 and 300 N, respectively, for one minute without fatigue (Kilbom et al., 1991). Since the 95th percentile value is only around 90 N (Greenberg and Chaffin, 1977), maximum grip force required should not exceed this value. However, if force exertion is required repeatedly the force requirements must be reduced. Thus, female and male subjects could use 40–50% of their maximum handgrip strength while cutting with plate-shears for one minute without fatigue.

Grip surface characteristics: The grip surface should be slightly compressible, nonconductive, and smooth (Konz, 1990). Compressible materials dampen vibration and allow better distribution of pressure. The grip material, however, should not be too soft otherwise sharp objects, such as metal chips, will get embedded in the grip and make it difficult to use. The grip material should not absorb oil or other liquids and should not permit conduction of heat or electricity. Wood or plastic are desirable handle materials (Wu, 1975). Foam rubber grip is preferred by users as it reduces the perception of hand fatigue and tenderness (Fellows and Freivalds, 1991). It, however, increases tool grip force. In general, metallic handles should be avoided or encased in a rubber or plastic sheath (Konz, 1990). Grip surface area should be maximized to ensure *pressure distribution* over as large an area as possible. Field and laboratory investigations suggest that excessive localized pressure sometimes causes pain that forces workers to interrupt their work. Pressure-pain thresholds of around 500 kpa for females and 700 kpa for males have been registered, the most sensitive areas being the thenar and the os pisiforme areas (Fransson and Kilbom, 1991). During maximal power grips these values are greatly exceeded. The *frictional characteristics* of the tool surface vary with the pressure exerted by the hand, with the smoothness and porosity of the surface and with contaminants (Buchholz et al., 1988; Bobjer, 1990). Sweat increases the coefficient of friction while oil and fat reduce it. When pinch force increases, the coefficient of friction is reduced. Adhesive tape and suede provide good friction when moisture is present, and would be good handle materials for reducing grip force requirements in workplaces with a lot of moisture (Buchholz et al., 1988).

In addition to the surface material, the pattern of the surface influences the perceived discomfort markedly, and is, therefore, a factor of significance in the acceptability of the tool. When the hand is clean or sweaty, the best (maximum) frictions were obtained with smooth or finely patterned surfaces (Bobjer, 1990).

Grip shape: The grip shape should be such that it maximizes the area of contact between the palm and the grip. Maximizing this area of contact leads to better pressure distribution and reduces the chances of forming pressure ridges or pressure concentration points. This is particularly important for tools that require a power grip. Generally, the tools available in the market have a cylindrical shape grip. Rubarth (1928) reported that cylindrical handles with rounded end maximized force. For a turning action, exerted on tools with cylindrical grips, Pheasant and O'Neill (1975) and Grieve and Pheasant (1982) reported that the torque is proportional to the handle diameter. Mital et al. (1985) and Mital and Sanghavi (1986), however, did not observe that; they found the torque increased with the grip diameter but not proportionately. When there is no turning action on the grip and yet torque is exerted on the workpiece (e.g., wrenches), torque varies with the lever arm (Mital and Channaveeraiah, 1988).

Pheasant and O'Neill (1975) reported that the shape of the handle is not relevant as long as the hand does not slip around it. The chances of hand slippage are reduced when a non-circular handle, such as rectangular or triangular, is used. The edges in such grips resist the slippage. Mital and Channaveeraiah (1988) compared cylindrical, triangular, and rectangular handles and reported that the torque exertion capability of individuals with triangular handle screwdrivers was greater than with cylindrical handle screwdrivers. For wrenches, torque exertion capability was maximized when cylindrical handles were used. According to Cochran and Riley (1986a), the thrust forces exerted with straight knives are as much as 10% greater with triangular handles as compared to cylindrical or rectangular handles. A handle perimeter of 110 mm is recommended for knives.

A T-handle is preferable for screwdrivers than the straight handle (Pheasant and O'Neill, 1975) as it not only prevents wrist deviations but also increases torque exertion by as much as 50%. According to Saran (1973), the T-handle should be slanted (60° angle) to allow the wrist to be straight and should be 25 mm in diameter. For knives, a pistol grip is preferable to a straight grip (Armstrong et al., 1982; Karlqvist, 1984). The angle of the grip should be 78° from the horizontal (Fraser, 1980).

Grooves, indentations and guards: Form fitting tools that have grooves for fingers, etc., are not recommended. The main problem with grooves is that they do not fit people; either the grooves are too big or too small. In both cases, pressure ridges are created, leading to nerve compression and impairment in circulation. Vertical grooves on the handle, provided by the designer to prevent the hand from slipping, also cut into the palm of the hand and create pressure ridges. Such grooves or flutes should be avoided. According to Bobjer (1984), Ergo in Sweden has developed screwdrivers that have 40 small axial grooves. These grooves are small enough to avoid pressure concentration but large enough to provide better friction between the hand and the grip. For small screwdrivers, used in precision work, regular size handles were used to provide better control.

In general, grooves and indentations are undesirable (Tichauer and Gage, 1977). Slight and uniform surface indentations, however, are desirable as they allow greater torque exertion capability. Pheasant and O'Neill (1975) reported that screwdrivers with cylindrical and knurled surfaces provided greater torque exertion capability than smooth surface cylindrical handles. The 50 mm diameter handle with knurled surface maximized torque exertion.

The primary purpose of the guard in front of the grip is to prevent injury when the hand slips or the hand and tool collide against a sharp or rigid surface. The guard in such cases prevents the hand from slipping and shields the hand against the impact. According to Cochran and Riley (1986b), guards of 1.52 cm height provide adequate safety. Greater guard heights improve safety only minimally.

3.2. Effective weight of tool

The effective weight is the weight of the tool that is supported by the worker. For manual precision tools it is usually so low as to not cause any problems. With many power grip tools (axes, hammers, saws), and with power tools (especially those with external power supply through pneumatic or electrical leads) weight is frequently a problem. In order to reduce fatigue, the tool should not weigh more than 2.3 kg (Greenberg and Chaffin, 1977; Eastman Kodak Company, 1983). However, if the center of gravity of a heavy tool is far from the wrist this weight limit should be further reduced. This is also corroborated by Johnson and Childress (1988), who observed that powered screwdrivers weighing about 1.12 kg or less do not produce significantly different magnitudes of EMG activity. The subjective assessment study of workers to determine the weight of the tool, conducted by Armstrong et al. (1989), also indicated that workers find tools weighing between 0.9 kg and 1.75 kg feel 'just right'. These weight recommendations apply to tools supported predominantly by the large muscle groups of the forearm. For fine precision work, where small muscle groups of the hand support the tool (e.g., dentistry and electronic assembly work), much lighter tools have to be used.

3.3. Type of tool

Considerably greater torque is exerted with wrenches (10 to 20 times more) than with screwdrivers as they use different muscles and lever arms (Mital, 1991). For screwdrivers, torque exertion capability generally increases with the handle diameter. For wrenches, the torque varies linearly with the lever arm. However, there are exceptions. For instance, more torque is exerted with a socket wrench than with a vise grip or spanner wrench or crescent wrench of about the same lever arm. This is expected to be caused by the type of grip and the nature of coupling between the tool and the workpiece.

3.4. Design of trigger

Many powered tools are activated by a trigger, operated either by the thumb, or one or several fingers. The trigger may have an on–off design, which usually does not cause problems. However, the trigger often has to be activated repeatedly or for prolonged periods of time, and this may cause problems as it requires some precision simultaneously with force exertion (holding and guiding the tool) (Tichauer, 1966b). In order to avoid fatigue, the tool should be designed such that it can be activated by either hand. While this will reduce the burden on one finger, the need to operate the trigger continuously by muscle force will still cause problems. Under such conditions, the trigger should have a latch (or some comparable locking mechanism) that can hold the trigger in place while the tool is in use. A slight dislocation of the latch will release the trigger and turn off the tool.

3.5. Special purpose

In general, special-purpose tools, even though costly, are faster. The savings in labor cost, due to shorter time per use, over the life of the tool more than offsets the high cost of the tool (Konz, 1990).

3.6. Human versus external power

Mechanical energy is generally cheaper than human energy. Besides, humans tend to fatigue much faster than machines. The efficiency of human power generation is also much lower than machine power generation. Therefore, power tools with external energy whenever possible (Konz, 1990).

3.7. Vibration characteristics

Vibrating tools can cause vasospastic disease (Raynaud's disease) and contribute to carpal tunnel syndrome (Wasserman and Badger, 1973). The seriousness of occupational diseases resulting from using vibrating hand tools for prolonged periods has been further emphasized by the National Institute for Occupational Safety and Health in its 1983 bulletin (NIOSH, 1983). Some researchers, however, doubt the existence of a direct relationship between vibration and carpal tunnel syndrome (Taylor, 1988a,b; Meagher, 1991). Silverstein et al. (1987a), however, assert on the basis of epidemiological evidence that the risk of CTD is 5 times increased in jobs inducing high force/high repetitiveness/vibration compared to twice in jobs inducing only high force/high repetitiveness. Thus, although there is no definite proof that vibration alone will cause CTD, in combination with forceful and repetitive work with tools vibration will most certainly aggravate the disorder.

It is generally recommended that vibrations in the critical range (40 to 130 Hz) should be avoided (Wasserman and Badger, 1973). The Technical Research Centre of Finland (1988) in its report on pneumatic screwdrivers and nut runners recommends that for vibration levels exceeding 126 dB, the exposure times be less than 8 hours. Lundstrom and Johansson (1986) reported that exposure to vibration between 2 and 200 Hz leads to acute impairment of tactile sensibility. Since, above 1000 Hz, resonances of the hand-arm systems do not occur (Iwata et al., 1971), it is important to avoid segmental vibration frequencies below 1000 Hz. In summary, the NIOSH (1983) recommendations – that: (i) jobs be redesigned to minimize the use of vibrating hand tools, (ii) powered hand tools be redesigned to minimize vibration, and (iii) engineering controls, work practices, and administrative controls be used to minimize vibration where jobs cannot be redesigned – are critical. The exposure to vibration can be reduced through a reduction in the driving force and the use of vibration damping materials (Andersson, 1990).

3.8. Duration and repetitiveness of use

The duration and repetitiveness of use of a hand tool profoundly increase the risk of occupational injury, either alone or in combination with factors discussed above (Hammer, 1934; Kurppa et al., 1979; Luopajarvi et al., 1979; Kuorinka and Koskinen, 1979; Cannon et al., 1981).

Silverstein et al. (1987a, b) observed that in work with a high degree of repetitiveness and high manual force exertions, the prevalence of carpal tunnel syndrome was 15 times higher than in jobs with low repetitiveness and low force exertions.

Sperling et al., 1991 proposed a set of criteria whereby the demands on time are divided into three categories:

low – less than one hour distributed over the entire day, or less than 10 minutes continuously or repetitively.

moderate – one to four hours distributed over the entire day, or 10–30 minutes continuously or repetitively.

high – more than 4 hours distributed over the entire day, or more than 30 minutes continuously or repetitively.

These researchers propose that high demands on time, in combination with high force or precision requirements, should be considered unacceptable. Combinations of moderate demands on time and high force or precision requirements, according to them, should also be analyzed.

Both duration and repetitiveness of work have been studied experimentally. Greenberg and Chaffin (1977) reported that 2.5 kg held in one hand leads to significant muscle fatigue within 20 min., even with a comfortable work posture. With extended or raised arms, or in persons with reduced muscle strength, time to fatigue is markedly reduced. In repetitive exertions, the duration of each exertion, and the ensuing rest period, influence the time it takes to reach subjective fatigue. Thus, one-minute exertions with one minute rest periods lead to fatigue within four hours if 5.6 kgs are held in one hand. If the weight of the object is 1.9 kg, one-minute exertions with 10 s rest periods lead to fatigue in 4 hours (Greenberg and Chaffin, 1977).

Mital and Channaveeraiah (1988) reported that if repeated exertions are made at a self-determined pace, the maximum torque that can be exerted with screwdrivers and wrenches declines by as much as 38% after 240 seconds. The decline in force exertion capability due to fatigue, thus, limits the endurance time.

Attempts have been made to model optimum work-rest regimes for intermittent isometric exercise (Rohmert, 1973; Milner, 1985; Dul et al., 1990). The criteria have been recovery heart rate and recovery of endurance, respectively. Recently, Bystrom and Kilbom (1991), using force response to electrical stimulation of the forearm muscles and local blood flow as criterion for recovery, studied time-to-recovery after sustained and intermittent handgrip exercise. At 25% MVC continuous exercise until exhaustion, recovery of force did not take place even after 24 hours. After 10 + 2 s (exercise and rest) intermittent exercise until half endurance time, recovery had not yet occurred after 60 minutes. Future use of such sensitive indices of recovery may lead to sharpening of the criteria previously used.

It is likely that with high demands on precision, time-to-fatigue is further reduced, both in sustained and intermittent exercise. In addition, the models used so far have been recovery of cardiovascular or muscular parameters. Consideration of these criteria does not necessarily provide protection from overexertion of connective tissue.

Several attempts have also been made to identify the maximal number of exertions per hour, or per day, that can be tolerated without fatigue or muscle-tendon disorders. In a tea-packing factory, Obolenskaja and Goljanitzki (1927) observed that a high rate of tenosynovitis of the upper extremities occurred in a group of workers who performed 7,000–12,000 hand movements per day. Luopajarvi et al. (1979) found a high prevalence of muscle-tendon disorders in assembly line packers performing up to 25,000 movements per day. Kuorinka and Koskinen (1979) also observed an exposure–response relationship between muscle tendon syndromes and number of pieces handled per year in a light mechanical industry. In these studies, exertions were repetitive but of short duration, and the injury was usually peritendinitis or tenosynovitis. It is not possible, based on these studies, to set limits of acceptable number of exertions per work day as the force in each movement will also influence the risk of injury. It can, however, be concluded that repetitive hand-wrist exertions distributed at a high rate over the work day introduces an increased risk of disorders.

3.9. Gloves

The primary purpose of gloves is to provide worker protection against injuries that may result from slipping (cuts, bruises, abrasions, etc.) or touching hot, cold, or toxic materials. Gloves protect by resisting sharp edges, splinters, extreme temperatures, sparks, electricity, and chips. Gloves are also used to reduce transmission of vibration energy by absorbing or attenuating it and thereby preventing Raynaud's syndrome (white finger, or dead finger, condition).

Gloves also have many disadvantages. In general, gloves interfere with a person's grasping ability and hand movements. Gloves also change the effective dimensions of the hand. The hand thickness may increase anywhere from 8 mm to 40 mm (Damon et al., 1966). The glove material over the hand tends to reduce manual dexterity (McGinnis et al., 1973; Nolan and Cattroll, 1977; Plummer et al., 1985). It is, however, possible to design gloves that provide good dexterity and manipulatory capabilities (Andruk et al., 1976; Gianola and Reins, 1976). Gloves also reduce feedback and its seams and edges could cause irritation at the contact points.

The influence of gloves on human performance is generally mixed. Weidman (1970) studied the influence of different kinds of gloves on manual performance and observed that performance times decreased by 12.5% when Neoprene gloves were worn. With terry cloth, leather, and PVC gloves, performance decreased by 36%, 45%, and 64%, respectively. Jenkins (1958), however, reported superior performance when operating small knobs while wearing gloves. Plummer et al. (1985) contradicted Jenkins's results and reported that the assembly of a 0.635 cm diameter bolt, nut, and washer took significantly longer time and resulted in more errors than those of 0.79 cm diameter and 1.27 cm diameter bolts. Desai and Konz (1983), however, found no difference in error rates in the tactile inspection task of hydraulic hoses. As a matter of fact, in some cases use of gloves has been reported to enhance performance. For instance, Bradley (1969a, b) reported improvement in occupational speeds for control operations. Riley et al. (1985) also reported an increase in friction and reduction in task strength requirements with gloves.

The force/torque exertion capability of workers wearing gloves has been of interest to a number of researchers. When wearing gloves, a fraction of the force generated by muscular contraction may be directed in maintaining the grip and may result in reduced force production. Reduction in grip (grasp) force when using gloves has been reported by several researchers (Hertzberg, 1973; Lyman, 1957; Swain et al., 1970; Sperling, 1980; Cochran et al., 1986; Sudhakar et al., 1988). According to Hertzberg (1973), grip strength declines by 20% when gloves are used. Reduction in strength may be as much as 30% or more, according to Lyman (1957) and Sperling (1980). Cochran et al. (1986) reported a reduction in grasp force due to gloves ranging from 7.3% to 16.8% when compared to no gloves. Sudhakar et al. (1988) also found the peak grip strength with rubber and leather gloves to be 10 to 15% lower compared to grip strength without gloves even when there was no significant difference in muscle activity between glove and no-glove conditions. Gloves also reduce the transmission of vibration energy (Goel and Rim, 1987).

The reduction in grip or grasp force, however, has not always been observed. For instance, Lyman and Groth (1958) reported that workers exerted more force with gloves than without gloves when inserting pins in a pegboard. Riley et al. (1985) also reported an increase in the maximum pull force, push force, wrist flexion torque, and wrist extension torque.

The investigations concerning force exertion capability of workers with gloves have simulated mainly those activities that require compressive forces (grip or grasping forces). For example, using pliers requires exertion of compressive forces. The exception is the study reported by Riley et al. (1985) in which actions with a knife are simulated. There are many industrial activities, however, that require a different kind of force exertion. For instance activities, such as maintenance and repair activities, that require the use of wrenches and screwdrivers. True, a compressive (grasping) force is needed to maintain a firm grip, but the main exertion is a rotational force at the periphery of the grip that when transmitted to the workpiece causes fastener devices, such as nuts and screws, to tighten or loosen.

Recently, Mital et al. (1992) used seven different hand tools and nine varieties of commercial gloves and measured peak volitional torques on a simulated workpiece. The peak torque and electromyogram (emg) of flexor digitorum profundus and supinator brevis were also recorded. The results indicated that muscle activity did not differ significantly between the glove and no-glove conditions and the peak torque exertion capability of individuals generally increased with gloves. The magnitude of torque exerted on the workpiece was different for different gloves. These results contradict some of the findings reported above.

Overall, gloves are a mixed blessing. They provide safety and comfort but occasionally at the cost of reduction in manual performance.

3.10. User considerations

Worker characteristics such as left/right handedness, gender and age, strength, technique, body size and posture profoundly influence both risk of injury and productivity in work with hand tools. Research in these areas is summarized below, especially with regard to torque exertion capability as reviewed by Mital (1991).

Left-handed and right-handed users. Approximately 10% of the population is left-handed (Konz, 1974). Yet tools are mostly designed for right-handed users. Since the preferred hand is stronger (about 7% to 20% – Shock, 1962; Miller, 1981), more dexterous (Kellor et al., 1971), and faster (Konz and Warraich, 1985), a left-handed user is at a disadvantage when using a right-handed tool. Besides the strength disadvantage, certain right-hand tools require a different action when used by the left hand (Capener, 1956). For instance, the right-hand scissors when used by a left-handed person requires a reversal of pressure upon the rings of the scissors. This leads to a tendency to twist it. The shearing surfaces thus lie in the vertical plane instead of the horizontal plane. In many instances, tools designed for right handers can not be used by left handers (Laveson and Meyer, 1976). Tools which are designed for use by either hand avoid such problems.

Worker gender and age. The grip strength of females is approximately 50 to 67% of male grip strength (Lindstrom, 1973; Konz, 1990; Chaffin and Anderson, 1984). Males, on the average can exert about 500 N force while females on the average are able to exert 250 N force. The difference in force is primarily due to females having smaller grip size and muscles. This means that tools designed on the basis of male strength and anthropometry will not accommodate females (Ducharme, 1975). Mital (1991) reported that the torque exertion capability of males with screwdrivers and wrenches is 10% to 56% greater than females. Thus, the tools designed for male dimensions and force capabilities will not permit easy integration of females into the workforce.

The grip strength is also significantly affected by age (Shock, 1962). By age 65, the decline in strength is about 20 to 40%. While no study of the effect of age on torque exertion capability with nonpowered screwdrivers and wrenches has been reported in the literature, it is logical to assume that variation in the age of the workforce would affect their torque exertion capability and should be taken into consideration in selecting tools and designing fastener devices.

Isometric strengths and anthropometry. Mital and Sanghavi (1986) observed a significant, but not very large, correlation between isometric shoulder strength and torque exertion capability (0.26 for males and 0.23 for females). In general, heavier and stronger individuals applied more torque. Other isometric strengths did not show appreciable correlations. Thus, it appears that shoulder strength is a limiting factor in torque exertion capability. Body heights, sitting or standing, were unimportant. A similar conclusion about anthropometric characteristics was drawn by Johnson and Childress (1988) in their study of powered screwdrivers.

Body posture. The effect of posture on torque exertion capability was studied in detail by Mital (1986) and Mital and Channaveeraiah (1988). Postures ranging from standing to lying-on-the-side were studied. It was concluded that minor variations in posture are unimportant. Extreme variations, such as between standing and lying on the stomach or leaning sideways from a ladder, however, cause large differences in torque exertion capabilities. Mital and Channaveeraiah (1988) have given detailed tables showing the distributions of torque exertion profiles of males and females in different postures with various wrenches and screwdrivers.

Technique and experience. Marras and Rockwell (1986) studied the force generated by experienced and inexperienced workers during the spike maul use. They observed that experienced workers generated almost twice as much force (136, 846 N) as the novice workers (64, 446 N). This large difference was due to the technique used by experienced workers (snapping action as opposed to sustained force application). Many tools, such as screwdrivers, however, do not permit snapping action and, therefore, the difference in the force/torque exerted by experienced and inexperienced workers is not expected to be as large but yet significant.

Reach distance. The torque exertion capability with wrenches and screwdrivers decreases linearly with the reach distance (distance between the front of the ankles and the workpiece) between 33 cm to 58 cm in the standing posture and between 46 cm to 71 cm in the sitting posture (Mital and Sanghavi, 1986).

Wrist orientation. The maximum torque with wrenches is exerted when the long axis of the wrench is kept horizontal. With screwdrivers, torque exertion is maximized when the wrist is rotated 90° counterclockwise from the prone position (Mital and Channaveeraiah, 1988).

4. Need for further research

Based on the above review, it appears that knowledge in the following areas is limited, thus limiting the development of comprehensive guidelines:
- Grip design and force exertion capabilities with different kinds of precision grips.
- Grip surface characteristics, for all types of grips but especially for gripping of contaminated handles.
- Hand sensitivity and occupational injury risk with localized pressure concentration while exerting for prolonged periods and/or repeatedly.
- Optimum weight of hand tools for precision as well as power.
- Occupational injury risk during work with high demands on precision.
- Optimum work-rest regime, with regard to short-term recovery and long-term musculoskeletal disorders.

References

Aghazadeh, A. and Mital, A., 1987. Injuries due to hand tools: Results of a questionnaire. Applied Ergonomics, 18: 273-278.

Andersson, E.R., 1990. Design and testing of a vibration attenuating handle. International Journal of Industrial Ergonomics, 6: 119-125.

Andruk, F.S., Shampine, J.C. and Reins, D.A., 1976. Aluminized Fireman's (Fire Proximity) Handwear: A Comparative Study of Dexterity Characteristics. Report no. US DOD-AGFSR-76-17, Navy Clothing and Textile Research Facility, Natick, Massachusetts.

Armstrong, T.J., 1983. An ergonomics guide to carpal tunnel syndrome. American Industrial Hygiene Association, Akron, Ohio, USA.

Armstrong, T.J., Punnett, L. and Ketner, P., 1989. Subjective worker assessments of hand tools used in automobile assembly. American Industrial Hygiene Association Journal, 50: 639-645.

Armstrong, T.J., Foulke, J., Joseph, B. and Goldstein, S., 1982. Investigation of cumulative trauma disorder in a poultry processing plant. American Industrial Hygiene Association Journal, 43: 103-116.

Ayoub, M.M. and LoPresti, P., 1971. The determination of an optimum size cylindrical handle by use of electromyography. Ergonomics, 14: 509-518.

Bobjer, O., 1984. Screwdriver handle design for power and precision. In: Proceedings of the 1984 International Conference on Occupational Ergonomics. Human Factors Association of Canada, Rexdale, Ontario, pp. 443-446.

Bobjer, O., 1990. Greppytors friktion samt upplevelsen av obehag vid beroring (Friction of grip surface and perceived discomfort upon touch). Preliminary report to the Swedish Work Environment Fund.

Bradley, J.V., 1969a. Effect of gloves on control operation time. Human Factors, 11: 13-20.

Bradley, J.V., 1969b. Glove characteristics influencing control manipulability. Human Factors, 11: 21-35.

Buchholz, B., Frederick, L.J. and Armstrong, T.J., 1988. An investigation of human palmer skin friction and the effects of materials, pinch force and moisture. Ergonomics, 31: 317-325.

Bystrom, S. and Kilbom, A., 1991. Low frequency fatigue as an index of fatigue and recovery after static and intermittent exercise. European Journal of Applied Physiology (in press).

Cannon, L., Bernacki, E. and Walter, S., 1981. Personal and occupational factors associated with carpal tunnel syndrome. Journal of Occupational Medicine, 23: 255-258.

Capener, N., 1956. The hand in surgery. The Journal of Bone and Joint Surgery, 38B: 128-151.

Chaffin, D.B. and Anderson, G.B.J., 1984. Occupational Biomechanics. John Wiley and Sons, New York, New York, USA.

Cochran, D.J. and Riley, M.W., 1986a. The effects oh handle shape and size on exerted forces. Human Factors, 28: 253-265.

Cochran, D.J. and Riley, M.W., 1986b. An evaluation of knife handle guarding. Human Factors, 28: 295-301.

Cochran, D.J., Albin, T.J., Riley, M.W. and Bishu, R.R., 1986 Analysis of grasp force degradation with commercially available gloves. Proceedings of the Human Factors Society 30th Meeting, Santa Monica, California, USA, pp. 852-855.

Damon, A., Stoudt, H.W. and McFarland, R.A., 1966. The Human Body in Equipment Design. Harvard University Press, Cambridge, Massachusetts, USA.

Desai, S. and Konz, S., 1983. Tactile inspection performance with and without gloves. Proceedings of the Human Factors Society 27th Annual Meeting, Santa Monica, California, U.S.A., pp. 782-785.

Ducharme, R.E., 1975. Problem tools for women. Industrial Engineering, September: 46-50.

Dul, J., Douwes, M. and Smitt, P., 1990. A work-rest model for static postures. Biomechanics Seminars, 117-124.

Eastman Kodak Company, 1983, Ergonomic Design for People at Work. Lifetime Learning Publications, Belmont, California.

Feldman, R.G., Hyland Travers, P., Chirico-Post, J. and Keyserling, W.M., 1987. Risk assessment in electronic assembly workers: Carpal tunnel syndrome. Journal of Hand Surgery, 12A: 849-855.

Fellows, G.L. and Freivalds, A., 1991. Ergonomics evaluation of a foam rubber grip for tool handles. Applied Ergonomics, 22: 225-230.

Fransson, C. and Kilbom, A., 1991. Sensitivity of the hand to surface pressure. Submitted.

Fransson, C. and Winkel, J., 1991. Hand strength: The influence of grip span and grip type. Ergonomics (in press).

Fraser, T.M., 1980, Ergonomics principles in the design of hand tools. Occupational Safety and Health Series No. 44, International Labour Office, Geneva, Switzerland.

Garrett, J., 1969a. References on hand anthropometry for women. AMRL Technical Report, 69-42. Army Medical Research Laboratory, Wright Patterson Air Force Base, Ohio.

Garrett, J., 1969b. References on hand anthropometry for men. AMRL Technical Report, 69-26. Army Medical Research Laboratory, Wright Patterson Air Force Base, Ohio.

Garrett, J., 1971. The adult human hand: Some anthropometric and biomechanical considerations. Human Factors, 13: 117-131.

Gianola, S.V. and Reins, D.A., 1976, Low Temperature Handwear with Improved Dexterity. Report no. 117, US Navy Clothing and Textile Research Unit, Natick, Massachusetts.

Goel, V.K. and Rim, K., 1987. Role of gloves in reducing vibration: An analysis for pneumatic chipping hammer. American Industrial Hygiene Association Journal, 48: 9-14.

Greenberg, L. and Chaffin, D.B., 1977. Workers and Their Tools. Pendell, Midland, Michigan, USA.

Grieve, D. and Pheasant, S., 1982. Biomechanics. In: W.T. Singleton (Ed.), The Body at Work. Cambridge University Press, Cambridge, pp. 142-150.

Hammer, A., 1934. Tenosynovitis. Medical Record, 353-355.

Herig, F., 1964. Bessere Arbeit durch bessere Griffe. Carl Marhold Verlags Buchhandlung Berlin, Charlottenburg.

Hertzberg, H., 1973. Engineering anthropometry. In: H. Van Cott and R. Kincaid (Eds.), Human Engineering Guide to Equipment Design. Superintendent of Documents, U.S. Government Printing Office, Washington, DC.

Hunt, L., 1934. A study of screwdrivers for small assembly work. The Human Factor, 9: 70–73.

Informationssystemet for Arbetsskador (ISA) Arbetsskador, 1986. Arbetarsskyddsstyrelsen, Stockholm, Sweden.

Iwata, H., Matsuda, A., Takahashi, H. and Watabe, S., 1971. Roentgenographic findings in elbows of rock drill workers. Acta Scholae Medicine of University of Gifu, 19: 393–404.

Jenkins, W.L., 1958. The superiority of gloved operation of small control knobs. Journal of Applied Psychology, 42: 97–98.

Johnson, S.L. and Childress, L.J., 1988. Powered screwdriver design and use: Tool, task, and operator effects. International Journal of Industrial Ergonomics, 2: 183–191.

Jonsson, B., Lewin, T., Tomsic, P., Garde, G. and Forssblad, P., 1977. Handen som arbetsredskap (The hand as a work-tool), Arbetsskyddsstyrelsen, Stockholm, Sweden.

Kao, H., 1976. An analysis of user preference toward handwriting instruments. Perceptual and Motor Skills, 43: 522.

Karlqvist, L., 1984 Cutting operation at canning bench – A case study of hand tool design. In: Proceedings of the 1984 International Conference on Occupational Ergonomics. Human Factors Association of Canada, Rexdale, Ontario, pp. 452–456.

Kellor, M., Kondrasuk, R., Iverson, I., Frost, J., Silberberg, N. and Hoglund, M., 1971. Hand strengths and dexterity tests. Sister Kenny Institute, Manual 721, Minneapolis, Minnesota, USA.

Kelsey, J.L., Pastides, H., Kreiger, N., Harris, C. and Chernow, R.A., 1980. Upper extremity disorders: A survey of their frequency and cost in the United States. CV Mosby Co., St. Louis, Missouri, USA.

Kilbom, A. and Ekholm, J., 1991. Handgreppsstyrka (Hand grip strength, Swedish with English summary). Stockholm MUSIC study group, Stockholm, Sweden.

Kilbom, A., Makarainen, M., Sperling, L., Kadefors, R. and Lindberg, L., 1991. Betydelsen av verktygsutformning och individfaktorer for prestation och trotthet vid arbete med platsax (Tool design and individual factors – Their effect on productivity and fatigue in work with plate shears – Swedish with English summary). Submitted.

Konz, S., 1974. Design of hand tools. In: Proceedings of the Human Factors Society 18th Annual Meeting. Human Factors Society, Santa Monica, California, USA, pp. 292–300.

Konz, S., 1986. Bent handle hammers. Human Factors, 28: 317–323.

Konz, S., 1990. Work Design: Industrial Ergonomics. Publishing Horizons, Inc., Worthington, Ohio, USA.

Konz, S. and Warraich, M., 1985. Performance differences between the preferred and non-preferred hand when using various tools. In: I.D. Brown, R. Goldsmith, K. Coombes and M.A. Sinclair (Eds.), Ergonomics International 85. Taylor and Francis, London, United Kingdom, pp. 451–453.

Kurppa, K., Warris, P. and Kokkanen, P., 1979. Tennis elbow. Scandinavian Journal of Work, Environment and Health, 5 (Suppl. 3): 15–18.

Kuorinka, I. and Koskinen, P., 1979. Occupational rheumatic diseases and upper limb strain in manual jobs in a light mechanical industry. Scandinavian Journal of Work, Environment and Health, 5 (supplement 3): 39–47.

Laveson, J.I. and Meyer, R.P., 1976. Left out 'lefties' in design. In: Proceedings of the Human Factors Society 20th Annual Meeting. Human Factors Society, Santa Monica, California, USA, pp. 122–125.

Lindstrom, F.E., 1973. Modern Pliers. BAHCO Vertyg, Enkoping, Sweden.

Lundstrom, R. and Johansson, R.S., 1986. Acute impairment of the sensitivity of skin mechanoreceptive units caused by vibration exposure of the hand. Ergonomics, 29, 687–698.

Luopajarvi, T., Kuorinka, I., Virolainen, M. and Holmberg, M., 1979. Prevalence of tenosynovitis and other injuries of the upper extremities in repetitive work. Scandinavian Journal of Work, Environment and Health, 5, Suppl. 3: 48–55.

Lyman, J., 1957. The effects of equipment design on manual performance. In: R. Fisher (Ed.), Production and Functioning of the Hands in Cold Climates. National Academy of Sciences, National Research Council, Washington, DC, USA, pp. 86–101.

Lyman, J. and Groth, H., 1958. Prehension forces as a measure of psychomotor skill for bare and gloved hands. Journal of Applied Psychology, 42: 18–21.

McGinnis, J.M., Bensel, C.K. and Lockhar, J.M., 1973. Dexterity afforded by CB protective gloves. Report no. 73-35-PR. US Army Natick Laboratories, Natick, Massachusetts.

Magill, R. and Konz, S.A., 1986. An evaluation of seven industrial screwdrivers. In: W. Karwowski (Ed.), Trends in Ergonomics/Human Factors III. North-Holland, Amsterdam, The Netherlands, pp. 597–604.

Marras, W.S. and Rockwell, T.H., 1986. An experimental evaluation of method and tool effects in spike maul use. Human Factors, 28: 267–281.

Massachusetts Medical Society, 1983. Occupational finger injuries – United States 1982. Morbidity and Mortality Weekly Report, 32: 539–591.

Meagher, S., 1991, Vibration: Causation of repetitive trauma disorders. In: W. Karwowski and J.W. Yates (Eds.), Advances in Industrial Ergonomics and Safety III. Taylor and Francis, London, United Kingdom, pp. 69–74.

Miller, G.D., 1981. Significance of dominant hand grip strengths in hand tools. Unpublished M.S. thesis, Pennsylvania State University, University Park, Pennsylvania, USA.

Milner, N.P., 1985. Modelling fatigue and recovery in static postural exercise. Ph.D. Thesis, University of Nottingham, Nottingham, United Kingdom.

Mital, A., 1986, Effect of body posture and common hand tools on peak torque exertion capabilities. Applied Ergonomics, 17: 87–96.

Mital, A., 1991. Hand tools: injuries, illnesses, design and usage. In: A. Mital and W. Karwowski (Eds.), Workspace, Equipment and Tool Design. Elsevier Science Publishers, Amsterdam, The Netherlands, pp. 219–256.

Mital, A. and Channaveeraiah, C., 1988. Peak volitional torques for wrenches and screwdrivers. International Journal of Industrial Ergonomics, 3: 41–64.

Mital, A. and Sanghavi, N., 1986. Comparison of maximum volitional torque exertion capabilities of males and females using common hand tools. Human Factors, 28: 283–294.

Mital, A., Kuo, T. and Faard, H.F., 1992. A quantitative evaluation of gloves used with non-powered hand tools in routine maintenance tasks. Ergonomics (in press).

Mital, A., Sanghavi, N. and Huston, T., 1985. A study of factors defining the 'operator-hand tool system' at the work place. International Journal of Production Research, 23: 297–314.

Napier, J., 1956. The prehensile movements of the human hand. Journal of Bone and Joint Surgery, 38B: 902–913.

National Institute for Occupational Safety and Health, 1983. Current Intelligence Bulletin 38: Vibration Syndrome. Cincinnati, Ohio, USA.

Nolan, R.W. and Cattroll, S.W., 1977. Evaluation of British and Canadian Conductive Rubber Heating Elements for Handwear: Preliminary Report. Report No. 77-24, Canadian Defense Research Establishment, Ottawa.

Obolenskaja, A.J. and Goljanitzki, I., 1927. Die serose Tendovainitis in der Klinik und im Experiment. Deutsches Z. Chir. 201: 388–399.

Patkin, M., 1969. Ergonomic design of a needleholder. Medical Journal of Australia, 2: 490–493.

Pheasant, S. and O'Neill, D., 1975. Performance in gripping and turning – A study in hand/handle effectiveness. Applied Ergonomics, 6: 205–208.

Plummer, R., Stobbe, T., Ronk, R., Myers, W., Kim, H. and Jaraiedi, M., 1985. Manual dexterity evaluation of gloves used in handling hazardous materials. In: Proceedings of the Human Factors Society 29th Annual Meeting, Santa Monica, California, USA, pp. 819–823.

Radonjic, D. and Long, C., 1971. Kinesiology of the wrist. American Journal of Physical Medicine, 50: 57–71.

Riley, M., Cochran, D. and Schanbacher, C., 1985. Force capability differences due to gloves. Ergonomics, 28: 441–447.

Rohmert, W., 1973. Problems of determination of rest allowances. Part II. Determining rest allowances in different tasks. Applied Ergonomics, 24: 158–162.

Roto, P. and Kivi, P., 1984. Prevalence of epicondylitis and tenosynovitis among meatcutters. Scandinavian Journal of Work, Environment, and Health, 10: 779–784.

Rubarth, B., 1928. Untersuchung zur Bestgestaltung von Handheften für Schraubenzieher und Ähnliche Werkzeuge (Investigations concerning the best shape for handles for screwdrivers and similar tools). Industrielle Psychotechnik, 5: 129–142.

Saran, C., 1973, Biomechanical evaluation of T-handles for a pronation supination task. Journal of Occupational Medicine, 15: 712–716.

Schoenmarklin, R.W. and Marras, W.S., 1989a. Effects of handle angle and work orientation on hammering; Part I: Wrist motion and hammering performance. Human Factors, 30: 397–411.

Schoenmarklin, R.W. and Marras, W.S., 1989b. Effects of handle angle and work orientation on hammering; Part II: Muscle fatigue and subjective ratings of body discomfort. Human Factors, 30: 413–420.

Shock, N., 1962. The physiology of aging. Scientific American, 206: 100–110.

Silverstein, B.A., Fine, L.J. and Armstrong, T.J., 1987a. Occupational factors and carpal tunnel syndrome. American Journal of Industrial Medicine, 11: 343–358.

Silverstein, B.A., Fine, L.J. and Armstrong, T.J., 1987b. Hand wrist cumulative trauma disorders in industry. British Journal of Industrial Medicine, 43: 779–794.

Sperling, L., 1980, Test program for work gloves. Research Report No. 1980: 18. Department of Occupational Safety, Division for Occupational Medicine, Labor Physiology Unit, Umea, Sweden.

Sperling, L., 1986. Arbete med handverktyg. Utveckling av kraft, anstrangning och obehag vid olika grepp och skaftdimensioner. (Work with hand tools. Development of force, exertion and discomfort in work with different grips and grip dimensions. Report in Swedish with English summary) Arbetarskyddsstyrelsen, Undersokningsrapport 1986: 25.

Sperling, L., Dahlman, S., Wikstrom, L., Kadefors, R. and Kilbom, A., 1991. Tools and hand function: The cube model – A method for analysis of the handling of hand tools. Paper presented at the 11th International Ergonomics Association Congress, Paris, France.

Sudhakar, L.R., Schoenmarklin, R.W., Lavender, S.A. and Marras, W.S., 1988. The effects of gloves on grip strength and muscle activity. In: Proceedings of the Human Factors Society 32nd Annual Meeting, Human Factors Society, Santa Monica, California, USA, pp. 647–650.

Swain, A.D., Shelton, G.G. and Rigby, L.V., 1970. Maximum torque for small knobs operated with and without gloves. Ergonomics, 13: 201–208.

Taylor, W., 1988a. Hand-arm vibration syndrome: A new clinical classification and an updated British standard guide for hand transmitted vibration. British Journal of Industrial Medicine, 45: 281–282.

Taylor, W., 1988b. Biological effects of the hand-arm vibration syndrome – Historical perspective and current research. Journal of the Acoustical Society of America, 83: 415–422.

Technical Research Centre of Finland, 1988. Evaluating and choosing pneumatic screwdrivers and nut runners. Occupational Safety Engineering Laboratory, Tampere, Finland.

Tichauer, E.R., 1966a. Some aspects of stress on forearm and hand in industry. Journal of Occupational Medicine, 8: 63–71.

Tichauer, E.R., 1966b. Gilbreth revisited. American Society of Mechanical Engineers, Publication no. 66-WA/BHF.

Tichauer, E.R., 1976. Biomechanics sustains occupational safety and health. Industrial Engineering, February, pp. 46–56.

Tichauer, E.R. and Gage, H., 1977. Ergonomic principles basic to hand tool design. American Industrial Hygiene Association Journal, 38: 622–634.

Terrell, R. and Purswell, J., 1976, The influence of forearm and wrist orientation on static grip strength as a design criterion for hand tools. In: Proceedings of the Human

Factors Society 20th Annual Meeting. Human Factors Society, Santa Monica, California, USA, pp. 28–32.

Ulin, S.S. and Armstrong, T.J., 1991. Effect of tool shape and work location on perceived exertion for work on horizontal surfaces. Proceedings of the International Ergonomics Association, pp. 1125–1127.

Ulin, S.S., Ways, C.M., Armstrong, T.J. and Snook, S.H., 1990. Perceived exertion and discomfort versus work height with a pistol-shaped screwdriver. American Industrial Hygiene Association Journal, 51: 588–594.

Viikari-Juntura, E., 1987. Interexaminer reliability of observations in physical examinations of the neck. Physical Therapy, 67: 1526–1532.

Wasserman, D.E. and Badger, D.W., 1973. Vibration and the Worker's Health and Safety. Technical Report No. 77, National Institute for Occupational Safety and Health, U.S. Government Printing Office, Washington, DC, USA.

Weidman, B., 1970. Effect of safety gloves on simulated work tasks. AD 738981, National Technical Information Service, Springfield, Virginia.

Wu, Y., 1975. Material properties criteria for thermal safety. Journal of Materials, 7: 575–579.

Yoder, T.A., Lucas, R.L. and Botzum, C.D., 1973. The marriage of human factors and safety in industry. Human Factors, 15: 197–205.

Equipment design for maintenance: Part I – Guidelines for the practitioner *

Sheik N. Imrhan
Department of Industrial Engineering, The University of Texas at Arlington, Arlington, TX 76019, USA

Target audience and uses

The guidelines for equipment design for maintenance described in this paper are intended for the practitioner. This includes all personnel involved in the design of equipment for maintenance, in the planning of maintenance activities in the workplace, and in the supervision of maintenance activities. Maintenance planners include management and supervisors.

In the design environment, these guidelines can be used to determine equipment design parameters so that maintenance tasks may be simplified and performed more efficiently. Supervisors and other planners of maintenance activities can also use these guidelines for allocating equipment to particular tasks, and for conducting education and training programs to technicians.

Data collection and analysis

Certain data are necessary to determine the effectiveness of maintenance operations and, hence, indirectly determine the effectiveness of equipment design. For example, excessive time taken to reach a failed unit may indicate that the arrangement and placement of components need improvement, or that the maintenance tools used may not be compatible with the particular machine (component) design, or even that the technician is lacking in skills. Data for the following performance parameters should be recorded:

(1) Performance times for individual maintenance tasks
   (i) Fault detection time
   (ii) Fault isolation time
   (iii) Access time to reach failed unit
   (iv) Removal time for failed unit
   (v) Repair time
   (vi) Replacement time for failed unit
   (vii) Functional check-out time to verify repair
(2) The number of failures within a specified time period
(3) Total operating time within that period
(4) Downtime within that period (scheduled and unscheduled)
(5) The type and number of errors made
(6) The type and number of injuries to personnel while performing maintenance tasks; and medical and other costs incurred from injuries.

Correspondence to: Sheik N. Imrhan, Department of Industrial Engineering, The University of Texas at Arlington, Arlington, TX 76019, USA.

* The recommendations provided in this guide are based on numerous published and unpublished scientific studies and are intended to enhance worker safety and productivity. These recommendations are neither intended to replace existing standards, if any, nor should be treated as standards. Furthermore, this document should not be construed to represent institutional policy.

The following individuals participated in the discussion of the earlier version of this guide. Their suggestions (written or verbal) were incorporated by the authors in this version: Roland Andersson, *Sweden*; Alvah Bittner, *USA*; Peter Buckle, *UK*; Jan Dul, *The Netherlands*; Bahador Ghahramani, *USA*; Juhani Ilmarinen, *Finland*; Asa Kilbom, *Sweden*; Stephan Konz, *USA*; Shrawan Kumar, *Canada*; Tom Leamon, *USA*; Mark Lehto, *USA*; William Marras, *USA*; Barbara McPhee, *Australia*; James Miller, *USA*; Anil Mital, *USA*; Don Morelli, *USA*; Maurice Oxenburgh, *Australia*; Jerry Purswell, *USA*; Jorma Saari, *Finland*; W. (Tom) Singleton, *UK*; Juhani Smolander, *Finland*; Terry Stobbe, *USA*; Rolf Westgaard, *Norway*, Jørgen Winkel, *Sweden*. The guide was also reviewed in depth by several anonymous reviewers.

These data should be available on computer records, in a form that makes them amenable to immediate summarization and statistical analysis. Up-to-date summaries and analyses should be available for comparison with past ones, and for evaluation with stated objectives. The data should also be in a format that facilitates computation of costs for all aspects of maintenance, and especially according to the categorization given above. Cost comparisons among the various categories above can help management not only to identify problem areas, but also to rank problem areas in order of importance. In this way management can gain better insight about the performance of equipment or system.

It may be impractical to gather time data for items (1i)–(1vii), in many actual work situations. However these times can be measured for simulated idealized conditions, and the resulting data used as a standard of reference; that is, the best the equipment or system can do. In fact, such data are often better for comparing different equipment designs, since contamination from confounding variables, such as delays, shall have been eliminated.

Criteria for determining the effectiveness of maintainability

These data (items (1)–(6) above) provide valuable parameters for evaluating the effectiveness of maintenance, system performance, or equipment performance. Indexes derived from them can also be used to assess the effectiveness of system or equipment design. Traditional indexes include: mean time to repair (MTTR); mean time between failure (MTBF); and availability (A). These are discussed in greater detail in Part II of these guidelines (Imrhan, 1992). Good designs imply small values for MTTR, large values for MTBF, and large values for A.

Problem identification

Problems can be recognized when the data for the above performance parameters or indexes show any undesirable deviation from expected values, assuming 'normal' or ideal conditions. For example, a large number of failures during operations, or low availability of a system, may be an indication that equipment may not have been properly maintained or designed. Poor maintenance could be identified from the cost of maintenance. Poor equipment design can lengthen the times taken to perform maintenance tasks. High rates of accidents and injuries may also indicate problems with the use and design of equipment.

All data should be compared with some reference standard that is considered good (or even ideal) for the specific type of maintenance tasks. Such a standard may be derived from previous data.

Solutions

Several aspects of human–equipment interaction in maintenance work must be considered when trying to find solutions to problems arising from improper equipment design. These are:
(1) Visual access – the ability of the technician (1) to see what he/she is doing, (2) to see what other team members are doing, (3) to engage in free communication by gestures, as in relaying non-verbal instructions, and (4) to see dangers in the working environment.
(2) Physical access – the ability of the technician to position the body, or part(s) of it within the work environment in the most advantageous way, in order to perform his/her task(s); and the ability to manipulate tools, etc. within the environment.
(3) Physical mobility – the ability of the technician to maneuver the body or part(s) of it (often with tools) within the work environment, in order to perform the task(s).
(4) Strength – the ability of the technician to generate adequate muscular forces via hand-tools and other equipment under specified conditions, for manual tasks, without experiencing overexertion.
(5) Muscular and physiological endurance – the ability of the operator to maintain a certain level of performance for a certain period of time, while using tools and other equipment.
(6) Cognitive and decision making demands – the ability of an operator to perceive and process information (mentally) from the maintenance environment, and to decide on

actions to be taken. Thus the way in which instruction manuals are written, the arrangement and design of controls, the design of equipment, and so on, can determine the speed and accuracy of the technician's work.
(7) Education and training – the ability of the operator to perform tasks successfully. This includes the ability to comprehend written and other instructions from maintenance manuals or other written materials, and to apply them to tasks.
(8) Safety – this is a wide area, but it involves the ability of operators to use equipment and perform tasks without exceeding their cognitive, mental or physical limitations. In effect, all of the seven factors stated above determine the level of safety in the maintenance environment.

Guidelines for equipment design that diminish or eliminate the ergonomic problems listed above are stated below.

Access

Visual access

Visual access involves features in the design which prevent physical barriers in the line of sight between the part being worked on (or the general work area) and the eyes.
– Have means for large heavy equipment to be moved easily and quickly from one position to another, so that the technician's line(s) of vision is not blocked.
– Mount parts so that, ideally, the operator is not forced to remove other parts to see the one he/she must work on.
– Provide mechanical aids which can reorient and reposition the machine in different planes of space, and at various heights above the floor, to prevent technicians having to adopt constrained fatiguing postures, when trying to see their work. In other words, reorient the object not the human body.
– Have apertures in walls, which can be sealed by hinged doors when maintenance is not being performed, so as to allow workers to communicate more easily.
– Ensure adequate illumination for the task, taking into consideration the limits of human visual capacities. Small parts require more intense illumination; and light contrasts are often helpful, providing that they are not too strong. Fluorescent light sources should therefore be preferred over point sources. The effects of these light sources can be enhanced by large diffusing screens, big radiating surfaces and shades (to prevent direct glare).
– Use visibility enhancing features such as mirrors, windows, auxiliary lighting, remote cameras, etc.
– Design automatic activation of internal light for safety, where accesses are in close proximity to danger zones (e.g. moving machinery, high voltages, etc.)

Anthropometric access

Physical accesses should accommodate at least 95% of the potential population of maintenance personnel; and should be designed according to available anthropometric data. Accesses should be designed by size and shape, according to the largest part of the body which must access the space, and according to body posture. The body posture is determined, to a large extent, by the task being performed. Dynamic (functional) anthropometry is more important than static anthropometry, in most of these situations. Additionally extra space must be allowed for the potential use of tools and special apparel (e.g. protective suits, gloves, etc.). The following considerations should be taken:
– Whole body access – for a single person or for more than one person.
– Segmental access – relating mostly to the arm or hand, with or without hand tools. One arm (or hand) and two arms (or hands) should be considered separately.
– Special apparel worn by the technician, such as gloves, winter clothing and other bulky protective clothing.
– Size and shape of openings to accommodate whole body, the arm, and the hand. The shape need not be symmetrical, though a symmetrical shape is easier to define and design. Moreover symmetry has other advantages: it allows greater variations in body orientation when accessing apertures.

Clearance dimensions for access for typical maintenance postures and for various body parts are shown in tables 1–4 below. Tables 2 and 3 also give data for adjustments to dimensions when winter clothing, gloves, or other heavy clothing must be worn. These data were taken from a number of sources, as discussed in Imrhan (1992).

Locating and orienting parts to aid physical access

- Locate maintenance controls in front of operator, within his/her immediate view and reach.
- Locate lines such as wires, cables, piping and tubing so that the time taken to retrieve, to setup and to use them can be minimized.
- Locate units so that, when they are being removed, they will follow straight line or slightly curved paths instead of sharp turns.
- Orient parts (seats, pins, etc.) in order to minimize time for repair, removal or insertion; and to avoid constrained body posture.
- Position components and sub-assemblies so that they can be reached (removed and replaced) easily and quickly. Avoid locations that are difficult to reach, such as under seats, inside recesses, behind other components, and so on.
- Locate the most frequently failing (or most critical) components so that they are the easiest to access.
- Mount parts in an orderly way on a flat surface, rather than one on top of the other, to make them more accessible.

Other physical access design features

- Design and locate bulkheads, brackets, and other units so that they will not interfere with removal or opening of the cover of units.
- Provide adequate means for easy movement of machine parts that must be moved from their installed positions. Cabinet racks, and roll out, slide out, or hinged racks are useful methods.
- Design equipment such that technicians are not forced to remove non-failed items, or other obstructions, in order to access the failed item(s). The time taken to do this is completely wasted.
- Design covers so that they can be easily separated from the units they protect, instead of removing the unit from the cover. This would also avoid any tendency for an operator to grasp parts in order to move the unit.
- Arrange parts in equipment so that removal of any replaceable unit will eliminate unnecessary opening of covers, panels, etc.
- Arrange parts so that one technician does not need to wait for another one to complete a task before beginning his or her task.
- Design units so that the removal and replacement of one component or subassembly should be accomplished with as little disturbance as possible to others.
- Allow sufficient space around parts so that the hand(s) can grasp, remove, manipulate and replace these parts (tables 2 and 3).

Task simplification

- Design to promote quick and positive identification of the malfunctioning part or unit.
- Design to minimize the number and complexity of maintenance tasks.
- Minimize the number and types of tools and test equipment required for maintenance, in order to avoid supply problems, frequent tool changes and the use of wrong tools.
- Standardize equipment (fastener sizes, threads, etc.) in order to minimize the number and type of tools needed for maintenance, and to eliminate supply logistics problems.
- Avoid using equipment parts that require special tools.
- Use interchangeable components within and across equipment to reduce supply problems – except where components are not functionally interchangeable.
- Minimize and standardize the number of lines used (electrical, hoses, etc.)
- Use lines that are easy to manipulate.
- Use a minimum number of fasteners necessary to maintain equipment integrity and personnel safety.
- Use snap-in retainers, latches, spring-loaded hinges, etc. to facilitate the removal and replacement of parts.

Personnel safety

The variety of features of equipment which can be considered unsafe is vast. Good equipment design incorporates features that will reduce harm to personnel and damage to equipment, from sources such as electrical shocks, cuts or bruises. Hazards include sharp or hard edges on objects, slippery floors and obstructions in walkways, objects falling from above, unstable heavy equipment, and so on.

- Locate components that conduct and retain electricity or heat, in positions where the technician will not place his/her hands.
- Locate controls and tools access points away from dangerous sources of voltages to avoid electrical shocks.
- Equip moving parts (on service equipment) with enclosures and screen guards to prevent inadvertent contact with the hand, or other parts of the body.
- Maintain good housekeeping. Clear pathways of objects (cables, handtools, etc.) to prevent personnel tripping and falling.
- Incorporate automatic shut-off design features (e.g. interlock) where personnel must enter high voltage enclosed areas.
- Avoid sharp edges on equipment to prevent abrasions, cuts, excessive pressure, etc., on personnel. Edges can be rounded, smoothed or covered with a protective material.

Handles and loads

Handles should be used to prevent muscular overexertion when units must be moved from one place to another, or be held without support.
- Provide handles on units that are too bulky, too heavy, or too small.
- Consider the need for space when designing handles. Where space between or within equipment is expected to be limited, handles should be removable and quick-fitting; or recessed; or of the hinged type that locks in the extended position.
- Orient handles on bulky or heavy equipment such that the wrist and forearm are kept in their natural positions, as closely as possible. This reduces musculoskeletal stress in the wrist.
- Position handles above the center of mass of the object, for stability.
- Design handles so that their dimensions conform to the anthropometry of the large male (ideally 95th percentile). A grip width of at least 4.5 inches and depth of at least 2 inches are required for bare hands. Increase these dimensions by 1-2 inches for hands with mittens.

Table 1

Selected clearance dimensions in typical whole body maintenance postures (adapted from Eastman Kodak Co., 1985).

Activity/access dimensions		(cm)	(in)
Width for walking	1 person	61	24
	2 persons abreast	137	54
	3 persons abreast	183	72
Height for walking	accommodates very tall person wearing thick-soled shoes and hard hat	203	80
Lying supine			
– elbow completely flexed	vertical (fore-aft)	46	18
	body length clearance	193	76
– elbow in mid-flexion	vertical	61	24
– elbow completely extended	vertical	81	32
Standing naturally	vertical	203	80
	horizontal (reach)	76	30
Standing with legs braced	vertical	203	80
	horizontal	102	40
Kneeling	vertical	122	48
	horizontal	177	46
Prone arm reach	lying down	243	96

Table 2

Selected clearance dimensions for hand and arm access (adapted from Eastman Kodak Co., 1985).

Access	Dimensions (cm)
Hand/arm entry/exit	
Both arms	75% depth of reach + 15 cm (6in)
Arm to elbow	11 cm (4.5 in) diameter
Arm to shoulder	13 cm (5 in) diameter
	Add 8 cm (3 in) for winter clothing
Hand with extended fingers	5 cm (2 in) × 10 cm (4 in) rectangular
Hand with clenched fist	10 cm (4 in) × 13 cm (5 in) rectangular
Hand holding small object	10cm (4 in) × 10 cm (4 in) square
Hand holding larger object	4.5 cm (1.75 in) clearance around object;
	Add 2 cm (0.75 in) for gloved hand
Push button access	4 cm (1.5 in) diameter
Two finger twist	9 cm (3.5 in) diameter

Table 3
Clearance dimensions for selected tool use in various wrist positions (adapted from Eastman Kodak Co., 1985).

Tool and task	Wrist position							
	Radial deviation		Ulnar deviation		Extension		Flexion	
	cm	in	cm	in	cm	in	cm	in
Turning screwdriver [20 cm (8 in.) length] or spinate wrench [15 cm (6 in.) length]	3.8	1.5	6.2	2.4	6.2	2.4	4.2	1.7
Grasping, turning, and cutting with needle-nosed pliers [14 cm (5 in.) length] or wire cutters [13 cm (5 in.) length]	5.3	2.1	7.3	2.9	4.6	1.8	6.8	2.7
Turning socket wrench [10 mm (3/8 in.) base, 7 cm (3.2 in.) shaft]	5.4	2.1	8.3	3.2	8.3	3.2	7.3	2.9
Turning allen wrench [5 cm (2 in.) length]	3.6	1.4	8.6	3.4	9.4	3.7	6.4	2.5

- Allow at least 2 inches clearance between handles and obstructions (adjacent objects).
- Use two handles where significant muscular effort is required (over about 25 pounds), so that they can be handled with both hands. Biomechanical stress on one side of the body will therefore be reduced by being spread over both sides.
- Use four handles (for two technicians) for loads between 40 to about 90 pounds.
- Use mechanical aids, not handles, for handling units greater than 90 pounds; or even for lighter loads, when constrained body postures cannot be avoided.

Minimizing errors

Equipment failures and wasted time from correcting errors can be reduced by designing out the sources of these errors. Among the most common error-related maintenance tasks are removing and replacing non-failed items, not removing and replacing failed items, failing to test components before assembly, and damaging items inadvertently. Cognitive and sensory problems are also common. Difficulties in reading or understanding instructional materials, and difficulties in detecting visual warnings and other information on equipment, may be due to poor visual display design.

Some of the guidelines for minimizing errors are similar to those that apply to 'designing for ease of maintenance'. Among the most important are:

- Design to make all parts accessible anthropometrically and visually (see sections on visual and anthropometric accessibility).
- Avoid crowded equipment bays and poor equipment layout.
- Avoid inappropriate placement of equipment and machines.
- Have adequate, properly written and properly designed maintenance instructions (not cumbersome).
- Avoid tasks requiring excessive muscular efforts (e.g. lifting heavy weights, manual exertions using wrenches in awkward postures, etc.)
- Ensure that adequate time is spent on task inspection and check-out.
- Separate controls for maintenance from controls for normal operations.

Table 4
Selected clearance dimensions for entering open-top vessels (adapted from Eastman Kodak Co., 1985).

Access	Dimension
Circular hatch	76 cm (30 in) dia
Rectangular hatch (horizontal entry)	61 cm (24 in) × 51 cm (20 in)
Square hatch 61 cm × 61 cm (vertical entry)	
	Add 10–20 cm (4–8 in) if heavy clothing is worn.

Labeling and coding

Labels and codes should conform to Human Factors/Ergonomics principles of visual display design. They should convey information which
- Helps to identify parts, units, etc.
- Helps to use the equipment appropriately.
- Warns maintenance personnel of hazardous conditions.
- Helps to distinguish among various areas (e.g. between test and service points).

Important design principles include:
- Use standardized color coding where labeling is necessary (e.g. red for danger), for quick discriminations or identification.
- Use colors only when necessary, and minimize the number of different colors. Too many colors are confusing.
- Add other codes for color defective personnel.
- Design labels on equipment large enough to be read easily. In good illumination characters on labels should make a visual angle of at least 12 minutes of arc at the furthest distance a technician is expected to read them. If illumination is expected to be diminished, the character size should be larger.
- Ensure that labeling and coding are consistent (e.g. red always means 'danger', and only red should be used for 'danger') so that instructions and procedures can be easily followed by maintenance personnel.
- Choose words and sentence structure to be compatible with the lowest level of reading and comprehension expected of maintenance personnel.
- Arrange instructions on labels in 'step by step' form. Avoid narrative.

Clearances between controls

To avoid errors in using (maintenance) controls, their spacing and arrangement should be within the anthropometric, sensory and cognitive capabilities of all technicians. The U.S. Air Force (Imrhan, 1992) recommends that clearances should be based on eight factors:
- The size of the control and the extent of its displacement
- The body member used (foot operated controls require greater separation).
- Whether one or two hands are expected to be used (two-hand operation implies greater spacing of controls).
- Sequential or simultaneous activation (simultaneous activation requires closer spacing).
- Sequential or random activation (random activated controls require greater spacing).
- Whether there is visual feedback (controls activated with no visual feedback require greater spacing).
- The use of gloves and heavy boots (greater spacing).
- Criticality of the control (critical controls require greater separation to prevent accidental activation).

Protection of equipment from damage

- Provide means for stabilization of parts (e.g. limit stops and braces) that are moved from their more stable installed positions.
- Protect equipment being maintained, from vibration (e.g. twist or push-to-lock mounts for small components; and shock mounts to protect against fine calibration interference).
- Protect sensitive adjustments physically, to prevent disturbance by maintenance personnel (e.g. collision with the body or with other equipment being transported).
- Use cables that are long enough to allow free movement, and locate them where they would not be pinched or bent.
- Separate adjacent components with sufficient space to avoid contact with test probes or other tools. This also applies to terminals to which wires are to be attached by soldering.
- Locate delicate components where they cannot be easily damaged by tools, etc.

Reference

Imrhan, S.N.,1992 Equipment design for maintenance: Part II - The scientific basis. International Journal of Industrial Ergonomics, 10: 45–52 (this issue).

Equipment design for maintenance: Part II – The scientific basis for the guide

Sheik N. Imrhan

Department of Industrial Engineering, The University of Texas at Arlington, Arlington, TX 76019, USA

Problem description

No machine can work forever, but all machines should be designed to perform their intended functions for as long as possible. However, machines and their components will fail, and must be repaired or replaced. The traditional concept and expectation among engineers are that a machine, or the system of which it is a part, must be designed for high reliability; that is, a high probability of successful performance. Efforts to enhance system reliability have traditionally concentrated on developing and using high reliability components and network systems.

System reliability can be enhanced not only by extending the failure times of new components but also by instituting preventive maintenance. Based on reliability theory to predict failure times, preventive maintenance methods have been used to service or repair components before they fail. This method improves system reliability by rejuvenating sources of expected failure. With the increasing sophistication and complexity of machines and systems, however, prediction of failures has become mind-boggling and very costly. This problem has been exacerbated by their complex relationships with both the environment and humans. Maintenance technicians must be better educated and trained than ever before. Training periods are becoming lengthier; and retraining for new equipment and methods of work have also become common.

For practical purposes two general kinds of maintenance can be distinguished – preventive maintenance and breakdown (corrective) maintenance. Preventive maintenance aims at maintaining equipment in good condition so as to obviate the need for breakdown maintenance. Parts may be serviced or replaced before they are expected to fail, so that equipment and system lifespan may be increased. This is one way of enhancing the reliability of both the equipment and the overall system. It has been compared with a person's health check by a physician (Nakajima, 1988). Breakdown (corrective) maintenance is performed after equipment has malfunctioned. The faulty equipment may be either repaired or replaced. This type of maintenance can be interpreted as a failure of preventive maintenance methods.

At one time it was thought that the rapid improvement in the reliability of machines and components would practically obviate the need for maintenance activities (Morgan et al., 1963; Crawford and Altman, 1972; and Oborne, 1981). Unfortunately this has not yet happened; and most likely will not happen in the foreseeable future. The growth of machine complexity has far outpaced the improvements in reliability techniques. Today the concept of 'maintenance' as a means to reduce downtime has assumed much

Correspondence to: S.N. Imrhan, Department of Industrial Engineering The University of Texas at Arlington, Arlington, TX 76019, USA.

greater importance compared to the concept of reliability of components. Preventive maintenance is being used as a method of improving overall system reliability.

Scope of the problem

Maintenance has always been a costly exercise. Costs arise mainly from the value of parts and maintenance equipment, training of personnel, and labor. Approximately 35% of the lifetime cost of a military system in the U.S has been spent on maintenance (McDaniel and Askrein, 1985). Maintenance of underground mine equipment in the U.S. accounts for 25%–35% of total mine operation costs, and has also accounted for one-third of all lost time injuries (Conway and Unger, 1991). Poor maintenance has also become a troubling problem. Errors are rife. It is estimated that maintenance errors account for 10% to 25% of maintenance time (Conway and Unger, 1991), and that 20%–25% of all failures are directly traceable to maintenance errors (Rasmussen, 1981). Military maintenance records also show that 4%–43% of all 'faulty' components sent back from the field for repair were actually good (Bond, 1987).

A number of factors have contributed to the overall increasing cost of maintenance. They include:
(1) The increasing technical complexity of modern equipment and systems.
(2) The great diversity of hardware choices made possible by computer aided designs (Adams and Peterson, 1988).
(3) The increasing and rapidly changing variability in equipment and system design.
(4) The competing and differing criteria in designing for operability and designing for maintenance. Usually the former is preferred over the latter.
(5) The continuing trend towards miniaturization and compactness (Tichauer, 1978). This is driven by the need to-lower the cost of production (and not necessarily the cost of maintenance); the need to enhance manipulation of machines during operation and transportation; and the need to satisfy the demands of a shrinking workplace size.
(6) The increasing cost to train and employ technicians at higher levels of skills to keep up with increasing technological sophistication and rapid changes in design.

An analysis of maintenance design limitations by the U.S. Bureau of Mines (Conway and Unger, 1991) identified four main problems that contribute to maintenance difficulties:
(1) Physical inaccessibility of components.
(2) Visual and physical inspection – due to crowding of components.
(3) Limited design fault isolation capability – due partly to limited or no designed-in fault diagnostic capabilities or effective failure indices.
(4) Lack of resources, especially of proper tools.

A consideration of previous guidelines (Morgan et al., 1963; Crawford and Altman, 1972; and Woodson, 1981) indicates that these types of problems were, and still are, among the major ones in almost all maintenance environments. Much trouble-shooting is still done by people (Bond, 1987), but it is clear that the ergonomic concept of designing to conform to human capacities and limitations has not yet been adequately applied to equipment design for maintenance.

Maintenance task sequence

Maintenance of machines and systems involves a number of operations which can extend time and cost if not properly designed for the capacities and limitations of the maintenance personnel. On the contrary the proper design of these operations and their associated equipment can reduce downtime and, hence, cost significantly. Once it has been recognized that a problem (malfunction) exists the following steps are typically followed:
(1) Detecting the fault(s)
(2) Isolating the fault(s)
(3) Accessing the part(s)
(4) Dissembling the machine
(5) Removing and repairing the part(s)
(6) Replacing the part(s)
(7) Reassembling the machine
(8) Testing and checking to verify successful repair

Any attempt at improving the efficiency of maintenance should attend to each of these operations.

Background

Literature review

Concepts

The primary objective of all aspects of maintenance, including equipment design, is to ensure that the system performs its intended function, at any time. The term 'availability' has been used to describe this property. It is defined as the probability that the system will be operating at any particular time (Crawford and Altman, 1972). Availability is related quantitatively to two other properties of equipment or system function. These are reliability and maintainability.

In the context of maintenance, reliability has been defined as the probability of failure-free system or equipment operation during a prescribed time period and under specified conditions. Reliability can be measured (Crawford and Altman, 1972) in terms of:

(i) The likelihood of successful operation for a specified period of time (usually 10^6 hours); or
(ii) The mean time between failures (MTBF) – the total operating time divided by the number of failures during that time period.

It must be emphasized that reliability measures can be misleading unless measurements are taken over a sufficiently long period of time, and under all conditions in which the machine (or system) is expected to operate.

Morgan et al. (1963) defines maintainability as follows: 'Maintainability is the degree of facility with which an equipment or system is capable of being retained in, or restored to, serviceable operation. It is a function of parts accessibility; internal configuration; use and repair environment; and the time, tools, and training required to effect maintenance'. Maintainability is therefore a function of equipment design. It can be measured as the mean time to repair the equipment (MTTR). It includes the time from the detection of a fault to the time when successful repair is verified. It can therefore be measured as the composite of the times for operations (1)–(8) above.

The relationship between availability (A), reliability (MTBF) and maintainability (MTTR) of equipment can best be expressed, quantitatively, as follows (Crawford and Altman, 1972):

(A) Availability = MTBF/(MTBF + MTTR)

MTBF can be also interpreted as uptime and MTTR as downtime, and the expression above becomes (Konz, 1990):

(A) Availability

= Uptime/(Uptime + Downtime)

= Uptime/Total Time

This type of availability is often referred to as 'inherent availability'. It does not include factors such as administrative delays, encumbrances, and unpredictables. It assumes near ideal conditions and can, therefore, be used as a standard of reference – the best the system can do. Crawford and Altman (1972) and Bond (1987) state alternative interpretations of availability, that include non-ideal factors.

The data from these indexes provide valuable parameters for evaluating the effectiveness of maintenance, the performance of a system, the performance of components of a system, or the effectiveness of equipment design.

Previous guidelines

A number of sources in the literature provide guidelines for equipment design (and workspace design) for maintenance. Among the two most comprehensive are Morgan et al. (1963) and Crawford and Altman (1972). Morgan et al. provide guidelines in relation to the nature of equipment:

(1) Units – unit location of components, mounting and assembly features, mounting bolts and fasteners, labeling and color coding.
(2) The design of unit covers and cases – structural characteristics, fasteners, and handles.
(3) Cables and connectors – cable routing, connecting features and disconnecting features.
(4) Equipment accesses – size and shape, safety features, and labeling and coding accesses.
(5) Equipment test points – functionally locating test points, physically loading test points, and labeling test points.
(6) Displays – location, scale markings and labeling.

(7) Controls – type, location, labeling and coding.
(8) Equipment space.
(9) Test equipment and bench mockups.

Crawford and Altman (1972) provide similar information in a more conceptual form, under the headings:
(1) Maintenance safety.
(2) Maintenance information.
(3) Handling and removal for replacement.
(4) Alignment and keying.
(5) Manual control layout.
(6) Workspace configuration.
(7) Accessibility – visual and anthropometric.

Even though these guidelines were intended for equipment existing at least two decades ago, the basic principles still apply today. Altman (1991), for example, provides an up-to-date discussion of the principles of design for maintainability, and his recommended guidelines are similar to those recommended two decades ago by Crawford and Altman, 1972.

A few recent survey studies of workspaces related to equipment design (Eastman Kodak Co., 1983; and Mital, 1991) have summarized data from several older sources, and have combined them with some more recent data. These guideline data and their original sources include:
(1) Physical (anthropometric) access for hatches (MIL-STD-1472B, 1974; and C, 1981).
(2) Aisle and corridor widths (MIL-STD-1472B & C; and Noble, 1982)
(3) Tunnels and catwalks (Woodson, 1965; and Damon et al., 1966).
(4) Walking clearance between obstacles (Pheasant, 1986; Thompson, 1972).
(5) Clearances for the arms and hands, including reach depth in apertures (Baker, McKendry and Grant, 1960; Kennedy and Filler, 1966; Woodson and Conover, 1964; DIN 31001, Pt. 1, 1983; and Clark and Corlett, 1984).
(6) Working postures such as lying down, kneeling, crawling, sitting and standing (Hertzberg, Emmanuel and Alexander, 1956; Rigby, Cooper and Spickard, 1961; Alexander and Clauser, 1965; Croney, 1971; Hertzberg, 1972; and Clark and Corlett, 1984).
(7) Safety distances for whole body reach over a barrier (Thompson and Booth, 1982).

Workspace is an important aspect of design for maintenance. It governs the ability to see, reach, remove, and replace parts; and to exert forces on handtools (Imrhan, 1991a). Any restriction on these functions can eventually translate into musculoskeletal stress, visual stress, poor performance, greater number of errors, extended worktime, and increased cost of maintenance. Inability to access parts has always been among the greatest difficulty in maintenance. A U.S. Bureau of Mines study found that the primary design limitation, for underground mining equipment, was inaccessibility to components (Conway and Unger, 1991). Limited workspace (for hand-arm access, manipulation of tools, visual access, etc.) is often imposed by poor equipment design. Good equipment design is therefore a solution to inaccessibility to components.

Several studies provide important experimental data that can be incorporated as ergonomic guidelines for designing equipment for maintenance. The data relate to workspace for the hand and arm; hand strength capabilities for torquing, pinching and gripping; and lifting strength capabilities.

Evidence indicates that there is an optimal workspace for tasks. Performance (strength exertion, control, time to complete task, and error minimization) is degraded below a certain limit but remains constant above that limit. Kama (1963 and 1965), for example, showed that the time taken to remove and replace a component decreased sharply as the aperture for the hand (access) increases up to a point, then levels off beyond that point. He also showed that the time increases as the depth of the aperture increases. Adams (1988) found that restricted hand clearance size decreases hand torque capabilities.

The ability to lift objects is also severely affected by restricted spaces. Restricted shelf opening spaces have been found to decrease maximal acceptable weight of lifts onto shelves significantly (Mital and Wang, 1989).

Awkward arm configurations during maintenance are often imposed by the orientation of the workpiece or by space constraints. This reduces manual muscular strength capabilities (Deivanayagam and Weaver, 1988; and Deivanayagam and Ratnavelpandian, 1991). Data from Adams (1988) indicate that the orientation of connectors to the right of the body's mid-saggital plane was better than either directly in front of or facing away from the subject (reverse frontal plane), for torque generation with the fingers. Data from Mital and Channaveeraiah (1988) also indicate that forearm orientation imposed by a particular type of task, and the type of hand-handle interface (from hand tools) influence manual peak torque exertion significantly. Equipment components and maintenance tools should, therefore, be oriented to allow the technician to capitalize on advantageous body postures and local skeletal configurations. Prudent location of components and the use of mechanical devices that can reorient units or subassemblies help the operator to achieve these advantageous body leverages.

Restricted handspaces impose extreme wrist joint deviations when using handtools. Grip and pinch forces are weakened considerably (Terrell and Purswell, 1976; and Imrhan, 1991b) and manual tasks become more strenuous. Limited hand space may also influence the type of pinch used. The strongest pinch is the lateral one (key pinch) but it requires more space than the commonly used pulp 2 pinch (with the thumb and index finger). Greater stress is therefore placed on the finger tendons when the pulp 2 must be used, instead of the lateral, for pinching or pulling with the fingers (Imrhan, 1991b, c). A little more clearance space can translate into less tendon strain and, eventually, decreased incidence of cumulative trauma disorders.

To aid maintenance, units that are too bulky or too small should be fitted with handles for lifting, lowering, or carrying (manual materials handling). Bulky or heavy units can impose significant musculoskeletal strain in the body; and small units may be difficult to grasp or hold, especially when gloves are worn by technicians. Altman (1991) provides guidelines for handles that are either bulky or small.

Lifting without handles reduces people's lifting capacity by about 15% (Bakken, 1983). This implies that musculoskeletal strain, for a given load, may be about 15% greater when lifting without handles compared to lifting with handles. This extra strain will decrease performance efficiency and, for heavy or bulky loads, can lead to overexertion injuries. Handles need not be fixed. Removable ones may be necessary where there are space limitations within and between equipment.

Need for future work

Representative anthropometric data

Most design recommendations are based on work from the U.S. Department of Defense (Mital, 1991). However, the exact differences between the industrial and military maintenance environments have not been established to permit accurate corrections to be made to military data, for industrial applications. Equipment clearances, for example, which depend on body dimensions are likely to be different, since anthropometric dimensions in the industrial population are likely to have greater variation than those of the military population. More representative anthropometric data from the civilian population are therefore needed.

Computer modeling

Traditional workspace design recommendations are almost exclusively 1- and 2-dimensional, and so are often inadequate for the real environments which are 3-dimensional. Dimensions representing volumes of space are the ideal data; and future efforts should be aimed at compiling 3-dimensional data. But first, traditional methods of workspace and equipment design for maintenance must be improved. They are too slow and limited in scope. Their use of mock-ups, anthropometric dummies, partial mannequins, stick figure mannequins, 2-dimensional drawing board mannequins, and manufacturers' drawings cannot adequately represent 3-dimensional space and user populations; they cannot be manipulated with the speed necessary to compare various alternatives; and they cannot respond quickly enough to rapid design developments (Rothwell, 1986).

The availability of personal and mini computers, and rapid software development in computer aided design (CAD) offer great hope in overcoming these limitations. Rothwell (1986) and Chaffin and Evans (1986) discuss the utility of CAD models in workspace design.

Computer simulation of tasks is an attractive option for the development of ergonomic equipment design data, because both field and laboratory studies are time consuming and difficult.

CAD models, with an anthropometric data base, can be used to derive equipment design parameters by analyzing the need for workspace and tool, and the demands of strength, range of motion and sensory functions (e.g. vision). Simulation of removal and replacement using different parts and tools may be estimated and compared more realistically and quickly. In effect, maintainability design evaluation can be made much less difficult at the design stage, by the design engineer (Quigley et al., 1991). Design quality will therefore be enhanced.

CAD programs can be used to analyze partial body models when specific areas of the body are of interest; or whole body models, for general workspace analysis. The human body may be modeled as a system of interconnected links (Calvert et al., 1982), with each link representing a body segment, with respect to one or more of the following: length; breadth; circumference; volume; density; mass; and internal modeling of bone, muscles, ligaments, skin and subcutaneous fat. Body surfaces and volumes can be modeled by 'enfleshing' stick figures. They are better than stick figures for analyzing clearances. The former are limited to distance-based analyses such as reach and visual interference (Lane, 1982). It is also possible to perform rotations and translations of the human graphical models to view the simulated maintenance person using particular equipment in workspaces, from a variety of angles.

Visual access can be analyzed by CAD models by relocating target objects and/or the human model within the environment (Rothwell, 1986). The ability of these CAD models to alter human dimensions facilitates the evaluation of special clothing and personal equipment (thick jackets, helmets, goggles, etc.) in clearance spaces.

Examples of CAD models

Crew Chief, developed by the U.S. air force, is perhaps the best computerized 3-dimensional model that simulates maintenance activities. Using a CAD system, this model creates a computerized maintenance person model (man or woman) representing a range of body sizes (Easterly and Ianni, 1991). It can evaluate proposed equipment and workplace designs before production while varying body posture, motion and strength characteristics. Hence ergonomic design parameters can be input into the model, making the final design parameters more representative of a maintenance person's characteristics. Graphic human animation enhances the interpretation of the outputs.

Other models deal more with the design of maintenance workspace. The capabilities of SAMMIE (System for Aiding Man/Machine Interaction Evaluation), the Graphic Man model and BUBBLEPERSON-TEMPUS were outlined by Chaffin and Evans (1986) and Galer (1987). SAMMIE allows the size of the human figure to be scaled to various population subgroups using segment link rates; heuristic rules can then be used to shape the body into realistic postures, allowing the determination of visual and reach envelopes. The Graphic Man model can be scaled to represent various population subgroups. A realistic graphic model of the human form can be produced for reach analysis; and biomechanical analyses can be performed. BUBBLEPERSON allows easy change of body segment size and shape, as well as manipulation of the total body form. This form can be used from a larger interactive graphic system called TEMPUS, to describe objects being handled as well as evaluating reach and physical body interference problems within a given workplace.

References

Adams, S.K., 1988. Hand grip torque for circular electrical connectors: The effect of obstructions. In: Proceedings of the Human Factors Society, 32nd Annual Meeting, Santa Monica, California, pp. 642–645.

Adams, S.K. and Peterson, P.J., 1988. Maximum voluntary handgrip torque for circular electrical connectors. Human Factors, 30(6): 733–745.

Alexander, M. and Clauser, C.E., 1965. Anthropometry of common working positions. Tech. Rpt. 65-73. Wright-Patterson AFB, Ohio: Aerospace Medical Research Labs.

Altman, J.W., 1991. Maintainability design. In: B.M. Pulat and D.C. Alexander (Eds.), Industrial ergonomics: Case studies. Industrial Engineering Institute and Management Press, IIE, Norcross, GA. pp. 297–315.

Baker, P., McKendry, J.M. and Grant, G., 1960. Supplement III – Anthropometry of one-handed maintenance actions. Tech. Rpt. NAVTRADEVCEN 330-1-3. Port Washington, New York: U.S. Naval Training Device Center.

Bakken, G.M., 1983. Lifting capacity determination as a func-

tion of task variables. Ph.D. dissertation, Texas Tech University, Lubbock, Texas.
Bond, N., 1987. Maintainability. In: G. Salvendy (Ed.), Handbook of human factors. John Wiley and Sons, New York, NY, pp. 1328–1354.
Calvert, T.W., Chapman, J. and Palta, A., 1982. Aspects of the kinematic simulation of human movement. IEEE Computer Graphics and Applications, 2(9): 41–50.
Chaffin, D.B. and Evans, S., 1986. Computerized biomechanical models in manual work design. In: Proceedings of the Human Factors Society 30th Annual Meeting, Santa Monica, California, pp. 96–100.
Clark, T.S. and Corlett, E., 1984. The Ergonomics of Workspaces and Machines: A Design Manual. Taylor and Francis, London, United Kingdom.
Conway, E.J.K. and Unger, R.L., 1991. Design of mining equipment for maintainability. In: B.M. Pulat and D.C. Alexander (Eds.), Industrial Ergonomics: Case Studies. Industrial Engineering Institute and Management Press, IIE, Norcross, GA, pp. 317–334.
Crawford, B.M. and Altman, J.W., 1972. Design for maintainability. In: H.P. Vancott and R.G. Kinkade (Eds.), Human Engineering Guide to Equipment Design. U.S. Department of Defense, Washington, DC, pp. 585–631.
Croney, J., 1971. Anthropometrics for Designers. Van Nostrand Reinhold Co., New York.
Damon, A., Stoudt, H.W. and McFarland, R.A., 1966. The Human Body in Equipment Design. Harvard University Press, Cambridge, Massachusetts.
Deivanayagam, S. and Weaver, T., 1988. Effects of handle length and bolt orientation on torque strength applied during simulated maintenance tasks. In: F. Aghazadeh (Ed.), Trends in Ergonomics/Human Factors V, Elsevier Science Publishers, North-Holland, pp. 827–833.
Deivanayagam, S. and Ratnavelpandian, K., 1991. Isometric strength model for tasks requiring access through openings. In: E. Boyle, J. Easterly, J. Ianni and S. Harper (Eds.), Human-Centered Technology for Maintainability: Workshop Proceedings, Armstrong Laboratory, WPAFB, Ohio.
DIN 31001, Part I, 1983. Safety distances for adults and children. Deutsches Institut fur Normung, Berlin.
Easterly, J. and Ianni, J.D., Crew Chief: Present and future. In: E. Boyle, J. Easterly, J. Ianni, M. Kodak and S. Harper (Eds.), Human-Centered Technology for Maintainability: Workshop Proceedings, pp. 31–38, WPAFB, Ohio.
Eastman Kodak Co., 1983. Ergonomic Design for People at Work: Volume I. Van Nostrand Reinhold Co., New York.
Galer, I.A.R. (Ed.), 1987. Applied Ergonomics Handbook. Butterworths, London.
Hertzberg, H.T.E., 1972. Engineering anthropology. In: H.P. Van Cott and R.G. Kinkade (Eds.), Human Engineering Guide to Equipment Design. U.S. Government Printing Office, Washington, DC, pp. 467–584.
Hertzberg, H.T.E., Emanuel, I. and Alexander, M., 1956. The anthropometry of working positions I. A preliminary study. WADC Technical Report: 54-520 (AD110-573). Wright-Patterson AFB, Ohio: Wright Air Development Center.
Imrhan, S.N., 1991a. Workspace design for maintenance. In: A. Mital and W. Karwowski (Eds.), Workspace, Equipment and Tool Design. Elsevier Science Publishers, Amsterdam, The Netherlands.
Imrhan, S.N., 1991b. Pinch strengths in various wrist positions. Applied Ergonomics (in press).
Kama, W.N., 1963. Volumetric workspace study: I. Optimum workspace configuration for using various screw drivers. Report No. AMRL-TDR-63-68(I), Aerospace Medical Research Laboratories, WPAFB, Ohio.
Kama, W.N., 1965. Volumetric workspace study: II. Optimum workspace configuration for use of wrenches. Report No. AMRL-TRD-63-68(II), Aerospace Medical Research Laboratories, WPAFB, Ohio.
Kennedy, K.W. and Filler, B.E., 1966. Aperture sizes and depths of reach for one- and two-handed tasks. Technical report no. AMRL-TR-66-27. Aerospace Medical Research Laboratories, Wright-Patterson Air Force Base, Ohio.
Konz, S., 1990. Work design: Industrial Ergonomics. Publishing Horizons, Inc., Worthington, Ohio.
Lane, N.E., 1982. Issues in the statistical modeling of anthropometric data for workplace design. In: R. Easterby, K.H.E. Kroemer, J.W. McDaniel and W.B. Askrein (Eds.), 1985. Report submitted to U.S. Airforce Aerospace Medical Research Laboratory, Wright Patterson AFB, Ohio.
McDaniel, J.W. and Askrein, W.B., 1985. Computer-aided design models to support ergonomics. Report submitted to U.S. AMRL, Wright-Patterson AFB, OH.
MIL-STD-1472B, 1974. Human Engineering Design Criteria for Military Systems, Equipment and Facilities. United States Department of Defense, Washington, DC.
MIL-STD-1472C, 1981. Human Engineering Design Criteria for Military Systems, Equipment and Facilities. United States Department of Defense, Washington, DC.
Mital, A. and Channaveeraiah, C., 1988. Peak volitional torques for wrenches and screwdrivers. International Journal of Industrial Ergonomics, 3: 41–64.
Mital, A. and Wang, L.W., 1989. Effects on load handling of restricted and unrestricted shelf clearances. Ergonomics, 32(1): 39–49. Study in hand/handle effectiveness. Applied Ergonomics, 6: 205–208.
Mital, A., 1991. Workspace clearance and access dimensions and design guidelines. In: A. Mital and W. Karwowski (Eds.), Designing for Maintenance. Elsevier Science Publishers, Amsterdam, The Netherlands.
Morgan, C.T., Cook III, J.S., Chapanis, A and Lund, M.W. (Eds.), 1963. Human Engineering Guide to Equipment Design. McGraw-Hill Book Co., Inc., New York.
Nakajima, S., 1988. Introduction to TPM: Total productive maintenance. Productivity Press, Cambridge, MA.
Noble, J., 1982. Activity and Spaces, Dimensional Data for Housing Design. The Architectural Press, London.
Oborne, D.J., 1981. Ergonomics at Work. John Wiley and Sons, New York.
Pheasant, S., 1986. Bodyspace: Anthropometry, Ergonomics and Design. Taylor and Francis, Philadelphia, PA.
Pulat, B.M. and Alexander, D.C. (Eds.), 1991. Industrial Ergonomics: Case Studies. Industrial Engineering Institute and Management Press, IIE, Norcross, GA.
Quigley, W., Baker, D. and Butdorf, D., 1991. Maintainability evaluation and integrated product development. In: E. Boyle, J. Easterly, J. Ianni, M. Korna and S. Harper

(Eds.), Human-Centered Technology for Maintainability: Workshop Proceedings, WPAFB, Ohio. pp. 122-133.

Rasmusen, J. and Rouse, W. (Eds.), 1981. Diagnosis of System Failures. Plenum Press.

Rigby, L.V., Cooper, J.I. and Spickard, W.A., 1961. Guide to integrated system design for maintainability. Tech. Rpt. 61-424, Aeronautical Systems Div., Wright-Patterson AFB, Ohio: U.S. Air Force.

Rothwell, P.L. and Hickey, D.T., 1986. Three-dimensional computer models of man. In: Proceedings of the Human Factors Society 30th Annual Meeting, Santa Monica, California, pp. 216-220.

Terrell, R. and Purswell, J.L., 1976. The influence of forearm and wrist orientation on static grip strength as a design criterion for hand tools. Proceedings of the 20th Annual Meeting of the Human Factors Society, Santa Monica, California, pp. 28-32.

Thompson, D. and Booth, R.T., 1982. The collection and application of anthropometric data for domestic and industrial situations. In: R. Easterby, K.H.E. Kroemer and D.B. Chaffin (Eds.), Anthropometry and Biomechanics: Theory and Application. Plenum Press, New York, New York, pp. 279-291.

Thompson, R.M., 1972. Design of multi-man–machine work areas. In: H.P. Van Cott and R.G. Kinkade (Eds.), Human Engineering Guide to Equipment Design. U.S. Government Printing Office, Washington, D.C. pp. 419-466.

Tichauer, E.R., 1978. The Biomechanical Basis of Ergonomics. John Wiley and Sons, New York.

Woodson, W.E., 1965. Human Engineering Guide for Equipment Designers. University of California Press, Berkeley, CA.

Woodson, W.E., 1981. Human Factors Design Handbook. McGraw-Hill Book Co., New York.

Woodson, W.E. and Conover, D.W., 1964. Human Engineering Guide for Equipment Designers (2nd edition). University of California Press, Berkeley, CA.

Designing warning signs and warning labels: Part I – Guidelines for the practitioner *

Mark R. Lehto
School of Industrial Engineering, Purdue University, West Lafayette, IN 47907, USA

Correspondence to: Mark R. Lehto, School of Industrial Engineering, Purdue University, West Lafayette, IN 47907, USA.
* The recommendations provided in this guide are based on numerous published and unpublished scientific studies and are intended to enhance worker safety and productivity. These recommendations are neither intended to replace existing standards, if any, nor should be treated as standards. Furthermore, this document should not be construed to represent institutional policy.

The following individuals participated in the discussion of the earlier version of this guide. Their suggestions (written or verbal) were incorporated by the authors in this version: Roland Andersson, *Sweden*; Alvah Bittner, *USA*; Peter Buckle, *UK*; Jan Dul, *The Netherlands*; Bahador Ghahramani, *USA*; Juhani Ilmarinen, *Finland*; Sheik Imrhan, *USA*; Asa Kilbom, *Sweden*; Stephan Konz, *USA*; Shrawan Kumar, *Canada*; Tom Leamon, *USA*; William Marras, *USA*; Barbara McPhee, *Australia*; James Miller, *USA*; Anil Mital, *USA*; Don Morelli, *USA*; Maurice Oxenburgh, *Australia*; Jerry Purswell, *USA*; Jorma Saari, *Finland*; W. (Tom) Singleton, *UK*; Juhani Smolander, *Finland*; Terry Stobbe, *USA*; Rolf Westgaard, *Norway*; Jørgen Winkel, *Sweden*. The guide was also reviewed in depth by several anonymous reviewers.

A. Intended audience

The guidelines presented in this paper are intended for use by practitioners with a strong background in human factors engineering. Such practitioners are likely to be actively involved in warning sign and label development in industrial or governmental settings, or assisting in the development of standards.

B. Application context

The complexity of designing effective warning signs and labels is indicated by the discussion in Part II pertaining to a scientific basis for design guidelines. The format, content, and mode of presentation of warning signs and labels are all potentially important influences. As emphasized there, much of the focus of traditional design solutions has been on format rather than content or the mode of presentation. This paper focuses to a much greater extent on these two latter issues. Consequently, attention is focused on the difficult area of human decision-making in real-world contexts.

The contents of this paper are applicable to solving warning sign and labeling problems both for inplant applications and the difficult problem of product safety in consumer settings. The focus is generic rather than application specific and strongly emphasizes effectiveness and methods of evaluating such. This, of course, increases the area of application to almost any setting in which warning signs or labels might be considered. This is not to say that these guidelines serve as a replacement for standards developed in other contexts which have specified the required content of warning signs or labels. Rather, the proposed guidelines provide a starting point for developing new warning signs and labels and re-evaluating past designs. As a final point it should be noted that the guidelines are also likely to be useful during the development of information signs, instructions, point of purchase displays, product inserts, tags, and other forms of information commonly provided with products or used in industry. While the focus has been on warning signs and labels, the evaluation methods and guidelines remain relevant in these other settings.

C. Problem identification

An initial point is that the provision of warning signs and labels must be recognized as a method for supplementing, rather than replacing, design solutions that focus on removing or reducing hazards in the workplace. In general, the accepted hierarchy of control from most to least effective is: (1) elimination of hazards, (2) containment of hazards, (3) containment of people, (4) training of people, and (5) warning of people (Lehto and Clark, 1990).

For the practitioner considering the development of a warning sign or label, the process of problem identification involves determining whether warning signs and labels should be provided. This can be difficult for reasons expanded upon in Part II (Lehto, 1991). The most obvious principle to guide such decisions is that warnings should be provided when they are likely to be effective. From a logical perspective this means that there must be a significant safety problem, in terms of incidence or severity of consequences. In certain cases this can be determined on the basis of accident statistics. In others, hazard analysis may reveal potential problems.

Furthermore, the safety problem must be caused by human behavior that can feasibly be modified by a warning sign or label. However, it should be recognized that warning signs and labels may be legally required even in cases where their effectiveness is highly questionable (i.e. labels describing the dangers of cigarette or alcohol use).

D. Data collection

As a necessary step early in the design process, task analysis should be performed to determine the critical information people need to know and the stage within the task when that information is best provided. The essential elements of such analysis are discussed in Part II. A large set of techniques have also been developed which are useful for evaluating the perception, comprehension, and behavioral effects of warning signs and labels once they have been developed.

D.1. Evaluating perceptual factors

Numerous approaches have been developed for evaluating aspects of sign and label perception. Such approaches include the evaluation of reaction time, tachistoscopic procedures, measures of legibility distance, study of eye movements, and the evaluation of accuracy or errors. By applying such methods insight can be obtained regarding both the perceptibility of signs and labels and the degree to which they attract attention.

Reaction time. Measures of reaction time have been used in many different contexts to evaluate warning signs. The basic assumption is that a symbol quickly reacted to is more salient (or attention demanding) than one that is reacted to slowly. Reaction time measures are particularly useful when quick reaction times are an actual requirement of the task (as in the driving of automobiles). Their value becomes questionable when reaction time is not of essence to the task. This latter point follows because reaction time is frequently not related to other measures of sign or label quality.

Tachistoscopic procedures. In this approach, a sign or label is presented to a subject with a tachistoscope for precisely timed intervals. After viewing the sign or label, subjects are then asked questions regarding its contents. Generally, the viewing times are very short, but longer viewing times may be used when the signs or labels are lengthy. By varying the presentation time, insight is obtained into how much time is required to correctly perceive elements of the sign or label. As such, these procedures provide a measure of how much attention is required to correctly perceive a sign or label.

Glance legibility is a variation of this approach originally used to compare textual and symbolic traffic signs at very short viewing times that has subsequently seen much application. However, as for reaction time measures, glance legibility is a questionable measure of symbol effectiveness when quick perception is not important. Several studies have found glance legibility to be unrelated to measures of comprehension or even other measures of legibility.

Legibility distance. Traffic signs have often been evaluated in terms of the maximum distance at which they can be recognized by representative subjects. Certain safety standards also recommend this approach. Legibility distance can be predicted accurately by computer programs that consider factors such as sign luminance, contrast, and visual angle. However, it traditionally has been directly measured. A common method requires the subject to move toward the sign beginning from a location at which the sign is not recognized. At the point where the sign is first recognized, the distance between the human subject and the sign is measured. Direct measurement can be inconvenient because the legibility distance of large objects is very long.

Eye movements. The tracking of eye movements during the viewing of signs or labels has also been proposed as a means of evaluating signs or labels. By doing so, insight can be obtained regarding how much attention people devote to specific elements of a sign or label. This approach is powerful because it can pinpoint areas of the sign or label which are hard to perceive or comprehend. It also illustrates the sequence in which items are viewed. A difficulty of this approach is that it generates a vast amount of data which can be difficult to analyze. Furthermore, the equipment used to collect eye movement data can be quite intrusive (older systems often require restraint of head movement), making it questionable whether subjects are following their ordinary forms of behavior when they are being observed.

Accuracy or errors. Perceptual errors also have been commonly used to evaluate elements of signs or labels. One of the standard approaches involves the development of confusion matrixes. In a confusion matrix, the various stimuli are listed in the same order on the x and y axes. Commonly, the x axis will correspond to the given stimuli, and the y axis will correspond to people's responses. Correct responses are on the diagonal line through the matrix described by cells for which the presented signal and elicited response are the same. All other cells in the matrix correspond to confusions.

Much work has been performed in which auditory and visual confusions between phonemes and letters are tabulated in confusion matrixes. The approach has also been used to document the confusions between automotive pictographs.

D.2. Evaluating comprehension

Much of the focus of standard-making organizations has been on developing signs and labels which can be easily comprehended. For example, ANSI Z535.3 includes provisions for (1) obtaining a representative target audience of at least 50 people which includes subgroups with unique problems, and (2) evaluating the meanings the test audience associates with the symbol. A minimum of 85% correct responses and a maximum of 5% critical confusions (defined as the opposite of correct) are required before a symbol is deemed acceptable.

Generic approaches for evaluating the comprehension of signs and labels include recognition/matching, psychometric scales, and readability indexes. Each approach will be briefly discussed below.

Symbol recognition/matching. Recognition/matching is a commonly applied technique for measuring the comprehension of symbols used within signs and labels. In a recognition task, the subject provides an open-ended response describing the meaning of a symbol. In a matching task, the subject selects the most applicable meaning from a list. Recognition is less commonly studied than matching because classifying open-ended responses from a large group of subjects is difficult. The approach does provide the advantage of measuring the meaning of symbols independently from other symbols.

A deficiency of both techniques, as normally used, is that they do not provide the contextual information commonly found within a task. One approach to alleviate this problem is to explicitly describe the context within which the symbol will be presented.

Message recall. The degree to which the message within signs or labels is correctly recalled sometime after viewing provides an additional measure of comprehension. In evaluating recall, consideration should be given to the task-related context and the time passed since the sign or label was viewed. Measures of recall in task-related settings more closely indicate the degree to which the message will be applied.

Psychometric scales. The rating of warning signs and labels on psychometric scales provides an alternative way of describing comprehended meanings. For example, after viewing a warning sign or label subjects could be asked to rate the probability of injury, the severity of injury, the importance of taking certain actions, the controllability of the hazard, the association between certain factors and the likelihood of an accident, etc. From these ratings various aspects of comprehension could then be inferred, using statistical approaches such as factor analysis, multidimensional scaling, or cluster analysis. A primary difficulty is that appropriate scales that denote relevant safety-related meanings need to be developed before this approach can be applied. The psychometric approach also focuses entirely on associations (a measure of declarative knowledge) and therefore is inadequate for describing the degree to which procedures are comprehended or learned.

The semantic differential is a variant of this approach which has been used to evaluate the meanings associated with pictographs and symbols and seems to be of potential value in a wide variety of other contexts. For example, it has been used to evaluate associations between signal words and the level of perceived hazard.

Readability indexes. A readability index describes the difficulty of written material in terms of word length, sentence length, or other variables. Many different readability indexes have been developed. This approach has been used to evaluate written statements of 'warning' and 'caution'. However, warning signs and labels often contain only terse fragments of sentences rather than prose, making readability indexes of little value.

D.3. Evaluating behavior patterns

As a final form of evaluation, there is a strong need to evaluate under realistic conditions the extent to which a warning sign or label influences behavior, simply because human behavior is difficult to predict. The ultimate objective of a warning sign or label is to change behavior, meaning that intermediate measures are of only secondary interest. Methods for evaluating behavior include experimental setups in which subjects use products under simulated conditions. Surveys and observations in the field are other commonly used approaches.

Field observations become especially important if the desired behavior is inconsistent with prevalent behavior patterns (i.e. the wearing of seatbelts). Under such circumstances, the simple presence of an observer or knowledge that an experiment is being performed may change the behavior patterns of the subjects. Furthermore, it is well known that people often claim to act in ways other than they really do.

E. Design solutions

In Part II of this paper, several generic guidelines are inferred from models and the available research on the effectiveness of warnings. These guidelines are: (1) Match the warning to the level of performance at which critical errors occur for the relevant population; (2) Integrate the warning into the task and hazard related context; (3) Be selective; (4) Make sure that the cost of compliance is within a reasonable level; (5) Make symbols and text as concrete as possible; (6) Simplify the syntax of text and combinations of symbols; (7) Make warning signs and labels conspicuous and each component legible; (8) Conform to standards and population stereotypes when possible.

In practice, each of these generic guidelines address specific aspects of the format, content, and mode of presentation of warning signs and label. Combinations of format, content, and mode of presentation correspond to particular design solutions. The relationship between these guidelines and design solutions will be expanded upon below.

E.1. Format

The format of a warning sign or label refers to physical aspects such as size, the presence of signal words and symbols, methods of color coding, typography and the arrangement of the sign or label's components. The two guidelines with the greatest relevance to specifying format are (7) Make warning signs and labels conspicuous and each component legible, and (8) Conform to standards and population stereotypes when possible. In application, the goal of making warning signs and labels conspicuous with legible components is relatively straightforward and will almost always be attained by conforming with labeling standards such as those summarized in Part II. Labeling standards also provide a means for conforming with, not to mention encouraging the development of population stereotypes regarding sign or label format.

Ways of satisfying format-related guidelines which seem to be justifiable on the basis of current research are summarized below.

(1) Provide a signal word or words in capital letters at the top of a warning sign or label which indicates the sign or label is a warning. Examples include the words ATTENTION, NOTICE, WARNING, CAUTION, and DANGER.
(2) Use lowercase for text within the sign or label.
(3) Text within labels should subtend a visual angle of at least 10 minutes of arc at the intended viewing distance. Consider larger visual angles if luminance conditions may be low or for special populations, such as the elderly. For signs, text should subtend a visual angle of at least 25 minutes of arc at the intended viewing distance.
(4) For short messages or signal words, wherein legibility is of primary concern, text characters should be sans serif. For more lengthy prose, consider serif fonts.
(5) Text characters should be dark against a light background.
(6) The strokewidth to height ratio of text characters should be between 1:6 to 1:10.
(7) The width to height ratio of text characters should be between 1:1 to 1:3.5.
(8) Provide for a brightness contrast of at least 50% between text and background. Don't depend on color contrast. If color is used, consider the energy spectrum of foreseeable lighting sources when specifying the color mix.
(9) Consider color coding schemes consistent with ANSI Z535.4. If color is used, consider the energy spectrum of foreseeable lighting sources when specifying the color mix.
(10) Make the sign or label large enough to avoid crowding of components, if possible. Each component must be legible.
(11) Consider the potential effects of an adverse viewing environment, including the presence of dirt, grease, and other contaminants; smoke or haze; and aging effects. Test the legibility under foreseeable conditions of this type. Consider means of replacing warning signs and labels when they become damaged.
(12) Consider the use of symbols when performance will be at a skill-based or rule-based level.
(13) Consider text if performance is likely to be at a knowledge-based level.
(14) Consider symbols rather than text if legibility is a major concern or viewing distances are long. However, if degradation due to contaminants is a likely problem, consider text.
(15) Consider less detailed pictographs or symbols if legibility is a primary concern.

E.2. Content

The content of a warning sign or label refers to the message itself, its level of abstraction, and syntactic structure. In specifying the content of a warning sign or label, several of the preceding guidelines become highly relevant. First, the *message* itself should satisfy guidelines (1) Match the warning to the level of performance at which critical errors occur for the relevant population, (3) Be selective, and (4) Make sure that the cost of compliance is within a reasonable level. Specific recommendations inferred from these guidelines and the discussion in Part II are:

(1) Messages should focus on critical errors which cause a significant safety problem. Avoid long lists of hazards. Avoid messages that describe hazards trivial or obvious to the intended audience. Such information is best provided on media other than warning signs or labels.

(2) Focus on developing messages for the two following types of error: (1) forgetting to perform an action ordinarily performed (i.e. the sign or label reminds), or (2) not knowing the consequences of performing or failing to perform some action.

(3) When performance is skill-based and the error purely involves the failure to perceive a condition or motor variability, provide a warning signal and consider training.

(4) When performance is rule-based and the error is caused by an incorrect or inadequate rule, determine whether the rule was originally developed on the basis of knowledge or judgement-based behavior. If it seems to be judgement based, focus on enforcement. If it seems to be knowledge based, determine whether a warning sign or label can be integrated into the task (i.e. will it place its message into short term memory at the time it is relevant). If it can, consider a sign or label containing the rule. Otherwise, focus on training or instructions.

(5) When performance is knowledge based and the error is caused by inadequate knowledge, determine the extent of the knowledge necessary to prevent the error. If the knowledge can be described with a small number of rules, consider a warning sign or label containing these rules. Otherwise, focus on training or instructions.

(6) When performance is judgement based and the error is caused by inappropriate priorities, evaluate the behavior pattern. If the behavior pattern appears to have significant value to the user (i.e. pleasure, comfort, convenience, etc.) or is likely to be entrenched, focus on enforcement. If the target audience is likely to be unfamiliar with the hazard or setting, consider a warning sign or label describing the hazard and the benefits of compliance.

(7) At all levels of performance, consider messages which minimize the cost of compliance. Make sure the desired form of behavior is within the user's capabilities.

(8) If a large number of potential messages are still present after applying these guidelines, other means of providing information, such as instruction manuals or training courses, should be considered to reduce the number of messages to be conveyed with a warning sign or label.

Given that a message satisfies the preceding screen, further recommendations pertain to the level of abstraction and are based on the guidelines (1) Make symbols and text as concrete as possible, and (2) Match the warning to the level of performance at which critical errors occur for the relevant population. Specific recommendations inferred from these guidelines and the discussion in Part II are:

(9) Consider pictographs instead of symbols if subjects are inexperienced.

(10) When performance is at a skill-based or rule-based level, consider brief messages that describe conditions or actions (focus on conditions if people know the relevant action). Also, consider symbols or pictographs instead of text.

(11) When performance is at a knowledge-based level, consider more detailed messages that describe both conditions and actions (i.e. rules or procedures). Use text rather than symbols.

(12) When performance is at a judgement-based level, consider messages that describe the hazard and the benefits of compliance. Consider citing highly credible sources.

(13) When the hazard is complex or occurs in many different manifestations, consider abstract text which better covers hazard contingencies rather than a long list of concrete examples.

(14) When knowledge or understanding of target population is low, consider concrete text.

(15) For all levels of performance, use words and symbols which people in the intended user population comprehend. Consider language, reading level, and cultural effects.

The *syntactic structure* of warning signs and labels is addressed by Guideline (6) Simplify the syntax of text and combinations of symbols. Specific recommendations derived from this guideline are:

(16) Use short simple sentences; complex conditional sentences, particularly those containing negations, should be avoided.

(17) Symbolic signs or labels should focus on describing conditions (i.e. flammable). With

few exceptions (i.e. a slash/bar to indicate negation) they should not combine multiple meanings or be used to describe complicated sequences of actions.

E.3. Mode of presentation

The mode of presentation refers to the location and task-specific timing of contact with a warning sign or label. Insight into the best mode of presentation can be obtained by considering guidelines (1) Match the warning to the level of performance at which critical errors occur for the relevant population, (2) Integrate the warning into the task and hazard related context, and (4) Make sure that the cost of compliance is within a reasonable level. Specific recommendations inferred from these guidelines include:

(1) The warning sign or label must be presented at a location and time at which the danger is still avoidable.
(2) The location and timing of presentation should be selected so as to minimize the cost or difficulty of compliance (i.e. a sign to wear goggles should be close to an available source of goggles).
(3) When performance is skill-based and the error involves the use of an inappropriate script, determine whether a way of interrupting the task and bringing attention to the label is feasible. If this can be done, consider a warning sign or label which describes the condition or prescribes an action, or both. If this cannot be done, consider a warning signal, consider training, and consider modifying the product so that activation of the inappropriate script no longer leads to danger. Note that this latter approach (i.e. eliminating the hazard) should always be considered, but becomes particularly critical in this situation.
(4) When performance is rule based and the error involves a failure to perceive a condition, determine whether a warning sign or label can be integrated into the task (i.e. will it place its message into short term memory at the time it is relevant). If it can, consider a sign or label. If not, consider using a warning signal.
(5) When performance is rule based and the error is caused by a missing rule, determine whether a warning sign or label can be integrated into the task (i.e. will it place its message into short term memory at the time it is relevant). If it can, consider a sign or label containing the rule. Otherwise, consider training or instructions.
(6) Attempt to present the warning sign or label at a time when the person has available attentional capacity.
(7) Avoid imbedding sign or label in a cluttered background.
(8) Consider means of providing positive feedback, if possible, for complying with the warning sign or label.

Reference

Lehto, M.R., 1992. Designing warning signs and labels: Part II – Scientific basis for initial guidelines. International Journal of Industrial Ergonomics, 10: 115–138 (this issue).

Designing warning signs and warning labels: Part II – Scientific basis for initial guidelines

Mark R. Lehto
School of Industrial Engineering, Purdue University, West Lafayette, IN 47907, USA

1. Problem description

Warning signs, labels, and other forms of safety markings have long been used in attempts to increase the safety of products and workplaces in situations where more fundamental methods of hazard control are infeasible. Concurrently, numerous guidelines for the design of warning signs and warning labels have been developed by standard-making organizations and regulatory agencies throughout the world. However, the existing guidelines have focused on perceptual issues and rudimentary aspects of comprehension. Almost no attention has been given to the critical issue of how to best assist human decision-making in the task-related context. There also are few objective criteria for deciding whether or not to apply an explicit warning label or sign which are both operationally applicable and scientifically justifiable. These significant shortcomings of the current guidelines severely hamper designers who are attempting to develop effective warning signs and warning labels.

The need for improved warning guidelines is further driven by the emergence of products liability litigation. In judging the adequacy of warnings, the courts themselves are (often naively) prescribing how warning signs and labels should be designed. Inspired by past experience with litigation, manufacturers and industry in general are now often applying legalistic or litigation preventive criteria such as 'warn against ALL hazards with explicit warning labels or warning signs'. The dangers associated with such an approach have previously been noted by Lehto and Miller (1986) and several other authors. One concern is that warning labels and signs may lose effectiveness as a means of influencing decision-making, if they 'cry wolf' too many times. Cognitive considerations also reveal that other less explicit warnings than those recommended in safety labeling and signage systems will be of greatest value in preventing accidents in certain situations (Lehto, 1991). Because of these potential problems and the increasing emphasis on the warning issue, there is a strong need for developing scientifically justifiable guidelines that (1) address all aspects of human behavior, including perception, comprehension, and decision-making, and (2) are operationally useful for determining when warnings should be provided, how they should be designed, and how their effectiveness should be evaluated.

2. Scope

The focus of this paper will be on synthesizing research findings and existing approaches to the design of warnings into useful guidelines for the sophisticated practitioner. However, since the approaches emphasized in safety signage and labeling systems are readily available in primary sources they will not be presented in detail here. Instead the focus will be on cognitive issues and evaluation methodologies.

Correspondence to: Mark R. Lehto, School of Industrial Engineering, Purdue University, West Lafayette, IN 47907, USA.

The paper begins by presenting background information relevant to the issue of providing guidelines to the designer of warning signs and labels. Within this section, existing safety signage and labeling systems will first be reviewed. Attention will then be directed toward describing models of the warning process and their important role in identifying and resolving warning sign and label-related design issues. This discussion will be followed by a brief description of research issues regarding warning effectiveness. Guidelines for the practitioner derived from a review of the literature and the earlier presented models are then given in the following section. Included is a discussion of when should warning signs and labels be provided, innovative methods of data collection, and design solutions. During the course of this discussion, often neglected factors will be emphasized such as (1) the level of human performance and its relation to information processing, (2) user knowledge and knowledge requirements, (3) the flow of information and the task-related context, (4) the cost of compliance, and (5) the need for selectivity.

3. Background

3.1. What is a warning?

Numerous methods of providing safety-related information to people are commonly used in industrial, consumer, and other settings. Commonly recognized approaches include: (1) safety markings, (2) safety signs and labels, (3) safety instructions and training, and (4) safety propaganda. Safety markings refer to methods of non-verbally identifying or highlighting potentially hazardous elements of the environment (i.e. painting step edges yellow, or emergency stops red). Safety signs or labels refer to a variety of informational displays which contain text or symbols conveying several forms of safety information. Other similar media include safety tags, product inserts, and point of purchase displays. Safety instructions and training refer to a wide variety of educational methods. Safety propaganda refers to methods of persuasion.

Given that numerous methods of providing safety-related information exist, it becomes important to clearly distinguish warnings from safety information in general. Along these lines, Lehto and Miller (1986) distinguish 'warnings' from safety information that 'informs', 'instructs', or 'persuades'. They state:

"warnings are specific stimuli which alert a user to the presence of a hazard, thereby triggering the processing of additional information regarding the nature, probability, and magnitude of the hazard. This additional information may be within the user's memory or may be provided by other sources external to the user".

As made clear by this definition, safety markings and other non-verbal forms of information, as well as signs and labels, can and often do act as warnings. A warning sign or label is simply a warning that as its *primary* function conveys using text or symbols "... a message intended to reduce the risk of personal or property damage by inducing certain patterns of behavior and discouraging or prohibiting certain other patterns of behavior" (Dorris and Purswell, 1978).

Based on this definition, warning signs and labels can be distinguished from numerous other types of signs and labels. Certain signs and labels (information signs) inform people of facts often critical to safety (i.e. locations of exits, fire extinguishers, first aid stations, emergency releases, etc.). Other signs and labels (instructions) describe how to perform tasks. A final group (propaganda signs or labels) attempt to persuade people to behave in certain ways. A complicating factor is that signs and labels often combine aspects of warning, informing, instructing, and persuading. Consequently, it becomes necessary to address the degree to which these additional roles (besides warning) are effectively performed by signs and labels in various contexts. Doing so will be a primary focus of this paper.

3.2. Safety signage and labeling systems

A large number of operationally applicable criteria exist within safety signage and labeling systems that describe how to develop a warning sign or label. Such guidelines are primarily provided by standard-making organizations that provide voluntary recommendations. Regulatory agencies also develop and mandate certain types of warnings as explicit requirements, while the courts through their decisions provide implicit requirements for warnings. Notable among standard-making organizations are: (1) international groups, such as the United Nations, the European Economic Community (EURONORM), the International Organization for Standardization (ISO), and the International Electrotechnical Commission (IEC), and (2) national groups, such as the American National Standards Institute (ANSI), the British Standards Institute, the Canadian Standards Association, the German Institute for Normalization (DIN), and the Japanese Industrial Standards Committee.

Other guidelines for warning sign and label design have been developed by (1) independent agencies such as the Underwriters Laboratories (UL) and the National Fire Protection Association (NFPA), (2) professional organizations such as the Society of Automotive Engineers (SAE), the American Society of Mechanical Engineers (ASME), the National Safety Council (NSC), and the American Society of Safety Engineers (ASSE), (3) trade associations such as the National Electrical Manufacturers Association (NEMA), the Chemical Manufacturers Association (CMA), and the Material Handling Institute, (MHI), and (4) corporations such as FMC (1985) and Westinghouse (1981).

In particular, safety signage and labeling systems provide recommendations for signal words and text, color coding, typography, symbols, arrangement, and hazard identification, among other factors (table 1). A commonly cited advantage of conforming with these systems is that they convey hazards in a rather stereotypical and explicit way. While this point appears to be true, there is substantial diversity in the provided design recommendations, which can become especially troublesome for manufacturers who market products in different countries. The following discussion will briefly summarize some important aspects of such systems and relate them to available research.

Signal words and text. Safety signage systems generally make use of particular signal words to signify levels of hazard. It is assumed that using such terminology will minimize possible misperception of the significance of warned-against hazards. Particular words which have been proposed, listed in decreasing order of their signified severity, are DANGER, WARNING, and CAUTION. Several commonly referenced sources including the FMC labeling system, use this terminology, while the ANSI Z35.1 standard uses the terms DANGER and CAUTION. One study (Bresnahan and Bryk, 1975) has addressed the perception of the terms 'danger' and 'caution' by industrial workers. Here it was found that greater levels of hazard were associated with the term 'danger' than with 'caution'. Lirtzman (1984) consistently found for both student consumers and chemical workers that the term 'danger' was associated with high levels of danger. However, the terms 'warning' and 'caution' were both associated with low danger and could not be distinguished. A simpler system might use a numerical risk scale, or the phrases 'extreme-danger', 'serious danger', and 'moderate-danger'. Hadden (1986) provides a strong rationale for developing consistent means of conveying the degree of risk, along these or similar lines.

Many safety standards (ANSI Z35.1, ANSI D6.1, ANSI Z129.1) and other standard sources (Westinghouse, 1981; FMC, 1980) also list examples of messages to be used in warning signs. The messages prescribed in such sources are often short fragments of sentences (phrases) that describe actions or conditions. It appears that such phrases should generally be easily understood. Multilingual concerns are also addressed by a subset of standards. For example, certain standards in the United States recommend text to be in both English and Spanish. Others recommend the use of symbols in multi-cultural settings. While the later approach has been successful in certain instances, as discussed further in the section on symbols, studies have shown that certain symbols are not well-understood across cultures (Easterby and Zwaga, 1976).

Color coding. Color coding methods, also referred to as 'color systems', consistently associate colors with particular levels of hazard. For example, red is used in all of the standards in table 1 to represent the highest level of danger. Orange is often used to identify hazard, Yellow caution, Green first aid, and Blue sources of safety information.

Some research has been performed evaluating stereotypical associations between different colors and the intended meanings. Bresnahan and

Table 1
Summary of recommendations within selected warning systems (based on Lehto and Miller, 1986; Lehto and Clark, 1990).

System	Signal words	Color coding	Typography	Symbols	Arrangement	Hazard ID
ANSI Z35.1 Specifications for accident prevention signs (to be replaced by ANSI Z535.2)	Danger Caution	Red Yellow	Sans serif typeface. All upper case or upper and lower case.	Symbols only as supplement to words	Defines Signal word, Message, Symbol panels (optional, attached to side of label)	Not specified
ANSI Z129.1 Precautionary labeling of hazardous chemicals	Danger Warning Caution Poison optional words for 'delayed' hazards	Not specified	Not specified	Skull-and-crossbones as supplement to words. Acceptable symbols for 3 other hazards types.	Label arrangement not specified; examples given	Provides guidance about how to select signal words
ANSI Z535.2 Environmental and facility safety signs (to replace ANSI Z35.1)	Danger Warning Caution Notice [general safety] [arrows]	Red Orange Yellow Blue Green as above; B&W otherwise per ANSI Z535.1	Sans serif, upper case, acceptable typefaces, letter heights	Symbols and pictographs per ANSI Z535.3	Defines Signal word, Word message, Symbol panels in 1–3 panel designs. 4 shapes for special use. Can use ANSI Z535.4 for uniformity.	Provides guidance
ANSI Z535.4 Product safety signs and labels	Danger Warning Caution	Red Orange Yellow per ANSI Z535.1	Sans serif, upper case, suggested typefaces, letter heights	Symbols and pictographs per ANSI Z535.3; also SAE J284 Safety Alert Symbol	Defines Signal word, Message, Pictorial panels in order of general to specific. Can use ANSI Z535.2 for uniformity. Use ANSI Z129.1 for chemical hazards.	Provides guidance
NEMA Guidelines: NEMA 260	Danger Warning	Red Red	Not specified	Electric shock symbol	Defines Signal word, Hazard, Consequences, Instructions, Symbol. Does not specify order.	Not specified

Standard	Signal words	Colors	Typography	Symbols	Layout	Other
SAE J115 safety-signs	Danger / Warning / Caution	Red / Yellow / Yellow	Sans serif typeface, upper case	Layout to accommodate symbols; specific symbols/pictographs not prescribed	Defines 3 areas: Signal word panel, Pictorial panel, Message panel. Arrange in order of general to specific.	Provides guidance
ISO Standard: ISO R557, 3864	None. 3 kinds of labels: Stop/prohibition, Mandatory action, Warning	Red / Blue / Yellow	Message panel is added below if necessary	Symbols and pictographs	Pictograph or symbol is placed inside appropriate shape with message panel below if necessary	Not specified
OSHA 1910.145 Specification for accident prevention signs and tags	Danger / Warning (tags only) / Caution / Biological Hazard, BIOHAZARD, or symbol / [safety instruction] / [slow-moving Vehicle]	Red / Yellow / Yellow / Fluorescent Orange/Orange-Red / Green / Fluorescent Yellow-Orange & Dark Red per ANSI Z53.1	Readable at 5 feet or as required by task	Biological hazard symbol. Major message can be supplied by pictograph (tags only). Slow-moving vehicle (SAE J943)	Signal word and major message (tags only)	Provides guidance
OSHA 1910.1200 [Chemical] Hazard Communication	Per applicable requirements of EPA, FDA, BATF, and CPSC; not otherwise specified		In English		Only as material safety data sheet	Provides guidance
Westinghouse Handbook; FMC Guidelines	Danger / Warning / Caution / Notice	Red / Orange / Yellow / Blue	Helvetica bold and regular weights, upper/lower case	Symbols and pictographs	Recommends 5 components: Signal word, Symbol/pictograph Hazard, Result of ignoring warning, Avoiding hazard	Provides guidance about how to select signal words

Bryk (1975) found that industrial workers appeared to associate the colors red and yellow with greater degrees of hazard than they did for the colors green and blue. Jones (1978) tested the importance of color cues in the comprehension of European road signs. Removal of the red color cue associated with signs that indicated hazard had no significant effect. A negative effect was noted, however, when the blue color cue for information signs was removed.

Typography. Explicit recommendations regarding typography are given in nearly all the systems. The most general commonality between the systems is in the recommended use of sans serif typefaces, based on legibility. Such recommendations are generally consistent with recommendations based on typographic research (i.e. Tinker, 1963). Standards also generally place emphasis on features of text such as text height, stroke width, contrast, or the use of capital letters. Lehto and Miller (1986) cite visual angle as the primary criteria influencing legibility, given adequate illumination. They note that a number of researchers recommend visual angles of 10 minutes of arc subtended by letters as adequate for legibility. However, larger visual angles, of up to 37 minutes of arc, have been recommended for poor viewing conditions. Recent work by Smith (1984) provides substantial insight into this issue. In a field study of over 2000 subjects, 90% legibility was found for letters or numerals at 9 minutes of arc. At 12 minutes of arc, the legibility was about 95%; at 16 minutes of arc, the legibility was 98%; at 25 minutes, the legibility was nearly 100%.

A strokewidth to height ratio of 1:8 has been recommended for black characters on white backgrounds and 1:13.3 for white characters on a dark background (McCormick, 1976). As noted by Lehto and Miller (1986), variations of at least 25% in the above strokewidth to height ratios can still result in good legibility. With regard to contrast, Lehto and Miller (1986) state that it is difficult to develop conclusive specifications from the literature. They note that problems in legibility seem to be minor as long as the luminance contrast ratio is greater than 50% and the illumination level is reasonably high. Color contrast provides a supplemental means of increasing legibility that may be useful when luminance contrast is low. However, color contrast may disappear under lighting conditions where the energy spectra is unbalanced and may not be perceived by color blind people.

Regarding the use of capital letters, studies have shown them to be somewhat more legible than lowercase under degraded viewing conditions (Miller et al., 1990). However, reading speed is often slower for capital letters (i.e. Tinker, 1963).

Safety symbols. Safety symbols have been developed as an alternative means of communicating safety messages to both literate and illiterate populations, often with the conviction that such symbols are more easily perceived and understood than written text. Some perspective is gained by considering the estimates given by Collins et al. (1982): (1) about 5 million individuals in the United States (as of 1976) reported difficulties in speaking or understanding English; and (2) between 2 and 64 million adults in the United States are functionally illiterate (this great variation, of course, reflects differing definitions of literacy). However, it has been well-recognized by standard-making organizations that symbol comprehension can be low. Because of such concerns, nearly all standards recommend that symbols and pictographs be supplemented with text. Other standards such as the proposed ANSI Z535.3 provide extensive recommendations regarding symbol configurations, their use, and development.

Recently, several substantial and well-known research efforts have evaluated the comprehension of safety symbols by industrial workers. Standard making groups such as the International Standards Organization (ISO) and the Occupational Safety and Health Administration (OSHA) have commissioned such work in the hope of developing easily comprehended symbols. In these studies, it has been found that very few symbols are universally comprehended. Consequently, effective symbols have been arbitrarily defined as those which are correctly interpreted by 85% or more of the evaluated population. Several of these studies have evaluated the comprehension of symbols commonly found in consumer and public settings. In particular, Easterby and Hakiel (1981) tested all known symbols pertaining to fire, poison, caustic, electrical, and general hazard. Approximately 4000 consumers participated in the survey. The comprehension of the best signs was only about 20%, when the criterion of correctness was stringent. When the criterion was lax, comprehension of the best signs increased to 50%. Markedly worse performance was observed for certain signs (5% or worse with the lax criterion). Collins and Lerner (1982) investigated 25 fire-safety signs for a sample of 91 subjects. Comprehension of the symbols varied from nearly zero to nearly 100%. Green and Pew (1978) studied the comprehension of 19 pictographs used in automobiles. Only 6 of the 19 symbols met the criteria of 75% recognition and 5% errors. Easterby and Zwaga (1976) surveyed public information signs, finding large variations in understandability.

Among those studies most focused on industrial settings, Collins et al. (1982) studied the comprehension of symbols used to convey 33 messages related to hazards, protective gear, first aid and emergency equipment, prohibited actions, and egress. The surveyed individuals consisted of 222 employees. Substantial variation was found for the evaluated symbols. For example, between 18% and 58% of the subjects correctly identified the meaning of at least some 'no exit' symbols. In contrast, between 90% and 100% of the subjects correctly identified the meaning of at least some 'eye protection' symbols. In a subsequent study, Collins (1983) studied 72 mine safety symbols conveying a total of 40 messages. The surveyed subjects were 267 miners located at 10 different mine sites. The results showed that 34 of the 40 messages were correctly interpreted by 85% or more of the subjects.

Other research clearly demonstrates that the context can influence the comprehension of symbols. For example, Cahill (1976) performed an experiment in which one group of subjects received a drawing of a vehicle cab within which the symbols were said to be used. A control group of subjects was given the symbols alone. The subjects who received contextual information, as conveyed by the drawing of the vehicle cab, correctly identified 62% of the symbols, while the subjects in the control group correctly identified 44% of the symbols. Galer (1980) evaluated the influence of a task-related context on the comprehension of signs (by truck drivers). A comprehension rate of 71% was found when no contextual information was given. Of those drivers who did not understand the sign, 37% were able to understand it when contextual information was given.

The above results indicate that measures of the comprehension of symbols, as normally obtained in surveys or tests, might be artificially low, since such approaches do not usually provide the contextual information found within real tasks. However, more recent research (Frantz et al., 1991) indicated that although people are quite good at inferring the generic meaning of the flame symbol, users of flammable adhesives may not correctly infer precautions associated with flammable vapors when they see the flame symbol in a use-related context. In any case, both findings (i.e. either increased or poor comprehension in actual use contexts) strongly support testing in real use contexts.

It does appear that many problems in comprehension may be due to a lack of standardization of design and a lack of opportunity for people to learn correct interpretations (Collins et al., 1982). The learning of symbols is an important issue in and of itself. Among related studies, Green and Pew (1978) found that the ease of learning the meaning of pictographs was not related to initial measures of comprehension. Cairney and Sless (1982a,b) found that most of the pictographs which gave trouble to their subjects during initial testing were readily learned. They also found high levels of retention (85% or better) for the majority of symbols when the subjects were retested one week later. Further research is needed to determine whether people get adequate opportunities to learn the meaning of symbols, not to mention remember them over long time intervals. For certain symbols this question can be answered affirmatively (i.e. 86% of a sample population in Canada reported seeing the flame symbol and 91% correctly interpreted its meaning – Gallup Canada, 1989).

A final point is that the use of symbols is frequently justified on the assumption that their meanings can be inferred across cultures. However, recent studies have shown that the mean comprehension of symbols varies greatly across cultures (Easterby and Zwaga, 1976; Cairney and Sless, 1982b). In particular, Cairney and Sless (1982b) found that culture-related effects resulted in profound misunderstanding of symbols by Vietnamese immigrants when compared to Europeans. Easterby and Zwaga (1976) found significantly varying degrees of comprehension between natives of different European countries for certain symbols.

Label arrangement. Substantial variation can be found in signage and labeling systems with regard to the recommended label arrangement. The proposed arrangements generally include elements from the above discussion and specify the: Image – graphic content, color; Background – shape, color; Enclosure – shape, color; and Surround – shape, color. Many of the systems also precisely describe the content and arrangement of the written text. Approaches often describe multi-panel formats (ANSI Z535.4, ANSI Z129.1, FMC) in which certain types of information are placed in particular panels.

There is experimental evidence documenting a relation between label or sign shape and hazard associations. One fairly consistent finding is that pointed shapes, such as diamonds, triangles pointing downward, or other regular figures with a vertex pointing downward, have greater hazard association values than shapes like rectangles oriented parallel to the ground or circles (Jones, 1978; Riley et al., 1982; Collins, 1983; Cochran et al., 1981). These effects might reflect stereotypes people develop from observing traffic signs. However, there is little if any evidence that particular label arrangements are critical determinants of sign or label effectiveness. In particular, two studies, Ursic (1984) and Dingus et al., (1991) have indicated that arrangement was not a significant determinant of sign or label effectiveness.

Hazard identification. Certain systems also provide guidance regarding methods of hazard identification within warning signs or labels. Often this takes the form of assigning signal words to particular levels of hazard.

3.3. Models of the warning process

While safety signage and labeling systems are useful in that they encourage the standardization of warning sign and label designs, it must be emphasized that in most applications the effectiveness of the designs proposed by existing systems is scientifically unproven. Part of the reason for this conclusion is that developed warning signs and labels are rarely evaluated in the context of actual use. The content of most warning signs and labels is also generally prescribed on the basis of practitioner experience or industry practice rather than scientific analysis. This is not to say that the traditional approaches are unstructured in their nature. In fact, the traditional focus of safety signage and labeling systems on perceptual issues and the more recent focus on rudimentary forms of comprehension reflects an organized, but incomplete, view of the warning process.

A more complete perspective can be obtained from cognitively oriented models that focus on the complicated process that takes place when an effective sign or label prevents accidents. In so doing, cognitively oriented models uncover and provide substantial insight into resolving design issues not addressed by safety signage and labeling systems. They also form a basis for evaluating the effectiveness of warnings. One such model describes the warning process as the sequential processing of information within a task- and hazard-related context; an extension of this model also considers the level of user performance. These models and the way people organize and apply knowledge have important implications, as expanded upon below.

Sequential processing model. Numerous researchers have described the warning process as a communication consisting of several information processing stages (Lehto and Miller, 1986; Lawrence, 1974; Schwartz and Driver, 1983; McGuire, 1980). In particular, Lehto and Miller (1986) describe the warning process as a sequence of eight stages: (1) exposure to the warning stimulus, (2) attention and active processing of the warning stimulus, (3) comprehension and agreement with the warning message, (4) retaining the message in memory, (5) retrieval of the message at the time it is relevant, (6) deciding to respond consistently with the message, (7) performing the response, and (8) adequacy of the response for preventing accidents.

This model implies that the effectiveness of a warning sign or label can never be greater than the probability of successfully completing any single step in the sequence. For example, consider the hypothetical situation where (1) 50% of the population will read the warning, (2) 50% of those individuals understand the warning after reading it, (3) 100% of those individuals, who read and understood the warning, will retain and retrieve the warning from memory when they need to, (4) 90% of those individuals will act in accordance with the warning after they retrieve it, and (5) that action is sufficient to avoid the accident 90% of the time. When effectiveness is defined as the probability of successfully completing the entire sequence, it becomes the product formed by multiplying the conditional probabilities of successfully completing each separate step. This results in an effectiveness of 0.20 for this example.

Recent research has confirmed such implications of this model. For example, Otsubo (1988) observed in an experimental study of warning labels designed to convince people to wear gloves while using a circular saw that 74% of the subjects noticed the warning, 52% said they read it, and 38% were observed to comply. For a jigsaw, 54% of the subjects noticed the warning, 25% read it, and 13% complied. Gomer (1986) found in a study of workers that when strong warnings of the need to use respirators were placed on bags of limestone, all of the subjects reported seeing the warning. However, only 3 of 14 subjects then requested a respirator after seeing the warning.

Levels of performance. Analysis of existing guidelines regarding the design of warning signs and warning labels reveals that they are expected to perform a wide variety of functions. These functions vary from simply alerting, to reminding, informing, instructing, and perhaps even persuading. This progression directly describes the roles of the traditional methods for providing warning information: (1) safety markings, (2) safety signs, (3) safety instructions and training, and (4) safety propaganda. The sequential model described above does not well distinguish between these

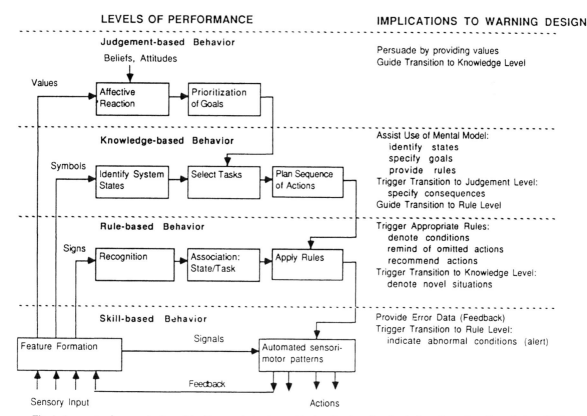

Fig. 1. A proposed conceptual model of human behavior and its implications for the design of warnings (from Lehto, 1991).

Table 2

Levels of operator and warning performance (from Lehto, 1991).

Levels of operator performance	Types of errors	Levels of warning performance
Belief/judgement based	Biases, inappropriate priorities	Persuading
Knowledge based	Fail to formulate the correct intentions, lost in problem space	Educating Assisting problem solving
Rule based	Missing or incorrect rule Omit step Forget/fail to perceive state	Recommending Reminding Informing
Skill based	Use wrong script, Fail to activate script Motor variability Fail to perceive signal	Feedback Alerting

forms of information and the differing ways they are processed.

To alleviate this problem Lehto (1991) developed a conceptual model which related these functions to a hierarchy of operator performance. This resulted in a model of human behavior (figure 1) which built upon Rasmussen's (1986) approach, by explicitly addressing judgement-based behavior. Belief or judgement-based levels of performance were placed at the top level, with the knowledge, rule, and skill-based levels of performance discussed by Rasmussen (1986) at the lower levels. At each of these levels, certain types of errors become prevalent (table 2).

The model further asserts that four fundamental forms of information are used during task performance: signals, signs, symbols, and values. Signals correspond to sensory data, signs indicate perceived or named states, and symbols are abstract constructs which can be formally processed in mental models (Rasmussen, 1986). Values represent deeply processed concepts associated with opinions, attitudes, or beliefs. As in Rasmussen's model, performance at each level is explicitly related to a particular form of information. At the skill-based level, signals trigger automated sensory-motor behavior. At the rule-based level, signs trigger the conscious application of rules. At the knowledge-based level, symbols are manipulated within mental models to identify system states, formulate goals, and plan sequences of task related activity. At the judgement-based level, goal priorities are developed on the basis of agreement or disagreement with presented values.

This model provides a fundamental way of classifying forms of warning information in terms of the depth of processing, by distinguishing between signals, signs, symbols, and values. As such, the model explicitly recognizes the value of many types of warnings including, but not limited to warning signs or labels. The model also provides substantial insight regarding the appropriateness and effectiveness of particular applications of warnings by directly relating each form of information to a specific level of performance. The primary insight is that the presented warning should be processable in a way that matches the level of task performance, i.e. as signals when task performance is at the skill-based level, as signs at the rule-based level, as symbols at the knowledge-based level, and as values at the judgement-based level.

It must further be recognized that shifts in the level of performance are to be expected both as part of task performance and over time as learning takes place. As shown in figure 1, each form of warning information naturally maps to particular functions, which often correspond to guiding transitions either up or down in the hierarchy of performance level. Transitions up the hierarchy occur as part of task performance, while transitions down the hierarchy correspond to longer term effects associated with learning or increased skill. For example, a warning at the skill-based level, by indicating an abnormal condition, may trigger a transition up the hierarchy to a rule or knowledge-based level of performance. On the other hand, a warning at the knowledge-based level may trigger the transition down the hierarchy to a rule-based level of performance in the future by teaching a rule. By considering user knowledge, attitudes, and beliefs in relation to shifts in the level of performance, substantial insight can be obtained regarding the role and effectiveness of warning signs and labels.

To begin with, a general distinction can be made between procedural and declarative aspects of people's knowledge. Procedural knowledge may take the form of scripts, rules, or problem solving strategies, corresponding respectively to the form of knowledge used when performing at a skill, rule, or knowledge-based level of performance. Declarative knowledge generally corresponds to the knowledge of system states, and as such is critical at each of the three lower stages of performance.

A script contains a set of procedures which are grouped together and followed during skill-based performance of a task. Numerous generic forms of behavior primarily involve the use of scripts (i.e. driving on a familiar road; opening of a container, removal, and application of a chemical compound; use of a hand or power tool; etc.). Errors at the skill-based level of performance are often perceptual-motor in nature, and may involve the failure to activate a particular script or the use of an incorrect script (table 2). From the view of a designer, it becomes important to determine whether people may possess inappropriate scripts. Of particular concern are scenarios involving changes in familiar settings or differences between products. In such scenarios, use of an existing script might lead to serious consequences (i.e. a person walking might not change his gait before stepping on a wet spot; a driver might not notice a new stop sign at a familiar intersection; a person who previously used a nonflammable adhesive without ventilating his work-area might do the same when using a flammable adhesive). Rhoades et al. (1990) provide several case studies of such errors.

Reducing or eliminating errors associated with inappropriate scripts is obviously of major concern. A primary difficulty is that when errors occur at the skill-based level, warning signs and labels will not ordinarily be effective, since reading is not part of skill-based behavior (i.e. information is processed only to the signal level). Consequently, in this difficult situation, the strategy of the designer must be focused on triggering a transition to a rule-based or higher level of performance, by some means other than the sign or label itself (see figure 1). Only at that higher level, will there be a possibility of the label being read. Along these lines, there is an emerging body of evidence that such strategies can increase the effectiveness of warning signs or labels. In particular, Frantz and Rhoades (1991) studied the effect of interrupting the skill-based performance of a simple task in which people filled filing cabinets. A warning label alone was an ineffective means of preventing the subjects from filling the top drawers first (13% compliance). The situation changed, however, when the top drawers were sealed shut with a strip under which a label was present (53% compliance) or when a cardboard obstruction displaying the label was placed within the drawer (73% compliance). The conclusion was that forcing the subjects to remove the strip or the cardboard obstruction before they could open the drawers caused a shift up to a rule-based or even knowledge-based level of performance, at which the warning label increased in effectiveness. Other studies (i.e. Dingus et al., 1991), not directly oriented toward evaluating this issue, have also demonstrated that interrupting skill-based behavior increases the effectiveness of warning labels to a major degree. The difficulty, of course, is that ways of interrupting skill-based behavior are not always available.

Rules refer to knowledge organized or describable as pairs of CONDITION(s) and associated ACTION(s). Important safety knowledge is often best described in terms of rules (i.e. if the floor is wet, then don't use an electrical tool). At the skill-based level of performance, such knowledge is embedded within scripts and may not be consciously available (i.e. it is compiled). At the rule-based level of performance, a person consciously perceives condition(s) and then performs the associated action(s). Note that such performance is generally routine and involves little thought. This, of course, reflects the fact that behavior often involves the rule-based coordination of skill-based activity (Rasmussen, 1986). Errors at the rule-based level commonly involve the failure to consciously perceive or retrieve from memory some condition (or state), which leads to omitting a critical action (table 2). For preventing such errors, a warning sign or label may be effective, but only if looking at the label is an ordinary part of the task at the stage in task performance at which the error occurs. Such warnings may be viewed as reminders which are consistent with existing rule-based behavior.

Errors also occur at this level when people don't know important rules or the ones they have are incorrect. In the situation where people don't know certain rules and are performing at a routine rule-based level, performance must be shifted to the knowledge-based level. As noted by Rasmussen (1986), inducing such shifts is difficult, often because people are overconfident in the adequacy of their routine behavior. A more problematic situation is where people possess incorrect rules or scripts which have moved down the hierarchy from knowledge-based or judgement-based reasoning (figure 1). The development of new rules or modification of the old rules governing certain forms of such behavior (i.e. diving in shallow water; failure to wear seat belts, failure to wear helmets, failure to wear eye protection, etc.) will often require a shift up to a judgement-based level of performance. Forcing such a shift is difficult; even intensive educational and persuasive programs often fail (Lehto and Miller, 1986).

Problem-solving strategies refer to methods people use in novel or unique situations when rules from previous experience are not available (Rasmussen, 1986). At the knowledge-based level of performance, people must infer appropriate actions by applying a problem-solving strategy. Much of this process involves the active seeking of information from external sources. As such, knowledge-based performance is prevalent during learning by inexperienced operators. Experienced operators, however, also exhibit similar levels of performance when working in exceedingly complex settings or during novel situations or emergencies. Errors at the knowledge-based level often correspond to the initiation of inappropriate actions, because of incorrect mental models or understanding of the problem to be solved. They may also correspond to an inability to adequately consider the side effects of actions taken earlier in the problem-solving sequence or a failure to shift to a judgement-based level when goal priorities are flawed.

Warning signs and labels intended for people performing at a knowledge-based level can be oriented toward stimulating needed transitions to a judgement-based level or assisting in the use of mental models during critical problem-solving processes (figure 1). Stimulating needed transitions to the judgement-based level might take the form of triggering reasoning directed toward evaluating the reasonableness of existing goal priorities (i.e. explaining why it is important to focus on attaining some goal). Assisting in the use of mental models can focus on at least three aspects of problem solving: (1) identifying states (what is currently true or likely to happen), (2) specifying goals (what to do), (3) providing rules (how to do it). Importantly, providing rules to apply in the form of recommendations may directly guide the transition from a knowledge-based to an appropriate rule-based level of performance.

Behavior at the knowledge-based level involves conscious problem solving directed toward attaining a goal. In this situation, it therefore seems people may actively seek out information from a warning sign or label. On the negative side, as noted by Rasmussen (1986), the majority of behavior is at either a skill or rule-based level. Furthermore, a tendency to seek out information may be outweighed by a profound lack of knowledge in certain instances. Certain people may not be capable of behaving safely even if they read a well-designed sign or label carefully. Simply put, a warning sign or label should not be viewed as a substitute for instruction manuals or training.

User attitudes and beliefs influence the processing of values at the judgement-based level of performance (Lehto, 1991). Such processing results in goal priorities which eventually influence lower levels of performance as behavior becomes more routine (figure 1). Attitudes and beliefs are distinguished from knowledge in that they are deeply held and not easily modified. Judgement-based performance represents the deepest level of information processing (i.e. the incoming information is processed as values which evoke an affective reaction). As such, judgement-based behavior occurs much less often than knowledge, rule, or skill-based behavior. It appears that beliefs and attitudes can be influenced significantly when people are unfamiliar with a topic (this situation, of course, corresponds closely to knowledge-based behavior). On the other hand, it is difficult to bring about changes in beliefs and attitudes; a more general result is that such efforts encourage people to justify or rationalize them. Changes generally occur when people experience highly significant events (i.e. an accident), causing them to reexamine their opinions or beliefs. As discussed by Lehto and Miller (1986) changes on the basis of persuasion are very unlikely when the proscribed behavior is perceived to have value (i.e. comfort, freedom, pleasure, etc.).

The clear implication is that warning signs and labels focused at influencing judgement-based behavior are unlikely to be effective, simply because it is hard to trigger such behavior in the first place and even if people process the warning label or sign to this degree, it is hard to persuade people to change. It is possible that warnings directed toward influencing judgement-based behavior will be effective when people are unfamiliar with a product or environment; however, such effects are undocumented. When there is a strong need for correcting errors in judgement, a focus on enforcement may be the most effective option (Lehto and Foley, 1989).

Conclusions. Analyzing behavior from the perspective of levels of performance provides substantial insight as to when warning signs or labels will be effective. In particular it may be concluded that (1) A warning sign or label is likely to have greatest influence when the behavior is at a knowledge-based level. This is not to say, however, that a warning sign or label can act as a substitute for training or more detailed instructions. (2) A warning sign or label will have no effect when behavior is at a skill-based level unless the behavior is somehow shifted to a higher level by something other than the sign or label itself. (3) A warning sign or label may be effective when performance is at a rule-based level, but only if looking at the label is an ordinary part of the task and the undesired behavior is not rooted in earlier judgement-based behavior. (4) A warning sign or label is very unlikely to be effective when performance is at a judgement-based level.

Because the majority of behavior is at either a skill or rule-based level (Rasmussen, 1986), signals may in fact be the most effective form of warnings in a wide variety of applications, and especially so when they are suitable for triggering transitions to rule or knowledge-based levels of performance (Lehto, 1991). Warning signals, as illustrated by the sounds of wood being fed into a table saw or by the markings on roads, can be particularly effective for skilled users of a product. Consequently, consideration should be given towards designing products which emit obvious signals when they are in a potentially hazardous state. For example, tires should squeal when speeds are excessive, poisonous chemical compounds should smell and taste bad, flammable objects should look or smell so, and so on. However, a critical issue is that the signals should be selective. This conclusion partially follows because an elementary analysis in terms of signal detection theory reveals that nonselective signals cannot convey information (Lehto and Papastavrou, 1991). Furthermore, continuous signals can be annoying and if they are too strong may in fact be hazards themselves (i.e. excessive noise levels can cause hearing loss).

3.5. Generic guidelines

Several generic guidelines regarding the design of warning signs and warning labels can be inferred from models and the available research on the effectiveness of warnings. These guidelines are: (1) Match the warning to the level of performance at which critical errors occur for the relevant population; (2) Integrate the warning into the task- and hazard-related context; (3) Be selective; (4) Make sure that the cost of compliance is within a reasonable level; (5) Make symbols and text as concrete as possible; (6) Simplify the syntax of text and combinations of symbols; (7) Make warning signs and labels conspicuous and each component legible; (8) Conform to standards and population stereotypes when possible; (9) Evaluate the developed warning sign or label in an actual use context.

Each of these guidelines are at a very general level, making them a superset of guidelines presented by previous authors (i.e. Christensen, 1983; Cunitz, 1981). Most importantly, they are based to a large extent on documented research and a modeling foundation. Satisfying these guidelines requires consideration of a substantial number of detailed issues during task analysis. With regard to this issue, Lehto and Miller (1986) have proposed a form of task analysis which first focuses on isolating the critical decision points within a task at which responses are required to avoid errors. At each critical decision point, the available forms of information are then evaluated.

The criteria proposed in the following discussion seem especially appropriate for analyzing the potential role of warning signs and labels at each of these decision points.

Match the warning to the level of performance at which critical errors occur for the relevant population

As an initial step in selecting the content of a warning sign or label, it becomes essential to determine the errors people make and the relationship between these errors and damaging events. Those errors which are critical in that they cause damages should then be evaluated in more detail and classified in terms of the level of performance at which they occur (table 2). For example, it appears that one of the more common errors leading to product-related accidents is that of using the wrong script (a skill-based error). Rhoades et al. (1990) provide several related case studies. In this situation, the person uses a product in a way grossly inappropriate for the given product but appropriate for a similar product which they are familiar with. Along these lines, a highly flammable adhesive might be used without adequate ventilation, in the same way a non-flammable adhesive was previously used. In such a situation, the process of opening the can and beginning application of the product is likely to be at a skill-based level with some rule-based control. Example warnings can easily be matched to errors occurring at particular levels of performance (table 3).

Given that the error can be matched to a particular level of performance, the preceding discussion on levels of performance provides insight into appropriate strategies for remedying the error. In summary form these are:

(1) When performance is skill-based and the error involves the use of an inappropriate script, determine whether a way of interrupting the task and bringing attention to the label is feasible. If this can be done, consider a warning sign or label which describes the condition or prescribes an action, or both. If this can not be done, use a warning signal, consider training, or consider modifying the product so the inappropriate script becomes adequate. Note that this latter approach (i.e. eliminating the hazard) should always be considered, but becomes particularly critical in this situation.

(2) When performance is skill-based and the error purely involves the failure to perceive a signal or motor variability, use a warning signal and consider training.

(3) When performance is rule-based and the error involves a failure to perceive a condition, determine whether a warning sign or label can be integrated into the task (see next sec-

Table 3

Example mapping between the warning information provided by a circular power saw and errors at different levels of performance (from Lehto, 1991).

Level of performance: Example error	Sensory channel		
	Visual	Auditory	Kinesthetic/ tactile
Judgement based: decide that guard is not important	disclaimer stating that if guard is removed the user is in violation of warranty and assumes all risk	–	–
Knowledge based: incorrectly install blade	instructions for use, repair and maintenance	–	–
Rule based: fail to react to unusual conditions	Label containing the statements: Do Not Remove Guard, Electrical Hazard	sound frequency and intensity indicates a performance problem	vibration intensity indicates a performance problem
Skill based: operate saw incorrectly	movement of blade visible	movement of blade audible	vibration of saw tactually sensible

tion). If it can, provide a sign or label. If not focus on using a warning signal.
(4) When performance is rule-based and the error is caused by a missing rule, determine whether a warning sign or label can be integrated into the task (see next section). If it can, provide a sign or label containing the rule. Otherwise, focus on training or instructions.
(5) When performance is rule-based and the error is caused by an incorrect or inadequate rule, determine whether the rule was originally developed on the basis of knowledge or judgement-based behavior. If it seems to be judgement based, focus on enforcement. If knowledge based, apply principle 4.
(6) When performance is knowledge based and the error is caused by inadequate knowledge, determine the extent of the knowledge necessary to prevent the error. If the knowledge can be described with a small number of rules, consider a warning sign or label containing these rules. Otherwise, focus on training or instructions.
(7) When performance is judgement based and the error is caused by inappropriate priorities, evaluate the behavior pattern. If the behavior pattern appears to have significant value to the user (i.e. pleasure, comfort, convenience, etc.) or is likely to be entrenched, focus on enforcement. If the target audience is likely to be unfamiliar with the hazard or setting, consider a warning sign or label describing the hazard and the benefits of compliance.

Integrate the warning into the task- and hazard-related context

The task which is being performed and the nature of the hazard together form a context within which the warning sign or label is present. Ideally, the sign or label will appear as an integral part of the task. Numerous examples can be found where warning signs or labels are well integrated into a task, as, for example, when a lockout tag is placed next to an activated lockout, when switches are labeled on a control panel, or when a sign placed on a door warns against entering. Warnings that are well-integrated into a task-specific context are the most likely to usefully exploit the human's knowledge, since such warnings can act as cues that trigger the retrieval of additional, and hopefully relevant, information from long-term memory. Well-integrated warnings also reduce the likelihood of overloading short term memory and the effects of distractions.

Inappropriate presentation schedules, by failing to present the information at the time it is needed, may force people to retain the needed information in consciousness. During this interval, the information may be lost because of distractions or the limited ability of people to retain information in short term memory (Lehto and Miller, 1986). In other more extreme situations, an inappropriately presented warning may require its recall from long term memory. This may be problematic, if the information within the warning is not well-learned or previously known, because people seem to be only marginally able to recall the information from warning signs or labels. In particular, Wright (1979) discovered that less than 10% of the purchasers of antacids remembered even a portion of an in-store warning label (placed next to the antacid display) when they were questioned while leaving the store. Harper and Kalton (1966) placed two posters in a coal mine and found that 52.3% of the subjects failed to recall and 26.7% failed to recognize either poster. On the other hand, Strawbridge (1986) found that shortly after performing a task, subjects were able to recall most of the information given in a product's warning label.

Part of the reason for poor recall of the information provided in signs or labels might be that people are at a skill-based level of performance when they view them. Several studies of drivers support this conclusion. In particular, Johansson and Backlund (1970) reported that the percentage of drivers who recalled a road sign after passing it varied from 21% to 79%, depending upon the particular sign; while Shinar and Drory (1983) found percentages of 4.5% and 16.5% during the day and night, respectively. Dramatically different results were obtained by Summala and Naatanen (1974), who found that motorists perceived 97% of the signs they passed when they were asked to look for the signs and report their presence to an investigator in the car. This discrepancy could well be due to the fact that drivers ordinarily are at a skill-based level of performance, but were forced to shift to a rule-based level in the latter study.

Be selective

Numerous studies have shown a need for selectively providing warning signs and labels. Failure to meet this criterion increases the chance that warnings will be ignored. One of the most obvious principles is that people are more likely to behave consistently with a warning sign or label if they believe the danger is large. This principle was confirmed in the experiment by Otsubo (1988), where subjects were more likely to behave in accordance with a warning label when the product was perceived as being more dangerous. Consistently, Perry (1983) summarizes a number of findings regarding the response of people to volcano, flood, and nuclear power plant-related warnings. The belief that real situational danger was present, as when officials or police warned them, was a very major determinant of behavior. If people didn't believe the warning (as when newspapers reported problems), they were much less likely to behave in accordance with the warning.

Familiarity and perceived importance also are likely to influence attention to product labels and signs. When Wright et al. (1982) investigated the reading of instructions, they found that the tendency to read instructions increased when people were unfamiliar with a product or when a product was perceived to be complex, unsafe, or expensive. Complexity and frequency of use were correlated significantly with the propensity to read instructions ($r = 0.47$ for the former and $r = -0.24$ for the latter). Consistently, Johansson and Backlund (1970) found that drivers were most likely to recall seeing traffic signs perceived as being important.

The number of items included is another factor which can influence the likelihood of ignoring warning signs or labels. In particular, Lirtzman (1984) presents a study in which the exposure time needed for subjects to identify specific hazards increased with the number of items listed on the label. Scammon (1977) further showed that subjects remembered important product-related information better when fewer items were listed on labels. It is consequently logical to selectively provide *critical* information in warning signs or labels.

Keep the cost of compliance within a reasonable level

As noted by Belbin (1956a) even if people understand and are able to recall safety precautions, they may behave in conflicting ways. From an economic view, this may happen because people heavily weight the cost of compliance. Illustrating such effects, Fhaner and Hane (1974) found that the perception of discomfort tended to outweigh people's knowledge of the effectiveness of seatbelts as a predictor of use. Other studies have shown that minimizing the cost of compliance increases the effectiveness of warnings. For example, Godfrey et al. (1985) found that only 3 out of 51 people obeyed a sign telling them to use an alternate exit fifty feet away, while 61 of 64 people obeyed a sign telling them to use the right instead of left door. Wogalter et al. (1988) found similar effects in a study where the convenience of obtaining gloves and masks for use in a chemistry lab task was manipulated. A more recent study (Dingus et al., 1991) of a common household cleaning product showed a dramatic increase in wearing gloves when they were provided along with the product (a change from 25% to 88%). By providing the gloves, it was assumed the cost of compliance was significantly decreased.

On the basis of such results, it seems clear that warning signs and labels should be designed to minimize the 'cost' of compliance. Generally the cost of compliance is most easily minimized by presenting the warning at a stage in task performance where taking the safety measure is most convenient. Along these lines, Wogalter et al. (1985) showed that the location of warnings in instructions can be a major determinant of effectiveness. In other situations it may be necessary to explicitly consider ways of decreasing the cost of compliance (i.e. by recommending an alternative action which is easier to perform). Alternatively, it may be feasible to increase the cost of noncompliance (i.e. by enforcement or by providing explicit rewards for compliance).

Make symbols and text as concrete as possible

Numerous studies of comprehension have revealed that both text and symbols are better comprehended and remembered when they are concrete rather than abstract (Lehto and Miller, 1986). Consequently, there is reason to use concrete, rather than abstract, words and symbols within warning signs and labels. Subject skill and experience in interpreting both symbols and text, however, plays a major role in determining the value of concreteness.

In particular, the need for concrete text increases when people's reading ability is low or when they poorly understand the subject they're reading about. For example, a naive subject may understand a concrete statement such as 'open all windows before using ...' better than a more abstract statement such as 'use in a well-ventilated area ...'. A single abstract statement, on the other hand, will often completely cover a large set of possible contingencies (i.e. for the above example, suppose there were no windows in a particular application scenario). In such situations, messages composed from concrete statements will often be lengthy and difficult to use because of space limitations.

Lehto and Miller (1986) summarize several results which indicate similar results for symbols and pictographs. In particular, naive subjects rarely are able to correctly identify abstract symbols (i.e. the biohazard or safety alert symbols used in the U.S.). More concrete pictographic symbols (i.e. a tongue of flame indicating flammability) are occasionally comprehended correctly by 85% or more of naive subjects. The conclusion is that pictographs should be used for naive subjects, but testing is of course necessary prior to implementation.

A final point is that when people are performing at a skill-based or rule-based level (i.e. familiar routine tasks), written text is unlikely to be read, meaning that symbols or very brief text describing a condition may be most effective. The opposite conclusion appears to be true when people are performing at a knowledge-based level, since in this situation they may actually be seeking information.

Simplify the syntax (grammar) of text and combinations of symbols

To be effective, the meaning of a warning sign or label must be correctly understood. Designers must realize that people often have difficulty comprehending syntax or grammar underlying both written text and groups of symbols. Regarding the first issue, writing text that poor readers can comprehend is not an easy task. Kammann (1975) cites what he calls the 'two-thirds rule'. The basic idea is that only two-thirds of written material will ever be comprehended if the material is at all complex. To alleviate such problems short simple sentences constructed in the standard subject-verb-object form are often recommended (Lehto and Miller, 1986). Negations and complex conditional sentences are especially likely to create comprehension problems (Wright, 1981). For signs in particular, Easterby and Hakiel (1981) found that descriptive signs (those that describe conditions) tend to be better understood than proscriptive signs (those that recommend against certain actions). Sell (1977) also recommends against proscriptive statements, based upon a literature review. Regarding the design of conditional sentences (Dixon, 1982) found that messages which recommended actions prior to describing the condition for which the action should be taken were preferable when few actions were feasible. Messages which first described the condition were preferable when many actions were feasible.

A second issue is that much of the information within many signs is encoded by the arrangement of the symbols or equivalently by a pattern, instead of by individual symbols. Flow charts, logical trees or decision tables are one method often used to organize symbols (Wright and Reid, 1973; Kammann, 1975; Green, 1982). While such approaches have been useful in instructions, they are rarely feasible for warning signs and symbols because of space limitations. Instead, the most common approach is to develop complex pictographs. There is a tendency for comprehension problems with such pictographs when they describe actions or combine multiple meanings (Cahill, 1976). In particular, Johnson (1980) discusses research where his group was unable to develop an easily comprehended pictorial method of describing to passengers (making emergency exits from planes) that they had to open either a door or a hatch. Cairney and Sless (1982b) found similar comprehension-related problems for pictographs. In particular, a symbol that indicated the availability of both gas and service for cars was commonly misunderstood. Jones (1978) provides an example of how even the prohibitive slash interacted with the type of symbol it was expected to negate. Interestingly, the message was correctly identified only 50% of the time, when the symbol to be negated was abstract (shape and/or color coding only). For more concrete symbols (i.e. an arrow, pedestrian, vehicle), the proportion of correct negations rose to 89.9%.

Make warning signs and labels conspicuous and each component legible

Several studies have shown that conspicuity related factors can influence the ability of people to correctly perceive safety signs. In particular, roadside advertisements (or other non-traffic-related signs) have been shown to influence the perception of traffic signs (Holahan, 1977; Boersema and Zwaga, 1985). Holahan was able to show that traffic accidents at a 'stop sign' increased with the presence of commercial signs. Boersema and Zwaga, in an experimental study, showed that the presence of advertisements interfered with the perception of routing signs. Godfrey et al. (1985) found that a large, very noticeable, and explicit sign warning of water contamination caused people to avoid using a water fountain to a much greater extent than did a small sign. Young and Wogalter (1988) found that using conspicuous text and icons for warnings within instruction manuals resulted in better recall of the warning information. Other studies, however, have shown weak or nonsignificant effects. Strawbridge (1986) found that highlighting warning messages in instructions increased reading from 71% to 83%, while Zlotnik (1982) found no significant differences in a similar experiment. Consistently, Ursic (1984) found that the use of a pictograph, the strength of a signal word, and the presence of capital letters all had no significant impact on memorability of the content of the warnings.

Regarding the issue of legibility, one consistent finding is that symbols are legible at greater distances than text (Jacobs et al., 1975), and are also less sensitive to degradation. However, pictographs (having a more detailed design) may be less legible than abstract symbols (Lerner and Collins, 1980). Numerous reference books (McCormick, 1976; Van Cott and Kinkade, 1972; Woodson, 1981; Westinghouse, 1981; FMC, 1980) and safety standards specify detailed legibility requirements (as discussed earlier). Although modern labels and signs will generally meet or exceed these specifications, exceptions do occur, as found in regard to the poor nighttime legibility of guide signs (Hahn et al., 1977), or the low conspicuity of exit signs in smoky buildings (Lerner and Collins, 1983). Difficult problems are also found in certain dirty environments where signs or labels become covered with oil, mud, or other contaminants and where space constraints limit the size of signs or labels.

Conform with standards and population stereotypes when possible

Despite the failure of safety signage and labeling systems to address many of the issues raised in this paper, it makes sense to conform with their recommendations for sign and label format. From a retrospective view, the guidelines proposed in this section provide a way for selecting applications and determining the necessary contents of labels and signs. Conforming with safety signage and labeling systems then provides an approach which helps minimize the need to evaluate the perception and comprehension of the warning signs and labels which are ultimately developed. Conforming with standards also allows the information to be conveyed in a stereotypical and explicit way, which of course ultimately poses advantages by increasing standardization.

On the other hand, it should be emphasized that there is little if any evidence that particular formats recommended in standards have documentable advantages in effectiveness. For example, as noted earlier, Ursic (1984) could find no influence of standard features recommended in safety signage standards (i.e. the use of pictographs, strength of signal words, or the use of capital letters) on the memorability of warning placards. Similarly, Dingus et al. (1991) found no effect due to changing the configurations of warning labels. It appears that the most significant influences on effectiveness are instead rooted in cognitive factors inherent in the guidelines presented here in this paper.

Evaluate the developed warning sign or label in an actual use context

Carefully considering each of the preceding guidelines during the design of a warning sign or label will in itself require evaluation well beyond that traditionally performed. In doing so, the developed warning will satisfy both perceptual and cognitive criteria essential for effectiveness. Nevertheless, the influence of the developed warning sign or label on both overt and covert forms of behavior should still be evaluated in the context of actual use. Methods of evaluation have already been covered earlier in this paper and are relevant for examining two issues: (1) the effectiveness of the warning, and (2) methods of improving the warning. Effectiveness is of course best determined by measuring the degree to which a warning sign or label modifies behavior as intended. Improvement of a warning sign or label, on the other hand, requires more detailed forms of evaluation in terms of the preceding guidelines. For example, by interrupting performance of the task at critical stages and asking the user relevant questions, insight can be obtained as to whether people consciously process and comprehend the information on the warning sign or label at the appropriate time in the context of actual use. If this is the case, the majority of the cognitively oriented guidelines will clearly be satisfied.

Several of the methods mentioned earlier can be used to evaluate applications of warning signs and labels in terms of the cognitively oriented guidelines. However, it must be emphasized that it can be difficult to determine whether the design objectives are adequately satisfied on the basis of these guidelines without carefully considering users and the application itself. For example, how large of a cost of compliance is reasonable, or at what point might we be too selective in providing warnings? Quite obviously, tradeoffs can occur (for example, when we are too selective in providing warnings, necessary information may be left out, while when we are not selective enough necessary information may be filtered because of information overload). Furthermore, the answers to these questions are context specific (for example, the appropriateness of the cost of compliance depends upon the situation). Consequently, there is a need for substantial research and development directed toward specifying the cognitively oriented guidelines in more detail for specific application areas and categories of users. Doing so would be a fruitful future direction for safety signage and labeling systems.

4. Research issues

It must be emphasized that few studies have evaluated warning signs and labels in real-world settings. Lehto and Miller (1986) found only two studies demonstrating any influence on behavior due to safety signs in industrial settings and of no such studies for warning labels. In the first study, Laner and Sell (1960) evaluated the influence of safety posters which told coal miners to hook slings. An average increase of 7.8% (a change from 37.6% to 45.4%) in hook-slinging behavior was observed when the safety posters were present. A second study, summarized in *National Safety News* (1966), evaluated the effect of placing posters on the steps entering an aircraft. The three posters considered were a picture of a man holding the rail, a picture of a man stumbling, and a picture of a man sprawled at the bottom of the stairs. During the course of the experiment, 2000 passengers were observed while entering an aircraft. A 6%, 13%, and 21% increase in railing-holding behavior was associated with the above posters, respectively.

Because so few studies have demonstrated the effectiveness of warning signs and labels in real-world settings, experts in the area have strongly questioned their value (i.e. McCarthy et al., 1984). Simply put, warning signs and labels will always be ineffective if they are filtered out, not understood, ignored, or when the actions prescribed are inadequate or beyond the capabilities of the receiver, as was shown earlier. The need to satisfy all of these conditions in a positive way makes the development of totally effective warnings a hopeless task. The point is that no matter how well a warning is designed, it still might be ineffective. Nevertheless, as indicated by the above-mentioned studies and other more recent research (Miller and Lehto, 1990; Dingus et al., 1991), warning signs and labels can be of value in certain situations.

Given the current state of research regarding warning effectiveness, it is difficult to go beyond determining when the numerous ways a warning can fail are not overwhelming in their negative impact. In other words we can develop fairly strong conclusions as to when warning signs and labels will not be effective, but determining when they will be effective is more difficult. Useful conclusions regarding both questions can be developed by synthesizing available research using the modeling framework presented earlier in this paper, and form the basis of the guidelines for the practitioner proposed in Part I. Although much can be inferred from such an approach, there is still a strong need for evaluating warning signs and labels, for simply knowing that a warning might be effective in a given situation is not sufficient.

Means of reducing the heavy burden associated with the evaluation of warning signs and labels are needed. This difficulty will only get worse when analyzing the issues associated with providing instructions, which is the next logical step. It appears that improved methods of task analysis that focus on describing user knowledge and user knowledge requirements may prove to be necessary for addressing this latter goal. Regarding the first question, a number of knowledge acquisition techniques have emerged which can be used to evaluate the knowledge people possess (Lehto et al., 1992). Knowledge acquisition approaches are yet to see significant application in the area of warning sign and label design, but seem particularly promising. For example, Tonn et al. (1990) were quite successful in using a frame-based knowledge acquisition tool to document risk beliefs. Other approaches include protocol analysis (Newell and Simon, 1972; Ericsson and Simon, 1980; Ericsson and Simon, 1984) and interviewing techniques. Both of the latter approaches have the disadvantage of generating massive amounts of data, which is often difficult to evaluate. They also poorly document aspects of skill-based performance, since much of such behavior is ordinarily at an unconscious level (Nisbett and Wilson, 1977).

Regarding user knowledge requirements, the NGOMS methodology proposed by Kieras (1989) for computer interface design and the SMESD proposed by Lehto (see Lehto et al., 1992) for consumer product safety applications seem relevant. NGOMS takes the form of describing the knowledge required to operate a particular system hierarchically in terms of goals and methods required to attain them. The methods themselves are hierarchically divided until they are described in terms of primitive operators. SMESD takes a similar approach to describing tasks, products, and user knowledge and was proposed for computer implementation. By describing the knowledge required to perform a task in this fashion, Kieras and Bovair (1986) have been able to predict the time required for people to learn to operate electronic equipment. They also have been able to predict the transfer of knowledge when people learn to use equipment similar to that they have used before. As such, the technique seems to provide a foundation for scientifically designing instructions, and perhaps warning labels for users at a knowledge-based level.

5. Conclusions

Safety signage and labeling systems provide an operational set of guidelines for how to develop a warning sign or label. Provisions are included for signal words, color coding, typography, symbols, arrangement, and hazard identification. A commonly cited advantage of conforming with these systems is that they convey hazards in a rather stereotypical and explicit way. They also provide a means for practitioners who are not necessarily experts in warning sign or label design to develop warnings which meet basic perceptual requirements. A number of techniques have also been developed for evaluating the perception, comprehension, and behavioral effects of warning signs and labels once they have been developed.

While traditional guidelines and evaluation techniques provide a useful set of tools to the practitioner, they inadequately address several issues of importance. In particular, practitioners need more sophisticated guidelines for selecting necessary applications of warning signs and labels, not to mention methods for evaluating their influence on decision-making. Current approaches have been based to a large extent upon the intuition of practitioners and/or a legalistic focus on warning against all hazards. A more scientifically justifiable approach is to focus upon user knowledge requirements and the flow of information within a task. Along these lines, emerging techniques based to large part upon the field of cognitive engineering provide a means for relating the form of the provided information to the level of user performance. The guidelines presented in this paper are developed on the basis of such an approach and provide a starting point for combining the positive elements of traditional approaches for designing warnings with a more cognitively oriented perspective.

References

American National Standards Institute, 1978. American National Standard Manual on Uniform Traffic Control Devices for Streets and Highways. ANSI D6.1.
American National Standards Institute, 1972. American National Standard Specifications for Accident Prevention Signs. ANSI Z35.1.
American National Standards Institute, 1987. Safety Color Code. ANSI Z535.1, draft.
American National Standards Institute, 1987. Environmental and Facility Safety Signs. ANSI Z535.2, draft.
American National Standards Institute, 1987. Criteria for Safety Symbols. ANSI Z535.3, draft.
American National Standards Institute, 1987. Product Safety Signs and Labels. ANSI Z535.4, draft.
American National Standards Institute, 1987. Accident Prevention Tags. ANSI Z535.5, draft.
American National Standards Institute, 1988. Hazardous Industrial Chemicals – Precautionary Labeling ANSI Z129.1.
Belbin, E., 1956a. The effects of propaganda on recall, recognition and behaviour I. The relationship between the different measures of propaganda effectiveness. British Journal of Psychology, 47 (3): 163–174.
Belbin, E., 1956b. The effects of propaganda on recall, recognition and behaviour II. The conditions which determine the response to propaganda. British Journal of Psychology, 47 (4): 259–270.
Boersema, T. and Zwaga, H., 1985. The influence of advertisements on the conspicuity of routing information. Applied Ergonomics, 16 (4): 267–273.
Bresnahan, T.F. and Bryk, J., 1975. The hazard association values of accident-prevention signs. Professional Safety, 17–25.
Cahill, M.C., 1975. Interpretability of graphic symbols as a function of context and experience factors. Journal of Applied Psychology, 60 (3): 376–380.
Cahill, M.C., 1976. Design features of graphic symbols varying in interpretability. Perceptual and Motor Skills, 42 (2): 647–653.
Cairney, P.T. and Sless, D., 1982a. Evaluating the understanding of symbolic roadside information signs. Australian Road Research, 12 (2): 97–102.
Cairney, P. and Sless, D., 1982b. Communication effectiveness of symbolic safety signs with different user groups. Applied Ergonomics, 13 (2): 91–97.

Caron, J.P., Jamieson, D.G. and Dewar, R.E., 1980. Evaluating pictographs using semantic differential and classification techniques. Ergonomics, 23 (2): 137–146.
Christensen, J.M., 1983. Human factors considerations in lawsuits. In: G.A. Peters (Ed.), Safety Law: A Legal Reference for the Safety Professional. American Society of Safety Engineers, Park Ridge, pp. 5–7.
Cochran, D.J., Riley, M.W. and Douglass, E.I., 1981. An investigation of shapes for warning labels. Rochester, New York, Proceedings of the Human Factors Society, 25th Annual Meeting, pp. 395–399.
Cole, B.L. and Jacobs, R.J., 1981. A comparison of alternative symbolic warning signs for railway level crossings. Australian Road Research, 11 (4): 37–45.
Collins, B.L., 1983. Evaluation of mine-safety symbols. Proceedings of the Human Factors Society – 27th Annual Meeting, pp. 947–949.
Collins, B.L. and Lerner, N.D., 1982. Assessment of fire-safety symbols. Human Factors, 24 (1): 75–84.
Collins, B.L., Lerner, N.D. and Pierman, B.C., 1982. Symbols for industrial safety. NBSIR 82-2485, Washington, D.C., National Bureau of Standards.
Cunitz, R.J., 1981. Psychologically effective warnings. Hazard Prevention, 17 (3): 5–7.
Dahlstedt, S. and Svenson, O., 1977. Detection and reading distances of retroreflective road signs during night driving. Applied Ergonomics, 8 (1): 7–14.
Dewar, R.E., 1976. The slash obscures the symbol on prohibitive traffic signs. Human Factors, 18 (3): 253–258.
Dewar, R.E. and Ells, J.G., 1977. The semantic differential as an index of traffic sign perception and comprehension. Human Factors, 19 (2): 183–189.
Dingus, T.A., Hathaway, T.A. and Huu, B.P., 1991. A most important warning variable: Two demonstrations of the powerful effects of cost on warning compliance. Proceedings of the Human Factors Society, 35th Annual Meeting, San Francisco, CA, pp. 1034–1039.
Dingus, T.A., Hunn, B.P. and Wreggit, S.S., 1991. Two reasons for providing protective equipment as part of hazardous consumer product packaging. Proceedings of the Human Factors Society, 35th Annual Meeting, San Francisco, CA, pp. 1039–1042.
Dixon, P., 1982. Plans and written directions for complex tasks. Journal of Verbal Learning and Verbal Behavior, 21: 70–84.
Easterby, R.S. and Hakiel, S.R., 1977. Safety labelling of consumer products: Shape and colour code stereotypes in the design of signs. A.P. Report No. 75, University of Aston in Birmingham.
Easterby, R.S. and Hakiel, S.R., 1981. Field testing of consumer safety signs: The comprehension of pictorially presented messages. Applied Ergonomics, 12 (3): 143–152.
Easterby, R.S. and Zwaga, H.J.G., 1976. Evaluation of public information symbols ISO tests: 1975 Series. A.P. Report No. 60, University of Aston in Birmingham, March, 1976.
Ericsson, K.A., and Simon, H.A., 1980. Verbal reports as data. Psychological Review, 87: 215–251.
Ericsson, K.A. and Simon, H.A., 1984. Protocol Analysis: Verbal Reports as Data. Cambridge, MA, MIT Press.
Fhaner, G. and Hane, M., 1974. Seat belts: Relations between

beliefs, attitude, and use. Journal of Applied Psychology, 59 (4): 472-482.
Fischer, P.M., Richards, J.W., Berman, E.J. and Krugman, D.M., 1989. Recall and eye tracking study of adolescents viewing tobacco advertisements. The Journal of the American Medical Association, 251 (1): 84-89.
FMC, Product Safety Sign and Label System, 3rd ed. Santa Clara, California, FMC Corporation, 1985.
Frantz, J.P. and Rhoades, T.P., 1991. The effect of warning location and task interruption on attention to and compliance with product warnings. Unpublished Technical Report, School of Industrial and Operations Engineering, University of Michigan.
Galer, M., 1980. An ergonomics approach to the problem of high vehicles striking low bridges. Applied Ergonomics, 11 (1): 43-46.
Gallup Canada, 1989. Consumer survey on the labelling and packaging of hazardous chemical products, Volume 1: Detailed findings. Report submitted to Product Safety Branch, Consumer and Corporate Affairs Canada.
Godfrey, S.S. and Laughery, K.R., 1984. The biasing effects of product familiarity on consumers' awareness of hazard. Proceedings of the Human Factors Society – 28th Annual Meeting. Human Factors Society.
Godfrey, S.S., Rothstein, P.R. and Laughery, K.R., 1985. Warnings: Do they make a difference. Proceedings of the Human Factors Society – 29th Annual Meeting. Human Factors Society, CA.
Gomer, F., 1986. Evaluating the effectiveness of warnings under prevailing working conditions. Proceedings of the Human Factors Society – 30th Annual Meeting. Human Factors Society, Santa Monica, CA pp. 712-715.
Green, P. and Pew, R.W., 1978. Evaluating pictographic symbols: An automotive application. Human Factors, 20 (1): 103-114.
Green, T.R.G., 1982. Pictures of programs and other processes. Behaviour and Information Technology, 1 (1): 3-36.
Hadden, S.G., 1986. Read the Label. Westview Press, Boulder, CO.
Hahn, K.C., McNaught, E.D. and Bryden, J.E., 1977 Nighttime legibility of guide signs. Research Report 50, Albany, N.Y., NY State Department of Transportation, pp. 1-64.
Harper, D.G. and Kalton, G., 1966. A study of the effectiveness of the National Coal Board's safety propaganda in Section 3 Safety posters. Unpublished Report.
Hicks III, J.A., 1976. An evaluation of the effect of sign brightness on the sign-reading behavior of alcohol-impaired drivers. Human Factors, 18 (1): 45-52.
Hodge, D.C., 1962. Legibility of a uniform-strokewidth alphabet: I. Relative legibility of upper and lower case letters. Journal of Engineering Psychology, 1 (1): 34-46.
Holahan, C.J., 1977. Relationship between roadside signs and traffic accidents: A field investigation. Research Report 54, Austin, Texas, Council for Advance Transportation Studies, pp. 1-14.
Hull, A.J., 1976. Reducing sources of human error in transmission of alphanumeric codes Applied Ergonomics, 7 (2): 75-78.
International Standards Organization, 1967. Symbols, dimensions, and layout for safety signs. ISO R557.

International Standards Organization, 1984. Safety signs and colors. ISO 3864.
Jacobs, R.J., Johnston, A.W. and Cole, B.L., 1975. The visibility of alphabetic and symbolic traffic signs. Australian Road Research, 5 (7): 68-86.
Johansson, G. and Backlund, F., 1970. Drivers and road signs. Ergonomics, 13 (6): 749-759.
Johnson, D.A., 1980. The design of effective safety information displays. In: H.R. Poydar (Ed.), Tufts University, Medford, Mass., Proceedings of the Symposium: Human Factors and Industrial Design in Consumer Products, pp. 314-328.
Jones, S., 1978. Symbolic representation of abstract concepts. Ergonomics, 21 (7): 573-577.
Kammann, R., 1975. The comprehensibility of printed instructions and the flowchart alternative. Human Factors, 17 (2): 183-191.
Kieras, D., 1989. A methodology for computer interface design. In: Handbook of Human-Computer Interaction.
Kieras, D. and Bovair, S., 1986. The acquisition of procedures from text: A production-system analysis of transfer of training. Journal of Memory and Language, 25: 507-524.
King, L.E., 1975. Recognition of symbol and word traffic signs. Journal of Safety Research, 7 (2): 80-84.
Klare, G.R., 1974-75. Assessing readability. Reading Research Quarterly, 10 (1): 62-102.
Laner, S. and Sell, R.G., 1960. An experiment on the effect of specially designed safety posters. Journal of Occupational Psychology, 34 (3): 153-169.
Lawrence, A.C., 1974. Human error as a cause of accidents in gold mining. Journal of Safety Research, 6 (2): 78-88.
Lehto, M.R., 1991. A proposed conceptual model of human behavior and its implications for the design of warnings. Perceptual and Motor Skills, 73: 595-611.
Lehto, M.R., Boose, J., Sharit, J. and Salvendy, G., 1992. Knowledge acquisition. Chapter 58 in: G. Salvendy (Ed.), Handbook of Industrial Engineering. John Wiley, New York, pp. 1495-1545.
Lehto, M.R. and Clark, D. 1990. Warning signs and labels in the workplace. In: A. Mital and W. Karwowski (Eds.), Workspace, Equipment and Tool Design. Elsevier Science Publishers, Amsterdam.
Lehto, M.R. and Foley, J., 1989. The influence of regulation, training, and product information on use of helmets by ATV operators: A field study. Proceedings of INTERFACE '89 – The Sixth Symposium on Human Factors and Industrial Design in Consumer Products. Human Factors Society, Santa Monica, CA, pp. 107-113.
Lehto, M.R. and Miller, J.M., 1986. Warnings: Volume I: Fundamentals, Design, and Evaluation Methodologies. Ann Arbor, MI, Fuller Technical Publications.
Lehto, M.R. and Papastavrou, J.D., 1991. A distributed signal detection theory model: Implications to the design of warnings. Proceedings of the 1991 Automatic Control Conference, Boston, MA, June 26-28, 1991, pp. 2586-2590.
Lerner, N.D., and Collins, B.L., 1980. The assessment of safety symbol understandability by different testing methods. PB81-185647, Washington, D.C., National Bureau of Standards.

Lirtzman, S.I., 1984. Labels, perception, and psychometrics. In: C.J. O'Connor and S.I. Lirtzman (Eds.), Handbook of Chemical Industry Labeling. Noyes Publications, Park Ridge, NJ.

McCarthy, R.L., Finnegan, J.P., Krumm-Scott, S. and McCarthy, G.E., 1984. Product information presentation, user behavior, and safety. Proceedings of the Human Factors Society – 28th Annual Meeting. Human Factors Society, Santa Monica, CA, pp. 81–85.

McCormick, E.J., 1976. Human Factors in Engineering and Design (4th edition). N.Y., McGraw-Hill, New York.

McGuire, W.J., 1980. The communication-persuasion model and health-risk labeling. In: L.A. Morris, M.B. Mazis and I. Barofsky (Eds.), Product Labeling and Health Risks. Banbury Report 6. Cold Spring Harbor Laboratory, pp. 99–122.

Miller, J.M., Lehto, M.R. and Frantz, J.P., 1990. Instructions and Warnings: The Annotated Bibliography, Ann Arbor, MI, Fuller Technical Publications.

National Electronic Manufacturers Association, 1982. Safety labels for padmounted switchgear and transformers sited in public areas. NEMA 260.

National Safety News, June, 1966.

Newell, H. and Simon, H.A., 1972. Human Problem Solving. Prentice-Hall, Englewood Cliffs, NJ.

Nisbett, R.E. and Wilson, T.D., 1977. Telling more than we can know: Verbal reports on mental processes. Psychological Review, 84: 271–276.

Norman, D.A., 1980. Errors in human performance. Report No. 8004, University of California, San Diego: Center for Human Information Processing.

Occupational Health and Safety Administration, 1985. Specification for accident prevention signs and tags. CFR 1910.145.

Occupational health and Safety Administration, 1985. [Chemical] Hazard communication. CFR 1910.1200.

Olson, P.L. and Bernstein, A., 1979. The nighttime legibility of highway signs as a function of their luminance characteristics. Human Factors, 21 (2): 145–160.

Otsubo, S.M., 1988. A behavioral study of warning labels for consumer products: Perceived danger and use of pictographs. Proceedings of the Human Factors Society – 32nd Annual Meeting, Human Factors Society.

Perry, R.W., 1983. Population evacuation in volcanic eruptions, floods, and nuclear power plant accidents: Some elementary comparisons. Journal of Community Psychology, 11: 36–47.

Pyrczak, F. and Roth, D.H., 1976. The readability of directions on non-prescription drugs. Journal of the American Pharmaceutical Association, NS 16 (5): 242–243, 267.

Rasmussen, J., 1986. Information Processing and Human-Machine Interaction. North-Holland, Amsterdam.

Rhoades, T.P., Frantz, J.P. and Miller, J.M., 1991. Emerging strategies for the assessment of safety related product communications. Proceedings of the Human Factors Society, 35th Annual Meeting, San Francisco, CA, pp. 998–1002.

Riley, M.W., Cochran, D.J. and Ballard, J.L., 1982. An investigation of preferred shapes for warning labels. Human Factors, 24 (6): 737–742.

Sanderson, P.M., 1988. SHAPA: An interactive environment for verbal protocol analysis. EPRL Technical Report EPRL-88-00, Engineering Psychology Research Laboratory, Department of Mechanical and Industrial Engineering, University of Illinois at Urbana-Champaign.

Scammon, D.L., 1977. 'Information load' and consumers. The Journal of Consumer Research, 4: 148–155.

Schwartz, V.E., and Driver, R.W., 1983. Warnings in the workplace: The need for a synthesis of law and communication theory. Cincinnati Law Review, 52: 38–83.

Sell, R.G., What does safety propaganda do for safety? A review. Applied Ergonomics, H77, 203–214.

Shinar, D. and Drory, A., 1983. Sign registration in daytime and nighttime driving. Human Factors, 25 (1): 117–122.

Sivak, M., Olson, P.L. and Pastalan, L.A., 1981. Effect of driver's age on nighttime legibility of highway signs. Human Factors, 23 (1): 59–64.

Smith, S.L., 1984. Letter size and legibility In: R. Easterby and H. Zwaga (Eds.), Information Design: The Design and Evaluation of Signs and Printed Material. John Wiley, Chicester, pp. 171–186.

Society of Automotive Engineers, 1979. Safety signs. SAE J115.

Society of Automotive Engineers, 1975. Safety alert symbol for agricultural, construction, and industrial equipment. SAE J284.

Society of Automotive Engineers, 1983. Slow moving vehicle identification emblem. SAE J943.

Strawbridge, J.A., 1986. The influence of position, highlighting, and imbedding on warning effectiveness. Proceedings of the Human Factors Society – 30th Annual Meeting. Human Factors Society.

Summala, H. and Naatanen, R., 1974. Perception of highway traffic signs and motivation. Journal of Safety Research, 6 (4): 150–154.

Tierney, W.J. and King, L.E., 1970. Traffic signing – Symbols versus words. Montreal, The Sixth World Highway Conference of the International Road Federation, October 4–10, 1970, pp. 1–40.

Tinker, M.A., 1963. Legibility of Print. Iowa State University Press, Ames, IA.

Tonn, B.E., Travis, C.B., Goeltz, R.T. and Phillippi, R.H., 1990. Knowledge-based representations of risk beliefs. Risk Analysis, 10 (1): 169–184.

Ursic, M., 1984. The impact of safety warnings on perception and memory. Human Factors, 26 (6): 677–682.

Van Cott, H.P. and Kinkade, R.G., 1972. Human Engineering Guide to Equipment Design. Washington, DC, U.S. Government Printing Office.

Van Nes, F.L. and Bouma, H., 1980. On the legibility of segmented numerals. Human Factors, 22 (4): 463–474.

Westinghouse Electric Corporation, 1981. Product Safety Label Handbook. Trafford, PA, Westinghouse Printing Division.

Wogalter, M.S., McKenna, N.A. and Allison, S.T., 1985. Warning compliance: Behavioral effects of cost and consensus. Proceedings of the Human Factors Society – 32th Annual Meeting.

Woodson, W.E., 1981. Human Factors Design Handbook, Information and Guidelines for the Design of Systems,

Facilities, Equipment, and Products for Human Use. New York, McGraw-Hill.

Wright, P., 1979. Concrete action plans in TV messages to increase reading of drug warnings Journal of Consumer Research, 6: 256–269.

Wright, P., 1981. 'The instructions clearly state ...'. Can't people read? Applied Ergonomics, 12 (3): 131–141.

Wright, P., Creighton, P. and Threlsall, S.M., 1982. Some factors determining when instructions will be read. Ergonomics, 25 (3): 225–237.

Wright, P. and Reid, F., 1973. Written information: Some alternatives to prose for expressing the outcomes of complex contingencies. Journal of Applied Psychology, 57 (2): 160–166.

Young, S.L. and Wogalter, M.S., 1988. Memory of instruction manual warnings: Effects of pictorial icons and conspicuous print. Proceedings of the Human Factors Society – 32nd Annual Meeting. Human Factors Society, Santa Monica, CA, pp. 905–909.

Zlotnik, M.A., 1982. The effects of warning message highlighting on novel assembly task performance. Proceedings of the Human Factors Society – 26th Annual Meeting, pp. 93–97.

Vision at the workplace: Part I – Guidelines for the practitioner *

Stephan Konz
Department of Industrial Engineering, Kansas State University, Manhattan, KS 66506-5101, USA

1. Audience

This guide is intended for practitioners who design workstations. The typical practitioner would have a job title of industrial engineer or ergonomist.

2. Problem identification

The problems will be identified through the judgment of the practitioner. Health and safety records and productivity records will not have the sensitivity to identify lighting problems.

Correspondence to: Stephan Konz, Department of Industrial Engineering, Kansas State University, Manhattan, KS 66506-5101, USA.

* The recommendations provided in this guide are based on numerous published and unpublished scientific studies and are intended to enhance worker safety and productivity. These recommendations are neither intended to replace existing standards, if any, nor should be treated as standards. Furthermore, this document should not be construed to represent institutional policy.

The following individuals participated in the discussion of the earlier version of this guide. Their suggestions (written or verbal) were incorporated by the authors in this version: Roland Andersson, *Sweden*; Alvah Bittner, *USA*; Peter Buckle, *UK*; Jan Dul, *The Netherlands*; Bahador Ghahramani, *USA*; Juhani Ilmarinen, *Finland*; Sheik Imrhan, *USA*; Asa Kilbom, *Sweden*; Shrawan Kumar, *Canada*; Tom Leamon, *USA*; Mark Lehto, *USA*; William Marras, *USA*; Barbara McPhee, *Australia*; James Miller, *USA*; Anil Mital, *USA*; Don Morelli, *USA*; Maurice Oxenburgh, *Australia*; Jerry Purswell, *USA*; Jorma Saari, *Finland*; W. (Tom) Singleton, *UK*; Juhani Smolander, *Finland*; Terry Stobbe, *USA*; Rolf Westgaard, *Norway*; Jørgen Winkel, *Sweden*. The guide was also reviewed in depth by several anonymous reviewers.

Visual problems at the workplace are unlikely to affect the worker's health although they may cause unnecessary stress on the visual system – sometimes resulting in headache and fatigue. Visual performance may be affected, resulting in increased errors and time/unit. In general, these three criteria (worker stress, increased errors, increased time) are rarely measured. Even if they are available for the present design, that doesn't tell you what they would be for good design.

3. Data collection and analysis

3.1. Hardware

The practitioner should have a:
- Color and cosine corrected light meter.
- Vision testing device for employee vision.
- Video camera and tape recorder to record the present work method.

3.2. Data collection

The practitioner should record the:
- Visual capability of the employees (often done by a plant nurse).
- Task visual requirements. For example, detail, on a second by second basis, exactly what the operator must do.
- Lighting characteristics (luminaire location and type, lamp type and size). Record room characteristics (size, shape, reflectances). Record illumination levels (lux) at different parts of the workstation.

3.3. Data analysis

The data is then compared to recommended values. For example does the individual have corrected vision? Have a color deficiency? Can the task be modified to reduce visual requirements? Is the lighting of the recommended quantity? quality?

There are a variety of computer programs available which will give predicted levels of illumination, after the user specifies room characteristics (shape, size, reflectances) and lighting characteristics (luminaire location and type, lamp type and size).

4. Solutions

4.1. Criteria

The cost of illumination is low in relation to labor cost. In general labor cost/sq foot of space is 100 to 300 times the illumination cost. That is, economics is not much of a constraint on good lighting ergonomics.

Visual problems can be due to deficiencies in the individual, the lighting or the task.

4.2. Individual deficiencies

In most cases, visual deficiencies can be corrected with corrective lenses – either spectacles or contact lenses. Spectacles with safety glass have an advantage (vs. contacts) of impact resistance. Spectacles with chemicals in the glass to reduce light reduce radiation only in the visible spectrum and pass UV radiation so welding glasses are needed in welding areas.

Contact lenses not only do not protect vs. impact but they also tend to seal the cornea and hold irritants against the eye. Do not use contacts (even inside a respirator) in areas with chemical fumes, vapors, splashes or dusty atmospheres.

Goggles with side shields can protect the eyes. Face shields protect the entire face vs. liquids and impact.

Table 1

Recommended target maintained illuminance (lux) for interior industrial lighting (IES, 1983). See table 2 for office examples.

Reference work plane	Type of activity	Illuminance category	Illuminance (lux)		
			Total factor		
			−3 or −2	−1, 0, +1	+2 or +3
General lighting throughout spaces	Public spaces with dark surroundings	A	20	30	50
	Simple orientation for short temporary visits	B	50	75	100
	Working spaces where visual tasks are only occasionally performed	C	100	150	200
Illuminance on task	Performance of visual tasks of high contrast or large size	D	200	300	500
	Performance of visual tasks of medium contrast or small size	E	500	750	1000
	Performance of visual tasks of low contrast or very small size	F	1000	1500	2000
Illuminance on task obtained by a combination of general and local (supplementary) lighting	Performance of visual tasks of low contrast and very small size over a prolonged period	G	2000	3000	5000
	Performance of very prolonged and exacting visual tasks	H	5000	7500	10,000
	Performance of very special visual tasks of extremely low contrast and small size	I	10,000	15,000	20,000

Note: See table 3 for factor calculations.

Table 2

Office task examples of illuminance categories for use with table 1 (IES, 1982). One asterisk indicates potential problems with veiling reflections; two asterisks indicate special problems with veiling reflections. It may be better to modify the task than to compensate for poor task characteristics by adding lighting.

Illuminance category	General task/activity	Detailed example
B	Copied tasks	Microfiche reader (**)
	EDP tasks	VDT screens (**)
	General and public areas	Hallways, mechanical rooms during operation, stairs, utility rooms
C	Drafting	Light table
	EDP tasks	Machine rooms
	General and public	Display areas, elevators, escalators, janitorial space, lobbies and lounges, reception area, rest rooms
D	Copied tasks	Mimeograph, Xerograph
	EDP tasks	Impact printer: good ribbon, ink jet printer, keyboard reading, machine rooms – active operations and tape storage
	General and public	AV areas, conference rooms, duplicating and offset printing areas.
	Handwritten tasks	#2 pencil and softer leads (*), ball point (*) and felt tip pen
	Printed tasks	8 and 10 pt type (*), typed originals, newsprint, glossy magazines (**)
E	Copied tasks	Ditto copy (*), moderate-detail photographs (**), Xerography (third or greater generation)
	Drafting tasks	Mylar with high-contrast media (*), vellum and tracing paper with high contrast (*), blue line and blueprints
	EDP tasks	Impact printer; poor ribbon, second carbon and greater, machine equipment when being serviced, thermal print (*)
	General and public areas	First aid, mechanical room equipment when being serviced, mail sorting (*)
	Graphic design and material	Graphs, moderate-detail photographs (*)
	Handwritten tasks	#3 pencil (*), handwritten carbons
	Printed tasks	6 pt (*), maps, typed second carbon and later, telephone books
F	Copied tasks	Poor thermal copy (*)
	Drafting tasks	Low-contrast media and hard graphite leads, low-contrast vellum and tracing paper, sepia prints
	General and public	Model making
	Graphic design and material	Color selection, charting and mapping, layout and artwork
	Handwritten tasks	#4 pencil and harder leads, nonphotographically reproducible colors

Table 3

Factors for illuminance table (table 1) (IES, 1983). If the task was in illuminance category D, the worker age was 'Under 40', speed/accuracy were 'Important', and task background reflectance was '30 to 70', then the components would be −1, 0, and 0; the total factor would be their sum, −1, so the middle column in table 1 would be used and the recommended illuminance would be 300 lux.

Variable	Factor		
	−1	0	+1
For illuminance categories A, B, C (general lighting throughout spaces)			
Occupants ages, year	Under 40	40–45	Over 55
Average weighted room surface reflectance (%)	Over 70	30–70	Under 30
For illuminance categories D, E, F, G, H, I (illuminance on task)			
Workers ages, year	Under 40	40–55	Over 55
Speed and/or accuracy	Not important	Important	Critical
Task background reflectance (%)	Over 70	30–70	Under 30

Table 4

Lamp characteristics of typical industrial lamps. Table courtesy of General Electric Lighting Business Group.

Type of lamp	Watts	Lumens/Watt		Lumen Maintenance (%)	Rated life (h)	Restrike (min)	Relative cost
		Initial	Mean				
High-pressure sodium	35–1000	64–140	58–126	90–92	24,000	1–2	Lowest
Metal halide	175–1000	80–115	57–92	71–83	10,000–20,000	10–15	Medium
Fluorescent	28–215	74–100	49–92	66–92	12,000–20,000+	Immediate	Medium
Mercury	50–1000	32–63	24–43	57–84	16,000–24,000+	3–6	High
Incandescent	100–1500	17–24	15–23	90–95	750–2000	Immediate	High

4.3. Lighting deficiencies

Lighting deficiencies will be divided into quantity and quality.

Quantity. There are two basic approaches to lighting workstations: uniform lighting and non-uniform (task) lighting. Uniform lighting provides maximum flexibility in arranging machines and workstations, eliminates the need to move lighting fixtures if the workstations are moved, and uses large lamps (which have higher lumens/watt). It tends to be energy inefficient. Task lighting advantages are lower lighting cost, more precise control of the light, and more esthetic appeal.

If uniform lighting is used (see Part 2 for more detail):
– Consider the spacing.
– Remember a distant light is dim.
– Reuse the light.
– Use efficient fixtures.

Tables 1, 2 and 3 give specific recommended amounts of lighting. The amount of light to use depends upon the worker age, the room reflectance and the required speed/accuracy.

Table 4 gives lamp characteristics. Lamps should be selected for their color characteristics and convenience as well as their cost.

Quality. In addition to light color and light orientation and variability, consider glare.

Direct glare reduction techniques include:
– Reduce brightness of the glare source.
– Reduce size of the glare source.

Indirect (reflected) glare reduction techniques include:
– Reduce amount of light from the source.
– Reorient the reflecting surface.
– Use matte surfaces.
– Put a filter (sunglasses) at the eye.
– Change the eye position.

4.4. Task deficiencies

The goal is to enhance the object so it emerges from its environment like an island from the sea.

Increase size. Some possibilities are:
– Increase the object size.
– Bring the object closer to the eyes.
– Use optical aids (magnification). Optical aids can be considered 'effort saving devices', not just 'defective vision correcting devices'.

Increase contrast. Camouflage consists in obliterating contrasts; the goal is anti-camouflage:
– Modify target color.
– Improve target contrast.
– Use spacing and grouping.

Increase time. Some possibilities are:
– Shoot sitting ducks. View stationary objects, not moving objects.
– Avoid machine-paced inspection.

Part 2 gives some recommendations for four specific applications:
– VDT areas,
– inspection,
– emergency lighting,
– security lighting.

Reference

Konz, S., 1992. Vision at the workplace: Part II – The scientific basis (knowledge base) for the guide. International Journal of Industrial Ergonomics, 10: 145–160 (this issue).

Vision at the workplace: Part II – Knowledge base for the guide

Stephan Konz

Department of Industrial Engineering, Kansas State University, Manhattan, KS 66506-5101, USA

1. Problem description

This paper provides information concerning good vision in the work environment. Visual problems can be due to deficiencies in the person, the place or the procedure. Some personal deficiencies can be corrected with lenses (contact lenses, spectacles). Place deficiencies can be corrected through: (1) Increasing the quantity of illumination or (2) Improving the quality of the illumination. The procedure (task) also can be improved.

Some situations requiring special attention are: (1) VDT areas, (2) Inspection, (3) Emergencies, and (4) Security.

2. Literature review

This article is a concise version of Chapter 19, 'Eye and Illumination', in Konz (1990).

Other sources are Frier and Frier (1980), Boyce (1981), Illuminating Engineering Society Handbook (1981, 1984) and Murdoch (1987).

VDT guidelines include: IES (1989), Human Factors Society (1987), and Chartered Institution of Building Service Engineers (Anon, 1989a,b).

Table 1 defines lighting units.

3. Equipment for the practitioner

3.1. Hardware

The practitioner will need a light meter to record the ambient light level. Meters should be color and cosine corrected. Color-corrected means the reading is corrected for the sensitivity of the human eye. Cosine-corrected means the reading will not be affected by holding the sensor area at an angle to the light flux. The better meters give a digital output and have a 'freeze' button so the last reading remains on the display even though the meter is moved.

Hardware also is needed to evaluate an individual's visual capability. Ferguson et al. (1974) reported that 69% of workers without glasses needed glasses and 37% with glasses needed a new prescription. There are great differences in accommodation and convergence ability so if the worker has prolonged close work (say 0.5 to 1.0 m), near vision should be checked. Some people also have color weakness. Color deficiency is related to the X chromosome. Since females have 2 X chromosomes, only about 0.4% of females are color weak while about 8% of males are color weak (Rushton, 1975).

Although a Snellen wall chart is a possibility, most firms use a testing device (Martin, 1989). Typical parameters include far and near point, peripheral and color vision; stereo depth perceptions are measured for some occupations.

Note that some tasks require tracking a moving target – dynamic visual acuity. There is little adverse effect on resolution if the target velocity is less than 30–40 degrees/s. There is little correlation between a person's static visual acuity and dynamic visual acuity. Dynamic visual acuity can be improved by training (Long and Rourke, 1989); athletes such as baseball players have better dynamic visual acuity than non-athletes (Rouse et al., 1988). Unfortunately at present there are no clinical devices to test dynamic visual acuity – just research apparatus.

In low light conditions (mesopic luminance),

Correspondence to: Stephan Konz, Department of Industrial Engineering, Kansas State University, Manhattan, KS 66506-5101, USA.

Table 1
Units and definitions of illumination.

Quality	Unit	Definition and comments
Luminous flux, lm	lumen	Light flux, irrespective of direction from a source.
Luminous intensity, I	candela	Light intensity within a very small angle, in a specified direction (lumen/steradian). Candela = 4π lumens.
Illuminance, dIm/dA dIn.d/A	lux	1 lumen/m^2 = 1 lux = 0.093 footcandle 1 lumen/ft^2 = 1 footcandle = 10.8 lux
Luminance	candela/m^2 also called nit	Luminance is independent of the distance of observation as candelas from the object and area of the object perceived by the eye decrease at the same rate with distance. 1 candela/m^2 = 1 nit = 0.29 footLambert 1 candela/ft^2 p = footLambert = 3.43 nits
Reflectance	unitless	Percentage of light reflected from a surface.

Typical reflectance

Object	%
Mirrored glass	80–90
White matte paint	75–90
Porcelain enamel	60–90
Aluminum paint	60–70
Newsprint, concrete	55
Dull brass, dull copper	35
Cardboard	30
Cast and galvanized iron	25
Good quality printer's ink	15
Black paint	3–5

Recommended reflectances

Object	%
Ceilings	60–80
Walls	40–60
Furniture and equipment	25–45
Floor	20–35

Munsell value	Reflectance (%)
10 =	100
9 =	100
8 =	78
7 =	58
6 =	40
5 =	24
4 =	19
3 =	6

Quality	Unit	Definition and comments
Luminance contrast	unitless	$C = \dfrac{\text{Luminance of brighter} - \text{Luminance of darker}}{\text{Luminance of brighter}}$
Wavelength	nanometers	The distance between successive waves (a 'side view' of light). Wavelength determines the color hue. Saturation is the concentration of the dominant wavelength (the degree to which the dominant wavelength predominates in a stimulus). Of the 60 octaves of the electromagnetic radiation, the human eye detects radiation in the octave from 380 to 760 nanometers.
Polarization	degrees	Transverse vibrations of the wave (an 'end view' of the light). Most light is a mixture; horizontally polarized light reflected from a surface causes glare.

static visual acuity declines with age. Sturr and Taub (1990) recommend low light vision tests for all automobile drivers over age 65.

3.2. Models and software

There are two basic approaches to lighting workstations: uniform lighting and non-uniform (task) lighting. Uniform ($\pm 15\%$ from mean) lighting arrangements (generally from the ceiling) provide maximum flexibility in arranging the machines and workstations in the area, eliminate the need to move lighting fixtures if the workstations are moved, and use large lamps (which have higher lumens/watt than small lamps).

Uniform ceiling lighting assumes: (1) There are no walls or partitions to obscure the light, and (2) The entire area should be illuminated at the level determined by the one task requiring maximum illumination. Since, in addition, many ceiling lights traditionally are connected to one switch, uniform ceiling lighting tends to be energy inefficient.

Advantages of non-uniform lighting (i.e., low general illumination level supplemented with task lighting) are lower lighting cost, more precise control of the light, and more aesthetic appeal.

3.2.1. Uniform lighting

For uniform lighting, the concept is to divide the zone to be lighted into three cavities. The ceiling cavity is the space above the lamps, the room cavity is the space between the lamp and the illuminated surface, and the floor cavity is the space between the illuminated surface and the floor. The basic equation, an identity, is:

$$I(A) = (N_1)(N_2)(L)$$

where I = Illuminance in the area, lux; A = Area illuminated, m^2; N_1 = Number of fixtures; N_2 = Number of lamps/fixture; L = Lumens/lamp. For example, to obtain 750 lux evenly in a 2500 m^2 area, and assuming fixtures with 1 lamp/fixture and each lamp emitting 22,500 lumens, then N_1 = 83 fixtures are needed.

However, the equation needs to be modified for three types of losses. The first loss, coefficient of utilization (CU), considers absorption of the light by the room surfaces. The second loss, lamp lumen depreciation (LLD), considers the loss of light output of a lamp with age. The third loss, luminaire dirt depreciation (LDD), considers the loss of light output due to dirt on the fixture.

The resulting equation is

$$I(A) = CU(LLD)(LDD)(N_1)(N_2)(L)$$

For example, assume the product of CU(LLD)(LDD) = 0.5; then the space would need 166 fixtures instead of 83.

Detailed values for CU, LLD, and LDD (as well as N_2 and L) can be obtained from lighting handbooks or from the various equipment manufacturers. Many computer programs are available to perform the calculations; the coefficients come with the programs. The user specifies the room dimensions, the reflectances of the walls, ceiling and floor, the type of fixture (also called luminaire), type of lamp, and the fixture location. The program then calculates the amount of light at any specified height above the floor throughout the room. The typical output is a contour diagram (or a table) of the light level on a plan view. The user then either accepts the design or enters a change and the program gives the new lighting levels. The programs allow the user to use iteration (trial and error) to achieve the desired goal.

Four guidelines for uniform ceiling lighting follow.

(i) *Consider the spacing.* Fixture manufacturers recommend maximum spacing/mounting height for each fixture. A 'checkerboard pattern' with the fixtures 'on the same color squares' is reasonably efficient. However, this arrangement tends to overlight the room center and underlight the room perimeter; to compensate increase the spacing in the center and reduce it on the perimeter. End-to-end fixtures are an inefficient pattern.

(ii) *Remember a distant light is dim.* Because of the inverse square law, the farther the light has to travel to the worksurface the more it spreads out and the weaker it is. Thus uniform ceiling lighting tends to be used with low ceilings or fixtures suspended below the ceiling.

(iii) *Reuse the light.* Direct light travels directly from the fixture to the work surface. However, indirect light (light which is first reflected from a surface) helps. To get more reflection, the reflectances of the walls, floors and ceilings should be high; generally this means using light colors.

(iv) Use efficient fixtures. Efficient fixtures trap only 5–10% of the light inside the fixture while other fixtures trap as much as 30% of the light. In addition, some fixtures can direct the light better than others. Fixtures with some light going up (uplight) tend to be better than those with zero uplight as they permit air flow through the fixture (which keeps it cleaner) and the light on the ceiling gives better luminance contrast for the people in the room.

Uniform lighting can also be achieved by indirect lighting. The lamps are in fixtures which bounce the light off the ceiling and walls and have no direct component; the fixtures can be floor mounted, wall mounted or ceiling mounted. The result is a 'soft fog' with minimum luminance contrast. Because each reflecting surface absorbs at least 20% (white surfaces) and up to 90% (dark cloth) of the incident light, indirect lighting inherently uses more power than direct lighting.

In general, people prefer a mixture of direct and indirect lighting.

For a specific fixture, the downward component is described by 'beam spread' (highly concentrating, concentrating, medium spread, spread, widespread). Wider beam spreads give more overlapping (better illumination on vertical surfaces and less dependence on a single lamp). At high mounting heights, use narrower beams. The shielding angle (the angle between a horizontal line and the line of sight at which the source becomes visible) should be greater than 25°, preferably approaching 45°. From linear sources (fluorescent and low pressure sodium), the light distribution tends to 'batwing' (be emitted at a 45° angle downward, as viewed from the tube end).

3.2.2. Non-uniform (task) lighting

From an ergonomic viewpoint the primary advantage of task lighting is precise control of the light. Almost all of the light is direct as the indirect contribution is minimized.

Just as there are computer programs for uniform general lighting, there also are programs to calculate the amount of light from single sources.

Task lights furnish an amount of light; however, in addition, task lights can be used to furnish a lighting orientation and to vary the amount of light in different parts of the room.

Orienting the light to sharpen or blur the surface texture or form of an object is called modeling. For example, bottles of liquid can be inspected with light through the bottom. Back lighting may be useful for transparent materials. Low-angle lighting helps detect surface flaws. Figure 1 gives examples of five different luminaire locations.

Orientation can be used to modify people's facial appearance. On the stage, light from below is used to make a person look 'evil'. Lighting of a speaker is best with a light 20° horizontally on each side of the face; vertically they should be above 45° to reduce glare for the speaker but at 30° to improve audience impression (Golden, 1985). To emphasize the brightly lighted speaker,

a   b   c   d   e

Fig. 1. Placement of supplementary light can use five techniques: (a) luminaire located to prevent reflected glare (reflected light does not coincide with angle of view), (b) reflected light coincides with angle of view (it is in the 'offending zone'), (c) low-angle lighting to emphasize surface irregularities, (d) large area surface source and pattern are reflected toward the eye, and (e) transillumination from diffuse source.

consider a dark background; fill light (even from a light-colored podium or podium light) also can help.

Light should be seen as a form of communication – evoking a perceptual response in addition to the visual performance response. Light communicates subjective impressions of the environment and also provides suggestions for behaviors.

For example, orient user attention with a higher illumination level than the surroundings. An extreme example is a spotlight on a stage; a subtle example is a display case in a store. Lighting can suggest a circulation pattern; when following a path, people tend to follow the brighter path.

In general, aesthetic lighting design should consider the design of shadows as well as the design of light – emphasizing asymmetry and variability. Without shade or darkness, light loses much of its aesthetic appeal. A cloudy overcast day is perceived as bland; the more appealing bright sunny day has bright highlights and sharp shadows.

The designer's use of light orientation and variability thus affects the aesthetic impression of the environment.

4. Guidelines

4.1. Recognition that a problem exists

Visual problems at the workplace are unlikely to affect the worker's health although they may cause unnecessary stress on the visual system – sometimes resulting in headaches and fatigue. Visual performance may be affected, resulting in increased errors and time/unit. In general, these three criteria (worker stress, increased errors, increased time) are rarely measured. Even if the values are available for the present design, that doesn't tell you what they would be for good design.

Thus an ergonomist can recognize potential areas for improvement by contrasting existing jobs with design guideline recommendations. Health and safety records and recorded productivity records do not have the sensitivity to identify lighting problems.

4.2. Data collection

Data should be collected concerning the individual's visual capabilities, the illumination techniques used at the workstation and the task requirements.

4.3. Data analysis

The data is then compared with recommended values. For example, does the individual have corrected vision? Does the individual have a color deficiency? Is the lighting of the recommended quantity? quality? Can the task be modified to reduce the visual requirements?

4.4. Solutions to problems

4.4.1. Criteria

The cost of illumination is low in relation to the labor costs. Annual lighting cost for 2000 h/year in the USA in 1989 for an office or manufacturing area (at 750 lux) was about $70/ft^2 or about $07/m^2 (GE, 1990).

In comparison, annual labor cost/ft^2, assuming a labor cost/hour of $12 (wages and fringes) and 1900 h/year and 1 worker/300 ft^2, is $77/ft^2. This is 110 times the lighting costs. Of course, the ratio will vary with labor cost, worker density and type of lighting but ratios usually are over 100 and often are over 300 (especially with task lighting). The point is that economics is not much of a constraint on good lighting ergonomics.

In general, performance generally increases with increased illumination up to a certain level; beyond that level there may be glare problems. See table 2 for recommended amounts of illumination. However, performance generally can be improved more by changing target size and contrast than illumination.

Visual problems can be due to deficiencies in the individual, the lighting or the task.

4.4.2. Individual deficiencies

As pointed out earlier, many people do not have perfect vision.

One problem is the eye's ability to focus an image on the retina. A near-sighted person (myopia) has a long eyeball – the light rays from a distant object converge in front of the retina.

The solution is a concave external lens to bring the rays farther apart on the eye lens. A farsighted person (hyperopia) has the opposite problem – a short eyeball; the solution is a convex lens. People with 'watermelon'-shaped eyeballs have unequal radii of curvature in two axes; the result, a line focus instead of a point focus, is called astigmatism. Astigmatism can be corrected by an external lens with unequal curvature. With increasing age, the lens becomes stiffer, resulting in a decreased accommodation range. Bifocals, trifocals or progressive focal length lenses extend the range of clear focus. (With bifocals, there are two focal length lenses; with trifocals there are three focal length lenses; with progressive, there is one lens with the focal length varying continuously from top to bottom).

Naturally there are other visual deficiencies such as color weakness, lack of convergence, etc.

Corrective lens can either be spectacles or contact lenses.

Spectacles can be made impact resistant but note that the glass will break if hit hard enough. Polycarbonate lenses are lighter than glass, do not break as easily but scratch easily. Spectacles with chemicals to reduce light (either permanently as in sunglasses or temporarily as in photogrey glasses) may present a safety hazard in welding areas. The chemicals cut off radiation only in the visible spectrum, while hazardous non-visible radiation passes through. Thus proper welding glasses must be used in welding areas.

Contact lenses do not protect the eye vs. impact as a well as spectacles can. They also tend to 'seal' the cornea surface; thus they should not be worn where people are exposed to chemical fumes, vapors, splashes or dusty atmospheres (even inside a respirator) as they tend to hold the irritant against the eye and not let the normal tears wash it out. Intense heat (e.g., furnaces) can also be a problem.

The eyes can also be protected with goggles and face shields. Goggles should have a side shield; dust goggles have a 'fuzzy cloth' next to

Table 2

Recommended target maintained illuminance (lux) for interior industrial lighting. See table 3 for office examples.

Reference work plane	Type of activity	Illuminance category	Illuminance (lux)		
			Total factor		
			−3 or −2	−1, 0, +1	+2 or +3
General lighting throughout spaces	Public spaces with dark surroundings	A	20	30	50
	Simple orientation for short temporary visits	B	50	75	100
	Working spaces where visual tasks are only occasionally performed	C	100	150	200
Illuminance on task	Performance of visual tasks of high contrast or large size	D	200	300	500
	Performance of visual tasks of medium contrast or small size	E	500	750	1000
	Performance of visual tasks of low contrast or very small size	F	1000	1500	2000
Illuminance on task obtained by a combination of general and local (supplementary) lighting	Performance of visual tasks of low contrast and very small size over a prolonged period	G	2000	3000	5000
	Performance of very prolonged and exacting visual tasks	H	5000	7500	10,000
	Performance of very special visual tasks of extremely low contrast and small size	I	10,000	15,000	20,000

Note: See table 4 for factor calculations.

the skin to give a better seal. Face shields protect the entire face against liquids and impact.

4.4.3. Lighting deficiencies

Lighting will be divided into quantity of illumination and quality of illumination.

Quantity of illumination. The USA Illuminating Engineering Society recommendations (1983) are given in table 2. Table 3 gives office examples for illuminance categories B-F. Table 4 shows which of the three factors to use; the factor depends upon the worker age, the room reflectance, and the required speed/accuracy.

Although sunlight can act as a supplement to

Table 3

Office task examples of illuminance categories for use with table 2. One asterisk indicates potential problems with veiling reflections; two asterisks indicate special problems with veiling reflections. It may be better to modify the task than to compensate for poor task characteristics by adding lighting.

Illuminance category	General task/activity	Detailed example
B	Copied tasks	Microfiche reader (**)
	EDP tasks	VDT screens (**)
	General and public areas	Hallways, mechanical rooms during operation, stairs, utility rooms
C	Drafting	Light table
	EDP tasks	Machine rooms
	General and public	Display areas, elevators, escalators, janitorial space, lobbies and lounges, reception area, rest rooms
D	Copied tasks	Mimeograph, Xerograph
	EDP tasks	Impact printer: good ribbon, ink jet printer, keyboard reading, machine rooms – active operations and tape storage
	General and public	AV areas, conference rooms, duplicating and offset printing areas.
	Handwritten tasks	No. 2 pencil and softer leads (*), ball point (*) and felt tip pen
	Printed tasks	8 and 10 pt type (*), typed originals, newsprint, glossy magazines (**)
E	Copied tasks	Ditto copy (*), moderate-detail photographs (**), Xerography (third or greater generation)
	Drafting tasks	Mylar with high-contrast media (*), vellum and tracing paper with high contrast (*), blue line and blueprints
	EDP tasks	Impact printer; poor ribbon, second carbon and greater, machine eguipment when being serviced, thermal print (*)
	General and public areas	First aid, mechanical room equipment when being serviced, mail sorting (*)
	Graphic design and material	Graphs, moderate-detail photographs (*)
	Handwritten tasks	No. 3 pencil (*), handwritten carbons
	Printed tasks	6 pt (*), maps, typed second carbon and later, telephone books
F	Copied tasks	Poor thermal copy (*)
	Drafting tasks	Low-contrast media and hard graphite leads, low-contrast vellum and tracing paper, sepia prints
	General and public	Model making
	Graphic design and material	Color selection, charting and mapping, layout and artwork
	Handwritten tasks	No. 4 pencil and harder leads, nonphotographically reproducible colors

Table 4

Factors for illuminance table (table 2). If the task was in illuminance category D, the worker age was 'Under 40', speed/accuracy were 'Important', and task background reflectance was '30 to 70', then the components would be −1, 0, and 0; the total factor would be their sum, −1, so the middle column in table 2 would be used and the recommended illuminance would be 300 lux.

Variable	Factor		
	−1	0	+1
For illuminance categories A, B, C (general lighting throughout spaces)			
Occupants ages, year	Under 40	40–45	Over 55
Average weighted room surface reflectance (%)	Over 70	30–70	Under 30
For illuminance categories D, E, F, G, H, I (illuminance on task)			
Workers ages, year	Under 40	40–55	Over 55
Speed and/or accuracy	Not important	Important	Critical
Task background reflectance (%)	Over 70	30–70	Under 30

artificial light in some situations, the choice generally is among artificial sources. Some people are strong advocates of 'daylighting' in place of artificial lighting. Under normal circumstances there is no physiological need for daylight in workplaces. However, close to the north and south pole, lack of daylight exacerbates depression; its appropriate name is Seasonal Affective Disorder (SAD). Usually the desire for daylight is associated with the desire for a view. There are many problems with daylighting, primarily due to the lack of control of the amount of light and the energy losses for air conditioning. In general, expect daylighting to cost more than artificial lighting. Figure 2 shows an innovative way of separating the lighting and view function of a window.

Table 5 gives some lamp characteristics. Which lamp to use depends upon cost, convenience and color.

About 90% of the cost is for energy with about 10% for the fixture, lamp, and lamp replacement. Lamp lumen output decreases with age. Therefore, the mean lumens/watt is a better measure than initial lumens/watt. Lumens/watt are higher for bigger bulbs; often the advantage is over 25%.

Convenience includes the replacement frequency and restrike time. All bulbs (except incandescent) tend to have lives over 10,000 h; mercury

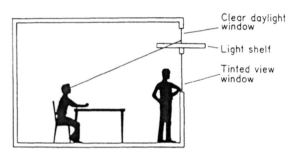

Fig. 2. Windows have both view and lighting functions. The tinted lower window permits a view while reducing direct glare. The clear upper window admits light but not glare to the occupant due to the light shelf. The light shelf also deflects light onto the ceiling deep into the room (Helms and Belcher, 1991).

Table 5

Lamp characteristics of typical industrial lamps (General Electric, 1990). Table courtesy of General Electric Lighting Business Group.

Type of lamp	Watts	Lumens/Watt		Lumen Maintenance (%)	Rated life (h)	Restrike (min)	Relative cost
		Initial	Mean				
High-pressure sodium	35–1000	64–140	58–126	90–92	24,000	1–2	Lowest
Metal halide	175–1000	80–115	57–92	71–83	10,000–20,000	10–15	Medium
Fluorescent	28–215	74–100	49–92	66–92	12,000–20,000+	Immediate	Medium
Mercury	50–1000	32–63	24–43	57–84	16,000–24,000+	3–6	High
Incandescent	100–1500	17–24	15–23	90–95	750–2000	Immediate	High

lasts especially long. If there is a power interruption, when the power resumes there is a delay (restrike time) before the lamp lights for sodium, mercury and metal halide lamps. Thus emergencies may require supplementary fluorescent or incandescent lighting.

Perceived color of an object depends upon the color sensitivity of the observer's eye, the surface color of the object, and the color of the light. With high-pressure sodium, color perception is poor and with low-pressure sodium, it is even worse. Don't use sodium sources where there is substantial viewing of faces (Lin and Bennett, 1983) as people look terrible. If fluorescent lamps are mounted among low pressure sodium lamps, the result is similar to high-pressure sodium – colors but not shades of colors are distinguishable. Metal halide lamps give good color (discriminate shades of colors) so they can be mingled with high-pressure sodium in high bays to get acceptable color and high lumens/watt. Fluorescent lamps not only permit good color rendition but permit you to specify the color. In the USA, cool white is the favorite; in Sweden warm white deluxe is the favorite. If you prefer a 'warmer' (more red) light than cool white, use cool white deluxe. Warm white gives the same color as incandescent. Colored light can also be used to enhance contrast in inspection (such as between gold, copper and copper oxide on printed circuits).

These general comments on color can be quantified with the color rendering index (CRI) of a lamp. After selecting a chromaticity range for its warmth or coolness, the CRI can be compared. On the 1–100 scale, high pressure sodium is 21, warm white fluorescent is 52, cool white fluorescent is 66, and daylight fluorescent is 79.

In inspection where colors need to be matched, it is necessary to specify the color of the ambient light as cool-white fluorescent, incandescent, high-pressure sodium, etc., as the perceived color varies with the lamp spectrum. Use lamps with a high color rendering index (CRI) for 'biological' inspection (meat, grain, cotton). The 'spectral power' of metal halide and sodium lamps is quite peaked so they are questionable for precise color discrimination. Metameric colors are colors which look the same under one light spectrum but different under another.

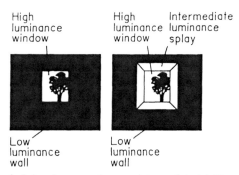

Fig. 3. Splayed recesses have an intermediate brightness between the window and the wall; this reduces luminance contrast.

Quality of illumination. In addition to light color (see previous paragraph) and light orientation and variability (discussed under task lighting), consider glare. Glare is any 'brightness within the field of view which causes discomfort, annoyance, interference with vision, or eye fatigue. Glare will be divided into direct and indirect glare.

Direct glare is caused by a light source within the field of view. Windows are the most common problem. One approach is to change the window brightness with louvers, curtains, awnings, etc. Another is to reduce window transmittance by using films or translucent glass; see figure 3 for another technique of reducing the contrast between a bright window and its surrounding wall. A third approach is to reorient the worker to face away from the window. A second source of direct glare occurs when inspecting backlighted objects such as microfilm readers; a solution is to use a mask to cover the portion of the screen not covered by the object. A third source of direct glare is lights, especially point sources with dark backgrounds. A solution here is to shield to at least 25° from the horizontal, although 45° is better; replace clear lamps with diffuse-coated ones. A light background and high ambient illumination level are preferable to a dark background and low ambient illumination – just think of vehicle headlights in daytime vs. at night. A fourth glare source is incandescent objects. Sliney and Wolbarsht (1980) describe in 1000 pages the safety problems of lasers, bright lights, welding arcs and hot metal.

Indirect (reflected) glare is caused by high luminance reflected from a surface. Reflected glare is horizontally polarized light.

One approach is to decrease the amount of incoming light so less is reflected. Lion (1964, 1968) reported better inspection performance when line sources (fluorescent) were used instead of point sources (incandescent). (For the same amount of light, a fluorescent lamp is less bright since it has a larger surface area). Another option is to obtain the light from multiple low luminance sources rather than a single high luminance source. The light source also can be filtered with a polarizer to minimize the horizontal reflection.

As a second approach, if the reflected glare is specular or directional (as in a mirror), reorient the reflecting surface or reposition the glare source so the glare misses the eye.

A third approach is to decrease reflectance through matte finishes. Some companies use curtains on the inside of inspection booths to reduce reflected glare. Car dashboards and 'satin finish' chrome are example matte finishes. Psychologically, people prefer matte blue, green or brown over matte grey.

A fourth approach is to put a filter at the eye – sunglasses. Sunglasses reduce luminance but not contrast so they reduce visual acuity. However, in most situations visual acuity is not critical as capability far exceeds task requirements; a possible exception is for very low luminance levels, close to the individual's sensitivity threshold. Thus people tradeoff a reduction of surplus luminance for an increase in comfort. Polarized sunglasses, which filter horizontally polarized light, improve both visual acuity and comfort. People working outdoors may want to use sunglasses marked 'General purpose'; they block at least 95% of ultraviolet B and 60% of ultraviolet A rays; sunglasses marked 'Cosmetic' only block 70% of UVA and 20% of UVB.

A fifth possibility is to move the operator's eye position. However, to avoid the glare, the operator may adopt a bad posture and thus get neck, shoulder or back pain.

4.4.4. Task deficiencies

The task can be improved by increasing object size, by increasing luminance contrast, and by increasing viewing time. If the object is the 'signal' and the environment is the 'noise', design the task with a high signal/noise ratio. The goal is to enhance the object so it emerges from its environment like an island from the sea.

Increase size. One possibility is to increase the size of the object. On transparencies for a meeting use 20 point size letters instead of 9 point (typewriter size). Size can be increased mechanically; examples are pantographs used in engraving or master/slave manipulators used in microscope work.

Another possibility is to bring the object closer to the eyes. This presents a conflict, however, because the hands usually are involved and holding objects close to the eyes becomes tiring. The arms can be supported with arm supports but the eye muscles (especially of older workers) become tired due to the effort of accommodation and convergence.

Optical aids are another possibility. When optics are used, visual clarity will decline due to light scattering in the lens and reduction of the visual field and depth of field will decrease; however the apparent size of the object will increase. Kruger et al. (1989) point out the advantage of magnifying devices in reducing accommodative strain; they also complain of the poor optical quality of many magnifying devices. If the target is to be magnified, the question is how much? Wei and Konz (1978) found $4 \times$ magnification better than $2 \times$ magnification but increasing the target size (after magnification) beyond 7 minutes of arc gave little benefit in their experiment. Smith and Adams (1971), using microscopes, reported time/correct inspection was minimum at 8–12 min of arc in their experiment. A lens (say $2 \times$) on a swinging arm with a local lighting unit can be used in many tasks. Magnifying devices tend to put the operator into a very fixed posture – a 'straightjacket'.

Reduce static muscle fatigue by building working breaks into their job every 20–30 minutes. For example, have them get supplies or records or carry completed units to a rack 10 m away. Magnification by projection (TV or optical comparators) reduces the fixed posture problems.

Another possibility is to put the magnification on the operator's head. A watchmaker's glass gives only monocular vision but there is no accommodation strain. It is also possible to use magnifying lenses in spectacle frames; usually the top half is omitted so the operator can have normal distance vision. VDT operators often have prescriptions ground to the focal length of their task (typically about 0.5 to 0.6 m) rather than a 'general-purpose' prescription. However depth of field is limited when the focal length is low (say 0.4 m); that is, the range of in-focus objects may only be from 0.2 to 0.7 m. The point is that glasses can be considered as 'an effort saving device' rather than just as a 'defective vision correction device'.

Sometimes reorienting the object will increase the visual size of a defect or improve contrast.

Increase contrast. The art of camouflage consists in obliterating contrasts; here we are concerned with anti-camouflage. Improved contrast can be either in color or luminance; it is especially critical if luminance contrast is less than 30%.

Seize any opportunity to modify target color. For example, color code file folders, tool handles, areas within a building, equipment, and worker clothing (to aid in identification of the type of worker). Use color in text and sign printing, photographs, and video displays. Some examples are 'bluing' a piece of metal before scribing lines on it, bluing bearings before fitting, staining a tissue in a microscope slide, and painting a white line on the pavement to differentiate the highway pavement and shoulder.

A simple reflector under the task often improves viewing. Thus use light table tops. Paint the interior cavities of machines white to improve visibility of objects being maintained through access ports.

For reading materials, use new printer ribbons, clear photo-duplications, pens not pencils, black pens not blue, felt tip pens not ball points, white paper for computer printouts instead of green, and white paper instead of brown or manila.

In inspection, an operator can inspect for object shape or object surface color. To detect shape, maximize object and background contrast. For example, if buttons are inspected for holes, make the table color contrast with the button color. When looking for shape, object orientation is critical. To demonstrate this, try to recognize a print of a face turned upside down.

However, to inspect object surface color, minimize the contrast of the object and the background. If you are sorting beans for color, sort on a table painted the color of a good bean. Pearls sold at retail are displayed on black velvet; the maximum contrast makes it difficult to detect color differences between pearls. When pearl merchants buy from each other, they display the pearls on white cloth to maximize color differences.

Spacing and grouping are helpful. For example, Fox (1977) reported coin inspection was more accurate when coins were presented in a systematic array than in a random array – the good coins presented a 'background' against which the defective coins stood out.

Increase time. Shoot sitting ducks. That is, work on or inspect stationary items rather than moving items.

A less desirable alternative is inspecting an object moving past the inspector. Cochran et al. (1973) found inspection performance was degraded at viewing time of 0.25 s/item but not 0.5 s. Keep angular velocity below 30–40°/s. Remove visual obstructions to maximize viewing time. If an inspector is viewing randomly spaced objects on a fixed velocity conveyor, there is a large variability in available viewing time. If, in addition, occasionally the inspector makes a motor movement (say to remove a defective item), the next 10 items may not be inspected at all.

A serious problem with some machine-paced conveyors or index tables is that the operator must make a positive action to prevent departure of a defective unit to the next station. If it is necessary to use machine pacing, have the operator make a positive action to send the units to the next station; this will increase physical activity and counter monotony, thus giving better quality. However, the best design is an operator-paced station, so the operator may allocate as much time as needed to each unit.

5. Special areas

5.1. VDT areas

Paper vs. screens. Until approximately 1980 the vast majority of tasks in offices involved working with paper; the paper tended to be relatively horizontal and more light improved the ability to do the task. However by 1990 there were over 40,000,000 Visual Display Terminals (VDTs) in the U.S.A. The screen is relatively vertical but more importantly ambient light washes out the contrast on the screen. At present, we have not yet gone to the 'paperless office' so lighting must be designed for both vertical screens and vertical paper (in document holders) and horizontal paper.

If the light is uniform throughout the area, this tends to be too much for the screen and too little for the paper. An alternative is task lighting. Because most VDTs are in offices, there is considerable interest in aesthetically pleasing solutions rather than just functional solutions.

Of the two tasks (paper and screen), the screen task tends to be the most difficult as the electronic characters are not as sharp as print, the letter/background contrast is worse, and there is some flicker. Reading rates are about 25% slower on screens than print (Gould and Grischkowsky, 1986). Thus if the lighting must be uniform, design for the screen since it is the more difficult task.

Light below 100 lux enhances screen legibility; above 500 lux enhances paper legibility.

For uniform lighting, the Human Factors Society (1988) recommends 200-500 lux. The IES recommends the light level for the paper task (see table 2) should not exceed 750 lux if there is uniform lighting.

If task lighting is used, then there can be relatively high illumination on the document (see table 2) and relatively low illumination on the screen – assuming the task light is properly directional. The general lighting should not be too low. The IES recommends 200-300 lux – assuming proper shielding of glare sources. Yearout and Konz (1987) reported that when VDT operators had general lighting of 350 lux supplemented by task lighting, the operators wanted their general view of the rest of the office to be much brighter – 770 lux – and to be visually interesting. Thus, for aesthetical reasons, avoid 'cavelike' offices for VDTs.

Operators prefer having the general lighting to be a mixture of indirect (uplight) and direct (downlight) rather than all downlight. An example recommendation is 60% indirect and 40% direct.

Reflections. Reflections on the screen can be diffuse or specular.

Diffuse (veiling) reflections are caused by ambient light as well as the phosphor. Diffuse reflections increase the luminance of both the screen background and characters, thereby reducing the contrast ratio. It is possible to improve the contrast with some types of screen filters. Their general characteristic is that the ambient light goes through the filter twice (in and back out) while the light from the characters goes through the filters once (out). Thus, although the luminance is reduced for everything, the luminance is reduced less for the characters and thus contrast is improved. If the filter is neutral density (gray), it reduces all wavelength energy equally. A color filter (the color of the phosphor) will pass most of the phosphor color but will not pass much of the ambient light (which typically is white); thus contrast is enhanced.

Specular reflection, a mirrorlike image, also occurs. Coatings such as quarter-wavelength filters (like the coating on camera lenses) and matte treatment (frosting) can reduce specular reflection. In addition remove bright objects from the environment. Detect these bright objects with the 'mirror test' – holding a mirror in front of the screen and looking at it from the operator's position.

One potential glare source is fixtures – especially in larger offices. (In smaller offices, the angle of the eye to the screen and the light from the fixture to the screen do not match.) The reflected luminance of the fixture seen on the screen needs to be reduced. This can be done by applying diffusers, prismatic lenses, polarizers, egg-crate louvers, and parabolic louvers to the light fixtures (IES, 1989). The screen also can be reoriented – tilting it vertically (to avoid windows). The entire workstation can be reoriented so there is no window behind the operator. If windows must be present in a room with VDTs, put them at the operator's side; windows placed in front of the operators cause problems due to the eye travel from the bright window to dim screen. Put blinds, shades or dark curtains on windows. Other solutions include replacing white shirts with dark shirts, replacing light colored table tops with dark table tops, finishing walls and other vertical surfaces in dark colors, moving bright objects on the shelves behind an operator, and placing a hood over the screen.

Another alternative (though not commonly used) is a negative contrast screen – dark characters on a light background. Specular reflections are less noticeable on the light background. Negative contrast screens may be more expensive, software may have to be modified, flicker may be more apparent, and it depletes the phosphor (IES, 1989). In some word-processing programs (e.g., WordPerfect), the user has an option of positive or negative characters and a choice of colors. For example, the user can have blue characters on a tint blue background with red highlighting or white characters on a blue background with yellow highlighting. Some 'graphics cards' permit additional enhancements such as bold face, italics and different font sizes.

See table 6 for a summary of measures to reduce screen reflections.

Luminance ratios. The Human Factors Society (1988) recommends the character be at least 7 times brighter than the screen background. The IES (1989) says the ratio (for both positive and negative displays) should be between 5:1 and 10:1.

Table 6

Measures for reducing screen reflections. The optimal combination of measures depends on the specific office, office layout, office equipment, and VDT (Helander, 1987).

Measure	Advantage	Disadvantage
At the source		
Cover windows		
Dark film	Reduces veiling and specular reflections	Difficult to see out
Louvers or mini-blinds	Excludes direct sunlight, reduces veiling and specular reflections	Must be readjusted in order to see out
Curtains	Reduces veiling and specular reflections.	Difficult to see out
Lighting control		
Control of location and direction of illumination	Reduces veiling reflections, may eliminate specular reflections	None
Indirect lighting	Reduces specular reflections, economy of office space by moving work stations closer	None
Task illumination	Reduces veiling reflection, increases visibility of source document	None
At the work station		
Move workstation	Reduces veiling and specular reflection	None
Tiltable screen	Reduces specular reflection	Readjustment necessary
Tilted screen filter	Eliminates specular reflection	Bulky arrangement for large screens
Screen filters and treatments		
Neutral density (gray) filter	Reduces veiling reflection, increases character contrast and visibility	Less character luminance
Color filter (same color as phosphor)	Reduces veiling reflection, increases character contrast and visibility	Less character luminance
Micro mesh, micro louver	Reduces veiling reflection, increases contrast	Limited angle of visibility; nonembedded filters get dirty
Polaroid filter	Reduces veiling reflection, increases contrast and visibility	Decreased character luminance
Quarter wavelength anti-reflection coating	Eliminates specular reflection	Expensive, difficult to maintain
Matte (frosted) finish of screen surface	Decreases specular reflections	Increases character edge spread (fuzziness, increases veiling reflections)
CRT screen hold	Reduces veiling and specular reflection	Difficult to avoid shadows on screen
Sunglasses (gray, brown)	None – contrast unchanged	Less character luminance and visibility
Reversed video	Reduces specular reflections	Increased flicker sensitivity
Screening of luminaires and windows	Reduces specular reflections	Might create isolated workplaces

Does the eye have a problem looking at a bright surface (such as a document) and then to a dim surface (such as a screen)? The HFS (1988) says this is not a problem and the brightness within a workstation can vary as much as 20 to 1 with no problem.

To minimize accommodation strain, the VDT screen should be at the 'dark focus' distance. Dark focus distance varies with individuals but a reasonable value is about 65 cm (range 50–80 cm) from the eyes (Jaschinski-Kruza, 1990; Akbari and Konz, 1991). This dark focus distance may be too far if the characters are small – for example being presented on a small screen.

5.2. Inspection

The general inspection task is: (1) search for the defect, (2) detect the defect, (3) judge if it is a defect, and (4) implement the decision.

Search and detection. Improve the search by ensuring the operator knows exactly what to look for by providing periodic updates on the probability and importance of various types of defects.

If the object is moving, it should move parallel to the shoulders, not perpendicular to the shoulders (Suresh and Konz, 1991).

As pointed out in the task deficiencies section, enough time must be allowed.

The amount of light can be estimated from table 2. General uniform lighting is unlikely to be sufficient. Be sure the task lighting is on the inspection point, not over the conveyor from which items are obtained. Ideally the inspector should have control of both the amount and orientation of the task lighting. Cushman and Crist (1987) have a good discussion of inspection lighting.

Judgment. Give the operator a precise quality standard so the enormous variability in inspector concept of standard can be reduced. Photographic standards, for each type of defect, giving one or two examples of acceptable and one or two examples of rejectable quality are excellent – especially for batch production and rare defects.

Each inspector will accept some bad items and reject some good items. The cost of each of these types of errors should be determined and a desirable ratio calculated for each type of error. Then, from a sample of each inspector's actual output, determine what the actual ratio is (Drury, 1988). With the inspector knowing that a test is being performed, have items with known defects be inspected along with good items. What percent of the defects were detected? The point of this information gathering is not to penalize individual inspectors but to obtain information on what is the inspector's mental quality standard, whether the lighting is sufficient, whether inspection devices are adequate, etc.

Implementation. Improve implementation by reducing social pressure on the inspector by the operator and supervision. If the inspection is at receiving inspection, penalize the vendor for defective items. Publicity is a good tool. For example, send a bill to the vendor's president requesting payment for rework of defective items. Much rework and repair is done 'under the table' to avoid attention from management. One of the primary advantages of the 'just-in-time' system is that it is more difficult to hide rework and repairs; thus there is enormous pressure to solve problems instead of hiding them.

5.3. Emergency lighting

Emergency lighting is designed to aid people to leave the building if the normal power fails. The Life Safety Code (1980) specifies 10 lux on the floor along the route people might take when leaving the building. Frier and Frier (1980) recommend 30 lux for congested or critical areas such as corridor intersections, tops of stairs, dangerous machinery, etc. The vertical surface around the exit door should have 20 to 50 lux. The recommendations in table 7 depend upon the hazard and the activity level.

The restrike problem of some lamps makes them unsatisfactory for emergencies, even with a

Table 7

Minimum illuminance (lux) levels for safety (IES, 1983). It is the absolute minimum at any time and at any location on any plane where safety is related to seeing conditions.

Hazards requiring Visual Detection	Normal activity level	
	Low	High
Slight	5	11
High	22	54

Table 8
Recommended area illuminances (lux) for security lighting, measured on the horizontal plane (Baker and Lyons, 1978).

Risk	District brightness		
	High (Adjacent main road lighting; lighted adjacent land; floodlighting)	Medium Adjacent (secondary road lighting)	Low (No adjacent lighting on adjoining property or nearby roads)
Extreme	20–30	10–20	5–15
High	10–20	5–15	2–10
Moderate	5–15	2–10	1–5

backup power system. Some solutions include emergency fluorescents, a quartz lamp in the high-intensity-discharge circuit, and dual-filament HID lamps (one filament is on 'standby').

Exit signs can be self-powered but many local codes permit the exit sign to be illuminated externally by the emergency lighting – this is less expensive than having both emergency lights and internally powered exit signs.

5.4. Security lighting

The purpose of security lighting is crime prevention. The light should discourage intruders (reduce 'offense') and improve detectability when entry is effected (improve 'defense'). If there is a guard or TV camera on-site, consider them as the 'audience' and the intruder as the 'actor'. That is, the audience should have a good view of the actor; conversely the actor should not be able to see the audience. The defender (the guard) should not be visible to the intruder. A guard station should be lit so it is not obvious whether it is occupied – it should be relatively dim inside and relatively bright outside. Design the 'stage lighting' to maximize glare for the intruder and minimize it for the guards, cameras, and neighbors in the plant vicinity.

If there is no on-site defender and you depend upon outside police observers, turn the lighting system around and light the building exterior.

Table 8 gives recommended security lighting. High-pressure sodium is the recommended source due to its low cost and compatibility with most TV cameras. Consider low light cameras for areas without good lighting.

Garages and parking spaces need about 10 lux. For garages, place the fixtures to illuminate the between car walk spaces rather than the area over the car stalls.

References

Akbari, M. and Konz, S., 1991, Viewing distance for VDT work. In: Proceedings of the 11th IEA Congress. Taylor and Francis, London.

Baker, J. and Lyons, S., 1978, Lighting for the security of premises. Lighting Research and Technology, 10(1): 10–18.

Chartered Institution of Building Service Engineers, London, 1989. The industrial environment.

Chartered Institution of Building Service Engineers, London, 1984. Areas for visual display terminals.

Cushman, W. and Crist, B., 1987. Illumination. In: G. Salvendy (Ed.), Handbook of Human Factors, chapter 6.3. Wiley, New York.

Drury, C., 1988. Inspection performance and quality assurance. In: S. Gael (Ed.), Job Analysis Handbook, chapter 6.5. Wiley, New York.

Ferguson, D., Major, G. and Keldoulis, T., 1974. Vision at work. Applied Ergonomics, 5(2): 84–93.

Frier, J. and Frier, M., 1980. Industrial Lighting Systems. McGraw-Hill, New York.

Fox, J., 1977. Quality control of coins. In: J. Weiner and H. Maule (Eds.), Human Factors in Work, Design and Production. Taylor and Francis, London.

GE, 1990. Lighting application bulletin: Industrial lighting. General Electric Corp.

Golden, P., 1985. The effects of lighting a public speaker upon observer impression. Lighting Design + Application, 15(12): 37–43.

Gould, J. and Grischkowsky, N., 1986. Does visual angle of a line of characters affect reading speed? Human Factors, 28(2): 165–173.

Helander, M., 1987. Design of visual displays. In: G. Salvendy (Ed.), Handbook of Human Factors, chapter 5.1. Wiley, New York.

Helms, R. and Belcher, M., 1991. Lighting for Energy-Efficient Luminous Environments. Prentice-Hall, Englewood Cliffs, NJ.

Human Factors Society, 1988, American national standard for human factor engineering of visual display terminal workstations. Human Factors Society, Santa Monica, CA.

IES, 1981. IES Lighting Handbook: Application Volume. Illuminating Engineering Society of America, New York.

IES, 1984. IES Lighting Handbook: Reference Volume. Illuminating Engineering Society of America, New York.

IES, 1989. The IES recommended practice for lighting offices containing computer visual display terminals. Illuminating Eng. Society of America, New York.

IES Industrial Lighting Committee, 1983. Proposed American national standard practice for industrial lighting. Lighting Design + Application, 13(7): 29–68.

IES Office Lighting Committee, 1982. Proposed American national standard for office lighting. Lighting Design + Application, 12(4): 27–59.

Jaschinski-Kruza, W., 1990. On the preferred distance to screen and document. Ergonomics, 33(8): 1055–1064.

Konz, S., 1990. Work Design: Industrial Ergonomics. Publishing Horizons, Scottsdale, AZ.

Krueger, H., Conrady, P. and Zulch, J., 1989. Work with magnifying glasses. Ergonomics, 32(7): 785–794.

Life Safety Code. An American National Standard, 1988. National Fire Protection Association, Boston.

Lin, A. and Bennett, C., 1983. Lamps for lighting people. Lighting Design + Application, 13(2): 42–44.

Lion, J., 1964. The performance of manipulative and inspection tasks under tungsten and fluorescent lighting. Ergonomics, 17(1): 51–61.

Lion, J., Richardson, E. and Browns, R., 1968. A study of industrial inspectors under two kinds of lighting. Ergonomics, 11(1): 23–24.

Long, G. and Rourke, D., 1989. Training effects on the resolution of moving targets – Dynamic visual acuity. Human Factors, 31(4): 443–451.

Martin, D., 1989. Visible difference. Occupational Health and Safety, 25–30.

Murdoch, J., 1985. Illumination Engineering. MacMillan, New York.

Rouse, M., DeLand, P., Christian, R. and Hawley, J., 1988. A comparison study of dynamic visual acuity between athletes and nonathletes. Journal of American Optometric Association, 59(12): 946–950.

Rushton, W., 1975. Visual pigments and color blindness. Scientific American, 64–74.

Sliney, D. and Wolbarsht, M., 1980. Safety with Lasers and Other Optical Sources. Plenum, New York.

Smith, G. and Adams, S., 1971. Magnification and microminiature inspection. Human Factors, 13(3): 247–254.

Sturr, J. and Taub, H., 1990. Performance of young and older drivers on a static acuity test under photopic and mesopic luminance conditions. Human Factors, 32(1): 1–8.

Suresh, A. and Konz, S., 1991. Movement and part orientation in paced visual inspection. Advances in Industrial Ergonomics and Safety III. Taylor and Francis, London.

Wei, W. and Konz, S., 1978. The effect of lighting and low power magnification on inspection performance. Proceedings of the Human Factors Society, CA: Santa Monica, pp. 196–199.

Yearout, R. and Konz, S., 1987. Task lighting for visual display unit workstations. In: Proceedings of IX Int. Production Engineering Conference, Cincinnati, OH, pp. 1862–1866.

Evaluation and control of industrial inspection: Part I – Guidelines for the practitioner[1]

Tim J. Gallwey

Department of Manufacturing and Operations Engineering, University of Limerick, Limerick, Ireland

1. Audience

These guidelines are intended to assist those people who are responsible directly or indirectly for jobs where there is a significant element of a viewing type of inspection. This occurs in a wide range of occupational activities where people examine such appearance factors as the quality of paint finish on motor cars, handling damage on fresh produce, texture of cloth, colour of plastics parts, shape of jam tarts, cavities in castings or machined surfaces, X-rays of patients or hip implants, corrosion of aircraft structures, or wear on rubbing surfaces. Early work on improving this activity was concentrated on manufactured parts, usually in metal working industries, and some of these can now be dealt with using machine inspection. However there remain numerous situations in maintenance work where inspection forms a major and crucial part of the job, and there are other non-traditional applications such as clerical work and medical procedures such as "key hole" surgery and mammograms. The areas to be addressed include information on inspectors' perception and decision making processes, statistical analysis procedures, job and workplace design requirements, appropriate measures of performance, and strategies to control and improve performance. Usually the targeted people work in some form of Quality Assurance job even though it may not be described officially in such terms.

2. Context for use of the guideline

In such inspection jobs, the characteristics of concern are difficult to measure (e.g. the sizes of irregular cavities) and they are difficult to define (e.g. how much damage is too much damage on a sheet of steel). They either cannot be assessed by machine or are prohibitively expensive to assess by machine. If machines are used in their assessment, a human has to interpret the results presented. Due to product variety, frequent changes are required in the inspection device and the standards for comparison are minimal or non-existent. A weighted

---

[1] The recommendations provided in this guide are based on numerous published and unpublished scientific studies and are intended to enhance worker safer and productivity. These recommendations are neither intended to replace existing standards, if any, nor should they be treated as standards. Furthermore, this document should not be construed to represent institutional policy.
The following individuals participated in the discussion of the earlier version of this guide. Their suggestions (written or verbal) were incorporated by the authors in this version: A Aaras, Norway; J.E. Fernandez, USA; A. Freivalds, USA; M. Jager, Germany; S. Konz, USA; S.Kumar, Canada; H. Krueger, Switzerland; K. Landau, Germany; A. Luttmann, Germany; A. Mital, USA; J.D. Ramsey, USA; M-J. Wang, Taiwan.

assessment may be required of all the characteristics in order to decide whether to accept or reject, in addition to individual assessment of the characteristics i.e. one or two may be borderline acceptable individually but not in combination. There are large numbers of possible deficiencies in the "object" inspected and many inspection standards to deal with, and agreement on what is acceptable is usually difficult and almost always only partial. These points indicate that inspection is a difficult job to carry out and control, and that it will be somewhat prone to errors and variations in performance. Despite this, sampling plans are devised on the assumption that inspectors will always make correct classifications, with the end result that the risks involved (producer's and consumer's) are often quite different from those supposedly designed into the plans. Nevertheless the job has to be carried out, and therefore it is necessary to determine what countermeasures to adopt to minimise these undesirable results. To achieve that it is necessary to understand what the job involves for the inspector.

The job can be broken down into three main stages: (1) the formation stage, (2) the processing stage, and (3) the deduction stage. In the formation stage, the inspector learns what types of deficiency to look for, what the various types look like, where they usually occur, and what is acceptable and rejectable for each. In the processing stage, the inspector searches the unknown item to find possible deficiencies, compares these against a mental image of the specification, and then exercises judgement to decide whether to accept or reject the item. In the deduction stage, the inspector has an idea in memory of various types of deficiencies rejected, the total number accepted, and perceives some value for the frequencies of deficiencies, trends for the different types of deficiency, and of the work rate achieved.

To achieve an improvement in inspection work requires an improvement in all three stages. The formation and deduction stages entail a heavy load on long term memory (LTM) plus a calculative part relating to deficiencies accepted and rejected, and the "costs" of errors. Most, if not all, of these two stages could be better handled by a hardware device to leave the inspector to concentrate on the processing stage for which the human is probably better in most cases. The lack of such hardware is a primitive state of affairs which maintains this job at the level of crude hand tools. As a result most inspection jobs go well beyond the abilities of human inspectors and they will remain so until this shortcoming is rectified. This document addresses the processing stage in particular.

3. Problem identification

3.1. Mechanics of the process

In this type of work, it is usually the case that the object contains a variety of marks, irregularities, discolourations, and other imperfections which are extraneous to its acceptability. It is also common experience that some deficiencies tend to occur only (or mainly) in certain areas, others in some other areas, and yet others occur anywhere on the object. In addition, the objects are usually large relative to the area that can be seen when the eyes are fixated on one spot; this area is called the visual lobe since it has the shape of a distorted ellipse.

The size of the visual lobe varies with the task characteristics. The larger it is, the quicker the deficiencies can be found and the higher the probability of detecting their presence. If the inspector fixates the eyes exactly on a possible deficiency, there is a high probability that its presence will be detected. As the fixation point falls further away (i.e. fixation eccentricity increases), so the probability of detection decreases, eventually to chance level. By finding the eccentricity in a variety of directions at which the chance level is reached, a measure of visual lobe for the inspector/job combination can be obtained. The limit of eccentricity for a high probability of detection (i.e. the area of the fovea) is about 2 degrees from the line of sight of the inspector.

If the object being inspected is a square with sides of length 364 mm, and if it is at a distance of 1 m from the inspector, its edges will subtend angles of 20° by 20° at the inspector's eyes as can be seen in this calculation:

$$\tan(\alpha) = 364/1000 = 0.364 \quad \text{and so } \alpha = 20°.$$

In such a case, to examine the object completely with foveal vision and making an exactly non-overlapping set of fixations, the inspector will have to make approximately 100 fixations (at a width of 2° there have to be 10 fixations per row and 10 per column). Hence such an object requires a visual search process. The manner in which the fixations are distributed will have a big effect on the success of the inspector and on the time required. In this way the processing stage of inspection of a single object can be seen as a job divided into a series of fixations, each of which is a single task consisting of the subtasks depicted in Table 1.

For each subtask some of the results go into LTM, either consciously or unconsciously. These combine with previous results on this object and other objects to help determine where subsequent fixations are going to be placed, whether or not to re-search an area, how likely it is that there is still an undetected deficiency on the object, and so on. When the perceived benefit of continuing to search the same object has fallen below some perceived value of utility, or time runs out, the inspector will abandon the search and accept the object. Such results, reject results, whereabouts the rejectable deficiency occurred, and so on, all will be added to the store in LTM for future reference. Similarly, current rejection rates and deficiency trends are perceived in some way and added to LTM. But all these mental processes are competing for the limited channel capacity available. If this capacity is exceeded, information will be lost so that the success of the inspection will be reduced. Therefore it is necessary to establish what these limits are and design the job so that they are not exceeded.

3.2. Effects of the process

The above description of the process demonstrates that there is a heavy mental processing load with a constant high level of mental concentration. These are complicated by the addition of difficult perception and decision tasks. The net result is a high potential for error. Accept decisions can be wrong because of: (1) not recognising the presence of a deficiency, (2) running out of time before finding it, (3) poor distribution of fixations, (4) inadequate perceptual ability, (5) poor recall of the inspection specification, or (6) a poor decision process. Reject decisions can be wrong only if a deficiency has been found; so the scope for error is slightly less. If objects are inspected a second time, by the first inspector or someone else, the decisions

Table 1
Subtasks contained in each fixation

Step	Subtask	Memory aspects
1	Examine the area fixated	Process info in STM
2	Decide what is the most likely deficiency there, if any	Use long term memory (LTM)
3	Select the appropriate visual image for that deficiency	Recall from LTM
4	Compare this image with the image transmitted from the eye	Use short term memory (STM)
5	Decide on the similarity or otherwise of these images	Process info in STM
6	Decide whether to accept or reject the object	Recall standard from LTM
7	If accept, decide whether or not there is enough time left to examine another area within the fixation	Examine STM, compare to LTM data.
8	If decided enough time left, decide which new area within the fixation to examine next	STM info, data from past in LTM
9	Exhaust the possibilities within the fixation	What was seen, where, use STM
10	Decide whether or not enough time left to fixate another area of the object	Examine STM and LTM
11	If decided enough time left, decide where to place the next fixation	Recall location of previous ones (STM), usual locations (LTM)
12	Move eyes to next fixation point, return to Step 1 and repeat the above	Recall location of previous ones (STM), usual locations (LTM)

may not be the same and have been shown, in a variety of situations, to be different.

The human has difficulty coping with the huge variety of demands made and the minimal availability of aids for performing the task. Similarly the time available is often insufficient (especially on machine-paced tasks), there are big individual differences between inspectors, and the skills required have been shown to be task-specific. However, with what we now know and understand about the tasks involved, we are able to achieve much better results. The problem then is how to apply our knowledge to particular industrial inspection situations to achieve such improvements.

For the employer, these errors show up as inadequate product being shipped to customers so that warranty claims, and customer complaints and returns, are higher than they should be. Alternatively, if the employer audits the quality of outgoing objects, the average outgoing quality level is lower than was thought to be the case, especially when these are based on the calculations used for sampling plans. Such audits also will reveal that acceptable objects are being rejected, and reworked or scrapped unnecessarily, and therefore adding unnecessary costs to the operation. But, to carry out such audits, it is necessary to obtain appropriate measures of the performance of the inspector, and of the system as a whole, if useful and reliable information is to be obtained.

4. Measurement and estimation

For more detailed information on many of the topics below see Gallwey (1998).

4.1. System performance

The first thing to measure is the success of the system in detecting deficient objects. For this purpose it is necessary to know the various types of deficiency and the rate of occurrence of each. Then a selection of objects needs to be made up with the appropriate proportions and variations of each, along with the required proportion of deficient free objects. At the same time all the deficient objects must be given some secret and invisible method of identification (e.g. a dye that is only visible under ultraviolet light, or metal that is magnetised), so that they can be recovered once introduced to the system. The test batch, and preferably several similar but different batches, are submitted to the inspectors along with their normal batches, and data are collected on the results of the inspectors' classifications of these objects. There is a school of thought that the inspectors should not be told about this process, as their performance might be affected if they know. On the other hand, there is the contrary opinion that such secret tests should not be carried out, as they may create a climate of suspicion, or because it is ethically wrong to perform such secret tests. If they are done, it is necessary to ensure that measures are obtained on the proportions of acceptable objects rejected, as well as the rejectable objects accepted.

A well known problem is that of "flinching" (when articles are only slightly deficient, inspectors are reluctant to incur the wrath of their fellow employees that results from classifying them as reject, so they are accepted). The result is that there is a high incidence of objects classed as being just acceptable and none classed as being just deficient. While this is most easily demonstrated on the dimensions of the objects, it can be shown as well on appearance issues. Data therefore should be collected in regard to such marginally deficient objects. The results should be plotted in a frequency histogram to show up any "spikes" or blanks in the spectrum of classification categories.

4.2. Inspector performance

The second thing to measure is the success, or otherwise, of the inspector on the job in question. From the performance data obtained from Section 4.1 above, the difficulty of this job for this inspector can be determined, by calculating a measure from the Theory of Signal Detection (TSD) (e.g. d'). It will also be possible to calculate from TSD a measure of the strictness of the inspector i.e. Beta. These measures will indicate whether the task is too difficult, or that the inspector has a wrong interpretation of what is required, in terms of trade-offs between correct actions and the inevitable errors that they will make.

4.3. Times and errors

The inspection process involves at least two parts, a visual search part and a decision part. The major time element is due to the search part. Data need to be collected on this, both for the time taken and for deficiencies missed. Classical visual search concerns situations where a "target" is always present and so time is the only measure of performance. To convert this to an inspection situation, other non-deficient objects need to be introduced. Thus the test inspector will not expect to find deficiencies all the time, and is therefore more likely to miss some of those that are there. Examples are needed of the different types of deficiency, with the different degrees of deficiency across the range found in practice. However there is a trade-off between time and errors, so time cannot be considered on its own. Therefore multivariate analysis is needed for properly valid analysis.

4.4. Measures of inspectors' abilities

Vision: Obviously inspectors need to be able to see what they are doing. They need to be tested to see that they have at least 6/6 vision. If colour is involved they need to be tested for colour vision; the simple way to do this is to use Ishihara Pseudo-Isochromatic plates. Mention has already been made of the size of the visual lobe; a test of this is desirable e.g. expose, for 300 ms, 35 mm slides with pairs of images at various eccentricities either side of the centre, and ask the subject to say if they were the same or different.

Mental processing: As the description above demonstrates this constitutes a critical part of the job. Three parts of the Weschler Adult Intelligence Test have been found useful. These are the Mental Arithmetic Test, Digit Span for Short Term Memory, and Digit Symbol (where as many symbols as possible have to be matched to corresponding digits in 90 s). In addition people differ in their ability to "dig out" particular shapes when they are mixed in among other shapes, patterns and lines; the Embedded Figures Test measures this and has been found useful in identifying people who are better in inspection work.

5. Analysis of the data

5.1. Procedures and testing

Use good experimental design to address the particular issues that arise in experiments with human subjects (prior experience in this type of work, the effects of learning during the task, balancing of orders to avoid order effects, whether the design should be within subjects or between subjects, and the use of a repeated measures design). Since there are large numbers of variables and conditions, a partial factorial design (such as Taguchi) should be used to perform a screening process to home in on the most relevant factors. Also, in analyses such as ANOVA, it is required that the data be Normally distributed. However, search times are exponentially distributed, so the times need to be transformed to their logarithm value. Similarly errors are expressed in proportions, which are Binomially distributed; these values need to be transformed to the arc $\sin((X)^{0.5})$ value.

Due to the large number of variables affecting inspection performance, the design needs to be "tightened" as much as possible. A good way to do this is to include the scores obtained from Section 4.4 as covariates, thus it probably ends up as a multivariate analysis of covariance design. However, in some experiments/studies, it is probably difficult or impossible to find any valid type of parametric test; in that case, nonparametric tests should be used.

Many of the analyses can be carried out using standard statistical analysis packages such as the Statistical Package for the Social Sciences (SPSS) or the Biomedical Computer Programs (BMDP) package or the SAS package. These contain some procedures for nonparametric tests but another possibility is to use the OMNIBUS package of Meddis (1984), which provides a unified approach to such tests.

5.2. Subjects

These should be representative of the typical population involved in the job. If the test task is in any way different from their normal duties, they should be given a thorough training in the new task, to reduce as far as possible the learning effect. Experience shows that this effect is extensive and failure to reduce it substantially results in experiments that do not produce significant effects.

6. Actions to solve the problems

6.1. Information to the inspectors

The inspectors need to receive regular feedback on their performance. Knowledge of results has been shown to be very effective in keeping them motivated and aware of what is happening. As part of this, it is necessary to ensure that they are aware of the particular deficiencies occurring at the time of their inspection, so that they are not keenly searching for the wrong things due to a persisting mindset from their preceding experience(s). To achieve this, there should be a preliminary inspection of incoming objects and the inspectors should then be told the types of deficiencies found. This is called feedforward and has been shown to be very effective. As a follow-up to this, it is good practice, as in many situations, to audit the results of the inspectors' work. The objects that are ready to be sent to the "customer" should be examined and the inspectors told what deficiencies they have missed i.e. feedback to them how successful they have been. This also has been very effective.

6.2. Inspection aids

The stimuli provided naturally by the objects inspected may be close to the perceptual limits of the inspector. In such cases they need to be enhanced e.g. by providing magnification for small sizes of deficiencies, using directional light to create shadows to make surface irregularities more visible, using image enhancement techniques to give the inspector a clearer view of an X-ray plate, or even transforming the stimulus from sound (say) to a numerical value e.g. a noise level meter. To assist decision comparisons, adequate and readily usable reference standards must be provided. Then the decisions can be made through relative judgements rather than absolute judgements.

The simplest form of aid is a folder detailing the requirements. Preferably these should be enhanced by including sketches of the different types of deficiency. They should also show several levels of deficiency for each type, e.g., definitely not acceptable, probably not acceptable, just acceptable, probably acceptable, definitely acceptable. Better than sketches would be colour photographs of each of these, and better still would be physical examples of each of them. One of the problems of folders is that of change control. Whenever there is a change in a specification there are serious difficulties in ensuring that the old document is withdrawn (at exactly the right moment) and replaced by the new one. Also, these documents tend to become "shop-soiled" very quickly with a loss in the usefulness and accuracy of the information. Ideal would be a multi-media set of aids and information available to the inspector at the press of a button, or click of a mouse key; only the copy of the specifications on the server has to be updated. Use of this can assist the "formation" stage described in Section 2 above.

There is a heavy memory load on an inspector, especially where there is a wide variety of products. It is necessary to remember where deficiencies usually occur, where they have occurred in recent batches, which types of deficiency have been most common lately, and so on. Rather than rely on the inspector's memory, which will probably be overloaded, use should be made of the computer to store this information as it arises. It could also be used to calculate trends and frequencies rather than rely on subjective impressions and memory of the current and recent experiences. The machine should perform these rather mechanical tasks of the "deduction" stage of Section 2 above. That will leave the inspector to concentrate on the "processing" stage, which must be done by a human.

6.3. Time and rest breaks

As the amount of time needed on any one object depends on whether or not it can be classified quickly, the inspector needs to be able to allocate time as necessary. This means that machine-paced inspection leads to the inspector having too much time on some occasions and too little on others, resulting in reduced utilisation and increased error rates. Hence all inspection should be self-paced. Enough time must be allowed for it to be done well. In addition to the time for actual inspection, there needs to be "enough" time for rest breaks at intervals of not more than 30 minutes (down to 15 minutes in some cases); the length of these depends on the nature of the task. It may well be sufficient to provide the inspector with "working rest", i.e., some other task to perform, such as minor materials handling or some type of administrative or clerical work. Such job enlargement has been shown to maintain attention and combat boredom.

An associated problem is that of visual strain or fatigue which usually results from strenuous fine work such as viewing poorly defined objects, working with inadequate lighting or flickering light, excessive contrast in the visual field (including glare), or optical aberrations of the person. Typical symptoms are discomfort of the eyes, difficulty in focusing, or headaches and general tiredness. Remedies include rectification of the adverse conditions and the provision of adequate rest breaks which should be more frequent as the conditions become more difficult.

6.4. Inspection strategies

As the frequency of deficiencies decreases, failure to detect them increases, perhaps due to a decrease in vigilance. An effective method of increasing the apparent rate is to use a two-stage strategy. A first-stage inspector rejects everything that is less than perfect; a second-stage inspector examines these at greater length to make the final decision to accept or reject.

The number of types of deficiency that the inspector is looking for is usually high (may be as much as thirty on quite simple objects, while reaching two hundred or more on complex ones). This represents an excessive load for most people. Thus many inspectors limit themselves to a subset, which is probably a number that constitutes the limit of their STM, say about five in most cases. Greater success is likely if this point is accepted, and if inspectors are requested to look only for the five main deficiencies currently occurring, as notified by the feedforward process. It is also better to search for one type of deficiency everywhere on the object, and then for a second one everywhere, and then a third, and so on, rather than trying to look for all types in all areas of the object. However this one-at-a-time approach is rather tedious, even when dealing with just 30 types. An instruction to search for the five most likely types, and then the next five most likely, and so on, would be a good compromise approach.

6.5. Selection of inspectors

It has been shown that selection tests are probably task-specific. The result is that there is no one test, or one specific set of tests, that can be used for selection of inspectors for all jobs. Instead it is necessary to carry out a detailed task analysis of the job in question; then a set of tests can be devised to look for the particular skills required. An aid to doing this is provided in Table 2 which presents a summary of findings reported more fully in Gallwey (1998). Factor analysis of the results of the tests probably will be needed to interpret them before the appropriate set of tests can be identified.

Ideally test cutoff scores ought to be specified here but this is not possible due to the task-specific nature already described. However, it should be noted that on some tests a low score is desirable (e.g. time on the Embedded Figures Test) and on others a high score is to be desired (e.g. number of symbols matched to numbers in the WAIS Digit Symbol). Therefore, when evaluating the results, it is necessary to ensure that, for this purpose, all the scores are adapted so that a high score represents better performance e.g. by using reciprocal values. It is really the composite score obtained from the selected test battery that is the criterion for selection, for example one obtained from multiple regression. In practice it is more likely to be a case of deciding how many inspectors one requires and then choosing that many from the top scorers on this composite value.

6.6. Training

As in many jobs it has been shown that good inspector training produces significantly better

Table 2
Guide to the choice of inspector selection tests

Characteristic to test	Selection test suggested for consideration
Short term memory	WAIS Digit Span (WAIS = Weschler Adult Intelligence Scale) Memory Drum (e.g. Lafayette No. 23012)
Speed of information processing	WAIS Digit Symbol WAIS Arithmetic
Use of secondary cues	Task itself or facsimile of it
Finding shapes intermixed with other shapes	Embedded Figures Test
Long term memory	Difficult to find a suitable test
Ability to make comparative judgements	IPI Precision (IPI = Industrial Psychology Inc., N.Y.) IPI Dimension IPI Blocks IPI Parts Embedded Figures Test Richardson's Quotient of Mental Imagery
Memory for past things	Difficult to measure
Ability to concentrate	Introvert on Eysenck's Personality Inventory Combined score on: WAIS Digit Span WAIS Digit Symbol WAIS Arithmetic
Skill in selecting area to search	Task itself or facsimile of it Artificial task involving similar stimuli and scanning patterns Perhaps IPI Parts, IPI Precision or Harris Inspection Test
Ability to judge time used	WAIS Digit Symbol IPI Precision
Detect deviations from standard shape or form or appearance	Flanagan Inspection Test (from Science Research Associates Inc., Chicago)
Visual lobe size	IPI Blocks, Dimension or Precision Tachistoscope test as for Gallwey's Lobe Probability
Mental imagery	Sheehan's Short form of the Betts' Quotient of Mental Imagery Gordon's test of Mental Imagery Control Richardson's combination of the above two
Personality	Eysenck's Personality Inventory

Note: These classifications are approximate only, as it is not possible in many cases to discern exactly why a test is highly correlated with good performance.

results. Such a programme should include at least the following:
- Items inspected are examples of the real thing or exact facsimiles of them.
- Trainees participate actively in the process.
- Immediate feedback is given in detail on each action of the inspector.
- Correct learning is emphasised rather than trying to correct mistakes.
- Right or wrong items used in training are clearly identified.
- Mistakes are corrected immediately.
- Level of difficulty is easy initially and then raised progressively to the highest level.
- A progressive-part approach is used as depicted in Table 3.
- Features of the step above are performed in restricted time to learn time judgement.
- Practice is provided with finer and finer discriminations of deficiency to train judgement.
- Paired-associates learning is used to train for comparisons against standards.
- Practice is provided with feedback and a monetary incentive for good performance.

Such training programmes require extensive preparation of specialist materials, are rather time consuming to carry out, and take much longer than those usually employed. This may initially appear to be a disadvantage but data in the literature show that the benefits can be very significant over the long term due to savings in unnecessary rework and scrap.

6.7. Workplace design

In order to facilitate inspector comfort, and to militate against muckuloskeletal injuries, inspection workplaces should be designed to ensure good postures as for any other industrial workplace. Most of the normal requirements apply equally to those for inspection and reputable textbooks such as Konz (1990) should be consulted. Chapter 7 of Harris and Chaney (1969) addresses organisation, instructions, workspace design, and controls and displays. Other chapters deal with general aspects of the inspection job and give much useful information which is too extensive to attempt to summarise here. An example of some of the special features required for a particular inspection station is that of Astley and Fox (1975) for which some details are given in Section 7.2 of the case studies below. It is probably the most detailed example of what issues need to be addressed on this aspect of the job and how they should be approached.

7. Case studies

As an aid to practitioners some examples of the application of some of these principles are given below. In all cases they involve inspection of real products with real inspectors. There are many examples available in the literature and those given here are just typical examples.

7.1. Machined parts for an electric motor (Chaney and Teel, 1967)

Average rate of defect detection was about 40%, with individual levels in a range from 25% to 80%. Defects which were to be detected in the study included mislocated holes, defective hole threads, measurement of paralellism and concentricity, and dimensions that were out of tolerance.

Table 3
Features of the progressive-part approach to training

Step	Action performed
1	One part of one element is mastered
2	A second part of the same element is mastered
3	These two parts are synthesised by practice together until mastered
4	A third part of the same element is mastered
5	The third part is synthesised by practice with the first two until mastered
6	This is continued until the whole element is mastered
7	A second element is mastered in the same way
8	The first two elements are synthesised by practice together until mastered
9	Further elements are added successively until the whole job is mastered
10	Learning is reinforced at the start and the end of each part by information on what is acceptable and what rejectable

Procedure: Test pieces were made up consisting of two brackets and two supports. Each had 100 characteristics to be tested where 34 characteristics were defective.

Inspectors: There were twenty six males, each with several years of experience, and they were divided into four groups matched in terms of inspection performance. One group was the control and received no change, one group received training by the supervisor, another group was given visual aids, and the final group was given both training and visual aids.

Training: This was given by means of four one-hour sessions of lectures, demonstrations and discussion on precision measurement, thread gauging, how to operate specific pieces of equipment, and how to interpret drawings.

Visual aids: These consisted of simplified drawings of the parts with the appropriate dimensions, tolerances, and inspection criteria for each characteristic on the drawing to minimise calculations and references to other material. Similar items were grouped on a page without any information on characteristics that were not to be inspected.

Evaluation: One month after completing the training scheme the inspectors were tested again but on a different pair of test pieces with the results shown in Table 4 below. Although the improvement percentages are impressive it should be noted that even in the best case only 60% of the defectives were detected. The combined treatments gave a 450% increase in detection of mislocated holes and resulted in 100% detection of threaded hole defects. Cost savings for these specific improvements were not directly obtainable but they were part of a two year programme of improvements of ergonomics in inspection costing $50 000. The total programme resulted in documented savings of $200 000 at 1966 prices.

7.2. Rubber seals for automobile brake cylinders (Astley and Fox, 1975)

The job was to detect surface defects by means of twenty eight female inspectors at final inspection. The seals were classified as acceptable or as either of two types of reject. To improve performance a detailed study of the work design was carried out.

Illumination: Intensity was 495 lux versus a requirement of 900, contrast between the seals and their immediate surrounds was 1 : 15 against a requirement of 10 : 3, and at 21 the Glare Index was much greater than the recommended value of 10. Glare was reduced by (1) reducing transmissibility of the roof windows by using a coating or by eliminating the windows, (2) replacing the light reflector fittings with diffuse fittings, and (3) decreasing reflectance of the walls and ceilings by applying more light absorbent paint. Bench top reflectance was reduced from 0.8 to 0.01 by replacing the cream plastic surface with green baize which changed the contrast of object : surround to 10 : 11.6. Intensity was improved by providing local illumination to supplement the general illumination to the level required.

Magnifier: The seals were viewed through a flattish magnifier glass mounted on a tiltable frame. It distorted the edge of the image if two seals were viewed side by side, which precluded the preferred design of handling two at a time with one in each hand. The focal length of the lens induced an eye-seal position which meant that some viewers did not view a magnified image. A series of experiments was carried out which showed that performance was significantly better with this combination: lens of 450 mm focal length, magnification of 1.5 ×, the 24 W lamp supplemented by two 6 W tubes to give an intensity of 970 lux at the seals. By a number of detailed optical calculations it was found that the eye-to-object distance should be in the range of 200–300 mm.

Bench and chair dimensions: To hold the seals at the distance given in the line above, the upper arm had to be held extended well forward of the trunk with considerable flexion of the elbow, or the head

Table 4
Fault detection rate for the groups

Group	Before (%)	After (%)	% Improvement
Control	36	34	Not significant
Training only	30	41	+ 32
Visual aids only	37	54	+ 42
Training + visual aids	36	60	+ 71

needed to be inclined sharply forward. To contain the arm and head positions within comfortable limits, the arm would need to be supported at a point midway between the elbow and the midpoint of the forearm. The underside of the bench top had to be high enough to allow larger inspectors to sit with the lower leg vertical to ensure comfort when sitting for long periods. There had to be sufficient leg space below the bench top to provide enough clearance for the inspector's knees, lower legs, and feet. Seat height had to be such that the inspector could support her arms on the upper surface but still have enough thigh clearance between the underside of the bench top and the seat. It also could not be so high that the small inspector's feet could not reach the floor and so inhibit blood circulation.

Body dimensions: The equipment was developed to accommodate the 5%–95% range of body size data for adult British females. These were used in conjunction with a simple biomechanical model using rigid segments pivoted at the major joints. Data were culled from the literature for maximum possible angular movement at the joints and for the angles that people find comfortable to maintain for long periods. Particular attention was paid to meeting dimensional needs of both the small and the large woman by plotting out the heights, reaches and joint angles.

Evaluation: The effects of these changes were evaluated after several months. The fault detection rate had gone from 92% to 90.5% and the percentage of good parts accepted from 97% to 97.2% but with 25% fewer inspectors. The seven inspectors released from this job moved to other inspection work in the company with annual savings of about £23 000.

7.3. Integrated circuit chips (Moclair and Gallwey, 1997)

There were approximately 20 different defect types to be detected by means of viewing through a microscope unit which allowed the inspector to choose whatever magnification level was necessary.

Materials: A batch of 50 "wafers" was specially engineered for the study to reproduce the range of defects normally encountered by the inspectors. The five most frequently occurring defects were reproduced ten times. On each of these there was a second defect. Thirty of them had the second five most frequently occurring defects reproduced six times each. The remaining ten defect types were reproduced twice on the last 20 wafers.

Inspectors: There were twenty four volunteers consisting of sixteen males and eight females. Ages ranged from 21 to 27 with a mean of 24, and all had normal or corrected vision. They all had post high school education. There were shift groups of 9 and 15 and all did the task twice, both on night shift and then on day shift.

Search performance: The effect on search performance of inspecting for only the five most frequently occurring types of defect was examined, versus searching for all 20 types.

Decision aids: Scanned images were provided of the minimum acceptable quality level of the defects by means of a 133 MHz Pentium PC with a fourteen inch colour monitor.

Selection tests: To reduce the large effect of individual differences, scores were obtained for each inspector on the Embedded Figures Test and on the WAIS Digit Span, Digit Symbol and Arithmetic tests. These were then used as covariates in the analysis.

Evaluation: The WAIS Digit Symbol and Arithmetic scores were significantly correlated with search time and there was a significant difference in search times between five and 20 types. The decision aid produced significantly greater accuracy and significantly different decision times, particularly with regard to 20 types of defect. But for decisions on the five types the results were confused by differences due to order effects between the day shift and night shift results. The WAIS tests had no significant effect in regard to the decision task, and scores on the Embedded Figures Test had no significant relationship with either task.

References

Astley, R.W., Fox, J.G., 1975. A study of human factors in inspection of rubber goods. In: Drury, C.G., Fox, J.G. (Eds.), Human Reliability in Quality Control. Taylor & Francis, London, pp. 253–272.

Chaney, F.B., Teel, K.S., 1967. Improving inspection performance through training and visual aid. Journal of Applied Psychology 51, 311–315.

Gallwey, T.J., 1998. Evaluation and control of industrial inspection: Part II – The scientific basis for the guide. International Journal of Industrial Ergonomics, 22(1–2): 51–65 (this issue).

Harris, D.H., Chaney, F.B., 1969. Human Factors in Quality Assurance. Wiley, New York.

Konz, S., 1990. Work Design: Industrial Ergonomics. 3rd ed., Publishing Horizons, Scottsdale, Arizona.

Meddis, R., 1984. Statistics Using Ranks: A Unified Approach. Blackhall, Oxford, UK.

Moclair, F.M., Gallwey, T.J., 1997. Memory load and decision aids in inspection of silicon wafers. Proceedings of 13th Congress of the International Ergonomics Association, Tampere, Finland, June 29th–July 4, 1, 226–228.

Evaluation and control of industrial inspection: Part II – The scientific basis for the guide[1]

Tim J. Gallwey

Department of Manufacturing and Operations Engineering, University of Limerick, Limerick, Ireland

1. Problem description

Rejection of acceptable objects, and acceptance of rejectable objects, are well known problems in all types of activities that contain some element of inspection. This part of the guideline provides the rationale for evaluating these errors and their associated effects, and for interventions that will help to reduce the risk of their occurrence.

An overview of the inspection problem is provided. It shows the types of problems encountered, highlights the types of corrective measures that have been tried, and reports the successes or failures achieved with them. To improve these processes it is necessary to obtain a good scientific understanding of the steps entailed, and the ways in which they can be modelled and evaluated. These will provide a quantitative basis for the evaluation of various inspection situations and countermeasures, and then propose some control actions.

2. Scope of the problem

Various researchers have investigated the effectiveness of inspection. These are summarised in Table 1 and show a sorry picture which belies somewhat the rosy view of some people involved in it. Even faults shown in radiographs of aircraft structural components have been missed during routine maintenance, so the problem is not confined to manufacturing industry. Clearly something needs to be done to investigate the problem in depth and to see whether or not it can be improved.

As a result of their improvements Harris and Chaney (1969) claimed an annual savings of $60 000 in one job, Drury and Sheehan (1969) saved $13 000 in another. Elsewhere the cost of quality was reduced from 5.5% of sales to 3.4% with a cumulative contribution to earnings of $75 million over 10 years (Industrial Engineering, 1978). Rae (1979) reported that the losses in British industry due to non-conformance of product were estimated to be at least £10 000 million. The US Office of the Secretary of Defence reported a rejection rate

---

[1] The recommendations provided in this guide are based on numerous published and unpublished scientific studies and are intended to enhance worker safety and productivity. These recommendations are neither intended to replace existing standards, if any, nor should they be treated as standards. Furthermore, this document should not be construed to represent institutional policy.

The following individuals participated in the discussion of the earlier version of this guide. Their suggestions (written or verbal) were incorporated by the authors in this version: A. Aaras, Norway; J.E. Fernandez, USA; A. Freivalds, USA; M. Jager, Germany; S. Konz, USA; S. Kumar, Canada; H. Krueger, Switzerland; K. Landau, Germany; A. Luttmann, Germany; A. Mital, USA; J.D. Ramsey, USA; M.-J. Wang, Taiwan.

Table 1
Summary of the results of investigations into the effectiveness of inspection

Reference	Thing inspected	Score (%)	Measure used
Tiffin and Rogers (1941)	Sheets of tin plate	79	Detection of surface defects
Hayes (1950)	Piston rings	10	Agreement of verdict between inspectors
		23	Change from first to second inspection
Jacobson (1952)	Wiring and solder	80	Average inspection accuracy
		45	Accuracy of worst inspector
R.M. Belbin (1957)	Ball bearings	63	Detection of faults
George (1963)	Ceramic articles	36	Consistency of decisions
Fox (1964)	Coins	55	Fault detection rate
Carroll (1969)	Printed circuit boards	73–95	Fault detection rate
Belbin (1970)	Tin cans	58	Fault detection rate
Astley and Fox (1975)	Rubber seals	92	Fault detection rate
Yerushalmy (in Wiener, 1975)	X-ray plates	20	Faults missed by radiologists
Mills and Sinclair (1976)	Knitwear products	50	Fault detection rate
Thornton and Matthews (1982)	Sauce pans and dutch ovens	62–75	Fault detection rate

of 5% on the products they purchased (Morawski, 1978); applied to the US Gross National Product at that time this would amount to $40 billion per year. Not all of these losses were due to failures of inspection, but they do give some idea of the potential magnitude of the effects of the problems highlighted here.

There are suggestions from time to time that these shortcomings can be overcome by what is referred to as "automation", whereby machines will take over the job. Chin (1982, 1988) has provided extensive bibliographies on this subject. Chin (1988) described some 660 papers on automated visual inspection, grouped into thirteen areas of activity. The bulk of them related to applications in the electronics industry, with little information on their success. Also, there was no comparison of their performance with those using human inspectors. From this, automated systems appear to be well suited to situations where exact geometrical shapes are involved e.g. on printed circuit boards. He reported that there was litttle research on the inspection of solder joints, and highlighted the difficulty of it due to the three-dimensional shape of the surface. It is particularly difficult due to involving such aspects as brightness, size, volume, surface area, and surface curvature, combined in the one job. In regard to the computational needs of computer vision, processing speeds of at least one billion operations per second are required.

Drury and Sinclair (1983) compared human inspection performance on metal cylinders with that of a machine. The inspectors were significantly better. Although the machine could locate most of the faults, human decision making on classifying them as acceptable or rejectable was more sophisticated. Persoon (1988) reported that automated systems that were developed to replace human inspectors were removed at times from production because of their poor performance. Hou et al. (1993) looked into the feasibility of hybrid systems which combined humans and machines to do the inspection. There were four alternatives: (1) Purely human inspection, (2) Computer search with human decision, (3) Computer search with combined computer and human decision making, and (4) Purely automated inspection. The results showed that any system which incorporated human inspectors outperformed the automated systems (Table 2), although the differences were quite small.

Despite increases in automation, figures are regularly produced to show that even in the US some 75% of manufacture is by batch production. Also, due to the mechanics of the process described in Section 3.1 of Part I (Gallwey, 1998), it is very difficult (if not impossible) to design and develop

Table 2
Mean sensitivity values for false alarms and miss rate of alternative systems

System	Mean
Purely human inspection	0.9494
Computer search – human decision	0.9573
Combined computer – human decision making	0.9291
Purely automated inspection – Template	0.8949
– Neural-net	0.8330

inspection machines. Humans can do a reasonable job right now. Affecting all these points is the fact that quality standards are rising all the time due to the rising expectations of the public, the effects of competition, and the pressure from consumer groups and regulatory agencies. The trend is towards more inspectors, while the number of shop floor workers decreases with automation. The net result is that inspectors are becoming an increasingly larger proportion of the total labour force and therefore increasingly important.

Simple common sense remedies are not available or appropriate, and traditional industrial engineering measures such as incentive payment schemes have not been successful. The only realistic avenue available is to conduct meaningful research to find out what the job entails; then apply the results to improve the design of the job and the work system.

3. Background

3.1. Inspector performance

The most commonly used measure is the probability of correct rejection (P_2), also called detection performance. But, on its own, this can be misleading since, by rejecting everything, a high value can be obtained at the expense of an increase in false rejects. So it is necessary to measure also the probability of correct acceptance (P_1); its complement $(1 - P_1)$ is the probability of false alarms. These measures can be combined by plotting P_2 and $(1 - P_1)$ on the y and x axes respectively of a unit square to give the Receiver Operating Characteristic (ROC) curve; see Fig. 1. Ideally P_2 should be

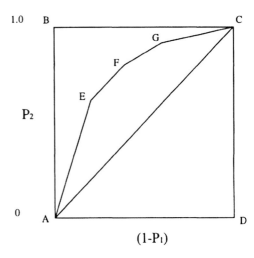

Fig. 1. Receiver operating characteristic (ROC) curve.

1.0 and $(1 - P_1)$ should be zero so that perfect performance is represented by the point B. Alternatively, pure chance performance will give constant values of 0.5 for both P_1 and P_2 which is represented by the diagonal AC. In practice, real performance is somewhere between these extremes (e.g., AEFGC). Hence the magnitude of the area of AEFGCD varies between 0.5 and 1.0. The larger it is the better is the inspector's performance, i.e., this gives a measure of the degree of difficulty of this task for this inspector.

The ROC curve above provides an overall measure of the inspector's classification performance. But it also can be used to measure the decision performance on its own, if P_1 and P_2 are for the decision part of the job only. In that case, it constitutes the general case for the Theory of Signal Detection (TSD) as it is nonparametric. The advantage of TSD is that it provides a measure of the difficulty of the decision separately from the measure of the bias of the inspector towards deciding reject or accept e.g. a "strict" inspector will be reluctant to accept an object unless it is very clearly non-deficient. The difficulty with this nonparametric TSD is that there is no generally approved measure of the bias. Thus it is more common to use the parametric form, even though it is a special case.

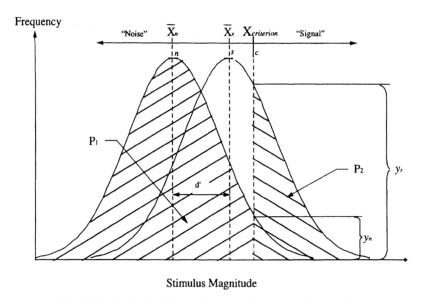

Fig. 2. Distributions and terms in theory of signal detection (TSD).

In parametric TSD, it is assumed that the stimulus magnitudes of the acceptable and rejectable objects form distributions; the task is analogous to detecting a telecommunications signal in a background of noise, where the noise comes from the extraneous features of the object described earlier in Part I; see Fig. 2. For convenience, it is assumed also that both distributions are Normal and that they have the same variance although this is merely for simplicity and is not essential to the theory. To simplify things further, it is assumed that the distributions are the standardised Normal with a variance of 1.0. In making a decision, it is assumed that the inspector chooses a criterion value of the stimulus magnitude and, if the stimulus is greater than this, the decision is that it comes from a signal (i.e., rejectable object). If less than this, the decision is that it comes from the noise, i.e., an acceptable object. The difference between the means of the two distributions represents the difficulty of the decision task, and it is measured in standard deviation units from the standardised Normal and is labelled as d'.

From Fig. 2, as the stimulus magnitude increases, the probability that it comes from a rejectable object (signal) increases while, at the same time, the probability that it comes from an acceptable object (noise) decreases. Hence these probabilities provide a direct measure of the inspector's bias. They are found from the ordinate values of the standardised Normal distribution. The bias measure then is defined as the ordinate of the signal distribution at the criterion value, divided by the corresponding ordinate of the noise distribution, to give a likelihood ratio, which is labelled as Beta.

In practice it is not necessary to go into the details of the actual distributions since the values of P_1 and P_2 can be obtained directly from the data collected, which can be arranged in a confusion matrix as shown in Table 3. From Normal distribution tables, P_1 and P_2 are used to obtain the z values that correspond.

$$P_1 = d/(b + d), \quad P_2 = a/(a + c),$$
$$d' = z(P_1) + z(P_2),$$

Beta = (Normal ordinate corresponding to P_2)/ (Normal ordinate corresponding to P_1)

Hence d' can be calculated. Similarly, P_1 and P_2 are used to obtain the corresponding ordinate values (Y_s and Y_n), and hence to calculate Beta. Once these values are available it is possible to determine whether a difference in detection performance is due to different degrees of difficulty

Table 3
Confusion matrix for the number of different types of decision of the inspector

Inspector's decision	True state of object inspected	
	Rejectable	Acceptable
Reject	a	b
Accept	c	d
Totals	a + c	b + d

among the people involved, or due to differences in their biases when making decisions. Differences in difficulty can be addressed by providing aids such as magnification, better lighting, or more time. Differences in bias can be addressed by better training or by better communication about the priorities of the organisation. A good illustration of how d' and Beta change with the level of defectiveness is given in Adams (1975).

3.2. The social environment

Some of the early work concentrated on this. McKenzie and Pugh (1957) found that wherever the functions of production and inspection were separated, there was a consistent pattern of conflict between them. However, Jamieson (1966) showed that when the inspectors were isolated there were fewer wrong classifications. He concluded that this was due to their greater group identity, and greater freedom to reject. To look at relations within a group Seaborne (1963) used a card sorting task. Three people worked together for two days, and then, for two further days, were joined by a fourth. Over all four days the latter received cards with a defect rate half that of the other three. In both of the test days, the test subject gradually raised the reject rate until it was almost the same as that of the other members. It is a good example of an inspector changing their Theory of Signal Detection (TSD) bias or criterion.

3.3. Provision of information

The simplest type of information is knowledge of results (KR). Wiener (1963) obtained superior detection performance on a vigilance task with it than without it. In an industrial inspection situation, Drury and Sheehan (1969) arranged for "oversight" inspectors to sample the products prior to 100% inspection. Their results were fedforward to the 100% inspectors i.e. they were alerted to the types of deficiency to expect (and their frequency) in the batch on its way to be inspected. This increased fault detection from 82% to 93%. On the inspection of glass products, Drury and Addison (1973) moved the sampling inspectors, in the finished goods store, back in the process to be closer to the 100% inspectors. This speeded up feedback of results to these inspectors (i.e. told them how well they had performed on a previous occasion); this increased d' from a mean of 2.50 to 3.16. Providing information to inspectors undoubtedly helps them to perform better.

3.4. Pacing and rest pauses

Colquhoun (1959) looked at the effect of rest pauses on a machine-paced job. He had two groups of inspectors working for an hour where the test group had a five minute break after thirty minutes. The results showed that the test group missed an average of 9.4 faults in the first thirty minutes and 9.9 in the second thirty; the others missed 17.6 and 44.1 respectively. In his coin inspection task Fox (1964) found that 15 minutes was the maximum length of time the inspectors could work before their detection efficiency began to decline. This was fortunate as it coincided with the length of time between their refillings of the work hopper.

Fox and Richardson (1970) obtained 34% detection with machine-pacing versus 70% with self-pacing. Three different speeds of machine-pacing were used by Smith and Barany (1970), and both types of inspection error increased directly with speed. On a simulated inspection task McFarling and Heimstra (1975) obtained a higher percentage of defectives detected in the self-paced condition. Inspection of chicken carcasses formed part of the Chapman and Sinclair (1975) study; from their findings they recommended that the inspectors should be changed to other work after 40 minutes of inspection work, to avoid a performance decrement. Coury and Drury (1986) had six pacing conditions in a complex decision-making inspection task. These had no effect on the probability of correct decisions but did affect response times significantly. In regard to the time available for inspection Geyer and Perry (1982) compared d' values obtained with viewing times of 1, 2 or 4 seconds. As time increased there was an approximately linear increase in d' from approximately 1.0 to 2.0 in one condition, and from about 1.7 to about 2.7 in another condition. It is reasonably safe to conclude from all these findings that self-pacing is much to be preferred in most cases (if not all) and that rest pauses (after 15–40 min) definitely help inspection performance.

3.5. Moving or stationary objects

Passing objects past the inspector on a conveyor is attractive to many in industry. Williams and Borow (1963) investigated angular velocities of 2, 4, 8, 16 and 31 degrees per second about the eyes of their subjects. Up to 8 degrees per second the search times were not significantly different, but performance deteriorated at the higher speeds. To simulate the inspection of sheets of steel Moraal (1975) used photographs in a slide projector for a static display. He used a movie projection of them to represent movement at 60 metres per minute. These two tasks gave d' values of 1.6 and 0.9 respectively. Chapman and Sinclair (1975) experimented with jam tarts moving at angular velocities of 3, 5.5 and 8 degrees per second. Correct rejection scores were 85%, 77% and 63%, respectively. Belbin (1970) noted that the eyes move in a series of jumps or saccades during which virtually no information is taken in (if any); the object on a conveyor moves some distance without being seen. He concluded that this is "...one of the least efficient forms of inspection". The data presented here support his contention and it is hoped that industry will see fit to abandon this type of inspection in the near future.

3.6. Reference standards

Kelly (1955) used paired comparisons to rank ten TV panels in order of defect seriousness. She placed them above the production line, to give a direct visual basis for decision comparisons (making it into a relative judgement task); this increased inspection accuracy from 51% to 90%. McKennel (1958) examined the task of grading wool fibres and found that the mean errors of the inspectors were consistently less when relative judgements were used instead of absolute judgements. In support, Van Cott and Kincade (1972) demonstrated that where absolute judgements are required humans can reliably make some 3–5 levels of discrimination; with relative judgements the number of levels can be in the hundreds (see Table 4 below) (Moclair and Gallwey, 1997).

3.7. Inspection strategies and job design

Colquhoun (1961) found that a higher proportion of faults was detected when a higher proportion of the batch was faulty. This "defect rate effect" has been explained as the finding of a defect acting as an arousal mechanism for the inspector. Therefore Colquhoun suggested a first-stage inspector to reject all items that are less than perfect, and then have these sorted subsequently by a second-stage inspector. As far as can be ascertained there has not been any experimental evaluation of this idea. Harris (1968) proposed the use of sampling plans (instead of 100% inspection) to give fewer but more intensive inspections as the quality level rises, but

Table 4
Comparison of relative and absolute judgements (Van Cott and Kincade, 1972)

Sensation	Number discriminable with relative judgements	Number discriminable with absolute judgements
Brightness	570 intensities of white light	3–5 with white light
Loudness	536 intensities at 2000 Hz	3–5 intensities of pure tones
Vibration	15 amplitudes on the chest	3–5 amplitudes

gave no data. However, Harris and Chaney (1969) reported that, on critical defects, the proportion detected rose from 45% after one inspection to about 93% after six inspections.

They also compared successive inspections of small areas for all defect types, against inspection of the whole product for one type of defect, and then the whole product for a second type, and so on. They reported that the second strategy "... was found to be 90% more effective". Purswell and Hoag (1974) reported that Waikar (1973) found an improvement from about 75% detection to about 80% when two inspectors made individual decisions rather than having one inspector. It increased to about 100% with three inspectors doing so. Su and Konz (1981) compared three strategies: (1) inspect for one defect type at a time, (2) inspect for all five defect types at a time, (3) inspect for similar items simultaneously. Condition (1) took significantly longer (22–87%, depending on condition). Decision errors were 16% less for this condition, but only when the decisions were "difficult" (Moclair and Gallwey, 1997). Whether or not these ideas are practical to implement is not clear, and there appears to be no evidence in the literature of their implementation in actual inspection jobs.

Thornton and Matthews (1982) changed the job by giving the inspectors a subsidiary task. They were inspecting sauce pans and dutch ovens, and the extra task was to sort the pans into three types, depending on the number and type of handle attachment studs. The fault detection rate rose from 65% to 89% during the experiment and was at 88% three months after the experiment ended. Savings were estimated to amount to $100 000 Canadian at 1981 prices, for zero cost.

3.8. Lighting and glare

On the inspection of "metal hooks" Drury and Sheehan (1969) tried illumination at 1950 lux on the first day, 975 lux on the second, 650 lux on the third, and inspector's chosen level on the fourth. There was no significant difference in performance level. But, in their study of inspection through microscopes, Smith and Adams (1971) reported a significant effect of illumination but did not measure the actual levels. Faulkner and Murphy (1973) reviewed lighting and pointed out that absolute levels are much less important than the use of directional light; it was demonstrated in two striking examples. (Their paper also gives many very useful ideas on how to provide appropriate lighting for a variety of specialist applications.) But Astley and Fox (1975) found it helpful to increase the level from 460 lux to 900 lux for the inspection of black rubber seals. They reduced the glare from 21 on the IES Glare Index (compared to a recommended maximum of 10 (IES, 1973)), by replacing the white background with one made from green baize, and they made the light directional. The exact effect of these particular changes could not be quantified as they were part of a whole series of changes. Chapman and Sinclair (1975) found illumination from 86 lux to 560 lux on jam tart inspection. They recommended that it be increased to comply with the IES Code and made directional, but were not able to quantify the effects individually of these changes.

3.9. Defect rate

Colquhoun (1961) had people inspect a set of discs, mounted on the six faces of a turret which rotated in a horizontal plane, and presented them in sequence. When the probability of a defect increased from 0.08 to 0.50, the detection efficiency rose from about 62% to about 79%. Supporting data were provided by Harris (1968) who obtained detection rates of 83%, 71% and 57% for fault probabilities of 0.16, 0.01 and 0.0025, respectively. Further support came from Fox and Haslegrave (1969). Their batches of screws had fault rates of 0.005 or 0.05 and gave a similar effect, provided the task was self-paced. Wallack and Adams (1969) found no difference in d' at 5%, 15%, 25% and 35% defective; this was probably because it measured the inspectors' decisions rather than their visual search processes. On subsequent analysis (Wallack and Adams, 1970) they found that the probability of correct rejection (P_2) increased as the percent defective increased. Drury and Addison (1973) experienced a decrease in detection rate as the fault rate decreased. The general picture is that detection rate decreases as fault rate decreases.

3.10. Inspection time distributions

Bloomfield (1972), amongst others, showed that search time was exponentially distributed, and Drury (1978) demonstrated how the cumulative exponential could be integrated into an overall inspection model. Gallwey (1980) obtained approximately Normal distributions of times when no fault was found, and Log-Normal when faults were found (i.e., search combined with decision). In a later experiment Gallwey (1982a) measured search and decision times separately on the same type of material as before. Decision times for the two types of fault were very close to an exponential distribution, but the Weibull shape parameters were 1.19 and 1.40 for the search times and 1.52 and 1.80 for the inspection times (i.e., totals of search time and decision time). These compare to a value of 1.00 for the exponential and 2.00 for the Log-Normal. More investigation is needed to clarify these issues before valid simulation models of the process can be built.

3.11. Width of vision

One investigation of the effect of this was by Erickson (1964). He used displays with 16, 32 or 48 objects from left to right which gave eccentricity angles of 3.6°, 4.8° and 6° about the visual axis. Peripheral acuity measured at the two smaller angles was significantly correlated with search times in displays of 16 or 32 rings and for 16 or 32 "blobs". At 6° almost all measures were not significant. Johnston (1965) used displays that subtended angles of 11.25°, 22.5° and 45°, and she measured the width of visual field of her subjects using a Brombach perimeter. The latter results were classed as large (1190–2080 square degrees) or small (638–1173 square degrees); those with the large field had a more effective search performance.

Bellamy and Courtney (1981) found a significant relationship between the area of the visual lobe of the inspector, and search performance. Gallwey (1982a) did something similar to Johnston by presenting 35 mm slides tachistoscopically. They consisted of a plotted search pattern, containing ten different punctuation marks, with a Left parenthesis situated at multiples of five spaces along the horizontal centreline, left or right of the central point. The probability of correct detection in this test was significantly related to the probability of wrong classification, and to the probability of search errors, in an inspection task. Courtney and Shou (1985) measured the visual lobe for six subjects searching an array of the letter X with a letter V embedded in it. Lobe measurements were made on eight axes passing through the fixation point and for two subjects these were mapped exhaustively. The boundaries were very irregular with some regions of insensitivity with the lobe. Both lobe shape and area were suggested as being important in visual search. The irregular boundaries and areas of insensitivity were put forward as explanations for the fact that some subjects do not locate a target despite repeated scanning.

3.12. Fault discriminability

Crossman (1955) devised his "confusion function" as a measure of discriminability. It was applied to a visual search task by Bloomfield (1972) and gave quite a good linear relationship with the fastest times obtained by each subject (as an estimate of reaction time). Engel (1974), in his search task, found a linear relationship between visual lobe size, and the logarithm of the difference between target and background diameters. On a CRT display, Monk (1976) used different luminances of his targets instead of size; there was a highly significant effect on Geometric Mean Search Time but he presented no mathematical formulation of the relationship. A similar device was used by Engel (1977) but he measured visual lobe size; it gave a linear relationship with the absolute difference in size between target and background spots. Gallwey and Drury (1977) found that the mean of the median search times was related in a linear fashion to Crossman's confusion function; it was calculated from the target/background differences in size. The equation was different for targets larger than the background than for targets smaller. Gallwey (1978) extended this to two different targets embedded in a mixture of ten different background items; there was a linear relationship between the logarithm of the search time and the logarithm of the target height.

3.13. Training

Eunice Belbin (1955) had objects on which the errors had been clearly marked up, and used them to build up a mental scale of "rightness" and "wrongness" before starting work on the job. The task was neither very easy nor very difficulty at the start and she aimed to achieve "correct learning" rather than permit a trial and error process. R.M. Belbin (1957) called for the grading of borderline defects to be identified by a code for training, and for critical defects to be identified. Chaney and Teel (1967) trained people for a measuring task. One group received no training, a second one received training only, a third received visual aids only, and the fourth received training and visual aids. The changes in correct detections for these groups were: −2%, +32%, +42% and +71%. Czaja and Drury (1981) applied and extended the ideas of Eunice Belbin by using a progressive-part training approach. They compared active with passive learning for young (20–35 years old), middle aged (36–50) and old (51–65) subjects. Active training gave much more successful inspection performance with significantly fewer search and decision errors; the advantage was consistent across the age groups. But, to ensure that there is good transfer of training, the conditions must be as similar as possible to those on the job. Gramopadhye and Wilson (1996) showed that subjects trained in quiet conditions, had lower d' scores when working in noise, than did those who were trained in a similar noise condition.

3.14. Memory

For information processing, the part of memory that is of interest is Short Term Memory (STM). Miller (1956) devised his "magical number seven, plus or minus two" which gives an approximate idea of STM capacity. But he also introduced the idea of "chunking", whereby several items can be grouped into a "chunk", and then STM capacity is seven or so of these. Yntema and Shulman (1967) required their subjects to recall the state of the variables for different numbers of categories. They had one category for each of six variables, one for every two variables, one for each set of three, and one for all six variables. The corresponding response errors were 25%, 39%, 44% and 54%.

Fitts and Posner (1967) state that running memory (which is probably the appropriate measure for inspection tasks) is limited to about 3 or 4 items. Gould and Carn (1973) used a visual search task where there had to be either 2 or 4 occurrences of a fault before an item could be rejected. They compared this to the case where only one had to be found. The subjects performed for 30 days and data over the last five days were analysed. There was a deterioration from 1 to 2 to 4 and the subjects reported that they could keep track of only two (sometimes three) different faults at once, even after 30 days of practice. Ainsworth (1982) studied the effect of searching for 1, 2, 3, 4 or 5 types of fault in the inspection of plastic parts moving on a conveyor. His analysis used d' as a measure and it fell successively from 2.7 to 2.5, 2.3, 1.9 and finally to 1.6 respectively (approximate figures), but this may contain a mix of the search and decision parts of the task. Gallwey and Drury (1986) increased the number of fault types, for which their subjects inspected, from 2 to 4 to 6. Response times increased notably and there was a significant increase in search errors. Decision and size judgement errors were unchanged. A strong case is made to limit the number of types actually searched for, to a figure between 3 and 5 (e.g. 5 was used by Moclair and Gallwey, 1997).

3.15. Selection

There is a school of thought that selection tests should not be used for choosing between potential new employees. The results of such tests cannot determine exactly whether or not a particular person will perform successfully on a particular inspection task. Hence there can be legal difficulties in relation to the fairness of using such tests. However, an employer may already have a number of people on the payroll when an inspection job becomes vacant. Before investing a lot of time and money in training such a new inspector it is advisable to take some measures to improve the probability that they will turn out to be able to do the job well.

For this purpose Harris (1964) devised a pencil-and-paper test which used symbols of items found on electronic circuit diagrams. It was validated on

65 inspectors who did four different kinds of jobs. Significant validity coefficients were obtained on three of the jobs. Reliability coefficients were calculated, by comparing scores on these jobs with scores on the two sheets of the test, and gave values of about 0.85. In Harris and Chaney (1969) two more groups of inspectors were quoted as having scored significant validity coefficients on the test. Wiener (1975) made a comprehensive review of inspector selection, and concluded that there was little "... in the current state of the art in inspection or laboratory studies of vigilance that can be put to work in the industry without further research". Gallwey and Czaja (1981) used the Harris Inspection Test in two quite large experiments. They were unable to obtain significant effects in relation to a variety of measures. It appears to be a test that works well retrospectively on existing experienced inspectors (where there has probably been some weeding out and self selection), but not so well proactively on identifying potential new inspectors. Gallwey (1982b) used it as one of a large battery of selection tests. The best two were a simplified facsimile of the actual task, and the combined score on the three subtests of the "Attention-Concentration" part of the Weschler Adult Intelligence Scale (WAIS) (see Table 2 of Part I). Of the other tests used, good results were obtained for lobe probability, the Embedded Figures Test of Witkin et al. (1971), and the Quotient of Mental Imagery of Richardson (1969).

Schwabisch and Drury (1984) examined the effect of cognitive style on performance in visual inspection. They used the Matching Familiar Figures Test (MFFT) to divide their thirty-nine subjects into: reflectives (longer times, fewer errors), impulsives (shorter times, more errors), fast-accurates (shorter times, fewer errors), and slow-inaccurates (longer times, more errors). The accurate groups (reflectives and fast-accurates) were significantly faster than the inaccurate groups (impulsives and slow-inaccurates) in the detection of certain flaws, and their size-judgement errors were fewer. But the inaccurates made fewer search errors.

Wang and Drury (1989) used three tasks consisting of a computer generated pattern of symbols, an unfilled circuit board, and a Colour Video Comparator for filled circuit boards. The skills and abilities important in the search subtask were found to be different for the different tasks i.e. they were task-specific. They also used a number of tests as possible predictors of inspection performance. Significant results were obtained for the Classifying Test, the WAIS digit span and digit symbol, IPI judgement, line and circle judgement tests, visual memory and PSI-memory tests, visual speed and accuracy, perceptual speed, and the Embedded Figures Test. They advised that it is first necessary to find out the specific mental requirements for a task, then to select a set of valid test items to measure them, and then devise a test battery.

3.16. Maintenance inspection

All forms of maintenance work involve large amounts of inspection of surfaces for wear, surface damage, cracks and discolouration, as well as listening for irregular or wrong noises, or rough running. Therefore they should benefit greatly from the findings of research into inspection work but there is little evidence of this in the scientific literature. A notable exception is that of Drury on aircraft inspection work such as reported in Drury and Lock (1992). Most of that report concerns such issues as socio-technical systems interventions, training, information systems design and error control. There is scope for a lot more.

3.17. Inspector error

Numerous examples of inspector errors have been cited in the literature above. An alternative source of information might be the research literature that relates to human error in general, such as the work of Rasmussen and others. However there does not appear to be anything in the literature to relate that work to the type of task covered here; for that reason those studies have not been included here. Many of the studies that are quoted here have been aimed at finding ways to reduce error magnitudes which strategy, while clearly necessary, has limited potential. When the errors have been reduced to the irreducible minimum (which is somewhat greater than zero), it is necessary to try to compensate for them. Drury (1978) has shown how this can be done in the design of statistical sampling plans.

These plans are devised by deciding on the level of producer's risk and consumer's risk, at the fraction nonconforming (p') of the product that corresponds to the acceptable quality level, and the rejectable quality level, respectively. For these two defined points on the Operating Characteristic curve, a particular plan is chosen according to other criteria. However, this procedure assumes that the inspector always makes perfect classifications of the product, and so the plans have to be modified as follows. If P_1 is the probability of correct acceptance of a good item, then $(1 - P_1)$ is the probability that it will be rejected. Similarly, if P_2 is the probability of faulty product being rejected, there is a probability of $(1 - P_2)$ that it will be accepted. Hence we can draw up the confusion matrix of Table 5. By totalling up its cell values the probability that material will be rejected is given by the effective fraction defective (p'_{eff}) where

$$p'_{eff} = (1 - P_1) + p'(P_1 + P_2 - 1).$$

If the sampling plan is devised using this p'_{eff} value, instead of p', valid plans will result. Otherwise, the actual producer's risk and consumer's risk that result, may be markedly different in practice from those used to design the plan.

3.18. Vigilance effects

A large number of papers on vigilance can be found in Mackie (1977). But Sinclair (1978), in an omnibus review of visual inspection research, stated that despite at least 1500 experiments concerned with human behaviour in vigilance and attention situations, there was still no generally accepted selection test for inspectors. He also criticised the nature of many of the experiments in terms of relevance to inspection. Drury (1980) also has cast doubt on the relevance of classical vigilance experiments for inspection work, pointing out that the classical conditions used are:
- The task is passive.
- The information rate is low.
- The environment is dull.
- Performance is prolonged.

It is questionable whether any of these conditions apply to typical industrial inspection jobs. Even so the idea of the "vigilance decrement" has taken hold (i.e., that performance deteriorates with time at work), but not without disagreement from others. In particular, Craig (1987) has argued persuasively that such a "decrement" is nothing more than the subjects adjusting to a much lower frequency of "signals" than they expected, what he calls probability matching, and his TSD data support this view. Matthews et al. (1993) performed three studies of individual differences in the performance of high rate vigilance tasks. Although they found some significant relationships to visual search performance they reported few correlates of decrement in perceptual sensitivity. At present it appears to be an open question.

3.19. Age effect

On the reading of micrometers, Lawshe and Tiffin (1945) found no correlation between age and accuracy, nor did Evans (1951) on the same task. But Jacobson (1952) found an increase in the accuracy of electronics inspectors up to the age of 34. Older inspectors in the study of Jamieson (1966) scored significantly fewer errors on electro-magnetic switches, and on faulty joints and connections on telephone exchange racks. Sheehan and Drury (1971) obtained a reduction of d' from about 3.1 at age 30 to about 2.4 at age 65 for the situation as found. After they introduced feedforward the values were raised to about 4.3 and 3.7, respectively. Czaja and Drury (1981) put their subjects into three groups: young (aged 20–35), middle (36–50) and old (51–65). Search time was the variable most affected

Table 5
Confusion matrix for inspector classifications of good and faulty product

Inspector classification	Good product	Faulty product
Accept	$(1 - p')P_1$	$p'(1 - P_2)$
Reject	$(1 - p')(1 - P_1)$	$p'P_2$
Totals	$(1 - p')$	p'

by age with mean times of 21.3 s, 25.5 s, and 33.4 s for the young, middle and old groups. It is safe to conclude that inspectors should be young for best performance.

4. Need for further research

Many of the major issues have now been researched to a fair extent (see also Gallwey, 1998), but it could be argued that, in most cases, greater depth is required to achieve the degree of generality and confidence with which one would be happy. In particular, eye movement research has been somewhat limited, due to the equipment and analysis difficulties that it entails; more work in this area should reveal a better understanding of the mental processes and performances of inspectors. For example, some results show that inspectors who make faster eye movements, also make fewer search errors. This apparently anomalous finding warrants further detailed study. Extensive work in eye movement research appears to have been done in the military domain, but very little of this has found its way into the publicly available scientific literature, and presumably it will remain that way, for some time at least.

The big area for future growth is that of maintenance and condition monitoring, and their associated techniques such as Non-Destructive Testing (NDT). The phenomenon of the "ageing aircraft" has become prominent in recent years; scares about aircraft failures in flight suggest that there are serious shortcomings in the existing procedures. These situations are manifestly different from those of traditional factories, and repetitive manufacturing, and will require extensive research investigations. Whatever applies to aircraft maintenance also applies, by and large, to other maintenance work, such as power generating plants (especially nuclear), chemical works, manufacturing plant and surface transport. As consumers demand higher levels of service and reliability, pressure is likely to be focused on these; this will require an extension and enhancement of the work described above. One area that appears to be growing in application is that of Acoustic Emission (Bray and McBride, 1992; Chuang and Chang, 1988; Martin, 1980; Spanner and McElroy, 1975), which uses sound as a means of NDT. It raises issues such as threshold levels, maintaining of attention, and signal-to-noise ratio, all of which need input from ergonomists.

Most of the work described in this document has dealt with visual inspection, but in many cases equivalent procedures are performed using other senses. In particular, hearing is used to gauge the smoothness, and sound, of the engines of luxury cars, and there is pressure on manufacturers of household appliances to ensure low noise from their products. In others the requirement is to listen for unwanted noises, or those that might suggest something unwanted, such as a failing bearing in maintenance. The usual method for these is to use the inspector's ear as the instrument. Ergonomics knowledge should be able to effect improved performance in these tasks.

Very little research appears to have been aimed at these situations. Many of the ideas used in visual inspection can be used for them or adapted to suit. For example, the Theory of Signal Detection can be applied to the task of listening for noises. The number of types of acoustic faults attended to at any one time should probably be limited to the inspector's Short Term Memory capacity, appropriate acoustic aids should be provided, and amplification is needed for weak sounds. There are also other tasks which use other senses. For example, the olfactory sense is important in the manufacture of cosmetics and perfumes, the taste sense in the roasting of coffee beans for making instant coffee, and the touch sense in the grading of wool (McKennel, 1958). Much more work needs to be done on these types of inspection task.

Research is needed to know which of the existing techniques can be applied successfully to these other situations, and/or how to adapt them to suit. Much more work is needed to develop adequate models to predict accurately the likely performance in a wide variety of situations. More must be done in regard to providing aids for the inspectors. New techniques need to be developed for those situations where existing ideas are found not to be successful. At present there is no clear indication of where to go on these issues but some indications may arise out of the work on aircraft inspection.

Acknowledgements

The author wishes to acknowledge the help of several reviewers whose suggestions and criticisms helped to effect a marked improvement in this document.

References

Adams, S.K., 1975. Decision making in quality control: some perceptual and behavioural considerations. In: Drury, C.G., Fox, J.G. (Eds.), Human Reliability in Quality Control. Taylor & Francis, London, pp. 55–69.

Ainsworth, L., 1982. An RSM investigation of defect rate and other variables which influence inspection. Proceedings of the Human Factors Society Annual Conference, pp. 868–872.

Astley, R.W., Fox, J.G., 1975. A study of human factors in inspection of rubber goods. In: Drury, C.G., Fox, J.G. (Eds.), Human Reliability in Quality Control. Taylor & Francis, London, pp. 253–272.

Belbin, E., 1955. The problems of industrial training. British Management Review 13, 165–173.

Belbin, R.M., 1957. New fields for quality control. British Management Review 15, 79–89.

Belbin, R.M., 1970. Inspection and human efficiency. Applied Ergonomics 1, 289–294.

Bellamy, L.V., Courtney, A.J., 1981. Development of a search task for the measurement of peripheral visual acuity. Ergonomics 24, 497–509.

Bloomfield, J.R., 1972. Visual search in complex fields: size differences between target discs and surrounding discs. Human Factors 14, 139–148.

Bray, D.E., McBride, D. (Eds.), 1992. Nondestructive Testing Techniques. Wiley, New York.

Carroll, J.M., 1969. Estimating errors in the inspection of complex products. AIIE Transactions 1, 229–235.

Chaney, F.B., Teel, K.S., 1967. Improving inspection performance through training and visual aids. Journal of Applied Psychology 51, 311–315.

Chapman, D.E., Sinclair, M.A., 1975. Applications of ergonomics in inspection tasks in the food industry. In: Drury, C.G., Fox, J.G. (Eds.), Human Reliability in Quality Control. Taylor & Francis, London, pp. 231–251.

Chin, R.T., 1982. Automated visual inspection techniques and applications: a bibliography. Pattern Recognition 15, 343–357.

Chin, R.T., 1988. Automated visual inspection. Computer Vision, Graphics, and Image Processing 41, 346–381.

Chuang, S.Y., Chang, F.H., 1988. Acoustic emission due to flight environment structural noise and crack growth. FZM-7620, General Dynamics, Fort Worth, Texas.

Colquhoun, W.P., 1959. The effect of a short rest pause on inspection efficiency. Ergonomics 2, 367–372.

Colquhoun, W.P., 1961. The effect of "unwanted" signals on performance in a vigilance task. Ergonomics 4, 41–51.

Courtney, A.J., Shou, C.H., 1985. Visual lobe area for single targets on a competing homogeneous background. Human Factors 27, 643–652.

Coury, B.G., Drury, C.G., 1986. The effects of pacing on complex decision-making inspection performance. Ergonomics 29, 489–508.

Craig, A., 1987. Signal detection theory and probability matching apply to vigilance. Human Factors 29, 645–652.

Crossman, E.R.F.W., 1955. The measurement of discriminability. Quarterly Journal of Experimental Psychology 7, 176–195.

Czaja, S.J., Drury, C.G., 1981. Aging and pretraining in industrial inspection. Human Factors 23, 485–494.

Drury, C.G., 1978. Integrating human factors models into statistical quality control. Human Factors 20, 561–572.

Drury, C.G., 1980. Personal communication.

Drury, C.G., Addison, J.L., 1973. An industrial study of the effects of feedback and fault density on inspection performance. Ergonomics 16, 159–169.

Drury, C.G., Lock, M.W.B., 1992. Ergonomics in civil aircraft inspection. Contemporary Ergonomics. Taylor & Francis, London, pp. 116–123.

Drury, C.G., Sheehan, J.J., 1969. Ergonomic and economic factors in an industrial inspection task. International Journal of Production Research 7, 333–341.

Drury, C.G., Sinclair, M.A., 1983. Human and machine performance in an inspection task. Human Factors 25, 391–399.

Engel, F.L., 1974. Visual conspicuity and selective background interference in eccentric vision. Vision Research 14, 459–471.

Engel, F.L., 1977. Visual conspicuity, visual search and fixation tendencies of the eye. Visual Research 17, 95–108.

Erickson, R.A., 1964. Relation between visual search time and peripheral visual acuity. Human Factors 6, 165–177.

Faulkner, T.W., Murphy, T.J., 1973. Lighting for difficult visual tasks. Human Factors 15, 149–162.

Fitts, P.M., Posner, M.I., 1967. Human Performance. Brooks/Cole Publishing, California.

Fox, J.G., 1964. The ergonomics of coin inspection. Quality Engineer 28, 165–169.

Fox, J.G., Haslegrave, C.M., 1969. Industrial inspection efficiency and the probability of a defect occurring. Ergonomics 12, 713–721.

Fox, J.G., Richardson, S., 1970. The effect of the complexity of the signal in visual inspection tasks. Report Eng/70/146, Department of Engineering Production, University of Birmingham, UK.

Gallwey T.J., 1978. The effect of visual lobe size on visual search time. Proceedings of 11th Annual Meeting of the Human Factors Association of Canada, pp. 5.1–5.9.

Gallwey T.J., 1980. Visual inspection for several fault types and its predictors. Unpublished Ph.D Dissertation, Industrial Engineering, State University of New York at Buffalo, NY.

Gallwey, T.J., 1982a. The distribution of visual inspection times. IEA'82, Proceedings of the 8th Congress of the International Ergonomics Association, Tokyo, Japan Ergonomics Research Society, pp. 574–575.

Gallwey, T.J., 1982b. Selection tests for visual inspection on a multiple fault type task. Ergonomics 25, 1077–1092.

Gallwey, T.J., Czaja, S.J., 1981. An evaluation of the Harris inspection test. Proceedings of the Human Factors Society Annual Conference, pp. 634–638.

Gallwey, T.J., Drury, C.G., 1977. Size discrimination in visual inspection. Proceedings of 10th Annual Meeting of the Human Factors Association of Canada, pp. 32–37.

Gallwey, T.J., Drury, C.G., 1986. Task complexity in visual inspection. Human Factors 28, 595–606.

George, R.T., 1963. Visual inspection strategy. International Journal of Production Research 2, 213–228.

Geyer, L.H., Perry, R.F., 1982. Variation in detectability of multiple flaws with allowed inspection time. Human Factors 24, 361–365.

Gould, J.D., Carn, R., 1973. Visual search, complex backgrounds, mental counters, and eye movements. Perception and Psychophysics 14, 125–132.

Gramopadhye, A., Wilson, K., 1996. Noise, feedback, and inspection performance. In: Advances in Occupational Ergonomics and Safety I, Proceedings of the XIth Annual International Occupational Ergonomics and Safety Conference, Zurich, 8–11 July, International Society for Occupational Ergonomics and Safety, Cincinnati, Ohio, pp. 660–665.

Harris, D.H., 1964. Development and validation of an aptitude test for inspectors of electronic equipment. Journal of Industrial Psychology 2, 29–35.

Harris, D.H., 1968. Effect of defect rate on inspector accuracy. Journal of Applied Psychology 52, 377–379.

Harris, D.H., Chaney, F.B., 1969. Human Factors in Quality Assurance. Wiley, New York.

Hayes, A.S., 1950. Control of visual inspection. Industrial Quality Control 6, 73–76.

Hou, T.H., Lin, L., Drury, C.G., 1993. An empirical study of hybrid inspection systems and allocation of inspection functions. International Journal of Human Factors in Manufacturing 3, 351–367.

Illuminating Engineering Society, 1973. The I.E.S. Code of Practice. The Society, London.

Industrial Engineering, 1978. News item. Industrial Engineering 10 (10), 58.

Jacobson, H.J., 1952. A study of inspector accuracy. Industrial Quality Control 9, 16–25.

Jamieson, G.H., 1966. Inspection in the telecommunications industry: a field study of age and other performance variables. Ergonomics 9, 297–303.

Johnston, D.M., 1965. Search performance as a function of peripheral acuity. Human Factors 7, 527–535.

Kelly, M.L., 1955. A study of industrial inspection by the method of paired comparisons. Psychological Monographs 69 (394).

Martin, G.C., 1980. In-flight acoustic emission monitoring. In: Proceedings of Conference on Mechanics of Nondestructive Testing, Blacksburg, Virginia.

Matthews, G., Davies, D.R., Holley, P.J., 1993. Cognitive predictors of vigilance. Human Factors 35, 3–24.

McFarling, L.H., Heimstra, N.W., 1975. Pacing, product complexity, and task perception in simulated inspection. Human Factors 17, 361–367.

McKennel, A.C., 1958. Wool quality assessment. Occupational Psychology 32, 50–60 and 111–119.

McKenzie, R.M., Pugh, D.S., 1957. Some human aspects of inspection. The Production Engineer 36, 378–386.

Miller, G.A., 1956. The magical number seven, plus or minus two: some limits on our capacity for processing information. Psychological Review 63, 81–97.

Mills, R., Sinclair, M.A., 1976. Aspects of inspection in a knitwear company. Applied Ergonomics 7, 97–107.

Moclair, F.M., Gallwey, T.J., 1997. Memory load and decision aids in inspection of silicon wafers. Proceedings of 13th Congress of the International Ergonomics Association, Tampere, Finland, 29 June–July 4, in press.

Monk, T.H., 1976. Target uncertainty in applied visual search. Human Factors 18, 607–611.

Moraal, J., 1975. The analysis of an inspection task in the steel industry. In: Drury, C.G., Fox, J.G. (Eds.), Human Reliability in Quality Control. Taylor & Francis, London, pp. 217–230.

Morawski, T.B., 1978. Economic models of industrial inspection. Unpublished Master's Thesis, Industrial Engineering, State University of New York at Buffalo, NY.

Persoon, E.H.J., 1988. Automatic Inspection. Proceedings of the First International Conference on Visual Search, Durham, pp. 205–210.

Purswell, J.L., Hoag, L.L., 1974. Strategies for improving visual inspection performance. Proceedings of the Human Factors Society Conference, pp. 397–403.

Rae, J.C., 1979. Establishing a cost-effective system of total quality. Quality Assurance 5, 85–88.

Richardson, A., 1969. Mental Imagery. Springer, Berlin.

Seaborne, A.E.M., 1963. Social effects on standards in a gauging task. Ergonomics 6, 205–209.

Schwabisch, S.D., Drury, C.G., 1984. The influence of the reflective-impulsive cognitive style on visual inspection. Human Factors 26, 641–647.

Smith, G.L., Adams, S.K., 1971. Magnification and microminiature inspection. Human Factors 13, 247–254.

Smith, L.A., Barany, J.W., 1970. An elementary model of human performance on paced visual inspection tasks. AIIE Transactions 2, 298–308.

Spanner, J.C., McElroy, J.W., 1975. Monitoring Structural Integrity by Acoustic Emission. American Society for Testing and Materials, 1916 Race Street, Philadelphia PA 19103.

Su, J.-Y., Konz, S., 1981. Evaluation of three methods for inspection of multiple defects/item. Proceedings of 25th Annual Meeting of the Human Factors Society, Rochester, NY, pp. 627–630.

Thornton, D.C., Matthews, M.L., 1982. An examination of the effects of a simple task alteration in the quality control inspection of cookware. Proceedings of the 15th Annual Meeting of the Human Factors Association of Canada, Toronto, pp. 134–137.

Tiffin, J., Rogers, H.B., 1941. The selection and training of inspectors. Personnel 18, 14–31.

Van Cott, H.P., Kincade, R.G., 1972. Human Engineering Guide to Equipment Design. US Government Printing Office, Washington, DC. 20402.

Waikar, A., 1973. Quality improvement using multiple inspector systems. Unpublished Master's Thesis, University of Oklahoma.

Wallack, P.M., Adams, K.S., 1969. The utility of signal-detection theory in the analysis of industrial inspector accuracy. AIIE Transactions 1, 33–44.

Wallack, P.M., Adams, K.S., 1970. A comparison of inspector performance measures. AIIE Transactions 2, 97–105.

Wang, M.J., Drury, C.G., 1989. A method of evaluating inspector's performance differences and job requirements. Applied Ergonomics 20, 181–190.

Wiener, E.L., 1963. Knowledge of results and signal rate in monitoring: a transfer of training approach. Journal of Applied Psychology 47, 214–222.

Wiener, E.L., 1975. Individual and group differences in inspection. In: Drury, C.G., Fox, J.G. (Eds.), Human Reliability in Quality Control. Taylor & Francis, London, pp. 101–122.

Williams, L.G., Borow, M.S., 1963. The effect of rate and direction of display movement upon visual search. Human Factors 5, 139–1546.

Witkin, H.A., Oltman, P.K., Raskin, E., Karp, S.A., 1971. A Manual for the Embedded Figures Test. Consulting Psychologists Press, Palo Alto, California.

Yntema, D.G., Shulman, G.M., 1967. Response selection in keeping track of several things at once. Acta Psychologica 27, 316–424.

Evaluation and control of hot working environments: Part I – Guidelines for the practitioner [*]

Jerry D. Ramsey [a], Thomas E. Bernard [b], Francis N. Dukes-Dobos [b]

[a] *Texas Tech University, Lubbock, TX 79409-2019, USA*
[b] *University of South Florida, Tampa, FL 33612-3805, USA*

1. Audience

The audiences for this guideline are the practitioner at a hot work site who is responsible for the heat management program, as well as the individual who is working in the heat. Environmental heat is a commonly occurring working environment; it is found at outdoor worksites, at indoor worksites with natural ventilation, at heat producing operations, and even in air-conditioned work areas where mechanical failure occurs in the cooling system or where the total heat load exceeds the capability of the cooling system. This guideline includes information on the way the human body stores heat, on potential illnesses from overexposure, methods for measuring and estimating the various components of heat exposure, limiting values of thermal exposure and work times in the heat, how these limits are affected by levels of acclimatization and clothing, and means of controlling heat exposure and heat stress.

2. Context for guideline use

Precise evaluation of heat stress is a complicated activity, but evaluation of heat stress for making decisions related to worker protection and productivity can be performed using simple and approximate guidelines as discussed in this paper. A fundamental assumption for heat stress decisions is that the workers themselves are adequately trained about heat stress, that they are practicing good heat stress hygiene, and that they have the opportunity to limit their exposure when they sense that the cumulative effects of thermal load on the body are reaching unsafe levels.

Heat stress may come from climatic heat, process heat, the workload (metabolic heat), or the use of additional/special clothing. Some combinations of these heat sources can produce exces-

---

[*] The recommendations provided in this guide are based on numerous published and unpublished scientific studies and are intended to enhance worker safety and productivity. These recommendations are neither intended to replace existing standards, if any, nor should be treated as standards. Furthermore, this document should not be construed to represent institutional policy.

The following individuals participated in the discussion of the earlier version of this guide. Their suggestions (written or verbal) were incorporated by the authors in this version: Arne Aaras, *Norway;* Fred Aghazadeh, *USA;* Roland Andersson, *Sweden;* Jan Dul, *The Netherlands;* Jeffrey Fernandez, *USA;* Ingvar Holmér, *Sweden;* Matthias Jäger, *Germany;* Åsa Kilbom, *Sweden;* Anders Kjellberg, *Sweden;* Olli Korhonen, *Finland;* Helmut Krueger, *Switzerland;* Shrawan Kumar, *Canada;* Ulf Landström, *Sweden;* Tom Leamon, *USA;* Anil Mital, *USA;* Ruth Nielsen, *Denmark,* Murray Sinclair, *UK*; Rolf Westgaard, *Norway;* Ann Williamson, *Australia;* Jørgen Winkel, *Sweden;* Pia Zätterström, *Sweden.* The guide was also reviewed in depth by several anonymous reviewers.

Table 1
Heat-related disorders

Heat exhaustion
Cause: Heavy sweating causing dehydration
Dehydration due to illness (vomiting, diarrhea, etc.)
Symptoms: General fatigue or weakness
Uncoordinated actions
Headache, thirst, weak pulse
Treatment: Rest in a cool environment
Fluid replacement
Prevention: Drink water or other fluids frequently
Add salt to food.

Heat stroke
Cause: Pre-existing illness (e.g., fever, flu, etc.)
Abnormal intolerance to heat stress
Excessive exposure to heat stress
Drug or alcohol abuse
Symptoms: Irrational, disoriented or unexpected behavior
Convulsions and/or unconsciousness
Hot dry skin
Treatment: Immediate, aggressive, and effective cooling
Transport immediately to an emergency medical facility
Prevention: Self-determination of one's heat tolerance limit
Maintain a healthy lifestyle

Heat syncope
Cause: Maintaining one work posture (e.g., standing or squatting)
Standing or sitting up quickly
Pooling of blood in the legs, away from head
Symptoms: When sitting or standing up:
    Faint, dizzy feeling ("gray out")
    Fainting ("black out") – brief, less than 30 seconds
Treatment: Rest laying down
Drink water or other suitable fluid
Prevention: Flex leg muscles several times before moving
Stand or sit up slowly

Heat cramps
Cause: Profuse sweating and hard work
Associated with an excessive loss of salts
Symptoms: Painful muscle cramps in legs, arms or abdomen
Cramps may occur during or after the work shift
Treatment: Drink water, preferably with 0.5% salt solution
Massage cramping muscles
Prevention: If hard physical work is part of the job, workers should add extra salt to food

Heat rash
Cause: Obstruction of the sweat glands caused by chronically wet skin
Symptoms: Itchy skin with small red spots
Unusually sensitive to radiant heat
Treatment: Keep out of heat until condition is healed
Seek treatment from a dermatologist
Keep the skin clean
Prevention: Keep skin clean, and periodically allow the skin to dry
Intermittent relief from the heat

sive heat strain. Because humans are homeothermic organisms, they can function very well in hot environments, and if there is good knowledge of heat illness symptoms and good information on the levels and limits of exposure, the total risk associated with work in the heat can remain at acceptable levels.

This guideline will be restricted to hot working environments. Thermal conditions which are warm/uncomfortable, cool/cold or have extreme surface contact temperatures will not be addressed in this document.

3. Problem identification

An understanding of the heat balance exchange between the human body and the environment is a fundamental ingredient in the evaluation of thermal effects, limits and controls. This can be simply represented by the following equation:

$$S = M \pm R \pm C - E$$

where, S = heat storage; M = metabolic heat; R = radiant heat; C = convective heat; and E = evaporative heat.

In this equation, heat storage $(S) = 0$ if the body is in heat balance or equilibrium with the work and environment. The metabolic heat (M) is always positive since it heats the body while the evaporative heat (E) is always negative since it removes heat from the body. The radiant heat (R) from solar or other hot surfaces is positive as a heat gain, although it is possible to have radiant heat loss toward cold surfaces of the environment. The convective heat (C) may be positive or negative based on whether the surrounding air is warmer or cooler than the skin. Other factors, such as conductive heat, external work, respiratory heat loss and insensible evaporative heat loss, are sometimes considered as separate components in the heat balance equation; however, their combined effect is typically small and usually neglected. Thermal balance and stability (S = 0), is the desired outcome of controlling heat stress conditions during prolonged worker exposure.

3.1. Heat-related disorders

Acute and prolonged exposures to heat stress may allow various heat illnesses to occur. Heat exhaustion typically includes symptoms such as headache, fatigue, fainting, profuse sweating and moist skin. Body temperature may be normal or slightly elevated. A person with these symptoms should be removed to a cooler area for rest and water intake (until symptoms cease). Heat stroke, on the other hand, is a life-threatening medical emergency with many of the same symptoms except the skin will usually be hot and dry because of central nervous system dysfunction. Body temperature will be above normal by several degrees. Prompt and aggressive cooling is required for heat stroke and may include removing clothing and applying water spray and ventilation to the body to reduce its temperature. If the person is conscious, then drinking water is also recommended. In any event, a person affected by heat stroke should be taken immediately to the nearest medical treatment facility.

Other heat-related disorders include heat cramps, heat rash and heat syncope. Heat cramps or muscle spasms are typically a result of body salt loss due to sweating. The drinking of salted water or electrolyte beverages and resting in a cool place are typically adequate to stop the cramps. For preventing any of the above illnesses, frequent intake of water and well-salted food is encouraged for anyone working in hot environments. Workers on salt-restricted diets should consult their personal physicians. The use of salt tablets or direct salt intake is not recommended. A summary of heat-related disorders, including causes, symptoms, treatment and prevention, is found in Table 1.

3.2. Personal factors

Increased risk of heat disorders is found with the worker who is a new employee, is unacclimatized or has recently returned from vacation. Similarly, personal risk increases after or during illness, with the use of drugs or alcohol, and for lower levels of general health and physical fitness. Special attention to these personal factors is warranted during heat waves, the hot part of the year, the hot time of the day, in the proximity of hot processes and when additional or special clothing is used. Reducing risk of heat disorders is the goal of the heat stress hygiene practices listed in section 5 of this guide.

4. Measurement and estimation

4.1. Thermal components

Many instruments are available for measuring the various components of heat, and there are several heat indices which have demonstrated utility in measuring hot environmental exposure. The Wet Bulb Globe Temperature (WBGT), however, is the most widely used and acknowledged index for use in evaluating hot occupational environments. The WBGT index combines the effects of the four main thermal components affecting heat stress: air temperature (as measured by the dry bulb thermometer $[T_a]$), humidity and air velocity (as indicated by the natural wet bulb thermometer $[T_{nwb}]$) and radiation (as indicated by the globe thermometer $[T_g]$). For indoor or sunless day exposures, WBGT can be determined from the following equation:

$$\text{WBGT} = 0.7 T_{nwb} + 0.3 T_g$$

For outdoors with solar load, WBGT becomes:

$$\text{WBGT} = 0.7 T_{nwb} + 0.2 T_g + 0.1 T_a$$

WBGT can be determined from taking measurements with a mercury in glass dry bulb thermometer for T_a, a thermometer with 1 inch of clean wicking over the mercury and the extension of the wick then placed in a glass of distilled water for T_{nwb}, and a thermometer enclosed in a 6 inch copper shell, painted matte black, for T_g. The T_a should be shielded, the T_{nwb} wick thoroughly wetted, and T_g given adequate time (around 20 minutes) for thermal stabilization. There are also several direct reading electronic instruments which are available on the market to determine WBGT. It is also possible to obtain a rough estimate of WBGT using a Botsball which

Table 2
Workload/metabolic heat produced (M) (Adapted from: J.L. Smith and J.D. Ramsey, Designing physically demanding tasks to minimize levels of worker stress, *Industrial Engineering*, Vol. 14, 1982.)

Level 1 – Resting / Less than 117W (100 kcal / h)

Level 2 – Light / 117–232W (100–199 kcal / h)
Sitting at ease: light handwork (writing, typing drafting, sewing, bookkeeping); hand and arm work (small bench tools, inspecting, assembling, or sorting light materials; arm and leg work (driving car under average conditions, operating foot switch or pedal).
Standing: drill press (small parts); milling machining (small parts); coil taping; small armature winding; machining with light power tools; casual walking up to 0.9 m/s (2 mph).
Lifting: 4.5 kg (10 lb), fewer than 8 lifts per minute; 11 kg (25 lb), fewer than 4 lifts per minute.

Level 3 – Moderate / 233–348W (200–299 kcal / h)
Hand and arm work (nailing, filing); arm and leg work (off-road operation of trucks, tractors, or construction equipment); arm and trunk work (air hammer operation, tractor assembly, plastering, intermittent handling of moderately heavy materials, weeding, hoeing, picking fruits or vegetables); pushing or pulling lightweight carts or wheelbarrows; walking 0.9–1.3 m/s (2–3 mph).
Lifting: 4.5 kg (10 lb), fewer than 10 lifts per minute; 11 kg (25 lb), fewer than 6 lifts per minute.

Level 4 – Heavy / 349–465W (300–399 kcal / h)
Heavy arm and trunk work transferring heavy materials, shoveling; sledge hammer work; sawing, planting, or chiseling hardwood; handmowing; digging; walking 1.8 m/s (4 mph), pushing or pulling loaded handcarts or wheelbarrows; chipping castings; laying concrete block.
Lifting: 4.5 kg (10 lb), 14 lifts per minute; 11 kg (25 lb), 10 lifts per minute.

Level 5 – Very Heavy / More than 465W (400 kcal / h)
Heavy activity at fast to maximum pace: ax work; heavy shoveling or digging; climbing stairs, ramps, or ladders; jogging, running, walking faster than 1.8 m/s (4 mph).
Lifting: 4.5 kg (10 lb), more than 18 lifts per minute; 11 kg (25 lb), more than 13 lifts per minute.

measures the Wet Globe Temperature (WGT). This is a small-size, rugged and easy-to-use device which has a reasonable correlation with WBGT, especially in non extreme environments.

4.2. Metabolic heat

The metabolic heat generated by a person increases as a function of the physical work performed. Metabolic heat can be estimated based on actual measurement of oxygen consumption of a worker, or estimated using detailed calculations and tabulations. For this guide, however, the metabolic workload can be estimated as either light, moderate or heavy, according to the typical work activities shown in Table 2. The estimate should represent the hourly time-weighted average workload at the hot workplace. This will provide an approximation of metabolic heat for use in determining heat exposure limits. Table 2 includes a category of "very heavy" work; however, it is unusual to find continuous work at this pace, since workers tend to adjust their work intensity and rest pauses during work with high metabolic demand, so that the "average" is at a more acceptable level.

4.3. Interpretation and use of heat limits

Fig 1 depicts WBGT limits for working in the heat as a function of the metabolic heat generated from the work. This recommendation from the National Institute for Occupational Safety and Health (NIOSH) for recommended heat stress exposure limits (REL) is basically the same as those suggested by the International Organization for Standardization, the American Conference of Government Industrial Hygienists, the U.S. military and others. The REL represents a threshold limit where there is onset of increased risk of heat distress. It applies to acclimatized workers who are healthy, physically and medically fit and are wearing regular lightweight work clothing. A ceiling limit proposed by NIOSH is also depicted in Fig. 1. Combinations of metabolic and environmental heat in excess of the ceiling limit require the worker to utilize heat protective clothing and equipment.

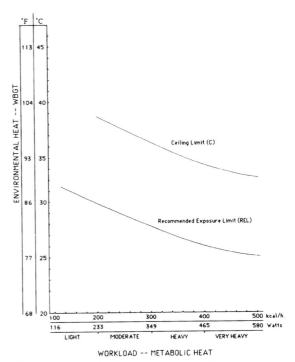

Fig. 1. Recommended heat stress exposure limits for acclimatized workers.

Both of these curves represent one-hour time weighted average (TWA) exposures. If the worker is exposed to "n" different temperatures during the hour (e.g., rest breaks in cooler areas), TWA would be calculated as follows:

$$\text{TWA(WBGT)} = [(\text{WBGT}_1) \cdot (\text{time}_1) \\ + (\text{WBGT}_2) \cdot (\text{time}_2) \\ + \ldots (\text{WBGT}_n) \cdot (\text{time}_n)] \\ /[\text{time}_1 + \text{time}_2 \ldots \text{time}_n]^{-1}$$

If the worker's job includes "n" different levels of workloads (metabolic heat, M) the TWA (M) would need to be calculated by substituting M for WBGT in the equation shown above. The limits of Fig. 1 then would be based on the time-weighted average values for both WBGT and M. This calculation provides a means of determining acceptable work/rest schedules for different combinations of environmental heat and metabolic heat.

4.4. Acclimatization

Acclimatization is the physiological adaptation which a person makes with repeated exposures to a set of thermal conditions. Acclimatization to a thermal stress will normally occur within 5–7 days of initial exposure to that combination of work and temperature. After acclimatization, the sweat glands excrete more sweat, removing more heat evaporatively. Also, the heart rate and body temperature tend to respond more appropriately to the level of stress, and thus the total risk of excessive heat strain is reduced. Persons new to a job, returning to the job after vacation or illness, or working at the onset of seasonal climatic changes should be allowed to acclimatize prior to assuming full work duties. The heat exposure limits of Fig. 1 should be adjusted as shown in Table 3 when determining limits for the unacclimatized worker. For example, the limits for unacclimatized workers should be lowered 1°C (1.8°F) for light work, 2°C (3.6°F) for moderate work and 3°C (5.4°F) for heavy work.

4.5. Clothing

Most recommendations concerning occupational heat stress are based on the premise of a worker wearing a single-layer clothing ensemble with trousers and a cotton work shirt, or the equivalent. Increasing the thickness (insulation characteristics) or the vapor/water barrier characteristics (permeability characteristics) of clothing can substantially affect the ability of a person to dissipate heat. The use of additional, special or protective clothing is a common phenomenon in today's workplace. The heat exposure limits of Fig. 1 should be adjusted as shown in Table 3 when determining limits for workers not in normal or regular work clothing. For example, the limits should be lowered: 2°C (3.6°F) if heavier clothing such as cotton coveralls or additional clothing such as a lab coat are worn; 4°C (7.2°F) for winter clothing, double cloth overalls or water barrier/vapor permeable clothing; 6°C (10.8°F) for light weight vapor barrier clothing such as PVC or coated TYVEK^R; and 10°C (18°F) for fully enclosed body suits. The completely enclosed suit will require its own cooling system if worn for longer than fifteen minutes or so.

5. Actions to control heat exposure and heat stress

The actions available for protecting heat exposed workers consist of an array of activities relating to the design of the job, the work and the environment. Some of the more important actions are summarized below.

5.1. Heat stress hygiene practices

(1) *Fluid replacement:* An adequate supply of cooled water should be available near the work site and workers should be informed of the necessity for frequent water intake and abundant salting of food during meals (not

Table 3
WBGT modification factors for acclimatization and clothing. The recommended exposure limits of Fig. 1 should be reduced by the amounts shown below

Unacclimatized		
Low workload level	117–232W (100–199 Kcal/h)	−1°C (−1.8°F)
Moderate workload level	233–348W (200–299 Kcal/h)	−2°C (−3.6°F)
Heavy workload level	349–465W (300–400 Kcal/h)	−3°C (−5.4°F)
Clothing		
Light work clothing		0°C (0°F)
Cotton coverall, jacket		−2°C (−3.6°F)
Winter work clothing, double cloth coveralls, water barrier		−4°C (−7.2°F)
Light weight vapor barrier suits		−6°C (−10.8°F)
Fully enclosed suit with hood and gloves		−10°C (−18°F)

salt tablets). Because thirst is not a sufficient drive for water replacement, workers should drink small quantities as frequently as possible. If work is to be performed in a drinking restricted area, drinking about one pint per hour of work before work begins will help meet the demands for water during the work.

(2) *Acclimatization:* Acclimatization is the adaptation of the body to chronic heat stress exposure. The ability to work increases and the risk of heat disorder decreases with acclimatization. Newly assigned employees and those recently returning from serious illness or long vacations and who are normally assigned to other than light work load tasks should not be required to perform at full-normal pace and full heat exposure until they have an opportunity to become acclimatized.

(3) *Self-determination:* One aspect of self-determination is limiting an exposure to heat stress. It is a responsibility of the worker and the foreman. In self-determination, the person terminates an exposure to heat stress at the first feelings of excessive heat strain. Serious injury can occur if the onset of symptoms is ignored. Another aspect of self-determination is reducing the effects of heat stress by lowering peak work demands and making the work demands lighter. For instance, when a fixed amount of work is assigned to a portion of the shift, peak demands can be reduced by stretching the work out over a longer period of time or taking more frequent breaks. For those working in crews, the pace should accommodate the least heat-tolerant worker. Self-determination is an important principle, but it should be noted that workers may not be very proficient in sensing the status of their own thermal storage.

(4) *Diet:* A well-balanced diet is important to maintain the good health needed to work under heat stress. Diets designed to lose weight should be directed by a physician who understands that the patient is working under conditions of heat stress. Salt intake as part of a normal diet is usually sufficient to meet the salt demands during heat stress work. Added salt may be desirable when repeated heat stress exposures are first experienced (i.e., before acclimatization). If salt is restricted by a physician's order, the physician should be consulted.

(5) *Lifestyle:* A healthy lifestyle is important to lowering the risk of a heat-related disorder. A worker should have adequate sleep and a good diet. Exercise helps. A healthy lifestyle also means no abuse of alcohol or drugs, which have been implicated in serious heat-related illnesses. In addition, exposures to heat stress immediately prior to work may increase the risk of a heat disorder at work.

(6) *Health status:* All workers should recognize that chronic illnesses, such as heart, lung, kidney, or liver disease, indicate a potential for lower heat tolerance and therefore an increased risk of experiencing a heat-related disorder during heat stress exposures. Any acute illness such as a cold, flu, fever, diarrhea, or vomiting should be reported to the individual's immediate foreman.

5.2. Engineering controls and personal protection

(7) *General ventilation:* Increased general ventilation or spot cooling can sometimes be used to reduce temperature at the workplace.

(8) *Exhaust ventilation:* Local exhaust ventilation at points of high heat or humidity production will help remove heat from the work area.

(9) *Local cooling:* Local or spot cooling of the worker may be an effective and energy efficient means of providing relief from heat exposure.

(10) *General cooling or refrigeration:* Evaporative cooling or mechanical refrigeration can be used to reduce the temperature of supply air and general work site temperatures.

(11) *Fans:* Personal cooling fans increase air velocity and evaporative and convective heat loss as long as air temperatures are below 35°C (95°F).

(12) *Radiant shielding:* Shielding by means of reflective screens, barriers, aprons or clothing

will interrupt line-of-sight radiant thermal exchange.

(13) *Excessive moisture reduction:* The relative humidity level can be reduced by repairing steam leaks, avoiding water on the floor or minimizing splashing or spraying from processes.

(14) *Isolation or change:* Isolation, relocation, redesign or substitution of equipment and/or processes should be considered as a means to reduce the thermal stress at the work site.

(15) *Personal clothing and cooling devices:* Personal cooling devices and/or protective clothing (e.g., circulating air systems, liquid cooling systems, ice cooling garments, evaporative cooling or reflective clothing) may help reduce heat stress in applications where other controls are limited.

5.3. Administrative practices

(16) *Training in heat stress and its control:* Each workplace where heat stress may occur should have all workers and supervisors trained in the recognition of heat strain and heat-related disorders, heat stress hygiene practices and countermeasures for controlling heat stress.

(17) *Medical:* Those exposed to extreme heat exposure should be medically evaluated before placement in this type of work and medically examined periodically thereafter.

(18) *Metabolic heat reduction:* Internally generated heat can be reduced by providing relief workers or other adjustments in the duration of the work period, the frequency and length of rest pauses, the pace and tempo of work, as well as the increased use of work-saving devices or mechanization.

(19) *Scheduled peaks:* When feasible, heavy work should be scheduled during the cooler parts of the work shift or on cooler days.

(20) *Rest areas:* The use of air-conditioned or cooler areas for rest and recovery will reduce heat storage in the worker, hasten recovery and increase heat tolerance.

(21) *Buddy system:* The risk of excessive heat exposure and heat-related illness will be reduced if the worker is under observation by a trained supervisor or fellow worker(s) who can detect early signs of excessive heat strain and also can respond to emergencies. Multiman work crews also provide for work-load sharing or transfer, if needed. In any event, working alone in heat near the Recommended Exposure Limit (REL) should be avoided.

The information in this document is for general guidance on assessing heat exposure and preventing heat-related illness at the workplace. If these methods do not provide a clear basis for hot work related decisions, other more analytical methods for measuring and assessing thermal balance should be used. Support for these activities is available from professionals in the field or from the literature, as described in Part II of this guideline. Employers with frequent employee exposure to hot work should develop their own comprehensive heat stress program which specifies all policies and procedures to be used in managing the risks of worker exposure to hot working environments.

References

Bernard, T.E., Dukes-Dobos, F.N. and Ramsey, J.D., 1994, Evaluation and control of hot working environments: Part II – The scientific basis (knowledge base) for the guide. International Journal of Industrial Ergonomics, 14(1–2): 129–138 (this issue).

Evaluation and control of hot working environments: Part II – The scientific basis (knowledge base) for the guide

Thomas E. Bernard [a], Francis N. Dukes-Dobos [a], Jerry D. Ramsey [b]

[a] *University of South Florida, Tampa, FL 33612-3805, USA*
[b] *Texas Tech University, Lubbock, TX 79409-2019, USA*

1. Problem description

Heat stress and the resulting heat strain are well recognized hazards in the workplace. This part of the guidelines provides the rationale for evaluation of heat stress and for interventions to reduce the risk of heat-related disorders.

Although tolerance to heat stress and the risk for heat-related disorders are affected by personal factors, the population risk increases as the level of stress increases. An overview of heat-related disorders is provided first to help the practitioner understand common causes and the rationale for most heat stress hygiene practices. The successful evaluation of heat stress requires a quantitative basis to make decisions about the level of heat stress present. The empirical basis of a scheme based on wet bulb globe temperature (WBGT) is presented. Finally, the rationale for basic heat stress controls is presented.

2. Heat tolerance and heat-related disorders

The three most obvious responses to heat stress exposures are increases in body temperature, heart rate and sweating. These are collectively heat strain. Core temperature increases as a direct response to the rate of work, and the increase is proportional to the work demands as a fraction of the individual's aerobic work capacity. Aerobic capacity is a measure of fitness and depends both on inherited traits and physical training. For a given rate of work in a cool environment, a less fit person will have a greater core temperature than a more fit person. If the climatic conditions reduce the ability of the person to dissipate the heat generated by energy metabolism, then the core temperature will increase further. The first step in dissipating the heat is the removal of metabolic heat from the working muscles and body core by blood flow through these tissues. The extra heat is delivered to the skin for dissipation to the environment. The additional demands for blood flow increase the cardiac demands on the person as reflected in heart rate. During exposures to heat stress, there is a tendency for skin temperature to increase, which means that the skin blood flow must increase further to achieve the same level of cooling of the body core, resulting in a further increase in cardiac demands. The ability of the person to meet these cardiac demands again depends on the level of fitness. A less fit person requires higher heart rates to maintain the same level of blood flow than a more fit person. Finally, sweat evaporation from the skin is the primary mechanism to dissipate heat from the skin to the environment. Usually the contribution of radiation and convection to the total heat load is small compared to metabolism, but when there is a net heat gain by radiation and convection, the requirements for sweat evaporation increase accordingly. The water and salt loss due to sweating is the price the body pays to maintain thermal equilibrium.

2.1. Heat tolerance and acclimatization

Heat tolerance is the ability of a person to physiologically adjust to a heat stress exposure. About 4% of the population can be described as heat intolerant; that is they do not thermoregulate well enough to work under most conditions of heat stress (Wyndham et al., 1972). The remaining population has varying abilities to tolerate heat stress that depend largely on personal factors such as fitness and acclimatization state. Table 1 is a list of several important personal risk factors that may indicate either a permanent or temporary inability to tolerate heat stress exposures. For those in good health, the single best predictor of heat tolerance is a history of heat tolerance. Beyond that, fitness or the capacity to support aerobic work is the next best predictor.

Acclimatization is an important physiological mechanism to enhance heat tolerance. During successive exposures to heat stress a person will exhibit progressively less physiological strain (lower core temperature and heart rate). This is illustrated in Fig. 1. The reduced strain is accomplished through an increased ability to sweat and thus enhance evaporative cooling. Another bene-

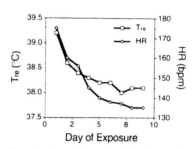

Fig. 1. Acclimatization as illustrated by changes in core temperature and heart rate with successive exposures to hot dry heat for two hours (from Lind and Bass, 1963).

fit of acclimatization is an improved ability to conserve salt (Lind and Bass, 1963).

Most of the benefits of acclimatization are seen after five successive days of exposure to heat stress. A necessary exposure to induce acclimatization is about one to two hours of simultaneous work and heat exposure per day. Longer periods or repeated periods in a day will not improve acclimatization. The usual recommendation for inducing acclimatization in an industrial environment is to work about 20% of the time the first day and increase the exposure time by 20% per day over the following days (NIOSH, 1986). As an alternative to formal acclimatization programs, Bernard has recommended reducing the expectations of workers and supervisors during periods of acclimatization and allowing more individual discretion of the exposures during the period of acclimatization (Bernard et al., 1991).

Acclimatization is lost during periods when there is no exposure to heat stress. As a rule of thumb, one day of acclimatization is lost for every two to three days a person is away from heat stress exposures for routine reasons (Stephens and Hoag, 1981). If the person is away due to an illness, then one day is lost for every day of illness.

2.2. Heat-related disorders

The major heat-related disorders are described in Part I. Most of the disorders are associated with fluid imbalances that are caused by loss of body water (and some salt) from sweating

Table 1
Personal risk factors associated with heat intolerance [a]

History of heat stroke or inability to acclimatize to heat
Chronic disease
Use of certain drugs (e.g., diuretics, cholinergics, beta-blockers, antihistamines, recreational, alcohol)
Obesity (> 25%/30% overweight for men/women)
Low physical fitness
Inability to sweat
Skin disease
Pregnancy (also potential risk to fetus)
Impaired mental capacity or ability to communicate

[a] These may represent temporary or permanent conditions. With regard to medical screening, careful evaluation of individual cases is important before work is restricted.

and improper replacement of the water and salt. Heat exhaustion is most often preceded by dehydration and is usually associated with unacclimatized workers. The loss in blood volume due to dehydration effectively reduces work capacity or fitness, causing fatigue as well as headache. Dehydration usually occurs because the sweat volume is greater than the volume of replacement fluids. It may also occur due to some illnesses that cause loss of appetite, vomiting or diarrhea.

Heat stroke in otherwise normal and healthy people results from a combination of excessive heat exposure and physical work. Workers who are physically fit and acclimatized may not have symptoms of heat exhaustion before becoming overheated. Dehydration, however, will make them more prone to overheating because sweating will decrease. On the other hand, if a person is overcome by heat exhaustion and is not removed from the heat, he or she may progress into a heat stroke. Workers who are strongly motivated are at greater risk of heat stroke because they may ignore the discomfort brought on by increased body temperature. By the time people reach a core temperature of 40°C, they lose their ability to make a reasonable decision to stop an exposure. While heat exhaustion is marked by normal to slightly elevated core temperature, heat stroke is marked by significantly elevated core temperature and the consequent loss of thermoregulatory control. The thermal regulatory control center in the brain can also be affected by acute and chronic illness and by drugs (prescription, over-the-counter and recreational including alcohol). With the loss of thermoregulation, the core temperature rises rapidly and there is increasing brain dysfunction that results in disorientation, collapse, unconsciousness and convulsions.

Heat syncope results from a transient (or sometimes sustained) hypotension. Dehydration will predispose a person to hypotension due to the loss of blood volume. In addition, the volume of the peripheral vasculature increases with heat stress. When there are rapid changes in posture (typically from sitting or kneeling to standing), blood flows quickly to the lower extremities before reflex responses redistribute blood volume to maintain an adequate blood pressure. The result may be dizziness or fainting.

Finally, heat cramps are associated with large losses of sweat where salt losses may be the critical element. In this case, water replacement may be adequate but salt replacement is not. Drinking lightly salted (0.5%) liquids alleviates the cramps, but some would argue that the dietary salt is sufficient to make up for salt losses.

2.3. Heat stress hygiene practices

From the discussion of heat-related disorders, it should be clear that fluid replacement is a fundamental heat stress hygiene practice. Under extremely hot conditions, sweating rates can reach a liter per hour over the working day and therefore fluid replacement must be as high as seven to eight liters per worker per day if dehydration is to be avoided. Drinking that amount can cause gastrointestinal discomfort, bloating and nausea. Acclimatized people are able to both sweat more and drink more without experiencing discomfort. The essential ingredient for fluid replacement is water. It has been demonstrated that people will drink more of flavored drinks than of plain water and this is one of the strongest arguments for commercial fluid replacement drinks. Another important factor in fluid replacement is frequency. Drinking about 200 ml (6 oz) every 15 minutes is better than trying to drink the equivalent volume once per hour. Because thirst is not an adequate driver, frequent drinking helps establish drinking as a habit and prevents bloating.

Acclimatization was included under hygiene practices rather than administrative controls in Part I. In this way, the worker through training understands the limits of working in an unacclimatized state and limits work expectations (especially compared to acclimatized workers). Slowing the pace of work and performing duties outside of a hot environment are ways to manage the heat exposures during acclimatization. It is important that supervision recognize the needs to reduce job demands during acclimatization and provide for reduced heat stress exposures.

Even with good fluid replacement and acclimatization, heat stress exposures can lead to extreme discomfort or the symptoms of heat-related disorders. When this occurs, the worker should be able to seek relief. Self-determination is a hygiene practice because the individual is the only one who is aware of extreme discomfort or the early symptoms of disorders. The worker must seek the relief before the early symptoms progress into an illness. Another aspect of self-determination is self-pacing of the work. Those workers with less cardiovascular reserve (lower fitness levels) can avoid peak demands through a leveling of the work. Management has a responsibility to encourage self-determination decisions.

With regard to diet, what is commonly recognized as a healthy diet is acceptable for those exposed to heat stress. Diets designed for weight loss should be undertaken with the direction of a physician who is aware of the heat stress exposures. A particular concern is dehydration, which may be the reason for apparent success in weight loss. Workers employed in hot jobs should be encouraged to salt their food more to make up for the extra salt loss. Unacclimatized workers may benefit from drinks that contain 0.1% salt.

Because any chronic disease may directly affect thermoregulation or the drugs prescribed for the disease may have an effect, the worker/patient must know to inform the primary care physician about exposures to heat stress. Acute illness may also affect thermoregulation. Because there may not be a physician directly responsible for treatment of a cold, flu, etc., the worker must inform the supervisor if he or she is well enough to be at work but suffering an acute illness.

3. Evaluation of heat stress

3.1. Limits based on physiological responses

A World Health Organization meeting on heat stress in 1969 set safe limits for controlling heat stress (WHO, 1969). It was agreed that core temperature should be kept below 38°C for chronic exposures to heat stress, allowing core temperature to reach 39°C for short times. The core temperature criterion is the one most often cited as the determinant of excessive exposures to heat stress (e.g., ACGIH, 1993; ISO, 1989a,b).

The first step to an evaluation scheme is to have a goal of controlling core temperature. Nielsen (1938) first described a relationship of core temperature at a constant rate of work in relation to environmental heat. There is a transition zone where body core temperature starts to increase with increasing environmental heat, which means that there is an increased risk that a worker under those conditions will be unable to maintain thermal equilibrium. Lind extended these studies to look at different rates of metabolism, and his studies are illustrated in Fig. 2 (Lind, 1963a). What is interesting about the Lind efforts is that he labelled the region in which there is little change in equilibrium core temperature with increasing environmental heat for a given rate of work as the prescriptive zone. That is, if work and heat are combined in such a fashion that it is in the prescriptive zone, thermal equilibrium is easily achieved and heat stress is not a significant problem. The upper limit of the prescriptive zone (as shown in Fig. 2) then becomes a threshold for excessive heat stress. It is clear that the limiting line depends on both the environment and metabolism. As metabolism increases, moving up through the lines, the upper limit of the prescriptive zone decreases. In other words, the heat gain from metabolism must be balanced by a greater capability to dissipate heat to the environment. The upper limit of the prescriptive zone was studied further and adjusted

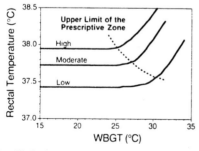

Fig. 2. Equilibrium core temperatures for varying environmental conditions at three work rates, and the upper limit of the prescriptive zone (after Lind, 1963a).

for a population of people that are representative of the general working population wearing light summer work clothing (Dukes-Dobos and Henschel, 1973).

3.2. Limits based on heat stress indices

Core temperature as well as the other components of physiological strain are outcomes of heat stress. Heat stress is described by environmental conditions and work demands as well as clothing. Numerous heat stress indices have been proposed based on physical factors of the environment, physiological strain, thermal comfort assessment, and rational heat balance. A review of these can be found in Beshir and Ramsey (1988). Rational evaluations of heat stress requires calculating the heat balance from $E_{req} = M + R + C$. From the heat balance, one can predict the evaporative cooling needed to maintain thermal equilibrium or to predict the rate of heat gain and/or time limits before recovery is required (Belding and Hatch, 1955; ISO, 1989b). The heat balance approaches require some detailed measurements and computations that are not easily accomplished. For this reason, simple empirical indices like the wet bulb globe temperature are desirable.

The wet bulb globe temperature (WBGT) is accepted as a reliable index to express the environmental contributions to heat stress (ACGIH, 1993; ISO, 1989a; NIOSH, 1986). The natural wet bulb is roughly indicative of the ability of the environment to support sweat evaporation while the globe temperature reflects dry heat exchange (radiation and convection). The weighting of wet bulb and globe temperatures is empirical and evolved from the work with Effective Temperature (Yaglou and Minard, 1957). It is an acceptable measure for industrial purposes because it is a good balance between predictive capabilities and relative simplicity.

Based on the work of Dukes-Dobos and Henschel (1973) on the prescriptive zone and the WBGT index, ACGHI published a Threshold Limit Value (TLV) for Heat Stress (ACGHI, 1973). The limit was a relationship between the environmental conditions described by WBGT and the rate of metabolism for workers in ordinary work clothes made from woven fabrics. The limits were expanded to include a Recommended Alert Limit and a Ceiling Limit in 1986 (NIOSH, 1986). The Recommended Alert and Exposure Limits were adopted by the ACGIH in 1990 as Threshold Limit Values for unacclimatized and acclimatized workers, respectively. In addition, the ISO promoted a standard using the WBGT as the environmental index using similar thresholds (ISO, 1989a).

An important assumption in the use of the WBGT-based thresholds is the validity of using hourly Time-Weighted Averages of work and rest to define total heat stress exposure. The validity was confirmed (Lind, 1963b).

3.3. Methods for assessment of metabolism

Evaluating heat stress requires information on metabolic heat as well as environmental heat. The methods to estimate metabolic rate include measurement of oxygen consumption, estimation by task analysis, estimation by activity analysis, and estimation from table look-up (Bernard and Joseph, 1994).

Measurement of oxygen consumption to assess metabolic rate is the *de facto* standard for assessing metabolism. An ISO standard describes techniques for assessing the oxygen demand of a task (ISO, 1989). The field measurement of oxygen consumption is accurate, but impractical.

Task analysis methods rely on the division of a job into individual tasks. The metabolic rate is estimated from previously developed relationships. For instance, the metabolic cost of lifting a box of known weight from the floor to a table of known height can be estimated (Garg et al., 1978). While the potential for assessment error is greater for the task analysis methods than for oxygen consumption, it is somewhat more practical.

Activity analysis is a group of techniques that borrow some of the features of task analysis with a less rigorous break-down of job elements. Two good examples are the one described in the ACGIH TLV for heat stress (ACGIH, 1993) and the Systematic Workload Estimation (SWE). (Tayyari et al., 1989). The ACGIH method breaks the metabolic rate into three components: basal metabolism, posture including walking and climbing, and work activities. The work activities are divided in two stages; the first stage describes the body involvement (e.g., hands, whole body) and the second stage describes the degree of effort from light to very heavy. For each activity in a job, the metabolic rate is estimated and then averaged by time weighting over an hour. The SWE uses a similar approach. First a code is assigned that indicates the class of activity (i.e., stationary, walking, extra exertion) and then a subclass code that depends on the amount of body involvement and degree of effort. The class and subclass codes are cross-referenced to a table that indicates the metabolic rate. While it was designed to record activities on a periodic basis for work sampling, it is applicable to an analysis of individual job activities. The skill levels and time demands are less than for task analysis or measurement of oxygen consumption. Activity analysis is a valuable method to quickly estimate metabolic rate with reasonable results.

Table look-up methods for assessing metabolic heat are a popular approach (ACGIH, 1993; Bernard et al., 1991; Eastman Kodak, 1986; ISO, 1985; Passmore and Durnin, 1955). The analyst seeks the best match between the job in question and standard activities for which the metabolic demand is known. A variation on table look-up is "category assignment". The purpose of category assignment is to place the job in one of three to five categories of metabolic demand (e.g., light, moderate, heavy or very heavy). There are usually descriptors of specific work that may be included in the category or broad descriptors of activities (e.g., light hand work or heavy work with the whole body). The range of metabolic demands that represent a category is about 115 to 175 W. Category assignment is the usual method for many heat stress management programs, and the threshold associated with each metabolic rate category has sufficient safety factors that broad characterizations of the metabolic rate do not place the workers at undue risk. In simplest form, table look-up methods are relatively easy, intuitive, and require little skill, but are the least accurate. An example table showing workload/metabolic heat produced is included in Part I (Smith and Ramsey, 1982).

There is a tendency to overestimate metabolic rate when using tables or categories as described in Part I. To avoid this, break the job down into tasks including break periods, assign a value to each task from the table, and then average the values together using a time-weighted average.

3.4. Effects of clothing

Clothing affects the transfer of heat from and to the body through insulation, permeability and ventilation. The insulating characteristics influence the transfer of heat by convection and radiation. As the insulation value increases, dry heat exchange through the clothing decreases. Permeability indicates the ability of water vapor to pass through the clothing. High permeability means that more sweat will be allowed to evaporate to support cooling of the person. Ventilation characteristics refer to the movement of air through openings in the clothing and through the material. High ventilation generally enhances cooling through evaporation. Conventional cloth work clothes have moderate insulation and permeability and good ventilation characteristics. There are a large number of alternative clothing ensembles to meet the needs for protective barriers against chemical, radioactive and biological contamination. High levels of contamination are usually answered by the requirement to wear vapor-barrier clothing with tight closures. This means that permeability and ventilation drop and insulation may increase. All of which serves to greatly increase heat stress for the same work demands and environmental conditions. To address this problem, vapor-transmitting fabrics are emerging as a major improvement in reducing heat stress and providing contamination protection. This all points to the need to incorporate clothing factors into the heat stress evaluation process.

Ramsey was the first to suggest fixed adjustments to the threshold values for different clothing ensembles (Ramsey, 1978). These ideas were extended by work supported by the Electric Power Research Institute (EPRI) (Bernard et al., 1991).

Finally, the ACGIH recommended clothing adjustments in the 1990 TLVs and have continued them since. The EPRI and ACGIH values are based largely on work by Kenney and others that maps the critical environmental conditions in which thermal equilibrium can be just maintained wearing different clothing ensembles (Kenney et al., 1988). These data were then used by ACGIH to correct the heat stress TLVs according to the thermal properties of the clothing being worn. For vapor-barrier clothing, an adjustment of 10°C WBGT was suggested (Paull and Rosenthal, 1987), but this may be high for lighter weight PVC or thin, coated fabrics.

4. Methods for self-assessment of heat strain

Following the evaluation scheme described in Part I, most of the work force wearing cloth work clothes will be well protected. The adjustment factors are the current best guess about the effects of clothing, but workplace practices based on them should be evaluated to assure that adequate protection is achieved. Vapor-barrier clothing poses the greatest risks to workers both directly from the clothing and from the fact that the work conditions may delay the worker from seeking relief when necessary. And under some circumstances, workers may work above the threshold values. The uncertainty of the threshold values for protective clothing and the intention to work above the thresholds means that self-assessment plays a critical role in stopping the exposure.

Self-determination as a hygiene practice means that a worker is free to stop a heat stress exposure at the onset of extreme discomfort or the symptoms of heat-related disorders. These are subjective decisions that require careful training and experience as well as a proper management climate. Even under these conditions, the reliability of subjective decisions may be poor (much less than 50%) (Honey, 1992).

NIOSH has stated that physiological monitoring is an appropriate means to assure worker protection, and there are a number of means available for this. Because core temperature is the guiding premise to protection, a reliable surrogate measure for core temperature is an obvious choice. Oral temperature is generally 0.5°C below rectal temperature. Oral temperature has been used to assess workplace exposures to heat stress (Fuller and Smith, 1981). Other alternatives involve the ear canal. A clinical device that uses an infrared sensor and a commercial heat stress monitor that uses a thermistor have been developed. They may be sensitive to external conditions and tend to drop faster than core temperature when work stops (Hower and Blehm, 1990; Wallace, 1992).

Heart rate can be easily recorded and monitored (Humen and Boughner, 1984). WHO suggested that the 8-hour average heart rate should not exceed 115 bpm (WHO, 1969). A fixed threshold alarm can be set on most sports-type heart rate monitors, and this should be set to no higher than 90% of the individual's maximal heart rate. Problems with this are premature alerts for peak activity and the potential to miss sustained high, but subthreshold, heart rates (Eastman Kodak, 1986). Goldman suggested a sliding scale of average heart rates for different time periods and Bernard has proposed a continuous assessment method that accounts for the trade-off between duration and average rate as well as adjustments for age-predicted changes in cardiac capacity (Goldman, 1980; Bernard and Kenney, 1994).

Recovery heart rate as an indicator of physiological strain was first proposed by Brouha (1960) and refined by Fuller and Smith (1981). It has been recommended that if recovery rates greater than 110 bpm are observed the rest breaks be extended or the work periods shortened (NIOSH et al., 1985). Bernard and Kenney (1988) further examined recovery heart rates and concluded that the heart rate one minute after stopping work and sitting can provide insight into the physiological strain. Simply stated, a one-minute recovery heart rate greater than 120 bpm represents significant physiological strain, while one-minute rates below 110 bpm represent low levels of strain.

Because either symptoms, core temperature or heart rate may be the critical factor to terminate a heat stress exposure, attention should be paid to all three (Bernard and Kenney, 1994).

5. Heat stress controls

Heat stress controls can be divided into two broad categories: General and Specific. General controls are those that should be implemented anytime there is a reasonable potential for heat stress on the job, that is conditions above the recommended exposure limit (REL) described in Part I. General controls are applicable across all heat stress jobs. Specific controls are directed to individual jobs, and are not readily applicable across all jobs.

5.1. General controls

General controls are the transcendent controls for managing the risk of excessive exposures to heat stress. They include training, heat stress hygiene practices and medical surveillance.

Training for heat stress exposures is like training for any other workplace hazard. It should be accomplished in a similar fashion and integrated with other workplace health and safety training. Topics should include the causes and physiological responses to heat stress, heat-related disorders including symptoms, first aid and prevention, and a description of heat stress hygiene practices.

Heat stress hygiene practices are described in Part I and the rationale for them is discussed above. Hygiene practices are crucial to the reduction of individual risk of heat intolerance and heat-related disorders. While the fundamental responsibility is on the individual, management has the responsibility of making sure that there are no barriers to the practice. For instance, fluid replacement stations must be near work areas and self-determination decisions must be respected.

Medical surveillance is the remaining general control. Preplacement and periodic physicals as well as monitoring of sentinel events are desirable. The physicals should cover both work and personal histories and include a comprehensive physical examination. The physician then should provide a written opinion about the capability of the person to work under conditions of heat stress (NIOSH, 1986). The success of the heat stress controls is seen in the absence over time of sentinel health events such as heat-related disorders, accidents, absenteeism, and chronic fatigue.

5.2. Specific controls

Specific controls are those that are dictated by individual job circumstances and the resources available for control. They follow the traditional hierarchy of engineering controls, administrative controls and personal protection.

Engineering controls are the first choice to reduce or eliminate heat stress as a hazard. While reduction of metabolic rate through mechanical assists or reallocation of work is the single most effective way to reduce heat stress, it may not be practical. If possible, changing clothing requirements to reduce heat stress is also effective – for instance, using vapor-transmitting fabrics instead of vapor-barrier for liquid contamination protection. Enhancing heat loss through ventilation is the next best alternative. Ventilation can be used to reduce ambient temperature and humidity. Ventilation systems can be permanent or temporary, may include mechanical cooling, and can be general or local. Thinking about ventilation systems should not be constrained by presumptions of cost; great improvements in productivity and reduced down time have been reported to more than offset the costs. Local air movement with fans can have some advantage if the air motion before fans is less than 0.5 m/s and the air temperature is less than 42°C. Under conditions of high radiant heat, shields, insulation and surface emissivity changes can reduce the radiant heat (AIHA, 1975).

Administrative controls are reasonable methods to control the risk. Generally they include planning the exposure time and distributing the exposure, scheduling the work to the lowest stress times, allowing adequate recovery, and permitting self-determination. While administrative controls are the most easily implemented and are most reasonable for nonroutine work, they tend to be more expensive than engineering controls in the long run.

Personal protection is appropriate when engineering and administrative controls are not feasible and as a back-up in case of failure. Most often personal protection means personal cooling, which can be in the form of air-circulating systems, liquid-cooling systems, and ice garments. There is a large variation in cost, capability and constraints with different cooling systems, and these must be carefully balanced with job demands. Reflective clothing is another form of personal protection against radiant heat. The selection also needs to match the job conditions and reduction in radiant heat load may or may not be accompanied by a greater loss of evaporative cooling.

References

American Conference of Governmental Industrial Hygienists (ACGHI), 1973. Notice of Intent to Establish Threshold Limit Values for Chemical Substances and Physical Agents in the Workroom Environment with Intended Changes for 1973. Cincinnati, ACGHI, pp. 58-68.

American Conference of Governmental Industrial Hygienists (ACGIH), 1993. Threshold Limit Values and Biological Exposure Indices for 1993-1994. Cincinnati: ACGIH, pp. 82-88.

American Industrial Hygiene Association (AIHA), 1975. Heat and Cooling of Man in Industry, 2nd ed. AIHA, Fairfax.

Belding, H.S. and Hatch, T.F., 1955. Index for evaluating heat stress in terms of resulting physiological strain. Heating, Piping and Air Conditioning, 27: 129-135.

Bernard, T.E. and Joseph, B.S., 1994. Estimation of work metabolism using qualitative job descriptors. American Industrial Hygiene Association Journal (under review).

Bernard, T.E. and Kenney, W.L., 1988. Heart rate recovery. American Industrial Hygiene Conference.

Bernard, T.E. and Kenney, W.L., 1994. Rationale for a personal monitor for heat strain. American Industrial Hygiene Association Journal (in press).

Bernard, T.E., Kenney, W.L., Hanes, L.F. and O'Brien, J.F., 1991. Heat-Stress Management Program for Power Plants. Electric Power Research Institute, Palo Alto, CA, NP4453L.

Beshir, M.Y. and Ramsey, J.D., 1988. Heat stress indices: A review paper. International Journal of Industrial Ergonomics, 3: 89-102.

Brouha, L., 1960. Physiology in Industry. Pergamon Press, London.

Dukes-Dobos, F. and Henschel, A., 1973. Development of permissible heat exposure limits for occupational work. American Society of Heating, Refrigerating and Air Conditioning Engineering Journal, 15(9): 57-62.

Eastman Kodak Company, 1986. Ergonomic Design for People at Work, Vol. 2. Van Nostrand Reinhold, New York.

Fuller, F.H. and Smith, P.E., 1981. Evaluation of heat stress in a hot workshop by physiological measurements. American Industrial Hygiene Association Journal, 42: 32-37.

Garg, A., Chaffin, D.B. and Herrin, G.D., 1978. Prediction of metabolic rates for manual materials handling jobs. American Industrial Hygiene Association Journal, 39: 661-674.

Goldman, R.F., 1980. Prediction of heat strain, revisited. In: F.N. Dukes-Dobos and A. Henschel (Eds.), Proceedings of a NIOSH Workshop on Recommended Heat Stress Standards. DHHS(NIOSH), Washington, DC, pp. 81-108.

Honey, K.M., 1992. The use of subjective responses to limit individual exposure to heat stress. MSPH Thesis, University of South Florida.

Hower, T.C. and Blehm, K.D., 1990. Infrared thermometry in the measurement of heat stress in firefighters wearing protective clothing. Applied Occupational and Environmental Hygiene, 5: 782-786.

Humen, D.P. and Boughner, D.R., 1984. Evaluation of commercially available heart rate monitors. Canadian Medical Association Journal, 131: 585-589.

International Standards Organization (ISO), 1989. Determination of metabolic heat production. ISO 8996, Geneva.

International Organization for Standardization (ISO), 1989a. Hot environments – Estimation of the heat stress on working man, based on the WBGT-index (wet bulb globe temperature). ISO 7243, Geneva. (Second edition.)

International Organization for Standardization (ISO), 1989b. Hot environments – Analytical determination and interpretation of thermal stress using calculation of required sweat rate. ISO 7933, Geneva.

Kenney, W.L., Lewis, D.A., Armstrong, C.G., Hyde, D.E., Dyksterhouse, T.S., Fowler, S.R. and Williams, D.A., 1988. Psychometric limits to prolonged work in protective clothing ensembles. American Industrial Hygiene Association Journal, 49: 390-395.

Lind, A.R., 1963a. A physiological criterion for setting thermal environmental limits for everyday work. Journal of Applied Physiology, 18: 51-56.

Lind, A.R., 1963b. Physiological effects of continuous or intermittent work in the heat. Journal of Applied Physiology, 18: 57-60.

Lind, A.R. and Bass, D.E., 1963. Optimal exposure time for development of acclimatization to heat. Federation Proceedings, 22: 704.

National Institute for Occupational Safety and Health (NIOSH), 1972. Criteria for a Recommended Standard for Occupational Exposure to Hot Environments. USDHEW, HSM 72-10296.

National Institute for Occupational Safety and Health (NIOSH), 1986. Criteria for a Recommended Standard for Occupational Exposure to Hot Environments. Revised 1986. USDHHS (NIOSH) Publication No. 86-113.

NIOSH, OSHA, USCG and EPA, 1985. Occupational Safety and Health Guidance Manual for Hazardous Waste Site Activities. DHHS(NIOSH), Washington, DC. pp. 85–115.

Nielsen, M., 1938. Die Regulation der Körpertemperatur bei Muskelarbeit. Skand. Arch. Physiol. 79: 193.

Paull, J.M. and Rosenthal, F.S., 1987. Heat strain and heat stress for workers wearing protective suits at a hazardous waste site. American Industrial Hygiene Association Journal, 48: 458–463.

Passmore, R. and Durnin, J.V.G.A., 1955. Human energy expenditure. Physiological Reviews, 35: 801–840.

Ramsey, J.D., 1978. Abbreviated guidelines for heat stress exposure. American Industrial Hygiene Association Journal, 39: 491–495.

Smith, J.L. and Ramsey, J.D., 1982. Designing physically demanding tasks to minimize levels of worker stress. Industrial Engineering, 14: 44–50.

Stephens, R.L. and Hoag, L.L., 1981. Heat acclimatization, its decay and reinduction in young caucasian females. American Industrial Hygiene Association Journal, 42: 12–17.

Tayyari, F., Burford, C.L. and Ramsey, J.D., 1989. Guidelines for the use of systematic workload estimation. International Journal of Industrial Ergonomics, 4: 61–65.

Wallace, D.O., 1991. Practical heat stress monitoring: A workplace demonstration of WBGT and personal temperature monitoring. MSPH Thesis, University of Utah.

World Health Organization (WHO), 1969. Health factors involved in working under conditions of heat stress. Technical Report Series 42. WHO, Geneva.

Wyndham, C.H., Strydom, N.B., Benade, J.S., van Renberg, A.J., 1972. Heat stroke risk in unacclimatized and acclimatized men of different maximum oxygen intakes working under hot humid conditions. Chamber of Mines Research Report No. 12/72. Chamber of Mines of South Africa, Johannesburg, South Africa.

Yaglou, C.P. and Minard, D., 1957. Control of heat casualties at military training centers. AMA Archives of Industrial Health, 16: 302–316.

Cold stress: Part I – Guidelines for the practitioner

Ingvar Holmér

Division of Work and Environmental Physiology, National Institute of Occupational Health, S-17184 Solna, Sweden

1. Introduction

This guideline is intended for persons involved in the design, planning and control of operations intended for cold environments as well as in the actual organization and performance of such work. The guideline provides recommendations on suitable and relevant methods for assessment and control of different types of cold stress as well as examples of measures for their alleviation and prevention.

---

* The recommendations provided in this guide are based on numerous published and unpublished scientific studies and are intended to enhance worker safety and productivity. These recommendations are neither intended to replace existing standards, if any, nor should be treated as standards. Furthermore, this document should not be construed to represent institutional policy.

The following individuals participated in the discussion of the earlier version of this guide. Their suggestions (written or verbal) were incorporated by the authors in this version: Arne Aaras, *Norway*; Fred Aghazadeh, *USA*; Roland Andersson, *Sweden*; Jan Dul, *The Netherlands*; Jeffrey Fernandez, *USA*; Matthias Jäger, *Germany*; Åsa Kilbom, *Sweden*; Anders Kjellberg, *Sweden*; Olli Korhonen, *Finland*; Helmut Krueger, *Switzerland*; Shrawan Kumar, *Canada*; Ulf Landström, *Sweden*; Tom Leamon, *USA*; Anil Mital, *USA*; Ruth Nielsen, *Denmark*, Jerry Ramsey, *USA*; Murray Sinclair, *UK*; Rolf Westgaard, *Norway*; Ann Williamson, *Australia*; Jørgen Winkel, *Sweden*; Pia Zätterström, *Sweden*. The guide was also reviewed in depth by several anonymous reviewers.

2. Application context

Recommendations given in these guidelines apply in a wide sense to occupational as well as leisure-type activities carried out under cold ambient conditions. They apply to continuous long exposures as well as intermittent, short term and single exposures. For the purpose of these guidelines a cold environment is defined as an environment under which greater than normal heat losses are anticipated and compensatory thermoregulatory actions required. Normal heat losses, hence, refer to what people normally experience during indoor living conditions (air temperature 20–25°C).

Cold stress implies a sequel of events of varied character. The guidelines apply to the direct and imminent thermal effects, their assessment and measures for prevention and control. The vast area of problems related to organisational, psychological, medical and ergonomic aspects are only briefly dealt with.

3. Definitions and terminology

Cold stress – General expression of the cooling power exerted by physical, climatic factors on human body tissues.

Physiological strain – General expression of the sequel of physiological responses evoked by the physical stress factors.

Cold stress index – Measure of cold stress determined by more or less complex empirical or analytical methods integrating climatic factors in a way relevant for their effect on the body.

4. Problem identification

Depending on the situation cold stress may be present at temperatures just below the normal indoor temperature. Naturally, the magnitude and severity of cold stress will be expected to increase with lowered ambient temperature.

Problems associated with occupational cold exposure comprise:
- thermal discomfort and pain sensation, in particular, from extremities,
- performance decrements due to cold hands, cold muscles or general cooling or caused by the hobbling effect of protective clothing (weight, bulk, friction, etc.),
- health effects in terms of cold injuries, initiation and aggravation of symptoms for cardiorespiratory diseases or elevated accident risks,
- special requirements for design and performance of work, construction of workplaces and design and use of equipment and tools, caused by the above factors.

4.1. Discomfort and pain

Thermal discomfort arises whenever excessive heat is lost from part of the body or the body as a whole. Severity of discomfort increases with magnitude of excessive heat loss and turns gradually into pain sensation. Individual variation in response is considerable and significant overlapping of discomfort and pain sensation may occur at moderate levels of cold stress.

4.2. Performance and work capacity

Fine movements of fingers and hands may suffer considerable deterioration in function, even at moderate levels of cold stress. Cooling of deeper tissues may also change physical and chemical properties of fluids and tissues, thereby impairing muscle function and aggravating losses in dexterity and power exertion.

With more pronounced extremity cooling, impaired conditions develop for the exertion of gross muscle movements, heavy physical work and control of walking. Capacity for prolonged, energetic work is reduced and more so when core temperature drops below 36°C. With rapid body cooling, as in cold water, work capacity degrades rapidly. In combination with impaired muscular efficiency a moderate effort may quickly turn into heavy work and, eventually, become exhaustive in a short time.

4.3. Health effects

Cardio-respiratory effects

Inhalation of very cold air cools the mucous membranes of the upper respiratory tract and may, eventually, cause irritation, microinflammatory reactions and provoke bronchospasm. Bronchospasm is a common reaction in the cold and is particularly pronounced in asthmatic persons and in people with hypersensitive airways. Airway cooling may provoke pain symptoms in persons with cardiovascular disorders.

Cold may increase blood pressure, acute or chronically. Persons with circulatory diseases (angina pectoris, Raynaud's disease, etc.) are more susceptible and prone to suffer from cold exposures.

Since many drugs act on the cardiorespiratory system their effect may interfere with thermoregulatory responses, for example medicine for hypertension. With ageing thermoregulatory adjustments become less efficient. A physician should be consulted for appropriate action, whenever the above cases can be suspected.

Local tissue damage ("white finger") may be triggered by exposure to cold after prolonged exposures to vibration from hand held power tools.

Cold injuries

The following types of cold injury are recognized (for explanations see Part II):
- Local cold injury
    Non-freezing cold injury

Frostnip (superficial freezing of skin tissue; white, pale spot)
Frostbite (deep, frozen tissue; hard and solid to touch)
- General body cooling
Hypothermia (subnormal body temperature)

With progress of body cooling, performance and physical work capacity deteriorate and mental confusion and impaired judgement develop. The person may not recognize the full danger of the situation. At this stage external assistance may be the only alternative to interrupt exposure.

4.4. Ergonomic requirements

Protection against cooling by necessity requires the wear of layers of clothing. Multi-layer clothing, handwear, and footwear, imply restrictions on mobility, gross muscle movements and dexterity. Accordingly, work tasks take longer time to finish in the cold and exert higher strain due to the extra cost of protection.

5. Assessment

The assessment of cold stress involves the ascertainment of a risk of one or more of the mentioned effects. Typically, Table 1 may be used as a first rough classification.

Information given in the table should be interpreted as signal to action. In other words, cold stress should be evaluated and controlled, if required. The table does not differentiate between different types of problems. At moderate temperatures problems associated with discomfort and

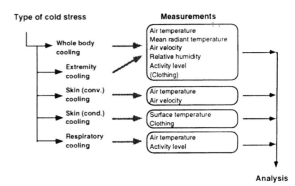

Fig. 1. Scheme for identification of cold stress and selection of appropriate measurements for its assessment. Evaluation methods are presented below.

function losses due to local cooling prevail. At lower temperatures the imminent risk of a cold injury as a sequel to the other effects is the important factor. For many of the effects discrete relationships between stress level and effect do not yet exist. It cannot be excluded that a cold problem may persist also outside the range of temperatures denoted by the table.

For practical purposes, the following types of cold stress should be identified and evaluated. Most likely, several, if not all, of them may be present in the typical case.
- Whole body cooling
- Local cooling such as
    Extremity cooling
    Convective skin cooling (wind chill)
    Conductive skin cooling (contact cooling)
    Cooling of respiratory tract

When a problem is present and needs assessment the strategy outlined in Fig. 1 may be applied. The problem is identified as one or more

Table 1
Alert scheme on the likelihood of appearance of one or more defined cold problems. Interpretation of marks: * low risk, ** moderate risk, *** high risk, **** very high risk, # only controlled, short exposures. Strong wind will shift marks by one or two columns for all work factors. Conditions assume normal measures of precaution

Work factor	< −30°C	−30 to −20°C	−20 to −10°C	−10 to 0°C	0 to 10°C	10 to 20°C
Fine, manual work	#	#	****	***	**	*
Light work	#	****	***	***	**	*
Moderate work	#	***	**	**	*	
Heavy work	***	**	*			

types of cold stress and the appropriate measurements are taken. Further details about data collection are given below.

6. Data collection

Depending on type of expected risk, different sets of measurements are required (Fig. 1). Procedures for data collection and accuracy of measurements depend on the purpose. A quick estimate may be useful as a first rough classification of problems and a basis for further action, be it a more detailed assessment or a preventive measure.

6.1. Quick estimate

A quick and rough estimate may be based on information about type and intensity of work, air temperature, wind speed and available personal protection.

Work is classified according to Table 2. Required clothing insulation (clo-value) is interpolated from the appropriate interval in Fig. 2 and matched with the estimated resultant clo-value of actual clothing (Table 3). If clothing provides less protection than required, exposure must be time limited. Recommended exposure time is estimated from Fig. 3.

In addition, the risk of local cooling effects is estimated by consulting Fig. 4 (finger cooling) and Table 4 (Wind chill).

6.2. Detailed assessment

A more accurate and reliable assessment requires a systematic collection of data for the relevant factors according to Fig. 1. Standard methods are available for climatic measurements

Table 2
Classification of activity level (metabolic heat production) in 6 classes. Modified from ISO 8996 (1)

Class 1 – Resting   < 117 W (< 65 W/m^2)

Class 2 – Very light   117–160 W (65–89 W/m^2)
 Sitting at ease: light handwork (writing, typing, drafting, bookkeeping); hand and arm work (inspecting, assembling or sorting light materials), arm and leg work (driving car under average conditions, operating foot switch or pedal).
Standing: assembling (small parts), finishing, packing (small parts).
Walking at ease: inspection.

Class 3 – Light   161–232 W(90–129 W/m^2)
 Sitting at ease: hand and arm work (small bench tools), arm and leg work (). Standing: drill press (small parts); milling machining (small parts); coil taping; small armature winding; machining with light power tools; casual walking up to 0.9 m/s ($\approx$ 3 km/h).
Lifting: 4.5 kg, fewer than 8 lifts per minute; 11 kg, fewer than 4 lifts per minute.

Class 4 – Moderate   233–348 W(130–195 W/m^2)
 Hand and arm work (nailing, filing); arm and leg work (off-road operation of trucks, tractors or construction equipment); arm and trunk work (air hammer operation, tractor assembly, plastering, intermitten handling of moderately heavy materials; pushing or pulling lightweight carts or wheelbarrows; walking 0.9–1.3 m/s (3–4.7 km/h).
Lifting: 4.5 kg, fewer than 10 lifts per minute; 11 kg, fewer than 6 lift per minute.

Class 5 – Heavy   349–465 W(196–260 W/m^2)
 Heavy arm and trunk work transferring heavy materials, shoveling; sledge hammer work; sawing, planting or chiseling hardwood; digging; walking 1.8 m/s (6.5 km/h), pushing or pulling loaded handcarts or wheelbarrows; chipping castings; laying concrete block.
Lifting: 4.5 kg, 14 lifts per minute, 11 kg, 10 lifts per minute.

Class 6 – Very heavy   < 465 W(> 260 W/m^2)
 Heavy activity at fast to maximum pace: ax work; heavy shoveling or digging; climbing stairs, ramps, or ladders; jogging, running, walking faster than 1.8 m/s (> 6.5 km/h).
Lifting: 4.5 kg, more than 18 lifts per minute; 11 kg, more than 13 lifts per minute.

Table 3
Examples of basic insulation values of clothing. Nominal protection level only applies to static, windstill conditions (resting). Values must be reduced with increased activity level (see text). Modified from ISO-TR 11079 (1)

Clothing-ensemble	clo
1. Briefs, short-sleeve shirt, fitted trousers, calf length socks, shoes	0.5
2. Underpants, shirt, fitted trousers, socks, shoes	0.6
3. Underpants, overall, socks, shoes	0.7
4. Underpants, shirt, coverall, socks, shoes	0.8
5. Underpants, shirt, trousers, smock, socks, shoes	0.9
6. Briefs, undershirt, underpants, shirt, overalls, calf length socks, shoes	1.0
7. Underpants, undershirt, shirt, trousers, jacket, vest, socks, shoes	1.1
8. Underpants, shirt, trousers, jacket, coverall, socks, shoes	1.3
9. Undershirt, underpants, insulated trousers, insulated jacket, socks, shoes	1.4
10. Briefs, T-shirt, shirt, fitted trousers, insulated coveralls, calf length socks, shoes	1.5
11. Underpants, undershirt, shirt, trousers, jacket, overjacket, hat, gloves, socks, shoes	1.6
12. Underpants, undershirt, shirt, trousers, jacket, overjacket, overtrousers, socks, shoes	1.9
13. Underpants, undershirt, shirt, trousers, jacket, overjacket, overtrousers, socks, shoes, hat, gloves	2.0
14. Undershirt, underpants, insulated trousers, insulated jacket, overtrousers, overjacket, socks, shoes	2.2
15. Undershirt, underpants, insulated trousers, insulated jacket, overtrousers, overjacket, socks, shoes, hat, gloves	2.6
16. Undershirt, underpants, insulated trousers, insulated jacket, overtrousers and parca with lining, socks, shoes, hat, mittens	2.6 3.4
17. Arctic clothing systems	3–4.5
18. Sleeping bags	3–8

and instrumentation (ISO 7726), for determination of activity level and metabolic heat production (ISO 8996) and for determination of clothing thermal charateristics (ISO 9920, prEN342 and prEN511). For detailed references see Part II.

Air temperature is measured with an instrument shielded for radiant heat fluxes (e.g. solar radiation). Mean radiant temperature can be cal-

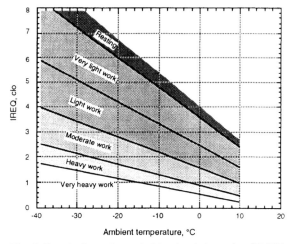

Fig. 2. Required, resultant clothing insulation value (IREQ) for different classes of activity (ISO-TR 11079). Values apply to conditions with no perceivable wind and heat radiation. For examples of activity see Table 2.

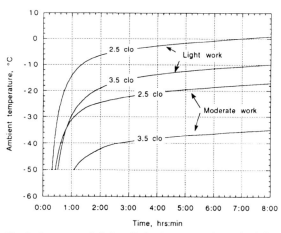

Fig. 3. Recommended time limit of exposure determined for two classes of activity and for two levels of basic insulation value of clothing. For examples see Tables 2 and 3.

Table 4
Cooling power of wind on exposed flesh expressed as a chilling temperature (t_{ch}) under almost calm conditions (wind speed 1.8 m/s). Values lower than −30°C represent a risk for frostnip or frostbite

Wind speed m/s	Actual thermometer reading (°C)										
	0	−5	−10	−15	−20	−25	−30	−35	−40	−45	−50
1.8	0	−5	−10	−15	−20	−25	−30	−35	−40	−45	−50
2	−1	−6	−11	−16	−21	−27	−32	−37	−42	−47	−52
3	−4	−10	−15	−21	−27	−32	−38	−44	−49	−55	−60
5	−9	−15	−21	−28	−34	−40	−47	−53	−59	−66	−72
8	−13	−20	−27	−34	−41	−48	−55	−62	−69	−76	−83
15	−18	−26	−34	−42	−49	−57	−65	−73	−80	−88	−96
20	−20	−28	−36	−44	−52	−60	−68	−76	−84	−92	−100

culated from measurements of globe temperature, air temperature and air velocity. Air velocity (wind speed) is measured as a 1 or 3 min average value. The contribution of travelling speed to the resulting wind speed should be estimated. For projecting and planning purposes information from a nearby, local weather station may be useful.

Estimation of the activity level (metabolic heat production) is required for some types of cold exposure. For practical purposes a rough classification of activity into six classes may be sufficient (Table 2).

Pertinent information must be obtained regarding variation in time of the climatic parameters, as well as of activity level and/or clothing.

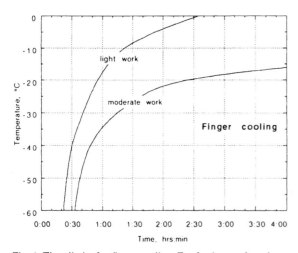

Fig. 4. Time limits for finger cooling. For further explanations see text. Total clothing insulation is assumed at 3.4 clo and total finger insulation is 2 clo. Wind speed is below 0.5 m/s.

Simple time-weighting procedures should be adopted for evaluating both activity level and exposure time.

7. Data analysis

In the following, procedures and methods for data analysis are presented.

7.1. Whole body cooling

On the basis of measurements or estimates according to Fig. 1 a required clothing insulation value is determined. This insulation value is calculated with the IREQ-method as described in Part II of this guideline. Fig. 2 provides required insulation (in clo) for different classes of activity (see Table 2). Once the requirement is determined, the IREQ-value is compared with the protection level offered by used or available clothing (Table 3).

Protection level is determined by the *resultant* insulation value of the clothing system, by air permeability of fabrics and by resistance to contact cooling of fabrics. These properties are measured according to prEN 342 (for clothing) and prEN 511 (for gloves) (1). Values may also be obtained from tables in the literature (e.g. ISO 9920 (1), Table 3). Such values are static values (basic insulation values) and must be corrected for presumed reduction caused by body motion and ventilation. Typically, no correction is made for resting level. Values of Table 3 are reduced by 10% for light work and by 20% for higher work levels.

When used or available clothing system *does not provide sufficient insulation* (lesser than IREQ), a time limit is calculated for the actual conditions. Fig. 3 shows examples of time limits for light and moderate work with two insulation levels of clothing. Time limits for other combinations may be estimated by interpolation.

7.2. Extremity cooling

The heat balance of hands and feet relies greatly upon heat input by warm blood from the body core, and input is usually high when activity is high. The curves in Fig. 4 may be used as basis for recommendations as to acceptable exposures of fingers. The curves represent two activity levels. Suitable handwear for winter conditions (2 clo) and, of course, adequate clothing, is assumed. For less effective protection exposure time is considerably shorter!

A similar set of curves should apply to toes. However, more insulation may be available for protection of feet, resulting in longer exposure times. Nevertheless, it follows from Figs. 3 and 4, that extremity cooling is probably more critical for exposure than whole body cooling.

7.3. Convective skin cooling

The Wind Chill Index (WCI) represents a simple, empirical method for assessment of cooling of unprotected skin (face). Table 4 provides equivalent cooling temperatures for various combinations of air temperature and wind speed. The table applies to active, well-dressed persons. A risk is present when equivalent temperature drops below $-30°C$ and skin may freeze within 1–2 minutes below $-60°C$.

7.4. Contact cold

Contact between bare hand and cold surfaces may quickly reduce skin temperature and cause freezing injury. Problems may arise with surface temperatures as high as 15°C. In particular, metal surfaces provide excellent conductive properties and may quickly cool contacting skin areas.
- Prolonged contacting with metal surfaces below 15°C may impair dexterity.
- Prolonged contacting with metal surfaces below 7°C may induce numbness.
- Prolonged contacting with metal surfaces below 0°C may induce frostnip or frostbite.

Other materials present similar sequence of hazards, but temperatures are lower with less conducting material (plastic, wood, foam).

7.5. Cooling of respiratory tract

Inhaling cold, dry air may cause problems for sensitive persons already at $+10$ to 15°C. Healthy persons performing light to moderate work require no particular protection of the respiratory tract down to $-30°C$. Very heavy work during prolonged exposures (hours) should not take place at temperatures below $-20°C$.

8. Preventive measures for alleviation of cold stress

Actions and measures for the control and reduction of cold stress imply a number of considerations during the planning and preparatory phases of work shifts, as well as during work.

Natural cold exposure comprises a large number of complex, often unpredictable exposure conditions. In contrast, artificial cold work is usually carried out in very constant climatic conditions, albeit temperature level may vary depending on type of business. Conditions are mostly constant over time and space. Problems and assessment methods are basically the same, as may be preventive measures. The appropriate assessment methods and preventive measures for the particular case should be selected by the practitioner.

The following section provides a list of pragmatic as well as more sophisticated measures for the alleviation of cold stress.

8.1. Work and environmental factors

Planning phase

The special problems and risks associated with work under cold conditions must be considered in the planning phase of projects. In particular the following recommendations should be tried:

- schedule work for a warmer season (for outdoor work)
- check if work can be done indoors (for outdoor work)
- allow more time per task with cold work and protective clothing
- analyze suitability of tools and equipment for work
- organize work in suitable work-rest regimens, considering task, load and protection level
- provide heated space or heated shelter for recovery
- provide training of complex work tasks under normal conditions
- check medical records of staff
- ascertain appropriate knowledge and competence of staff
- provide information about risks, problems, symptoms and preventive actions
- seperate goods and worker line and keep different temperature zones
- care for low velocity, low humidity and low noise level of the air-conditioning system
- provide extra manpower to shorten and/or reduce exposure
- select adequate protective clothing and other protective equipment

Preparation for work
Before every work shift
- check climatic conditions at onset of work
- schedule adequate work-rest regimens
- allow for individual control of work intensity and clothing
- select adequate clothing and other personal equipment
- check weather and forecast (outdoors)
- prepare schedule and control stations (outdoors)
- organize communication system (outdoors)

Exposure phase
During the actual work shift a number of observations and controls allows safe management of working conditions.
- provide for break and rest periods in heated shelter
- provide for frequent breaks for hot drinks and food
- care for flexibility in terms of intensity and duration of work
- provide for replacement of clothing items (socks, gloves, etc.)
- protect from heat loss to cold surfaces
- minimize air velocity in work zones
- keep workplace clear from water, ice and snow
- insulate ground for stationary, standing work places
- provide access to extra clothing for warmth
- monitor subjective reactions (buddy system) (outdoors)
- report regularly to foreman or base (outdoors)
- provide for sufficient recovery time after severe exposures (outdoors)
- protect against wind effects and precipitation (outdoors)
- monitor climatic conditions and anticipate weather change (outdoors)

8.2. Individual factors

Behavior
Individual behavior is critical to control of cold exposure. With knowledge and experience unneccesary stress can be avoided and efficient operation maintained.
- allow for time to adjust clothing
- prevent sweating and chilling effects by making adjustments of clothing in due time before change in work rate and/or exposure
- adjust work rate (keep sweating minimal)
- avoid rapid shifts in work intensity
- allow for adequate intake of hot fluid and hot meals
- allow for time to return to protected areas (shelter, warm room) (outdoors)
- prevent wetting of clothing from water or snow
- allow for sufficient recovery in protected area (outdoors)
- report on progress of work to foreman or base (outdoors)
- report major deviations from plan and schedule (outdoors)

Clothing
- select clothing you have previous experience with
- with new clothing, select tested garments

- select insulation level on the basis of anticipated climate and activity
- care for flexibility in clothing system to allow for great adjustment of insulation
- clothing must be easy to don and doff
- reduce internal friction between layers by proper selection of fabrics
- select size of outer garment to make room for adjustment of the insulative middle layer
- use multi-layer system
  inner layer for microclimate control
  middle layer for insulation control
  outer layer for environmental protection
- inner layer should be non-absorbent to water, if sweating cannot be sufficiently controlled
- inner layer may be absorbent, if sweating is anticipated to be none or low
- inner layer may consist of dual-function fabrics, in the sense that fibers in contact with skin are non absorbing and fibers next to the middle layer are absorbing water or moisture
- middle layer should provide loft to allow stagnant air layers
- middle layer, preferably, should be form-stable and resilient
- middle layer may be protected by vapor barrier layers
- garments should provide sufficient overlap in the waist and back region
- outer layer must be selected according to additional protection requirements, such as wind, water, oil, fire, tear or abrasion
- design of outer garment must allow easy and extensive control of openings at neck, sleeves, wrists, etc, to regulate ventilation of interior space
- zippers and other fasteners must function also with snow and windy conditions
- buttons should be avoided
- clothing shall allow operation, also with cold, clumsy fingers
- design must allow for bent postures without compression of layers and loss of insulation

Handwear
- mittens provide the best overall insulation
- mittens should allow fine gloves to be worn underneath
- prolonged exposures requiring fine hand work, must be intercepted by frequent warm-up breaks
- pocket heaters or other external heat sources may prevent or delay hand cooling
- sleeve of clothing must easily accommodate parts of gloves or mittens – underneath or on top
- outer garment must provide easy storage or fixing of handwear when taken off

Footwear
- boots shall provide high insulation to the ground (sole)
- sole shall be made of a flexible material and have an anti-slippery pattern
- select size of boot so it can accommodate several layers of socks and an insole
- ventilation of most footwear is poor, so moisture should be controlled by frequent replacement of socks and insole
- control moisture by vapor barrier between inner and outer layer
- allow boots to dry completely between shifts
- legs of clothing must easily accommodate parts of boots – underneath or on top

Head protection
- flexible headwear comprises an important instrument for control of heat and whole body heat losses
- headwear should be windproof
- design should allow sufficient protection of ears and neck
- design must accomodate other types of protective equipment (e.g., ear muffs, safety goggles)

Face and respiratory protection
- face mask should be windproof and insulative
- no metallic details should contact skin
- significant heating and humidification of inspired air can be achieved by special breathing masks or mouth pieces

8.3. Equipment

Tools and equipment
- select tools and equipment intended and tested for cold conditions

- choose design that allows operation by gloved hands
- prewarm tools and equipment
- store tools and equipment in heated space
- insulate handles of tools and equipment

Machinery
- select machinery intended for operation in cold environments
- store machinery in protected space
- prewarm machinery before use
- insulate handles and controls
- design handles and controls for operation by gloved hands
- prepare for easy repair and maintenance under adverse conditions

Education and training
- provide education and information on the special problems of cold
- provide information and training in first-aid and treatment of cold injuries
- test machinery, tools and equipment in controlled cold conditions
- select tested goods, if available
- train complex operations under controlled cold conditions

Reference

Cold stress: Part II – The scientific basis (knowledge base) for the guide. International Journal of Industrial Ergonomics, 14(1–2): 151–159 (this issue).

Cold stress: Part II – The scientific basis (knowledge base) for the guide

Ingvar Holmér

Division of Work and Environmental Physiology, National Institute of Occupational Health, S-17184 Solna, Sweden

1. Problem description

The major hazard associated with cold exposure is the potential for excessive body heat losses. As a sequel to abnormal heat loss, tissues may cool down and result in more or less irreversible damage to bodily functions. Depending on the nature and magnitude of cold stress, problems may vary from perceived thermal discomfort to death from hypothermia. The main problems associated with cold exposure are the following:
– thermal discomfort and pain sensation – in particular, from the extremities;
– impaired manual performance, caused by the cold hand and/or by the gloved hand;
– impaired mobility and operational capacity due to weight and bulk of clothing and/or environmental conditions (ice, snow, etc.);
– deterioration in physical work capacity with muscle and body cooling;
– risk of cold injury with extreme exposures;
– initiation and aggravation of symptoms associated with certain diseases;
– increased effort and exertion due to the hobbling effect of clothing and other protective equipment;
– increased accident risks associated with above factors;
– special requirements for design of work, work stations, walkways, tools, equipment, signal and alarm systems, caused by the above factors.

2. Scope of the problem

Cold stress may be present at temperatures at or just below the comfort zone, in particular with sedentary work (Fanger, 1970). Naturally, the lower the ambient temperature, the higher becomes the cold stress. The actual response to a given level of cold stress depends largely upon the capacity of evoked behavioral and thermoregulatory adjustments. In other words, the resulting cooling effect determines the severity of response. An understanding of the physics of heat exchange and the subsequent modeling of body heat balance, thus becomes an important instrument for assessment of cold stress.

Cold climates prevail in large parts of the world. In particular, subarctic and Arctic areas are cold. Except for tropical areas natural cold climates may prevail for shorter or longer periods during the year. In desert areas the hot day is often followed by a cold night due to negative radiant heat balance at night. Artificial cold climates (e.g., cold stores) are common and are found also in tropical countries. Water immersion, in most conditions, entails a cold stress problem. The particular aspects of cold water exposure is out of the scope of this guideline. A recent review of the problems associated with occupational cold exposure in Sweden (Anon, 1989) found that about 25% of the worker population experienced some problems with cold and about 5% experienced significant problems. In Sweden this 5% corresponds to approximately 250,000 workers.

Statistical evidence for the various types of cold effects is difficult to obtain. First, few of the systems surveying occupational health problems recognize cold as a primary risk factor. Second, when an accident occurs it is often explained by more obvious causes (slipping, clumsy handling, etc.) and cold, per se, may not be regarded as the primary factor.

3. Background

3.1. Cold effects

In general, older persons appear to be less tolerant to cold (Wagner and Horvath, 1985). In terms of general cooling, the insulation provided by the thicker subcutaneous fat layer seems to render women more tolerable to cold stress (McArdle et al., 1992; Wagner and Horvath, 1985), in particular, in cold water (Keatinge, 1969).

Cold may have direct effects on psychological and physiological functions as well as on health and well-being. In addition there is a "cost of protection", in a sense that the use of protective measures to prevent cooling often comprises increased physical and mental effort. Work becomes more cumbersome and tedious. Equipment and material become more difficult to handle.

Cold discomfort and pain

Extensive research provides a bulk of evidence for the increased discomfort experienced by people exposed, also to moderate thermal environments. Discomfort is largely a function of whole body or local heat balance (Fanger, 1970; McIntyre, 1980). Depending on temperature, activity and clothing, as well as the presence of thermal asymmetries, persons may experience cold discomfort at temperatures around +20°C.

Cold sensation and pain are well related to mean skin temperature for whole body cooling (Iampietro, 1971) and local skin temperature for local cooling (Enander, 1982). Local temperature is also the critical factor for receptor function. Blocking of peripheral sensory receptors seems to occur at local skin temperatures around 7–8°C (de Jesus et al., 1973; Fagius et al., 1989; Vanggaard, 1975).

Work capacity

Cooling of tissues slows down metabolic processes and retards the signaling and processing capabilities of the nervous system. As skin tissue and extremities are most prone to cooling, degradation of mechanoreceptor and neuro-muscular function easily develops. Deprivation of sensory functions impairs neuro-motor function and, accordingly, affects the ability to initiate and control movements.

Manual performance. Manual performance is likely to suffer considerable losses when hands and fingers are cooled and several investigators have addressed this problem (see reviews by Ellis et al., 1985; Enander and Hygge, 1990; Fox, 1967). Fine motor functions may be impaired already at finger temperatures around 30–31°C. Gross motor function and muscular power are reduced at hand temperatures below 20°C (Gaydos and Dusek, 1958; Litchfield, 1987; Lockhart et al., 1975; Rogers and Noddin, 1984).

Mental performance. Mental performance is more complex and may comprise different types of functions from a simple reaction time test to problem solving. Effects of cold are less well understood. Sometimes results of similar investigations show opposite effects. Nevertheless, brain cooling causes significant mental dysfunction (MacLean and Emslie-Smith, 1977; Wilkerson et al., 1986). Short exposures may actually result in improvement of performance, at least for some types of tasks. A possible explanation would be compensatory reactions due to increased arousal (Enander and Hygge, 1990). Thus, clearly sub maximal performance requirements are prone to be less affected than near-maximal. The longer the exposure (hours), the greater becomes the cooling of tissues and the more prominent would be the expected performance loss (Jokl, 1982; Lockhart et al., 1975; Meese et al., 1981).

Physical work capacity. When cooled, muscles perform less well, as previously mentioned. Action potential of the motor unit becomes delayed, is reduced in magnitude and progresses more slowly (Vanggaard, 1975). In addition, circulatory capacity degrades due to increased peripheral resistance and direct effects of cold on heart function (MacLean and Emslie-Smith, 1977; Popovic and Popovic, 1974). The overall expected effect is that sufficient circulatory support cannot be maintained to muscles performing at high intensity for prolonged periods. Physical work capacity, defined as maximal aerobic power, is reduced (Bergh, 1980; Holmér and Bergh, 1974). The energy cost performing sub maximal work may rapidly increase with falling muscle and core temperature (Holmér and Bergh, 1974). Apparently, exhaustion may develop rapidly parallel to reduction in maximal aerobic power and increase in sub maximal energy cost. This could be a plausible explanation to many cases reported in magazines and newspapers of sudden failure to cope with the conditions (MacLean and Emslie-Smith, 1977).

Muscular power diminishes with lowered tissue temperature (Bergh, 1980) and becomes less efficient (Oksa et al., 1993). This effect has implications for many critical exposure conditions. Survival and similar equipment must not require extensive effort to operate.

Health effects

Respiratory system. It is well-known that inhalation of cold air may induce broncho-constriction in sensitive persons (McFadden et al., 1985; O'Cain et al., 1980). Normal subjects usually cope with cold air inhalation without major problems, at least down to −30°C and at moderate work intensities (Horvath, 1981). Athletic endurance events in many countries are canceled when ambient temperature conditions drop below −20°C.

Cardiovascular effects. Cold may have significant cardiorespiratory effects. Thermoregulatory peripheral vasoconstriction induces an elevated blood pressure. In addition facial cooling may trigger a significant transient increase in blood pressure (LeBlanc et al., 1976). Cold may evoke and/or aggravate symptoms associated with different types of cardiovascular diseases (Harjula, 1980). In particular, persons with angina pectoris often feel discomfort and pain with cold exposure (Lassvik, 1979). Apparently, airway cooling plays an important role for the promotion of symptoms.

Recently, strong correlation between cold weather (low temperature and strong wind) and number of deaths from acute coronary heart disease have been reported (Gyllerup, 1992). The underlying mechanism is not clear, but an integrated action of many factors associated with adverse cold conditions could present a plausible explanation.

Cold injuries. Several investigators have addressed the subject and the reader is referred to a number of reviews on the subject for details (Granberg et al., 1991; MacLean and Emslie-Smith, 1977; Wilkerson et al., 1986).

The imminent danger of cold environments is the risk of getting a cold injury. The actual freezing of tissues results in frostnip or frostbite. Frostnip is a local freezing of the superficial skin tissue, causing a white, pale spot. Rapid rewarming prevents any significant persistent damaging effects. Frostbite defines a condition when deeper tissues freeze and form ice crystals in the tissue. The skin is hard and cannot slide on top of underlying layers. Severity and damage depend on tissue temperature, duration, intensity, extension and rewarming procedures.

Long exposures of extremities at low tissue temperature, albeit above °C, may result in a non-freezing cold injury. Low temperature, immobility of extremities, moisture and poor hygiene are predisposing factors. The main damaging effect appears to be vascular and cellular dysfunction, eventually irreversible.

Hypothermia is present when core temperature drops below 35°C. A sequel of physiological and psychological reactions are observed when body temperature drops to this level and lower. A general feature is the incapacitation of the individual in terms of both mental abilities and work capacity. The victim is almost entirely dependent on external help to relieve stress and recover. Further body cooling may progress into unconsciousness and paralysis of most bodily functions. Cardiorespiratory functions reduce to a minimum. A core temperature of 28°C is recognized as the level with significantly increased risk of cardiac fibrillation. The care and rewarming of deeply hypothermic victims must be done with hospital equipment and by persons with special medical training. Transportation must take place with extreme care to prevent cardiac stress.

Cost of protection

The hobbling effect of clothing and other kinds of protective equipment is well known and documented (Belding, 1949; Endrusick et al., 1991; Lotens, 1987; Reed et al., 1988; Rodahl, 1991; Rogers and Noddin, 1984; Siple, 1945; Teitlebaum and Goldman, 1972). The contradictory requirements for cold protection and for mobility and dexterity are difficult to accommodate in clothing and may neccessitate other solutions, e.g., work organisation and auxilary heating. The increased energy cost of working with multi-layer clothing due to internal friction between layers (Teitlebaum and Goldman, 1972) and to weight (Duggan and Haisman, 1992; Goldman, 1975) implies that a defined work task elicits a higher relative work load (earlier development of fatigue) or takes longer time to perform (reduction in work pace to maintain same relative work load). The same conditions apply to manual work and the effect of gloves (Rogers and Noddin, 1984; Sperling et al., 1983; van Dilla et al., 1949).

3.2. Cold stress

Can cold stress be defined and quantified and what relationships exist between such values and the above mentioned effects?

Cold stress is defined as the integrated effect of a combination of climatic factors affecting body heat exchange. The effects described previously are more or less directly caused by cooling of tissues. It is likely, then, that expressions of cold stress that relate to its cooling effect on the body have high validity. The integration of two or more climatic factors into more or less complex mathematical descriptions of their cooling effect is a useful approach. Cold stress indices, parallel to the heat stress indices, have been proposed, although they are few.

Cold stress may be classified according to various types of cooling. The following types can be defined and will be discussed in the next section.
- Whole body cooling
- Extremity cooling
- Convective cooling
- Contact cooling
- Airway cooling

Assessment of cold stress

Whole body cooling. Whole body heat balance is a function of metabolic heat production and the various forms of heat losses. Several equations have been proposed for the prediction of heat balance (Burton and Edholm, 1955; Holmér, 1984b; Steadman, 1984). They all account for the important effect of clothing, but use different expressions for calculation of heat losses. The IREQ-method (Holmér, 1988) has been developed along the lines of similar models for heat stress (ISO-7933, 1989) and thermal comfort (ISO-7730, 1984) and is published as a test standard (ISO/TR-11079, 1993).

Cold stress is calculated as a required clothing insulation (IREQ) for maintenance of body heat balance at defined levels of physiological strain. When required insulation cannot be met by selected clothing ensemble, a duration limited exposure is calculated based on a defined cooling (heat debt) of the body.

An analysis of reported field studies based on ISO-TR 11079 showed a good agreement between predicted IREQ and observed values for clothing insulation (Holmér, 1989). Nielsen (1992) observed for mild cold stress (+5 to +10°C) that IREQ slightly overestimated required or "worn" thermal insulation. The main problem with any type of predictive model appears to be a sufficiently accurate estimation of metabolic rate (Kähkönen et al., 1992).

Extremity cooling. Human extremities, in particular digits, are prone to suffer high rates of heat loss (Jaeger et al., 1977; van Dilla et al., 1949). Their temperature depends on the balance between local heat loss and the heat input by warm blood from the body core. Thus, not only the actual protection of the extremity, but also whole body protection and metabolic heat production are determining factors. This complex interaction has been investigated by many groups (Hellstrøm et al., 1970; Kruk et al., 1990; Livingstone et al., 1984; Marshall, 1972). In addition the thermal status of the body upon exposure is important (Allan et al., 1979). The cooling event has been modeled by Shitzer et al. (1991) for a (cylinder-shaped) finger. A first validation of the predictions appears to yield reasonable values for the time of finger temperature to drop to a defined level.

Convective skin cooling. The direct action of wind on bare, warm skin causes considerable heat losses and may endanger the local heat balance. The Wind Chill Index (WCI) (Siple and Passel, 1945) provides a rough estimate of this cooling effect. It is widely used for the assessment of the freezing risk of unprotected human skin. It is also proposed as a test standard by ISO (ISO/TR-11079, 1993). The advantage of the original WCI seems to be a fair agreement with practical experience. However, the rationale behind the WCI has been questioned (see Dixon and Prior, 1987). Indeed, the index may overestimate cooling of bare skin at low to moderate wind speeds and underestimate cooling at very high wind speeds (> 20 m/s).

Contact cold. ACGIH (1990) recommends action to be taken when metal surface temperatures below $-1°C$ may be contacted by unprotected skin. This may be a sufficient limit value for very short contacts (less than a few seconds). As soon as the contact period is prolonged, skin temperature may rapidly approach the temperature of the metal surface (Chen, 1991). Pain is not a reliable warning signal, because of the rapid response and the interaction with the blocking of superficial receptors at $+7°C$ and below (Chen, 1991).

Cooling of respiratory tract. Inhaling cold, dry air causes a considerable local cooling of the nasal mucosa and the upper respiratory tract (Hartung et al., 1980; Jaeger et al., 1980; Togias et al., 1988). With nasal breathing a certain amount of moisture and heat is regained by the mucous membrane. With mouth breathing and, in particular, at high ventilation rates cooling may extend deep into the airways and provoke epithelial inflammations (McFadden, 1983; Togias et al., 1988). In sports events of endurance type a lowest temperature of about $-20°C$ is often recommended. Nasal or combined breathing at low activity levels allows a much lower air temperature (-40 to $-50°C$) before onset of respiratory distress symptoms (Horvath, 1981).

4. Measures for prevention and alleviation of cold stress

Measures for prevention and alleviation of cold stress, principally, should involve actions to control body heat balance by, for example,
- reduction of heat losses;
- increase of internal heat production;
- supply of external heat;
- improvement of the ergonomics of cold work.

Reduction of heat losses, typically, encompasses use of protective clothing, protection of work place, insulation of surfaces in contact with parts of the human body, control of exposure time and other measures. It is evident that measures such as shorter work time and protection of work space from wind, reduce cold stress. Critical factors for cold protective clothing such as thermal insulation, air permeability, moisture and water permeability may be readily tested by standardized methods (prEN-342, 1992; EN-511, 1994). Thus, objective tests form the basis for selection of appropriate protection against defined levels of cold stress. In addition, a rational and cautious behavior is required for the continuous adjustment of personal protection to type of work, rate of work and ambient conditions. Wetting of clothing is detrimental to its insulative capacity and must be avoided (Belding et al., 1947; Belding,

1949; Gavhed et al., 1991; Holmér, 1984a). Protective clothing against cold, be it overalls, handwear, footwear or headgear, also must conform with ergonomic requirements of work (mobility, dexterity, field of vision, etc.) (Gonzalez et al., 1989; Lotens, 1987; Santee and Endrusick, 1988; Santee et al., 1988). Priorities between requirements influence the possible ways of action for control of cold stress.

Increase of heat production by muscular work is a behavioral adjustment of significant power. However, it may be restricted by requirements of work and by the physical work capacity of the individual.

Supply of external heat is a common solution to many cold problems. Warm-up breaks in a heated room or shelter or in front of a fireplace are standard requirements in cold work. Experience and/or rational calculations may form the basis for the establishment of appropriate work–rest regimens. Local heating by radiation, convection or conduction may be used with stationary work places and is particularly useful in cold stores and fridge houses (Lund Madsen and Saxhof, 1979; Nagasaka et al., 1987; Vanggaard et al., 1988). Much attention has been paid to the use of auxiliary heating of gloves (Goldman, 1969; Lockhart and Kiess, 1971; Vanggaard et al., 1988). However, the problems of durability and sufficient battery power at low weight remain.

Ergonomic measures reducing or eliminating the hobbling effect of protective equipment and the associated effort and exertion should be considered. Such measures include the selection of appropriate clothing, equipment and tools, training under adverse conditions, as well as organisation of work. Despite the introduction of new, promising fabrics and lightweight garments many problems remain (Reed et al., 1988). Protection of feet and hands is a common problem, in particular when activity is low. Electrically heated clothing, socks or gloves may help, but suffer from bulk, weight and low capacity (when battery operated) (Haisman, 1988; Kempson et al., 1989; Scott, 1988). However, more systematic compilation of experience and research in this field may help in the development of solutions for some applications (Oakley, 1984; Rosenblad-Wallin, 1988; Santee and Endrusick, 1988; Santee et al., 1988).

Organisation of work with adequate work–rest regimens and warm-up breaks is important for control of exposure. The length and frequency of breaks or rest periods may depend on work tasks, but should be designed for effective thermal recovery (Forsthoff, 1983; Kleinöder, 1988). In most cases, longer and less frequent breaks are more efficient than many short breaks.

Equipment and tools must be adopted for cold conditions. Handles, bars and controls should be covered by insulating material to prevent contact cooling when used at subzero temperatures (ACGIH, 1990; Chen et al., 1993). In addition, they must allow operation by gloved hand. Powered hand tools may be equipped with heated handles.

5. Need for further research

Cold stress accommodates a number of different physical stress factors and a number of distinct effects on human function, health and performance. Despite the bulk of knowledge continuously gained during the years, many problems still remain unsolved and many hypotheses require further validation. Much of the controversy of results may depend on the fact that the independent variable (cold stress) was not properly or similarly defined. Research on cold stress should focus on

– systematic, quantitative studies of cold effects and their relation to measures of exposure;
– validation of methods and models for prediction of cold stress;
– improved modeling of the physical factors and their interaction;
– more systematic studies of the cooling effect on skin, extremities and whole body;
– effect of clothing during various types of transient conditions;
– better and easy-to use methods and instrumentation for field use;
– ergonomic measures for improved function in the cold of clothing, equipment and tools.

References

ACGIH, 1990. Cold stress. In: Threshold limit values for physical agents in the work environment. Edited by ACGIH, New York.

Allan, J.R., Gibson, T.M. and Green, R.G., 1979. Effect of induced cyclic changes of deep body temperature on task performances. Aviat. Space Environ. Med., 50: 585–589.

Anon, 1989. Levnadsförhållanden. SCB, Rapport no. 61.

Belding, H.H., Russel, H.D., Darling, R.C. and Folk, G.E., 1947. Thermal responses and efficiency of sweating when men are dressed in arctic clothing and exposed to extreme cold. Am. J. Physiol., 149: 204–222.

Belding, H.S., 1949. Protection against dry cold. In: L. Newburgh (Ed.), Physiology of Heat Regulation and the Science of Clothing. Saunders, Philadelphia, pp. 351–367.

Bergh, U., 1980. Human power at subnormal body temperatures. Acta Physiol. Scand., Suppl. 478: 1–39.

Burton, A.C. and Edholm, O.G., 1955. Man in a Cold Environment. Edward Arnold, New York. 273 pp.

Chen, F., 1991. Contact Temperature on Touchable Surfaces. Luleå University, Sweden.

Chen, F., Nilsson, H. and Holmér, I., 1993. Cooling responses of finger pad in contact with an aluminum surface. Am. Ind. Hyg. Assoc. J. (in press).

de Jesus, P.V., Hausmanowa-Petrusewicz, I. and Barchi, R.L., 1973. The effect of cold on nerve conduction of human slow and fast nerve fibers. Neurology, 23: 1182–1189.

Dixon, J.C. and Prior, M.J., 1987. Wind-chill indices – A review. The Meteorological Magazine, 116: 1–17.

Duggan, A. and Haisman, M. F., 1992. Prediction of the metabolic cost of walking with and without loads. Ergonomics, 35: 417–426.

Ellis, H.B., Wilcock, S.E. and Zaman, S.A., 1985. Cold and performance: The effects of information load, analgesics, and the rate of cooling. Aviat. Space Environ. Med., 56: 233–237.

EN-511, 1994. Protective gloves against cold. Comité Européen de Normalisation, Brussels.

Enander, A., 1982. Perception of hand cooling during local air exposure at three different temperatures. Ergonomics, 25: 351–361.

Enander, A. and Hygge, S., 1990. Thermal stress and human performance. Scand. J. Work Environ. Health, 16, suppl. 19: 44–50.

Endrusick, T.L., Santee, W.R., DiRaimo, D.A., Blanchard, L.A. and Gonzalez, R.R., 1991. Physiological responses while wearing protective footwear in a cold-wet environment. Proceedings of Fourth International Symposium on the Performance of Protective Clothing, Montreal, Canada, 130 pp.

Fagius, J., Karhuvaara, S. and Sundlöf, G., 1989. The cold pressor test: Effects on sympathetic nerve activity in human muscle and skin nerve fascicles. Acta Physiol. Scand., 137: 325–334.

Fanger, P.O., 1970. Thermal comfort. Danish Technical Press, Copenhagen.

Forsthoff, A., 1983. Arbeit in −28°C. Verlag Dr. Otto Schmid, Köln.

Fox, W.F., 1967. Human performance in the cold. Human Factors, 9: 203–220.

Gavhed, D., Nielsen, R. and Holmér, I., 1991. Thermoregulatory and subjective responses of clothed men in the cold during continuous and intermittent exercise. Eur. J. Appl. Physiol., 63: 29–35.

Gaydos, H.F. and Dusek, E.R., 1958. Effects of localized hand cooling versus total body cooling on manual performance. J. Appl. Physiol., 12: 377–380.

Goldman, R.F., 1969. Factors in the design and use of clothing. Proceedings of Denver, Colorado meeting of the American Society of Heating, Radiation and Air-conditioning Engineers, July.

Goldman, R.F., 1975. Predicting the effects of environment, clothing and personal equipment on military operations. Proceedings of the 11th Commonwealth Defence Conference on Operational Clothing and Combat Equipment, India.

Gonzalez, R.R., Endrusick, T.L. and Santee, W.L., 1989. Thermoregulatory responses in the cold: Effect of an extended cold weather clothing system (ECWCS). Proceedings of Fifteenth Commonwealth Defence Conference on Operational Clothing and Combat Equipment, Canada.

Granberg, P.-O., Hassi, J., Holmér, I., Larsen, T., Refsum, H., Yttrehus, K. and Knip, M., 1991. Cold physiology and cold injuries. Arctic Medical Research, 50: 1–160.

Gyllerup, S., 1992. Cold as a risk factor for coronary mortality. University of Lund.

Haisman, M.F., 1988. Physiological aspects of electrically heated garments. Ergonomics, 31: 1049–1063.

Harjula, R., 1980. Cardiorespiratory effects of cold and exercise in health and coronary heart disease. Institute of Occupational Health, Helsinki.

Hartung, G.H., Myhre, L.G. and Nunneley, S.A., 1980. Physiological effects of cold air inhalation during exercise. Aviat. Space Environ. Med., 51: 591–594.

Hellstrøm, B., Berg, K. and Vogt Lorentzen, F., 1970. Human peripheral rewarming during exercise in the cold. J. Appl. Physiol., 29: 191–199.

Holmér, I., 1984a. Heat exchange by sweating in cold protective clothing. In: Aspect médicaux et biophysiques des vêtements de protection. Edited by Centre de Recherche du Service de Santé des Armées, Lyon, pp. 306–311.

Holmér, I., 1984b. Required clothing insulation (IREQ) as an analytical index of cold stress. ASHRAE Transactions, 90: 1116–1128.

Holmér, I., 1988. Assessment of cold stress in terms of required clothing insulation – IREQ. International Journal of Industrial Ergonomics, 3: 159–166.

Holmér, I., 1989. Relevance of required clothing insulation, IREQ, for the assessment of cold environments. Proceedings of Man–Thermal Environment Systems, Sapporo, Japan, pp. 190–193.

Holmér, I. and Bergh, U.J., 1974. Metabolic and thermal

response to swimming in water at varying temperatures. J. Appl. Physiol., 37: 702–705.

Horvath, S.M., 1981. Exercise in a cold environment. Exercise and Sports Sciences Reviews, 9: 221–263.

Iampietro, P.F., 1971. Use of skin temperature to predict tolerance to thermal environments. Aerospace Medicine, 42: 396–399.

ISO-7730, 1984. Moderate thermal environments – Determination of the PMV and PPD indices and specification of the conditions for thermal comfort. International Standards Organisation, Geneva.

ISO-7933, 1989. Hot environments – Analytical determination and interpretation of thermal stress using calculation of required sweat rates. Internationl Standards Organisation, Geneva.

ISO/TR-11079, 1993. Evaluation of cold environments – Determination of required clothing insulation (IREQ). International Standards Organisation, Geneva.

Jaeger, J.J., Deal, E.C., Roberts, D.E., Ingram, R.H. and McFadden, E.R., 1980. Cold air inhalation and esophageal temperature in exercising humans. Med. Sci. Sports, 12: 365–369.

Jaeger, J.J., Sampson, J.B., Roberts, D.E. and McCarroll, J.E., 1977. Characteristics of human finger cooling in air at 0°C. Physiologist US, 20: 47.

Jokl, M.V., 1982. The effect of the environment on human performance. Appl. Ergonomics, 13: 269–280.

Keatinge, W.R., 1969. Survival in cold water. Blackwell, Oxford, 131 pp.

Kempson, G.E., Clark, R.P. and Goff, M.R., 1989. The design, development and assessment of electrically heated gloves for protecting cold extremities. Ergonomics, 31: 1083–1091.

Kleinöder, R., 1988. Ergonomische Gestaltung von Kältearbeit bei −30°C in Kühl- und Gefrierhäusern. Bundesanstalt für Arbeitsschutz, Fb 562.

Kruk, B., Pekkarinen, H., Harri, M., Manninen, K. and Hanninen, O., 1990. Thermoregulatory responses to exercise at low ambient temperature performed after precooling or preheating procedures. Eur. J. Appl. Physiol., 59: 416–420.

Kähkönen, E., Nykyri, E., Ilmarinen, R., Ketola, R., Lusa, S., Nygård, C.-H. and Suurnäkki, T., 1992. The effect of appraisers in estimating metabolic rate with the Edholm scale. Appl. Ergonomics, 23: 186–192.

Lassvik, C., 1979. Angina pectoris in the cold. Effects of cold environment and cold air inhalation at exercise tests. Linköping university.

LeBlanc, J., Blais, B., Barabe, B. and Cote, J., 1976. Effects of temperature and wind on facial temperature, heart rate and sensation. J. Appl. Physiol., 40: 127–131.

Litchfield, P., 1987. Manual performance in the cold: A review of some of the critical factors. J. Roy. Nav. Med. Serv., 73: 173–177.

Livingstone, S.D., Nolan, R.W. and Cattroll, S.W., 1984. Effect of increased body clothing insulation on hand temperature in a cold environment. Defence Research Establishment, Ottawa.

Lockhart, J.M. and Kiess, H.O., 1971. Auxiliary heating of the hands during cold exposure and manual performance. Human Factors, 13: 457–465.

Lockhart, J.M., Kiess, H.O. and Clegg, T.J., 1975. Effect of rate and level of lowered finger surface temperature on manual performance. J. Appl. Physiol., 60: 106–113.

Lotens, W.A., 1987. Clothing design for work in the cold. Arct. Med. Res., 46: 3–12.

Lund Madsen, T. and Saxhof, B., 1979. An unconventional method for reduction of the energy consumption for heating of buildings. Proceedings of XVth International Congress of Refrigeration, Venezia, pp. 32–43.

MacLean, D. and Emslie-Smith, D., 1977. Accidental Hypothermia. Blackwell, Oxford.

Marshall, H.C., 1972. The effects of cold exposure and exercise upon peripheral function. Arch. Environ. Health, 24: 325–330.

McArdle, W.D., Toner, M.M., Magel, J.R., Spina, R.J. and Pandolf, K.B., 1992. Thermal responses of men and women during cold-water immersion. Eur. J. Appl. Physiol., 65: 265–270.

McFadden, E.R., 1983. Respiratory heat and water exchange: physiological and clinical implications. J. Appl. Physiol., 54: 331–336.

McFadden, E.R.J., Pichurko, B.M., Bowman, H.F., Ingenito, E., Burns, S., Dowling, N. and Solway, J., 1985. Thermal mapping of the airways in humans. J. Appl. Physiol., 58: 564–570.

McIntyre, D.A., 1980. Indoor Climate. Applied Science, London, 443 pp.

Meese, G.B., Kok, R., Lewis, M.I. and Wyon, D.P. 1981. The effects of moderate cold and heat stress on the potential work performance of industrial workers. Performance of tasks in relation to air temperature under eight environmental conditions. Proceedings of CSIR research report 381/2.

Nagasaka, T., Hirata, K., Nunomura, T. and Cabanac, M., 1987. The effect of local heating on blood flow in the finger and the forearm skin. Can. J. Physiol. Pharmacol., 65: 1329–1332.

Nielsen, R., 1992. Development of work clothing for cold workplaces (In Danish) (Udvikling av arbeidsbekledning for kolde arbeidsmiljöer). Lab. for Varme og Klimateknik, Danmarks Tekniska Högskola.

O'Cain, C.F., Dowling, N.B., Slutsky, A.S., Hensley, M.J., Strohl, K.P., McFadden, E.R.J. and Ingram, R.H.J., 1980. Airway effects of respiratory heat loss in normal subjects. J. Appl. Physiol., 49: 875–880.

Oakley, E.H.N., 1984. The design and function of military footwear: A review following experiences in the South Atlantic. Ergonomics, 27: 631–637.

Oksa, J., Rintamäli, H., Hassi, J. and Rissanen, S., 1993. Gross efficiency of muscular work during step exercise at −15°C and 21°C. Acta Physiol. Scand., 147: 235–240.

Popovic, V. and Popovic, P., 1974. Hypothermia in Biology and in Medicine. Grune & Stratton, New York.

prEN-342, 1992. Protective clothing against cold. Comité Européen de Normalisation, Brussels.

Reed, L., Osczevski, R.J. and Farnworth, B., 1988. Cold weather clothing systems: Recent progress and problems for future research. In: Handbook on Clothing. Biomedical Effects of Military Clothing and Equipment Systems. Edited by NATO's AC/243, Brussels, pp 5.1–10.

Rodahl, K., 1991. Working in the cold. Arct. Med. Res., 50: 80–82.

Rogers, W.H. and Noddin, E.M., 1984. Manual performance in the cold with gloves and bare hands. Perceptual and Motor Skills, 59: 3–13.

Rosenblad-Wallin, E.F.S., 1988. The design and evaluation of military footwear based upon the concept of healthy feet and user requirement studies. Ergonomics, 31: 1245–1263.

Santee, W.R. and Endrusick, T.L., 1988. Biophysical evaluation of footwear for cold-weather climates. Aviat. Space Environ. Med., 59: 178–182.

Santee, W.R., Endrusick, T.L. and Wells, L.P., 1988. Biophysical evaluation of handwear for cold weather use by petroleum (pol) handlers. In: F. Aghazadeh (Ed.), Trends in Ergonomics/Human Factors V. Elsevier Science Publishers B.V., Amsterdam, pp. 441–448.

Scott, R.A., 1988. The technology of electrically heated clothing. Ergonomics, 31: 1065–1081.

Shitzer, A., Stroschein, A., Santee, W.E., Gonzalez, R.R. and Pandolf, K.B., 1991. Quantification of conservative endurance times in thermally insulated cold-stressed digits. J. Appl. Physiol., 71: 2528–2535.

Siple, P.A., 1945. General principles governing selection of clothing for cold climates. Proc. of the Am. Philosophical Society, 89(1): 200–234.

Siple, P.A. and Passel, C.F., 1945. Measurements of dry atmospheric cooling in subfreezing temperatures. Proc. American Philosophical Society, 89: 177–199.

Sperling, L., Jonsson, B. and Holmér, I., 1983. Handfunktion och handskydd vid arbete med handskar. Arbetarskyddsstyrelsen, 171 84 Solna, 1983: 30.

Steadman, R.G., 1984. A universal scale of apparent temperature. J. Climate and Appl. Meteorology, 23: 1674–1687.

Teitlebaum, A. and Goldman, R.F., 1972. Increased energy cost with multiple clothing layers. J. Appl. Physiol., 32: 743–744.

Togias, A.G., Proud, D., Lichtenstein, M., Adams III, G.K., Norman, P.S., Kagey-Sobotka, A. and Naclerio, M., 1988. The osmolality of nasal secretions increases when inflammatory mediators are released in response to inhalation of cold, dry air 1–4. Am. Rev. Respir. Dis., 137: 625–629.

van Dilla, M.A., Day, R. and Siple, P.A., 1949. Special problems of the hands. In: R. Newburgh (Ed.), Physiology of heat regulation. Saunders, Philadelphia, pp. 374–388.

Vanggaard, L., 1975. Physiological reactions to wet-cold. Aviat. Space Environ. Med., 46: 33–36.

Vanggaard, L., Gonzalez, R.R. and Breckenridge, J.R., 1988. Physiological aspects of auxiliary heating and cooling. In: Handbook on Clothing. Biomedical Effects of Military Clothing and Equipment Systems. Edited by NATO's AC/243, Brussels, pp. 9A.1–5.

Wagner, J.A. and Horvath, S.M., 1985. Influences of age and gender on human thermoregulatory responses to cold exposures. J. Appl. Physiol., 58: 180–186.

Wilkerson, J.A., Bangs, C.C. and Hayward, J.S., 1986. Hypothermia, Frostbite and other Cold Injuries. The Mountaineers, Seattle, Washington.

Noise in the office: Part I – Guidelines for the practitioner [*]

Anders Kjellberg [a], Ulf Landström [b]

[a] *National Institute of Occupational Health, S-17184 Solna, Sweden*
[b] *National Institute of Occupational Health, P.O. Box 7654, S-90713 Umeå, Sweden*

1. Introduction

Noise in offices rarely constitutes a risk for hearing damage. The dominating problem is that the noise may be a source of annoyance and disturbance and that it may impair performance. The aim of this guideline is to identify the principal noise sources in offices and to point at possible ways of dealing with them to avoid these undesirable effects. The guideline does not aim to give detailed descriptions or recommendations of the different techniques available for noise reduction. In most respects these methods are the same as those used when the aim is hearing conservation and they have been amply described in many other contexts. The aim is rather to point out the main types of measures available for the solution of different noise problems in offices and to suggest highest acceptable levels of noise from different sources.

The guidelines are not written primarily for the acoustic expert but for the many other groups that are involved in the planning and hygienic assessment of office environments.

Four types of noise sources have been judged to be critical in these work places:
- External sources (e.g. traffic noise),
- Fixed installations, building noise (e.g. ventilation noise),
- Office machines (e.g. copying machines and printers),
- Office activities (e.g. conversations).

2. Problem analysis, general strategies and methods

Office noise may affect task performance, but such effects are very difficult to assess outside the laboratory. In practice, therefore, the point of departure for the analysis of the noise problem has to be the annoyance expressed by the office workers. The first step in an analysis of a noise problem in an office is to identify the critical noise sources. In this analysis the employees

---

[*] The recommendations provided in this guide are based on numerous published and unpublished scientific studies and are intended to enhance worker safety and productivity. These recommendations are neither intended to replace existing standards, if any, nor should be treated as standards. Furthermore, this document should not be construed to represent institutional policy.

The following individuals participated in the discussion of the earlier version of this guide. Their suggestions (written or verbal) were incorporated by the authors in this version: Arne Aaras, *Norway*; Fred Aghazadeh, *USA*; Roland Andersson, *Sweden*; Jan Dul, *The Netherlands*; Jeffrey Fernandez, *USA*; Ingvar Holmér, *Sweden*; Matthias Jäger, *Germany*; Åsa Kilbom, *Sweden*; Olli Korhonen, *Finland*; Helmut Krueger, *Switzerland*; Shrawan Kumar, *Canada*; Tom Leamon, *USA*; Anil Mital, *USA*; Ruth Nielsen, *Denmark*, Jerry Ramsey, *USA*; Murray Sinclair, *UK*; Rolf Westgaard, *Norway*; Ann Williamson, *Australia*; Jørgen Winkel, *Sweden*; Pta Zätterström, *Sweden*. The guide was also reviewed in depth by several anonymous reviewers.

themselves are the principal source of information. Often the noise from one source is considered to be much more annoying than other noises. In such cases, measures should be concentrated on this critical source. Measures against other noise sources are likely to be without effect on overall noise annoyance.

The second step is to determine what makes the noise annoying. Different technical methods may be necessary to use in this analysis. In many cases, the problem is just one of too high sound level, and a measurement of the equivalent sound level should be sufficient. The A frequency weighting has to be used to make it possible to relate the result to noise standards and legislation, which generally are based on dB(A) values. However, such an evaluation is in many cases insufficient and may sometimes be misleading as an indicator of expected annoyance reactions. In the technical assessment of the noise exposure and noise annoyance it may be necessary also to take into account parameters such as the frequency characteristics of the noise and its temporal pattern. The A-weighted equivalent sound level is particularly likely to lead to an underestimation of probable annoyance reactions when the noise contains strong low frequency components and pure tones (especially high frequency tones), and when the noise is intermittent, repetitive or in other ways shows strong fluctuations.

The A-weighted equivalent sound level should not be used for comparisons between noise levels at different work places or from different machines when the sounds to be compared have very different spectral shapes. In such cases the B- or D-weighting may be more appropriate, but there is not one filter optimal for all types of noise. In such cases a satisfactory comparison requires the use of a more complex assessment method like Zwicker's loudness. When comparing sources that emit noise with similar spectral shapes the choice of frequency weighting is of little consequence.

The technical methods have to be supplemented by analyses of non-technical factors that might be of importance for the degree of annoyance caused by the noise: the tasks at hand and their vulnerability to masking and distraction, the relation between the tasks and the noise source (the "necessity" of the noise and the individual possibilities of controlling the noise), the informational content of the noise (speech vs. other types of sounds) and the attitudes towards the noise source from other points of view.

General principles and strategies to follow in the acoustical design of offices are treated in VDI 2569 01.90 and ISO standards ISO/DIS 11690, Part 1 and 2.

3. General concepts and suggestions

Measures directed against the annoying noise in offices often follow the same principles as those against noise that constitutes a risk for hearing damage. Thus, measures shall in most cases primarily be directed at the source, secondly upon the transmission of the noise and thirdly upon the receiver. In office environments, however, the last alternative, i.e. the use of ear protectors, can almost never be recommended. Furthermore, the fact that the office workers themselves constitute a principal noise source, puts limits on the possibilities to take actions against the source of the noise.

Since noise annoyance is only partly explained by the physical characteristics of the noise, other measures than noise reduction may also be of importance. Three important principles to follow in this context are:

Noise annoyance is partly determined by the task at hand, and it is therefore particularly important to be protected from excessive noise during the performance of certain tasks. This is primarily true of tasks that require speech communication and other tasks in which noise may impair the perception of important auditory information. Furthermore, during work that puts high demands on information processing and concentration, noise tolerance is lower than during routine tasks.

Avoid exposing workers to "unnecessary" noise. One consequence of this principle is that one should avoid exposing workers to noise from machines and other sources that are unrelated to their own work.

Be careful to inform workers about the measures that have been taken against noise and the costs of any further improvement of the noise situation. The conviction of the worker that not much more could be done against the noise at reasonable costs, is likely to be as important as a fairly large reduction of the sound level.

4. Guidelines for different noise sources

4.1. External noise

Measures taken by the work place against noise from external noise sources (e.g. traffic and industries) can for obvious reasons seldom be directed at the source of the noise. Instead the main way of treating these problems must be to avoid undesired transmission of the noise to the office by acoustic treatment of the ceilings, doors, windows, etc. The low frequency character of much of the external noise creates special problems, since it is very difficult to interfere with the transmission of such noise. Furthermore, in smaller offices unfavourable room resonances may sometimes appear in cases of extremely low frequencies external noise.

At external noise levels below 45 dB(A) very few of the employees generally are disturbed by the noise, whereas this is likely to be true for half of the exposed at levels around 55 dB(A). When the external noise is intermittent or its level and quality varies greatly, the adverse reactions are likely to be strengthened.

Taking all these aspects into consideration, the exposure to external noise inside the office should preferably not exceed the level of 50 dB(A). Lower levels may be required when the noise is highly variable and when it contains strong low frequency components.

4.2. Ventilation noise

There are effective methods for the reduction of ventilation noise although the measures necessary are often expensive, especially when they are the result of inadequate acoustic planning and, thus, have to be carried out in a completed building. This is unfortunately a very common situation in offices. Measures can be directed at the fan, the ventilation canal systems or the tools for inlet and outlet of the air. The machine can be stabilised and the rotation frequency may be increased to generate higher sound frequencies which are easier to reduce by absorbing techniques. The ventilation noise can also be reduced by stabilising the canal system or through installations of silencers in the canal. Changes of the size and model of the tools for the inlet or outlet air may also sometimes have a positive effect.

The ventilation noise in the offices should preferably not exceed the level of 35 dB(A). In offices designed for work demanding high degrees of concentration, a maximum level of 30 dB(A) is suggested.

4.3. Office machines

Directing action against the source is most often rather easily accomplished with office machines. In most cases it is possible to solve the problem by changing the machine to one with different design or operating techniques. The transmission of the noise can be obstructed with hoods and screens, or by moving the machine to another room. Modern PC's are seldom a source of complaint. However, the noise from the cooling fan may be very annoying and should not exceed 40 dB(A).

The noise from office machines should not exceed the level of 50 dB(A). In cases of tonal components in the noise the level should not exceed 45 dB(A). Fluctuations exceeding 15 dB shall be avoided. When the work requires a high degree of concentration and complex information processing, the level should be reduced as much as possible, and should not exceed 40 dB(A).

4.4. Office activities

Co-workers constitute a principal source of noise in offices, and irrelevant speech is generally the most disturbing of the sounds produced by the co-workers. Therefore, the most effective measure against noise complaints often is to restrict the number of people working in the same room. However, in many cases such a change is neither possible nor desirable.

The degree of annoyance caused by irrelevant speech is a consequence of its informational content rather than its sound level. Therefore, the aim must be to create conditions where irrelevant speech is unintelligible, while preserving good speech communication conditions.

The intelligibility of speech depends on the relation between the speech level and the level of the background noise and it may be assessed by computing the Articulation Index (AI) for the irrelevant speech. An AI value of 0.1–0.2 or lower should be aimed at for such speech. It is most often preferable to accomplish this by reducing the speech level at the point of the listener, but the effect could also be attained by increasing the background noise level.

In closed offices it is generally possible to guarantee a sufficient reduction of speech from adjacent rooms by choosing wall and ceiling material with satisfactory sound transmission characteristics.

In the open-plan office the level of the irrelevant speech may be reduced in two ways. First, the distance between the speaker and the receiver may be increased, i.e. the distance between those workers who seldom need to communicate should be maximised, and that between those needing to communicate should be reduced to avoid the need for raising the voice. Secondly, the propagation of speech beyond the intended listeners should be minimised. The main way of achieving this reduction is to treat the ceiling, floor and walls with sound absorbents. A reverberation time of 0.3 sec, or, in bigger offices, a 5 dB fall-off with each doubling of the distance, should be aimed at for the main speech frequencies. In acoustically treated rooms further improvements may be achieved by screens. The size of the screen, its sound absorption and transmission are the characteristics most critical for its effectiveness as an attenuator of speech sounds. The Noise Reduction Coefficient of the screen should be more than 0.50 (average absorbed part of the incident acoustical energy in the 250–2000 Hz octave bands) and a Sound Transmission Class of 20–25 (a classification based on the transmission loss in dB in the 125–4000 Hz frequency range) most often is sufficient.

Other noises in the office mask irrelevant speech and therefore may have beneficial effects on noise disturbance. Artificial constant background noise ("sound conditioning") may also be installed to attain such an effect. This involves a delicate balance between two objectives: masking efficiency and acceptability of the noise in itself. However, there is no consensus about the possibility to achieve such a balance. When used, the level of such a masking noise should definitely not exceed 48 dB(A).

During tasks which are very vulnerable to distraction the use of earplugs might be considered in exceptional cases. Arrangements might also be made to make it possible to perform such tasks in a quiet one-man room.

The disturbance from noise of speech communication is another main reason for noise annoyance. As long as the ambient sound level does not exceed 55 dB(A) it should seldom be necessary to raise the voice to avoid speech interference when the distance between speaker and listener is no more than 2.5 m. Neither should telephone conversations between parties be any problem at this level of background noise.

Reference

Kjellberg, A. and Landström, U., 1994. Noise in the office: Part II – The scientific basis (knowledge base) for the guide. International Journal of Industrial Ergonomics, 14 (1–2): 93–118 (this issue).

Noise in the office: Part II – The scientific basis (knowledge base) for the guide

Anders Kjellberg [a], Ulf Landström [b]

[a] *National Institute of Occupational Health, S-17184 Solna, Sweden*
[b] *National Institute of Occupational Health, P.O. Box 7654, S-90713 Umeå, Sweden*

1. Introduction

Noise is one of the most widespread environmental problems in working places including offices. It is true that office workers seldom run the risk of developing hearing damage, but noise certainly may create serious problems at noise levels far below those at which such damage may occur. However, systematic research on such problems has been sparse in work places; it has generally dealt with community noise problems. The development of the open-plan office in the sixties and seventies accentuated the noise problem and led to some research (see review by Sundstrom, 1986), but noise is still given a very modest role in models developed for the analysis of health and well-being of office workers (see review by Hedge, 1989). Therefore, most knowledge of the effects of noise at the lower levels encountered in offices emanates from research that has dealt with problems in residential areas and from laboratory studies.

1.1. The scope of the noise problem

In Sweden about 25% of the white collar workers consider themselves exposed to noise and the number exposed is especially high in groups working in education and health care (Statistics Sweden, 1991). Keighley (1970) let 2000 office workers rate how satisfied they were with different aspects of their working environment and found that noise was the decidedly worst-rated feature. Nemecek and Grandjean (1973b) found that 90% of the workers in landscaped offices were disturbed by noise, and that 35% of them were severely disturbed. Corresponding figures for climactic conditions were considerably lower. In a later study (Nemecek and Turrian, 1978a) including different types of offices they found that 62% mentioned noise as the most disturbing aspect of their working environment. Similar results have been reported in other studies (Brookes, 1972).

There are also reasons to believe that dissatisfaction with the noise conditions may have a considerable impact on the general satisfaction with the job and the working environment. Thus, Sundstrom (1986) reports a study by himself and co-workers of noise and job satisfaction before and after moving to a new office. They found that when the new office led to increases in noise from people talking, telephones and typewriters, general job dissatisfaction also increased, whereas decreases of those noises had the opposite effect. In the study by Klitzman and Stellman (1989) self-rated noise exposure was the factor that showed the highest correlation with general satisfaction with the office. They also showed that after control of the effects of demographic variables, occupation and psychosocial conditions, noise was still significantly associated with job satisfaction and psychological well-being.

1.2. The aim and organisation of the review

The aim of the present paper is to review research relevant for the understanding of the noise problems in offices. Systematic work place studies of most aspects of these problems are lacking. The review therefore to a large degree rests on laboratory experiments and studies of community noise problems. Obviously, the results from such studies are only partly applicable to the working environment. However, together they may point out general principles applicable to most contexts in which noise is met and, hopefully, lead to a general understanding of noise problems and the critical conditions for their appearance. Such an understanding may be of great practical value, considering that noise and work conditions in offices change rapidly. Knowledge of the specific noise problems met in present day offices may soon become obsolete.

After some general comments on the office as a work place, the main types of noise are identified and characterised. This is followed by a review of the different types of effects, mainly subjective and performance effects, which are likely to constitute problems in offices.

2. The office as a working place

Spatial characteristics and working conditions vary widely between offices, and they, thereby, constitute a heterogeneous group of work places with respect to noise exposure and noise problems. The size of the office can vary between a few square meters including one or two workers to several hundreds of square meters with hundreds of workers engaged in different tasks. Activities also vary immensely both between and within offices. The number of employees, the size of the room and the tasks performed are all factors that have fundamental effects on the noise problems in offices (Hay and Kemp, 1972b; Nemecek and Grandjean, 1973b).

In some cases the offices have been located in buildings or areas originally not designed for office work. This situation is rather common in industries where the offices can often be found in extremely noisy areas. Also in other contexts external sources, such as road and air traffic, may be important to consider and keep apart from internally produced noise in the assessment of the noise situation.

Office technology and, thereby, the character of office work has undergone dramatic changes in the last decade and further changes can be expected in the near future. The rapid change in the nature of offices means that the accounts of noise exposure levels and other noise characteristics as well as guidelines for the office environments soon become dated.

3. Noise in the office environments

Generally, the noise levels emitted from individual machines installed in offices have been lowered compared to the situation in the 70's and 80's. At the same time the technological development has led to an increase in the total number of noise sources found in offices. Taken together this development has not led to any appreciable reduction of the total noise level in offices.

The introduction of the landscaped office in the 60's and 70's seriously changed the noise situation for the worse by increasing the exposure to noise originating from room activities.

The noise emitted from sources installed in the building is another great problem in many office environments. Ventilation noise has been an increasing problem in the last decade due to a notable lack of insight into noise problems among the installers of ventilation systems. Disturbance from outdoor noise sources is another example of noise problems in offices which depend on remarkable shortcomings of the building technique.

The subjective importance of different noise sources in offices has been treated in several studies. Nemecek and Grandjean (1973a,b) found in a study of noise in landscaped offices that co-worker's talk was the noise that most often caused complaints. This source of disturbance was mentioned by 46% of the workers, whereas 25% mentioned office machines, 19% telephones, 7% the coming and going of people and only 3% external noise. Langdon (1966) (as quoted by Sundstrom, 1986) also found co-workers to be the most often mentioned noise source in offices and the frequency of complaints increased with the number of persons sharing the room. Among the office workers that took part in the study by Nemecek and Turrian (1978a) 52% considered conversations the most disturbing noise, except when the windows were opened in which case traffic noise was judged as worse. In the landscaped offices studied by Boyce (1974) telephone signals and conversations were the most frequently mentioned sources of noise disturbance. Brookes (1972) reports similar results. In more recent Swedish studies (Landström et al., 1991a, 1992) which primarily investigated smaller offices, conversations were also the most mentioned type of noise, but many office workers also pointed at the ventilation system as an important source of disturbance.

Data on noise sources and noise annoyance have with very few exceptions been collected with interviews or questionnaires in which the worker has done retrospective judgements of the noise at work. One exception is a study by Purcell and Thorne (1977), who used a diary in which the workers recorded and described all episodes during a day when they were disturbed by noise. They also found that conversations constituted by far the most frequent source of disturbance. Other frequently mentioned sources were typewriters, a drink vending machine, telephones, general office noise and sudden changes of the noise, most often from office machines. Thus, there seem to be four main types of noise in offices: Traffic noise and noise from other outdoor sources, building noise (primarily the ventilation system), noise from office machinery (incl. telephones) and noise resulting from office activities (conversations, coming and going, etc.).

General principles and strategies to follow in the acoustical design of offices are treated in a VDI standard (Verein Deutsche Ingenieure, 1990) and two proposed ISO standards (International Standardization Organization, 1992a,b).

3.1. Outdoor noise sources

Neighbour road traffic is the most important outdoor noise source in most office environments. In industries the noise generated by the production may also reach offices. Air traffic, railways, construction work, ports and military activities may also contribute to the outdoor noise.

Due to the insulation properties of the walls, windows, etc., the outdoor noise inside the offices is often characterized by relatively high levels in the low frequency range. The noise is often rather constant in spectral content and is often of an intermittent character (e.g. road traffic noise, air plane noise). The hygienic standards which have been suggested for road traffic and air plane noises have therefore often taken both the levels and the number of events into consideration (Crocker, 1978a,b).

This kind of noise problem is of course best solved by taking the outdoor noise level into consideration in the construction of outer walls and the choice of windows.

3.2. Building noise

A number of permanent installations in the building constitute important noise sources in many offices, e.g. ventilation systems, heating installations, and elevators. Building noise often has many features common with outdoor noise. The ventilation noise is often described as a major source of annoyance and disturbances in both smaller office rooms and landscaped offices (Landström et al., 1991a, 1992).

The building noise transmitted into the offices is often of a low frequency character since the insulation of the building is more effective against higher frequencies. The level and spectral content of the noise is often relatively stable. The problem with building noise is complicated by the prominent risk of transmission through the building frames (structure borne noise).

Ventilation noise is most often broad band where the pressure level increases with decreasing frequency (Hay and Kemp, 1972a; Landström et al., 1991a; Leventhall, 1988). Sometimes, however, tones are present, the frequency of which is primarily determined by the number of revolutions per minute of the fan. According to a recent study (Landström et al., 1991a) carried out on 155 offices the dB(A) values of the ventilation noise varied between 30 and 45 dB(A). In most of the offices the dB(A) level exceeded 40 dB(A). The results of the investigation also indicated that the pressure levels at frequencies below 50 Hz fell below the hearing perception thresholds at all workplaces. This result indicates that infrasound is seldom the cause of adverse reactions to ventilation noise, since the sound frequencies that cannot be heard obviously do not influence the subjective response to the noise. This is important, since infrasound may have partly different effects on human beings than sounds at higher frequencies (Landström, 1987; Landström and Byström, 1984; Landström et al., 1983; von Ising et al., 1980)

Also for these types of noise the most effective measures can be taken if the problem is considered in the planning of the building and its ventilation system. Measures against noise from an existing ventilation system can be directed at the fan, the canal system or the tools for inlet and outlet of the air. The fan may be stabilized or the frequency of rotation can be changed; higher frequencies are generally easier to reduce. Silencers may be installed in the canals, which also can be stabilized. In some cases good results can be obtained just by changing the size, model or position of the inlet and outlet tools in the room.

3.3. Office machines

Noise levels generated from different types of office machines can vary between almost inaudible sounds from laser printers and VDU units at levels below 30 dB(A) to matrix printers and duplicating machines above 70 dB(A) (Costa et al., 1984; Ericsson Information Systems, 1984; Grandjean, 1988; Hay and Kemp, 1972b; Ying, 1988).

Noise emissions (dB(A)) from different types of office machines have been analysed by among others Grandjean in a study from the late 80's (Grandjean, 1988). The following noise levels were recorded from the different types of machines used by them. Matrix and daisy wheel printers: basic noise (73–75 dB(A)), peak levels (80–82); matrix printer with hood: peak levels (61–62); ink jet printer: basic noise (57–59), peak levels (60–62); laser printer: not measurable; cooling fan of VDT: 30–60 dB(A); old type writer: 70 dB(A); modern electronic type writer: 60 dB(A). Similar values have been reported in a number of other studies.

There are a great number of alternative measures which could be taken against noise from office machines, and the appropriate one varies beween the different types of machines. A machine may be changed for a less noisy model, a new technique may be used to fulfill the same function, the machine may be removed to another room or some technique for sound insulation (screens, hood) may be used (Hubert, 1978).

There are a number of ISO standards or proposals for such standards which treat noise from office machines (International Standardization Organization, 1988a,b,c).

3.4. Office activities

As mentioned above, conversations generally seem to be the most disturbing kind of noise in offices. The importance of co-workers as a noise source is also evident from the data presented by Langdon (1966) showing the relation between the number of persons in the office and the frequency of noise complaints. The percentage of complainants rose from six percent in the one-man rooms to 47 percent in rooms with 9–15 workers. Further increases of the number of persons did not affect the frequency of complaints.

Nemecek and Grandjean and others have found that the correlation between sound level and annoyance is very weak, and that one reason for this weak relation is that informational content is a more critical aspect of speech (Nemecek and Grandjean, 1973a). A primary aim therefore must be to reduce the intelligibility of the irrelevant speech, i.e. to lower its signal/noise ratio. A problem is that this must be accomplished without endangering good speech communication conditions in the office.

Conversations seem to be the main problem, but other sounds emanating from office activities may also be disturbing, e.g. sounds generated by different types of manual work, closing doors, walking and transportation.

In a one-man office noise problems of this kind are solved by proper sound insulation of the room. In the open-plan office measures taken against these types of noise must be directed at the sound propagation. This is mainly accomplished by acoustical treatment of the ceiling, walls and floor. In addition, screens are necessary to reach levels acceptable for the involuntary listener. Moreland has analysed the influence of the speaker-screen distance and three screen characteristics (its size, sound absorption and transmission) on the intelligibility of speech coming from the other side of the screen (Moreland, 1988). As expected a larger screen was found to be more effective than a smaller one. He also concluded that the least important of the three screen characteristics was how well it absorbed sounds, given a Noise Reduction Coefficient of at least 0.50, i.e. at least an average of 50 percent of the incident acoustical energy should be absorbed by the screen in the 250-2000 Hz octave bands. The diffraction of speech sounds over the screen was reduced as the distance between speaker and screen was decreased, but on the other hand more sound then was transmitted through the screen. A Sound Transmission Class of 20 was found to be sufficient except when the distance between the speaker and screen was very short. (Sound Transmission Class is a classification of the effectiveness of barriers to reduce the sound level based on the sound transmission loss in the 125-4000 Hz third octave bands, see ASTM, 1975.) He also found that it is important to minimise the gap between screens and between the screen and the floor.

Under free field conditions the sound level is lowered at a rate of 6 dB with each doubling of the distance, although in practice the reduction rate generally is lower and less regular (West, 1973). Another way of reducing these problems may therefore be to increase the distance between employees with little need of direct communication.

3.5. Overall noise level in offices

Several machines as well as other noise sources contribute to the noise in offices. The average equivalent noise level in different types of offices and work places using different machines are described by Costa et al. (1984). Depending upon the size of the room and the number of people in charge, the mean equivalent sound level (Leq) may vary between 60 and 90 dB(A).

Costa et al. (1984) also found, as expected, that the spectral composition of the recorded noises notably varied from one office to another. The frequency spectra were in some cases characterised by higher peaks corresponding to the human voice frequencies (telephone exchange offices). In some cases the noises were dominated by higher frequencies above 3000 Hz (coding machines and typewriters). As for the frequency composition, the broadband as well as the tonal character may vary widely.

4. Different noise effects

Hearing damage is very seldom a problem in offices. (The only study indicating that such a risk sometimes may exist is one by Hohensee (1990), who found that the sound level in the headphones of a dictating machine yielded Leq's above 85 dB(A) for many of the typists.) Furthermore, there are no data to support that the noise met in offices is likely to lead to any non-auditory health effects or to any sustained physiological responses (see reviews by, e.g. deJoy, 1984; Smith, 1991; and Van Dijk, 1990). Therefore, two main types of noise effects have to be considered in offices: Subjective responses (annoyance, disturbance) and behavioral effects, primarily performance effects. In the following sections these two types of effects are discussed. The effects are described as are the conditions (noise characteristics and other factors) determining the strength of the effects. The problem of interference with communication, which may be an important one in offices, is treated in connection with both the subjective and performance effects.

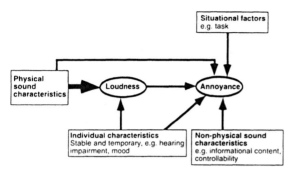

Fig. 1. The relations between loudness and annoyance and the various factors affecting these subjective responses to noise.

4.1. Subjective responses

4.1.1. Different subjective qualities

Most field studies and a large part of laboratory studies of the non-auditory effects of noise have dealt with the subjective responses to noise. In laboratory experiments one has most often been interested in determining the loudness of sounds with different physical characteristics. Loudness is a non-evaluative characteristic only referring to the subjective intensity of the sound. Loudness can with good accuracy be predicted from the physical sound qualities of the sound. When studying the response to noise at a work place one is generally not interested in the loudness of the noise but wants a judgement of the unwantedness of the sound, most often termed annoyance. This judgement is not only determined by the physical characteristics of the noise, but also by many other aspects of the noise, the situation and the individual.

Fig. 1 shows the relation between four classes of variables, i.e. physical and non-physical sound characteristics, individual characteristics and situational factors, which affect loudness and annoyance. It should not be viewed as a model for noise annoyance; that would, for example, require a subdivision of the four classes of variables and a specification of the relation between these variables. The figure shows that loudness to a very high degree is determined by the physical noise characteristics (indicated by the broad arrow). Furthermore, loudness is a major determinant of annoyance. However, the effect of the physical characteristics on annoyance are not wholly mediated by their effects on loudness; it may for example also be a result of the efficiency of the noise as a masker of speech. The influence of the different factors affecting annoyance will vary from one case to another, and nothing general, therefore, could be said of their relative importance.

The subjective responses are not limited to loudness, annoyance or other direct evaluations of the noise. Noise may also influence a person's subjective state, i.e. his mood, and this effect is not necessarily attributed to the noise.

The following four sections treat the different classes of variables affecting the subjective response to noise.

4.1.2. The effects of physical noise characteristics on the subjective response to noise

The basic psychoacoustics, i.e. the relation between the physical qualities of the noise and the subjective response, are common knowledge and have been summarised by many authors (Carterette and Friedman, 1978). Therefore, these relations are only treated briefly, stressing the exceptions to the general rules of thumb.

Sound level. It is generally agreed that the loudness of a sound in most cases is doubled with each 10 dB increase of the sound pressure level. The most important exception to this rule is low frequency noise. When for example a 100 Hz tone is raised 10 dB it is experienced as 4–5 times louder (Scharf, 1978). Such functions have also been determined for more evaluative aspects of the subjective response to noise like noisiness and annoyance with similar results (Kryter, 1985). Obviously, loudness is a major determinant of annoyance. However, when different types of sounds are compared, for example from different sources with different temporal and frequency characteristics, other factors than loudness may be of greater importance.

In reports of noise surveys in residential areas, the effect of noise level is often not described in terms of degree of annoyance but in terms of the percentage of the exposed who complain or report a certain degree of annoyance. A typical finding in studies of traffic noise is that the percentage of annoyed people, above a threshold level, increases by about 10 percent for each 5 dB increase of the sound level (Schultz, 1978). Landström et al. (1992) made a corresponding analysis of annoyance ratings from a group in which fifty percent were office workers. They found the same increase of the percentage "rather annoyed" above 50–55 dB(A) and a slower increase at lower levels. This indicates that other factors than sound level may be more important determinants of noise annoyance at the lowest exposure levels.

Landström et al. (1991a) compared ventilation noise annoyance among offices where the level varied between 33 and 40 dB(A). They found that low frequency ventilation noise at levels close to 40 dB(A) may contribute markedly to noise annoyance and that a ventilation noise at a 5 dB lower level was considerably more acceptable. The study, thus, indicates that a ventilation noise level below 35 dB(A) should be aimed at.

Landström et al. (1992) obtained a correlation of 0.49 between individual annoyance ratings and equivalent sound levels at the work places. However, when this analysis was restricted to office workers the correlation was extremely low and non-significant, which is in line with the result of several other studies (Cavanaugh et al., 1962; Nemecek and Grandjean, 1973b). There are two reasons for this weak relation between annoyance and sound level. One is that the sound level generally is rather low and varies within a rather narrow range in this group. In such cases it is to be expected that other factors beside the sound level (see below) will be of great importance for noise annoyance. Furthermore, the overall sound level is an adequate measure only when the critical noise source is a major determinant of this level. This is not always the case. Thus, an intermittent noise may, for example have a very small influence on the equivalent sound level, but still be the principle cause of the annoyance.

In the study by Nemecek and Turrian (1978b) the outdoor noise was found to be very critical. By comparing the indoor and outdoor noise disturbances, it was found that the indoor noise was more disturbing at levels below 52 dB(A). Outdoor noise at levels above 52 dB(A) was more disturbing than the corresponding levels of the indoor noise. The disturbance from the outdoor noise was related to its sound level, whereas no such relationship was found for the indoor noise. Hay and Kemp (1972b) also studied the relation between outdoor noise levels and disturbance in landscaped offices. The external noise was generated from factory sites and from different types of traffic. At levels below 45 dB(A) at the most 4% of the employees were dissatisfied with the noise coming from the outside. At levels above 55 dB(A) 55% or more of the people were dissatisfied. Both studies thus show a relatively high acceptability of the outdoor noise at levels below 50 dB. Together these studies indicate that different types of noise have to be evaluated in different ways.

When assessing the acceptability of noise from a certain source, the level of this noise must be related to the background noise. Thus, Hay and Kemp (1972a) concluded that not more than 15% of employees were likely to complain of the noise from office machines as long as it did not raise the overall level by more than 6 dB.

It should also be mentioned that in some situations a higher sound level is preferred to a lower one. One reason for this is that a raised level may lead to a masking of especially annoying sounds (see *Signal–noise ratio*, below). One example is that a person in some situations may choose to raise the sound level by turning up the radio. This alternative is often used in cases where a monotonous task is performed against a background of monotonous noise, e.g. in lorries (Landström et al., 1988), but also in offices (Landström et al., 1991a).

Research on the level dependency of the annoyance response has to a large extent treated noise from single sources and therefore provides a weak basis for determining what is the highest overall sound level acceptable in an office, although there is some data supporting that the level has to be raised above 50 dB(A) to lead to a marked increase of the number of complaints. No studies have been reported on the relation between the annoyance caused by separate sources and the total annoyance caused by noise in work places. However, studies of community noise (Berglund et al., 1981; Vos, 1992; Yano and Kobayashi, 1990) have shown that total annoyance is equal to the maximum annoyance caused by a separate source of noise when the other sources cause considerably less annoyance. Therefore, in such cases, a lowering of the overall sound level will not reduce annoyance if it is accomplished without an attenuation of the critical noise. In other cases, where the separate sources are about equally annoying, a certain summation of annoyance takes place, which makes the overall level a more relevant criterion for evaluating the measures taken against the noise.

Finally it should be noted that for some noise sources, especially speech, other qualities than the sound level are likely to be more important. As long as the unwanted speech is intelligible, a lowering of the level has little effect on the annoyance it may cause (see below).

Frequency. There is general agreement on the overall shape of the equal loudness curves, which describe the differential hearing sensitivity of man to sounds of different frequencies. Thus, the maximum sensitivity lies within the 500–5000 Hz frequency range, the sensitivity decreases rapidly in the lowest frequency band and few people hear any sounds above 15000 Hz. The equal loudness curves become flatter at higher sound pressure levels, i.e. the frequency dependency decreases when the level is raised. The frequency weighting filters which are normally used when noise is measured are based on such curves. However, it is to be noted that the same filter, usually the A filter, is used irrespective of sound level. This filter, yielding readings denoted as dB(A), is based upon the equal loudness curve at a very low level (40 dB) where the difference between the sensitivity to different frequencies is large. This means that the dB(A) value is virtually unaffected by the low frequency components of the noise. A consequence of this, which is confirmed by several studies, is that the dB(A)-value may grossly underestimate the subjective impression of noise containing strong low frequency components, e.g. ventilation noise (Kjellberg and Goldstein, 1985; Persson et al., 1985). The A-weighting therefore may lead to incorrect conclusions when for example two machines which emit noise with very different frequency composition are compared. Kjellberg and Goldstein (1985) found that the B- and D-weighting in such cases was less misleading. However, no frequency weighting can be satisfactory without taking into consideration the level-dependency of the equal loudness curves. Zwicker's loudness is an assessment method which fulfills this requirement, and which also takes other factors which determine loudness into account (Zwicker, 1987).

Extremely low frequency sounds, infrasounds, may be generated by the ventilation system in offices. Such sounds may be very annoying, provided that the level exceeds the threshold for hearing. It is therefore important to relate the levels in these frequency bands to the hearing threshold curve. However, it is rarely the case that the low frequency part of the noise reaches this level and the critical frequencies for the adverse effects of the ventilation noise in offices therefore generally are above 50 Hz (Landström et al., 1991a).

Another critical aspect of the frequency characteristics of the noise is the presence of pure tones in the noise, which is common in the noise from office machines. It has been repeatedly demonstrated that annoyance often is increased when the tone is clearly audible, i.e. when the signal–noise ratio is high enough, but the magnitude of this effect has varied between studies (Goulet and Northwood, 1973; Hellman, 1982; Kryter and Pearson, 1965; Ollerhead, 1973; Scharf and Hellman, 1979; Åkerlund et al., 1990). One has generally obtained a more pronounced effect for higher frequency tones. This was confirmed by an experimental study by Landström et al. (1991b), who found that a 100 Hz tone added nothing to the annoyance ratings of a ventilation noise. Bienvenue et al. (1988) found that annoyance was further increased by the presence of two tones. Obviously the effect of tones on annoyance is influenced by many parameters, and none of the suggested methods for tone corrections appears to be universally applicable (Hellman, 1982; Scharf and Hellman, 1979).

Exposure time. It is commonly believed that in a long term perspective people will adapt to noise in the sense that they become less annoyed by it. In fact, most evidence from residential areas argues against such an effect (Weinstein, 1982) at least after the first few weeks in a new environment. Kjellberg et al. (1992) found no relation between time in the same office and noise annoyance, whereas Hay and Kempf (1972b) report a lowering with time of the percentage dissatisfied by the noise. A study by Weinstein in a dormitory (1978) even indicates that the annoyance of those who initially are most annoyed by the noise undergoes the opposite change; their annoyance increases with exposure time.

In a shorter term perspective the question is whether one habituates to a constant noise, e.g. ventilation noise, in the office. A common description of the response to that kind of noise is that one rather seldom thinks of it, but that it is very pleasant when it is turned off (Landström et al., 1991a). This indicates some degree of habituation. However, as a study by Kjellberg and Wide (1988) shows, the absence of direct subjective responses to such a noise, does not rule out the possibility that the person is negatively affected by the noise. It is quite possible that an effect of noise on mood or performance is attributed to other factors than the noise.

Temporal variability. Several studies indicate that temporal variability generally increases the annoyance response. Thus, Keighley (1970) in a large survey of noise problems in offices found that the mean rated acceptability of the noise in an office could be reasonably well predicted from the average sound level and a peak index which was based on the number of times the noise exceeded mean level with a certain value. The importance of variability was also demonstrated by Nemecek and Turrian (1978a), who found that the standard deviation of the noise level was a better predictor of noise annoyance in offices than the equivalent noise level. Similarly, Hay and Kempf (1972b) showed that a better prediction of annoyance was obtained if the difference between the 90th and the 10th percentile of the noise level distribution was considered in addition to the equivalent level.

Similar results have been reported in studies of the community response to traffic and aircraft noise (Björkman, 1991; Björkman et al., 1992).

Laboratory evidence of the importance of intermittency is given by e.g. Schulz and Schönpflug (1982), who gave their subjects the opportunity to terminate a variable traffic noise for short periods during work with tasks of different difficulty. At the lowest noise level only peak sounds were audible. They found that subjects shut off the lowest noise (30–40 dB) as often as the loudest one (70–80 dB) and more often than the noise at an intermediate level. This means that an intermittent noise was regarded as more disturbing than a variable and continuous noise at a higher equivalent sound level.

Another type of intermittent noise is the repetitive noise produced by e.g. printers. This type of noise is also generally considered more annoying than continuous noise at the same equivalent sound level. Two experiments by Byström et al. (1992a,b) indicate that the acceptable level of a repetitive noise during mental work lies about 6 dB lower than for continuous noise.

When assessing ambient noise level in offices it is important also to take its variation into consideration. Smith (1975) concluded that a variation of more than 18 dB should be avoided.

Signal–noise ratio. The most obvious effect of noise is that it may mask speech and other sounds which one wants to hear. The mean sound level of a normal conversation where the persons stand one meter from each other is 55–65 dB(A) with a variation between different speech sounds of about 25 dB. For good communication the noise level should be at least 6 dB below the speech level, but speech may be intelligible even when the signal-to-noise ratio is negative. When the ambient sound level exceeds 60 dB(A) it is necessary to raise the voice to avoid speech interference. Wholly satisfactory communication at a distance of 2.5 meters between speaker and listener requires a noise level of at most 45 dB(A). In rooms where it is important to have perfect communication using normal voice levels it is, thus, necessary to have an even lower ambient noise level. Methods for the assessment of the effects of noise on speech intelligibility are given in

ISO/TR 3352 (International Standardization Organization, 1974). Reviews of research and methods in this area are given by e.g. Lazarus (1990) and Pearsons (1990).

Several studies have shown that noise is especially annoying when noise masks speech or other auditory information which is considered important (Kjellberg et al., 1992; Vanderhei and Loeb, 1977; Williams et al., 1969). Effortless speech communication is essential for many tasks in an office, and a noise that impairs speech intelligibility will therefore be considered very annoying. As shown by Nemecek and Turrian (1978a) such problems occur in offices. They found that 35% of the office workers reported that the noise often impaired speech communication.

Methods have been developed for assessing the communication conditions from the level and frequency characteristics of the background noise, e.g. the Articulation Index (AI) (American National Standards Institute, 1969) and the later Modified Articulation Index (International Standardization Organization, 1990). This index is calculated from the difference between speech level and noise level in different frequency bands, which are weighted with respect to their importance for speech intelligibility. The value of the AI varies between 0 and 1 and stands for the fraction of unrelated words predicted to be correctly identified in the noise. For good communication the AI should be 0.7 or higher (Kryter, 1985).

However, as noted above, masking may sometimes have a positive effect, by making unwanted sounds less intrusive. Faint intermittent sounds, like dripping water, may become very annoying when heard in an otherwise very quiet environment. In such cases a continuous masking noise would probably decrease the annoyance. The same is true of speech masking. Keighley and Parkin (1979) refer to an unpublished study of their own in which they found that the highest levels of satisfaction with the noise climate were found in offices with relatively high background levels, presumably because this noise prevented the overhearing of conversations. This effect naturally also operates in the other direction; by preventing others from hearing our conversations the noise also protects our privacy. A complicating factor is that the natural response to an increased background noise is to raise the voice when the background level is raised beyond about 48 dB(A) (Hamme and Huggins, 1972). However, this compensation generally is incomplete; a 10 dB increase of the noise level normally only makes the talker raise the voice by about 3 dB, which thus has a rather slight effect on the masking effect of the background noise (Van Heusden et al., 1979).

The Articulation Index may also be used to evaluate noise privacy for the speaker and the disturbing effects of irrelevant speech for the listener (Herbert, 1978; Smith, 1975; Warnock, 1973). In this context an AI of 0.1 or lower (Smith, 1975) generally is recommended.

There are several examples of work places where one has installed artificial constant background noise ("sound conditioning") to mask annoying noise, primarily speech. To be effective the masking noise spectrum should decline as a function of frequency with about 5 dB/octave (Smith, 1975). However, there is no consensus on the beneficial effects of such installations (Keighley and Parkin, 1979; Warnock, 1973). Keighley and Parkin (1979) tested twelve different types of artificial background noise but found that the lower the level of the artificially introduced sound was, the better was the overall rating of the office noise. They furthermore concluded that masking efficiency probably only can be attained by levels above 46 dB(A) but that such levels often would be regarded as unacceptable by themselves, at least by a large minority. Warnock (1973) used three levels (45, 48 and 51 dB(A)) in his first experiment and found that the masking noise system was unequivocally rejected at all levels. He concluded that this might have been the effect of the fact that the staff did not express any great desire for privacy, and that their tasks were not vulnerable to distractions from irrelevant speech. Warnock also cites a study by Hegvold (1971) in which the employees were instructed to adjust the level of the masking noise and where it was found that the level finally settled on was 48 dB(A). Völker, from an extensive experience of sound conditioning in offices, maintains that good results have been reached in many cases, but does not report any systematic testing of the effects on the employees (Völker, 1978, 1979). Warnock concludes that 48 dB(A) is the highest level that is likely to be accepted by the employees and Völker (1987) mentions 47 dB(A) as the optimal level of a masking noise. For smaller offices Smith (1975) recommends a highest level of 45 dB(A).

Music has been introduced in some offices with the primary intention of improving the working climate and motivation, but also to work as a masking sound. A study by Nemecek (1984) indicates that music may have beneficial masking effects, but also that 50% of the employees were at least somewhat annoyed by the music.

Obviously, it is very difficult to find an artificial masking noise that is an effective masker and which is not regarded as unacceptable in itself. The problem is further complicated by the fact that the individual responses to such noise vary widely, probably in part as a result of the character of one's job (see below).

4.1.3. The effects of non-physical noise characteristics on the subjective response to noise

In studies of noise annoyance in residential areas physical noise measures, like the dB(A) level, have generally been found to explain a minor part of the interindividual variance in annoyance (Griffiths and Langdon, 1968; McKennell, 1963). There is a lack of comparable data from occupational environments, but in a study from our own group (Landström et al., 1992) we were able to explain about 25% of the variance in a group exposed to noise levels below 85 dB(A), but practically nothing of the variance among the office workers. These results may partly be explained by shortcomings of the annoyance and the physical noise measures, but it is also evident that non-physical noise characteristics are of great importance. Unfortunately, research on the influence of such factors on the subjective response is rare and virtually non-existent within the work environment field.

Informational content. As mentioned above, sound level is a poor predictor of annoyance where irrelevant speech is the main noise problem (Nemecek and Turrian, 1978a; Nemecek et al., 1976). Rather the intelligibility of the irrelevant speech seems to be the main factor determining how disturbing it is. This is why the addition of a masking noise sometimes may have favorable effects.

Sounds other than speech may of course also carry information, which makes them more or less acceptable. The sound may for example tell a person that something is wrong with a machine, or it may be associated with pleasant or unpleasant previous events. However, knowledge about such factors may explain individual differences in the response to a sound, but are generally of little help in the planning of an office environment.

Predictability and controllability. Results from stress research indicate that an unpredictable and uncontrollable stressor generally yields a stronger stress response than a predictable and controllable event (Thompson, 1981). A predictable stressor offers greater possibilities to prepare oneself for the stressor, and the predictability also implicates that there are periods during which the person does not have to be prepared for the stressor. Accordingly, an expected noise should be less annoying than an unexpected one. Similarly, the person who operates a machine and, thus, controls its noise should be less annoyed by it than are other people exposed to the same noise. However, Kjellberg et al. (1992) found no significant difference in the annoyance response to noise from one's own machines and that from other sources. Although it might have been possible to reveal such an effect with a more sensitive study design, the result indicates that the effect cannot be very large.

Attitude to the noise source and other aspects of the working environment. The response to noise is also influenced by the attitude towards the source of the noise. McKennel (1980), for example found that the persons least annoyed by the noise from the Concorde aeroplane were those who had the strongest patriotic feelings for the Concorde project. Likewise, Sörensen (1970) managed to reduce the complaints about aircraft overflights by creating a more positive attitude towards the air force.

It is reasonable to assume that the same mechanisms are at work also in occupational settings. Thus, one would for example expect a secretary who does not like word processors to be especially irritated by the printer noise. However, no study of such effects has been reported.

It is also likely that the overall evaluation of the work and the working environment may influence the evaluation of the noise at the work place. However, systematic studies of such effects have only been reported from residential areas (Jonah et al., 1981; Weinstein, 1980).

A conclusion from these studies is that when the basic problem is one of attitudes towards the noise source or to other aspects of the working conditions, a reduction of the sound level is likely to have little effect on the noise problem at the work place.

Aspiration level and the "necessity" of the noise. In a work shop a major part of the noise may be regarded as an unavoidable consequence of the activity, whereas the same noise would be considered unnecessary in an adjacent office. This is probably the main reason why a noise deemed acceptable in a workshop would be regarded as unbearable in an office. Kjellberg et al. (1992) asked their subjects to judge the possibility to lower the noise level at their work place. These judgements were found to be closely related to their annoyance rating; persons who thought noise could very well be reduced were more annoyed than the others. The differences in annoyance ratings between those who estimated the possibilities as good and those who thought there was no real possibility for an improvement corresponded to a difference in exposure level of about 10 dB (these calculations were based on annoyance ratings that had been corrected for differences in exposure levels). Schönpflug and Schulz (as quoted by Sust, 1987) in a field study in offices found that even very high noise levels were tolerated when the noise was a result of the employee's own work. This might be a conjoint effect of several factors. Thus, the noise may contain information about the work process (performance feedback), the person has a certain amount of control over the noise source and the noise may be viewed as an almost unavoidable consequence of the task. Such an effect of the connection between task and noise on annoyance has also been demonstrated in the laboratory (Munz et al., 1971).

A conclusion from these studies is that a main goal should be to keep the employees protected from noise from sources that have no direct connection with their own job. Some types of noise, primarily ventilation and traffic noise, are regarded as extraneous to the job by virtually all office workers. From this point of view, it should be especially important to reduce these types of noise.

4.1.4. The effects of situational factors: Ongoing activity and noise annoyance

It is often assumed that the annoyance response depends on the task in which one is engaged. In one respect this is definitely true; one is certainly more disturbed when the noise masks auditory information required for the ongoing activity. However, few systematic studies have been reported on the importance of other task characteristics than sensitivity to masking effects. Nemecek and Turrian (1978a) found that persons with higher positions in the office were more easily disturbed by noise than other employees. They interpreted this result as an indication of heightened vulnerability to noise during work with more complex tasks. Boyce (1974) and Nemecek (1984) report similar results, whereas Hay and Kempf (1972b) found no differences between occupational groups. The noise disturbance diaries collected by Purcell and Thorne (1977) indicated that calculation tasks were the ones most sensitive to noise disturbance. Landström et al. (1993) determined annoyance thresholds and tolerance levels during work with a simple reaction time task and a more difficult reasoning task. These levels were found to be 6 dB lower during the more complex task. In the series of experiments reported by Kjellberg and Sköldström (1991) subjects rated their noise annoyance during work with different tasks. Among other tasks they used the same ones as Landström et al. (1993) and also obtained differences in rated annoyance corresponding to six dB. However, during work with a somewhat more complex reaction time task, annoyance was as high as during proof-reading and reasoning tasks. They also found that the most annoying sound–task combination was irrelevant speech during work with a verbal task. Such a combination, unfortunately, is very common in offices.

Obviously, noise tends to be more annoying when performing more complex tasks, and this is also generally considered in the planning of work places. However, the studies also indicate that a rather small part of the differences between work places in noise tolerance levels is explained by such task differences.

4.1.5. Individual differences in the reponse to noise

It is obvious that the same noise elicits widely different responses in different persons. Much research has been devoted to these individual differences and has mainly concerned three questions: Are there any stable individual differences regarding the subjective response to noise? If so, is it a case of specific noise sensitivity? And, finally, what characterises the noise sensitive persons? A comprehensive review of these studies is given by Jones and Davies (1984).

Concerning the first question, the low stability of annoyance ratings indicates that they to a large degree reflect transient states rather than stable traits (Griffiths and Delanzun, 1977). Furthermore, there are reasons to doubt that the stable differences in annoyance reflect a specific noise sensitivity. Thus, Weinstein (1980) found substantial correlations between the evaluations of different aspects of the neighbourhood. He also found that a critical tendencies scale (not including any noise questions) predicted noise annoyance as well as the noise sensitivity scale. Similarly, Kjellberg et al. (1992) and Nemecek and Grandjean (1973b) found that noise annoyance was closely related to dissatisfaction with other aspects of the working environment. Thus, noise sensitivity as a specific and stable trait has little support in data.

Research attempting to identify characteristics of the persons most annoyed by noise has not been very successful. The very inconsistent results indicate that noise sensitivity has no clear relation to demographic variables like sex and age (Jones and Davies, 1984). The only group that has been found to clearly deviate from others in their response to noise are those with a hearing impairment, who most often are more annoyed by noise than others. One reason for this is that the impairment is followed by strengthened masking effects, which means that much lower noise levels become annoying because of their interference with speech intelligibility (Aniansson et al., 1983).

In conclusion, the research on individual differences shows that one could expect widely different responses to the noise at a work place and that it is not possible to predict these differences from other known individual characteristics. One important exception, which should be considered at work places, is that persons with a hearing impairment are likely to be more sensitive to noise.

4.2. Performance effects and other behavioral effects

The research on the effects of noise on performance presents no simple and consistent picture of the effects of noise. Thus Kryter (1985) concluded that there are no significant non-auditory adverse effects of noise on performance, whereas the conclusion in another review covering the same period (Davies and Jones, 1985) was that recent research "suggested that a much wider range of mental functions than hitherto supposed is influenced by noise". Broadbent (1979) in a review from 1979 concludes that no clear effects on performance have been demonstrated of noise below 95 dB. Recent research has demonstrated that noise may affect performance at much lower levels. One reason for this has been a shift of interest from sensori-motor tasks (e.g. reaction time and vigilance tasks) to verbal tasks, which also are the ones that are of primary interest in the office context.

The effects of noise on performance have been treated within several theoretical contexts. The performance effects may be viewed as a result of masking, distraction, changes of arousal level or changes of the strategy chosen for the performance of the task.

4.2.1. Theoretical explanations of performance effects

Masking. It is an indisputable fact that the performance of any task that involves auditory cues may be deteriorated by noise. Poulton (1977) extends this to be true of almost all reported negative effects of continuous noise. Thus, according to him, these effects may be explained by masking of auditory feedback or of inner speech ("you cannot hear yourself think in noise"). Broadbent (1978) has convincingly argued against these ideas, showing that they leave many effects unexplained. However, Poulton must be credited for calling attention to the importance of masking of auditory cues also in tasks that are not primarily of an auditory nature. In offices the masking of speech is the main masking problem, which could affect work performance.

Distraction. Any change in the environment and, thus, any change of noise (raised or lowered level, changed quality) may lead to an attentional response generally called the orienting reflex. This involves a redirection of sense organs towards the noise source, but also a series of physiological responses lasting one or a few seconds (Graham, 1979). Such distractional effects appear at the onset and offset of noise but may also be important for the negative effects of variable and meaningful noise (e.g. irrelevant speech).

An important aspect of this response is that it habituates, provided that the noise is found to be of no importance for the individual. This habituation is faster for low level noise and when the intervals between the changes are short. Habituation is generally slower to sounds that carry information, or when the information is of no relevance for the receiver.

Purcell and Thorne (1977) demonstrated that irrelevant speech and sudden changes of the noise in offices may interrupt a chain of thought, for example a computation, and thereby impair performance.

Arousal level, allocation of attention. The effects of noise on performance have traditionally been treated within the framework of arousal theory, sometimes supplemented by the hypothesis that attention becomes more selective at high arousal levels. According to this view the effects of noise on performance are seen as consequences of changes along a unidimensional arousal continuum. Generally noise has been supposed to raise the arousal level and, thus, the effects on performance are the result of overactivation which results in a too selective intake of information. However, it should be noted that there is reason to believe that repetitive or continuous noise, especially low frequency noise, sometimes makes people sleepy (Bohlin, 1971; Hartley, 1973; Landström et al., 1983). When the radio is turned on during work it is partly to overcome such an effect of the noise most often in combination with a monotonous task.

In offices exposure to sounds to which one does not habituate may lead to overstimulation effects. The opposite problem may also be a real one: ventilation noise and other constant low frequency noise may lower the wakefulness level.

Strategy choice and compensatory processes. The unidimensional arousal theory is contradicted by several experiments in which combinations of noise and other arousing or de-arousing factors have been studied. To account for such results more complex arousal models have been proposed (Broadbent, 1971), but have not been tested systematically. Another problem for explanations of the performance effects in terms of arousal is that small changes in the task or the experimental situation may alter the results dramatically. This indicates that few of the performance effects are unavoidable consequences of the noise exposure in itself. Such effects have been one of the arguments for a "strategy choice hypothesis", which states that noise primarily influences the way in which a person chooses to carry out a task (Jones, 1990; Schönpflug, 1983; Smith, 1983, 1991). This means, for example, that attentional selectivity is not regarded as an inevitable consequence of the arousing effects of noise. Noise is rather viewed as one of many factors that may increase the mental load and decrease the capacity available for the performance of the task. In such a situation one strategy is to direct the available information processing resources towards the most important aspects of the task at the expense of the less important ones. In some tasks such a strategy may be very successful and leave performance unaffected.

A more selective intake of information is, therefore, an example of a way of coping with a situation in which the normal way of performing the task requires more resources than those available. However, in most tasks the performance level generally lies well below the person's maximum level. This means that it is generally possible to uphold the performance level in spite of the noise by exerting extra effort. At least this is true for a limited time. A negative effect of noise on performance then is attained if the load exceeds the level that can be compensated for, or, if for some reason it is not deemed to be worth the extra effort required to keep performance at the level that is normal in a quiet situation.

Thus, an unchanged performance level does not necessarily mean that the environment is without any effect; the performance may have been attained at a higher cost than otherwise would have been the case. This means that performance efficiency, defined as the ratio between performance and its costs, may be affected by noise rather than performance in itself. Another consequence of this view is that work should be more fatiguing in noise than under quiet conditions. The extensive experimental research on after-effects of work in noise (see below) indicates that this also is the case.

The issues of strategy and compensatory mechanisms are discussed at length by Schönpflug (1982, 1983), who also presents data to support this view of the effects of noise on performance.

This emphasis on strategy effects does not mean that there are no performance effects that are almost unavoidable consequences of the noise exposure. This is primarily true of those caused by masking effects, but it is also conceivable that some of the effects on short-term verbal memory are a result of direct and unavoidable interference with a stage of the information processing.

4.2.2. Field studies of safety and efficiency

Early reports on the effects of noise reduction claimed dramatic increases of productivity (Berrien, 1946; Wisner, 1967). However, all these studies as well as the more recent studies on the effects of noise on accidents and productivity have dealt with factories with very high noise levels (Broadbent and Little, 1960; Noweir, 1984; Wilkins and Acton, 1982). No systematic studies seem to have been made of noise effects on performance in offices, although claims have been made that such effects do occur (Warnock, 1973).

In studies of office workers only subjective ratings of performance impairment have been reported (Kjellberg et al., 1992; Landström et al., 1991a; Nemecek and Turrian, 1978a). According to Nemecek et al. (1976) this judgement is unrelated to noise level. Landström et al. (1991a) found that about 20% of the office workers in their studies thought that the ventilation noise made their tasks more difficult. Lindström and Vuori reported similar opinions on noise effects in an open plan office (1984). Purcell and Thorne (1977) asked their office workers to record in their diaries more detailed information about how noise affected their performance. They generally described the effects as disturbances of attention (distraction, reduced concentration and interruption of trains of thought).

4.2.3. Noise effects in different tasks

A vast amount of laboratory research on noise and performance has been reported. Most of this research, however, is of small or no relevance for the noise problems in offices; either the noise levels have been far above those in offices, or the choice of tasks has made it difficult to generalise the results to office settings. Thus, a large part of this research has used levels above those at which hearing protection is required at work, and the tasks studied have very often been reaction time and vigilance tasks. Sometimes the ambition has been to generalise results to industrial and military settings, but often the aim has not been to apply the results to any practical noise problem. The following review is limited to studies that bear some relevance for the noise problem in offices, i.e. studies in which the tasks engage abilities that might be of importance for office work.

Reading comprehension and similar verbal tasks.

Recent research has demonstrated that noise may affect performance at levels previously considered harmless. One reason for this has been a shift of interest from sensori-motor tasks to verbal tasks. Among those tasks the most relevant ones are the reading comprehension tasks.

Proof-reading requires abilities that are in one way or another of importance in many other office tasks. One component of this task is to detect typographical errors, misspellings and other errors that may be identified just from a look at single words. In this respect the task is similar to several other clerical tasks and requires a rather superficial processing of the text. However, the task may also involve the detection of contextual errors like grammatical or semantic errors, or missing words, i.e. errors that require comprehension of the text that require a deeper processing and understanding of the text. Thus, proof-reading is also an obvious example of a task in which one may adopt several strategies with different emphasis on performance speed and the detection of typographical or contextual errors. If noise affects the strategy choice it is necessary to describe all three aspects of task performance to understand the effect.

The first studies of proof-reading in noise were reported by Weinstein (1974, 1977). The first of these experiments (Weinstein, 1974) used an intermittent printer noise at a level varying between 55 and 70 dB(A). No effect of noise was found on performance speed or on the identification of simple typographical errors, but noise lowered the detection frequency of the more complicated contextual errors. In the second study (Weinstein, 1977), which gave similar results, the noise was an intermittent presentation of news items from a radio broadcast.

Schwabe (1982) had his subjects work for three hours during exposure to three levels of traffic noise (48, 58 and 68 dB(A)). The percentage of identified errors and speed was registered. Only typographical errors were present in the text, which was presented on a computer screen. In line with Weinstein's results no effects of noise were obtained.

Jones et al. (1990) studied the effects of irrelevant speech on proof-reading, distinguishing between typographical and contextual errors. They found that the effect was independent of speech level (between 50 and 70 dB(A)) and that only meaningful speech impaired error detection. Finally, in contrast to the studies by Weinstein, only the identification of typographical errors was impaired by the speech. Speed and contextual errors were unaffected. One conclusion, thus, is that noise and irrelevant speech constitutes a load on the reader, but that this may manifest itself in different ways dependent on what aspect of the task is given priority.

Martin et al. (1988) used another approach to the problem. They let their subjects read text passages in different noise conditions (including white noise, speech, random speech and different kinds of music) and tested the comprehension of the text by asking questions about the content. Again meaningful speech caused the greatest impairment of comprehension, but meaningless speech and music with lyrics were also found to impair comprehension.

Calculation. The diary study by Purcell and Thorne (1977) indicated that calculation tasks were the ones most sensitive to noise disturbance according to the employees themselves. Bhatia et al. (1991) exposed their subjects to a 40 dB factory or corridor noise and divided subjects into low and high noise sensitivity using Weinstein's noise sensitivity scale (Weinstein, 1978). They found that arithmetic performance was impaired at this extremely low noise level, although only in the high sensitivity group.

Memory. The effects of noise on verbal memory have been treated in many studies, which, however, generally have been of greater relevance for information processing theories than for the understanding of noise effects at work. In recent years a large part of this research has dealt with the impairment of serial verbal short-term memory by irrelevant speech, which has been found to be an unusually consistent effect (see review by Jones and Morris, 1992). Some main conclusions from this research are that the effect is independent of speech level (at least between 55 and 95 dB(A)) and that it appears also when the speech is incomprehensible. Furthermore, it is unlikely that distraction effects could explain the impairment, which rather seems to depend on direct interference with memory processes. The effect seems to be limited to a very special type of memory task, which has very few direct counterparts outside the laboratory. However, it is possible that the effect in this task reflects a disruption of working memory by irrelevant speech, which may be of relevance in several other tasks with a high memory load.

Incidental learning. Hockey and Hamilton (1970) instructed their subjects to remember words presented on different parts of a screen. This task was accomplished at least as well in noise as in a control condition. However, when subjects were unexpectedly asked to remember the location of the words more errors were obtained in the noise group. Other studies of such incidental learning have obtained the same result (Cohen and Lezak, 1977; Davies and Jones, 1975), although it should be noted that Niemi et al. (1977) failed to replicate the results obtained by Hockey and Hamilton. Thus, these studies provide further evidence that noise may cause an increased concentration of attentional resources on those aspects of the tasks that are deemed to be most important.

Complex office types of tasks. Schönpflug and Schulz made extensive analyses of office tasks and developed laboratory simulations of more complex office tasks. Many of the laboratory experiments have studied the effects of noise on these types of tasks, although noise effects have not been their central issue. Noise has rather been viewed as one of several environmental loads in the office that might influence task performance. The aim of these studies has been to analyse how such a load influences the way such tasks are performed; what effect does the noise have on the strategy used in the task and on the way information is picked up and processed? Sust (1987), Schönpflug (1983) and Schulz and Schönpflug (1982) review these studies.

In a typical task in these studies, subjects are presented with a problem of a type common for office workers (e.g. the application of a loan) and have to make a choice among alternative decisions. To reach a correct decision the subject needs supplementary information, which he has the possibility to order. Such tasks of varying difficulty have been studied during exposure to different levels of noise. A main conclusion from these studies is that noise constitutes a load while performing these tasks and that subjects cope with this load in two ways. They either try to compensate for the noise load by working harder, or they make more risky decisions.

Schönpflug and Schulz (1982, 1979) found that noise (50–60 or 70–80 dB(A) varying traffic noise) typically increased the number of repeated calls for the same information. This was interpreted as a compensatory measure for overcoming increased problems of keeping information in memory. This tendency was strengthened when the task was performed under time pressure. Under time pressure noise also increased the number of erroneous decisions, i.e. the compensation for the noise effects was sometimes incomplete. Performance efficiency (defined as the relation between number of correct solutions and the number of calls for information), was impaired already at the 50–60 dB level, but was more pronounced at 70–80 dB.

Similar results were reported by Schulz and Battmann (1980), who also found that noise had a larger effect on subjects with lower capacity for the task. Battman and Schönpflug (1986) studied the effect of exposure to variable noise during different phases of a time planning task. They found that when noise peaks coincided with critical phases of the task, the quality and speed of performance were impaired at an equivalent sound level of 64 dB(A).

4.2.4. After-effects of work in noise

One indication of the increased effort exerted during work in noise is the after-effects of work in noise, which has been demonstrated in a long series of laboratory experiments. In their influential book *Urban Stress*, Glass and Singer (1972) reported a series of studies, which indicated that performance may be impaired also *after* a period of noise exposure. A review of more recent studies is given by Cohen (1980). The tasks that seem to be most sensitive to these after-effects of noise do not primarily measure the capacity to perform, but rather the motivation to perform well, for example endurance in insoluble tasks, or very dull or extremely difficult tasks. In contrast to the acute effects, after-effects have been found to be very reliable. It is also to be noted that an acute effect during exposure is not a prerequisite for these after-effects.

The after-effects are most evident after exposure to uncontrolled or unpredictable noise, whereas they are less dependent on the noise level. In fact, after-effects of noise should not be regarded as due to noise in itself but rather as due to a general influence of uncontrollable and unpredictable stressors (Cohen, 1980).

In the only study demonstrating after-effects of noise at work the group was exposed to extremely high noise levels (Kjellberg et al., 1991). Nothing therefore is known of the likelihood of office noise yielding such effects.

4.2.5. Conclusions: Performance effects in the laboratory and the office

The effects of office noise on performance have to our knowledge never in a convincing way been demonstrated in a real office environment. Conclusions concerning the risk for such effects therefore must rest upon laboratory studies. These studies indicate that even moderate levels of noise may impair performance of many tasks performed in offices, and that the risk is higher for more complex types of tasks.

This is not to say that performance in real office tasks is also likely to be impaired by the noise, but it is a strong indication that noise constitutes a load when performing these types of tasks. The way in which a person chooses to cope with the increased load caused by the noise, however, is highly dependent on the situation. In some situations one is likely to accept a deteriorated performance, whereas in other situations everything will be done to maintain the performance level. Therefore, it is very difficult to make generalisations from the laboratory to the work place regarding the performance effects of lower level noise. If no effect was found in the laboratory, it is still possible that such effects would appear at the work place. During a short experiment where performance is being monitored continuously, the subject may choose to try to overcome the load created by the noise. This strategy might be less likely in the real work situation. Therefore, it is quite possible that tasks that have been found to be insensitive to noise in the laboratory, may be affected at work places. Of course the opposite relationship is also possible; an effect that has been found in the laboratory might not show up at work, where sometimes the motives for avoiding failures are stronger.

In conclusion, noise has been found to impair performance of office tasks in the laboratory, which indicates that noise is likely to make the task more demanding also under other circumstances. However, it is not possible to judge whether these increased demands are likely to result in impairments also at the work place. The costs of noise may instead be paid by the worker in the form of increased fatigue.

4.2.6. Critical physical noise characteristics for the performance effects of noise

In the studies of performance effects the interest has been concentrated on analyses of how performance is affected by noise and in which tasks such effects are likely to appear. In contrast to the research on subjective responses to noise, rather little research has been devoted to studies of the effects of noise characteristics.

Sound level. Noise level is the only physical noise characteristic that has been studied to any extent in connection with performance effects. Broadbent (1979) concluded that negative effects on performance were only to be expected at very high levels (> 90–95 dB(A)). Since then researchers have become increasingly aware of the complexity of the matter, and no general conclusions about the critical level now seem possible; it obviously depends on the nature of the task, the experimental setting and the choice of subjects and their experience. Evidently, the critical level is sometimes very low, as shown by Kjellberg and Wide (1988) and Hygge (1988), who both obtained performance effects of continuous noise at 51–53 dB(A). The issue is further complicated by the problem to generalise from laboratory to real working conditions where the duration of exposure and work is so much longer.

Frequency. Only Broadbent (1957) has specifically treated the effects of sound frequency on performance and found that serial reaction time performance was more sensitive to high than to low frequency noise. However, it seems likely that the effects of low and high frequency noise on performance are mediated by different mechanisms. Low frequency noise probably gets its effect by lowering the wakefulness (Landström, 1987; Landström et al., 1983), an effect that takes time to develop. Such an effect should be most evident in repetitive, non-demanding tasks and should primarily lead to slower work (Landström et al., 1993). Higher frequency noise is more intrusive and affects performance by competing for attentional resources.

Exposure time. Generally the effect of noise has been found to increase as a function of time on task. This is not necessarily an effect of exposure time in itself but is perhaps better described as an effect of working for a long time in noise, i.e. an interaction effect. However, in two studies subjects were exposed to noise for a longer or shorter period before the task (Hartley, 1974; Smith and Broadbent, 1985). These studies indicate that there may also be an effect of exposure time in itself.

Temporal variability. Intermittent noise has been found to impair performance more often than continuous noise (Smith, 1985). Schulz and Schönpflug (1982) in their study of a variable traffic noise at different levels found that the error rate was doubled in the low level, intermittent noise, condition as compared to that in a quiet control condition and only 25% lower than in the high level, continuous noise, condition. Wieland (1981) found that an intermittent noise, but not a continuous one, made subjects work less efficient in a complex planning task.

Signal–noise ratio. Obviously noise may impair the performance of tasks that require speech comprehension. This effect of noise and the methods available for the evaluation of speech communication problems have been described in many handbooks (Bailey, 1989; van Cott and Kinkade, 1972; Webster, 1984). The noise in offices seldom reaches levels that interfere seriously with word recognition, but this does not necessarily mean that the masking effects are of no importance. Speech comprehension may require the person to exert more effort and diminish the spare capacity necessary to process the information. Thus, Rabbit (1966) found that noise impaired the memory of words, although the words had been correctly identified.

Other sounds than speech may also carry information, which is important for the performance of a task. Thus, auditory cues may for example tell us whether a machine works as it should or whether an operation has been triggered as intended. Poulton brought forward subtle and until then disregarded examples of masking affecting performance (Poulton, 1976, 1977, 1978, 1979). However, only rarely do office tasks depend on such auditory feedback.

4.2.7. Other factors influencing the performance effects of noise

Of the factors mentioned in connection with the subjective responses only the individual differences and the information content and predictability of the noise has been the subject of research.

Informational content. As mentioned before, the informational content of the speech probably is the main determinant of its intrusiveness. It is true that in one very special type of task, serial verbal memory tasks, one has found incomprehensible speech sounds to be equally intrusive (Jones and Morris, 1992). However, such tasks are of limited validity for real life office tasks, and more important therefore is that the meaning of the speech is of critical importance for the negative effects of speech on reading comprehension tasks (Jones, 1990).

Predictability and controllability. Intermittent noise has most often been found to have more detrimental effects on performance than continuous noise and this effect seems to be accentuated when the time schedule is unpredictable (Kohfeld and Goedecke, 1978). The opposite effect of variability may be obtained if the subject is well acquainted with the noise and the time schedule is reasonably predictable (Warner and Heimstra, 1971).

In the one study that treated the effect of subjects having control over the noise (Blechman and Dannemiller, 1976), performance was less disrupted when subjects had, or rather thought they had, such control.

Individual differences. The research on individual differences in the vulnerability of performance to noise presents a very unclear picture. There are indications that the part of the individual differences in noise vulnerability is even less stable than that found regarding the subjective responses to noise (Wilkinson, 1974). The differential effects of noise on performance thus probably rather reflect the temporary state of the persons than stable individual characteristics.

The degree of experience with the task at hand is a possible basis for individual differences, which is of more interest in the present context. Apparently, noise is more likely to impair performance of complex than simple tasks. The prevalent explanation of this finding is that the additional load caused by the noise is more critical in a task which demands a great part of the available capacity, than in a task which leaves much spare capacity. This implies that one should be most vulnerable to noise in the early stages of training of a task. This is so since one important effect of training is that one learns to perform the task with the use of a smaller part of the total capacity. This notion gets some support by Schulz and Battmann (1980), who found a larger noise effect on subjects with a lower capacity for the task.

4.2.8. Other behavioural effects of noise

Many studies indicate that we behave differently towards each other in noisy environments than otherwise. One obvious reason for this is that noise makes speech communication more difficult, but this does not explain all social effects of noise. Thus, a series of field experiments shows that people tend to be less helpful in noise (a review of these and other studies of social effects is given by Cohen and Spacapan, 1984). The noise levels used in these studies, however, have in all cases been far above what could be expected in an office, and nothing, therefore, is known about possible social effects in such contexts.

4.3. Effects of noise reduction

Many studies have been performed to evaluate the effectiveness of different measures for reducing noise at work places. However, surprisingly few systematic studies have been reported of the subjective or behavioral effects of noise reduction on the office workers. This means that there is no real basis for predicting the effects of a certain reduction of the noise level in an office; differences between annoyance levels in offices with different noise levels provide an unsatisfactory basis for such predictions. This is indicated by studies of community noise in which it has been shown that a dose–response relationship based on noise and annoyance levels in different areas provides no valid basis for conclusions about likely effects of a reduction of the noise level. Thus, a 5 dB difference between two residential areas is likely to result in a ten percent higher frequency of complainants (Schultz, 1978), but this does not mean that a 5 dB reduction of the noise level leads to a ten percent lowering of the number of complainants. In fact, the typical result has been a larger effect on annoyance than was to be expected from the dose–response curves (de Jong, 1990; Raw and Griffiths, 1990). It is likely that this also is the case in offices.

The only study known to us of the performance effects of a noise reduction at a work place is the frequently cited study by Broadbent and Little (1960). They found that a noise reduction led to substantial decreases of the rejection percentage in a film production plant where the noise levels were very high. Only anecdotal evidence exists for an effect of noise reduction on office productivity (Makley, 1985).

A related question is what happens when the ventilation system or another noise source is turned off. A common answer to a question about the ventilation noise is that one seldom thinks of it during the day, but that it feels very pleasant when it is turned off at night. Kjellberg and Wide (1988) simulated this situation in the laboratory. During work with a reasoning task a ventilation noise was gradually increased until it reached 51 dB(A). After a period of constant noise it was suddenly turned off. In a control group the noise was turned on at the same point of time. None of the subjects thought that the noise was part of the experiment, and few had noticed it before it was turned off. Still, learning rate was slower in the noise group, and the number of errors was markedly reduced when the noise was turned off. Thus, these results indicated an immediately increased efficiency after the noise was turned off.

4.4. Research needs

There is a general need for work place studies on the effects of noise levels that do not constitute a risk for hearing damage. For such research to be relevant to the issue of improving noise conditions in offices, the analysis of the noise and other factors influencing the annoyance response must be more sophisticated than has been the case in most studies. Analyses that are restricted to the relation between equivalent sound levels and overall ratings of noise annoyance are of very limited value as a basis for improvements of the noise conditions in an office. There seem to be two fruitful alternatives to such a strategy. One would be to study the response to a specific noise source. In such studies there seem to be good possibilities to specify dose–response relationships. The other way would be to recognise the complexity of the problem and include other noise characteristics as well as other critical conditions in the analysis.

The measurement of noise and noise annoyance may also be done in more sophisticated ways than hitherto. The equivalent sound level is a too crude measure of exposure and should be replaced by measures taking account of e.g. temporal characteristics of the noise. Annoyance has generally been assessed with simple rating scales with verbally defined steps. One problem with such ratings is that their reliability generally is rather low. This problem may be remedied by the construction of annoyance indices based upon several questions. Another problem is that the scales are uncalibrated, and that, thus, the same rating does not stand for the same level of annoyance for all persons. Berglund and Berglund (1987) has developed a method for calibrating ratings of this kind. This method, however, has never been used in occupational contexts.

Performance studies have seldom had the primary aim of evaluating the risk of performance impairments due to noise in offices. Neither the tasks, nor the noise or other exposure conditions have been chosen with this purpose in mind. Therefore, there is a lack of studies using realistic tasks during long periods of exposure to noise at levels and of a character met in offices. Performance studies in real offices present many problems, but the increased computerisation of office work may make such studies more realistic. It should now be possible to collect detailed data reflecting both productivity and distraction during different noise conditions, which, preferably, should be arranged in the same office.

A lot of work is done to improve noise conditions in offices, but systematic evaluations of the effects in terms other than technical are almost completely lacking. Research reports therefore offer little guidance for the practitioner on how much annoyance could be expected to be reduced as a result of a certain improvement of the noise conditions.

References

Åkerlund, E., Kjellberg, A. and Landström, U., 1990. Annoyance in working environments due to low frequency noise with or without tones. In: U. Sundbäck (Eds.), Proceedings of Nordic Acoustical Meeting 90. Luleå University of Technology, Luleå, pp. 165–170.

American National Standards Institute, 1969. American national standard methods for the calculation of the articulation index. ANSI S3.5-1969. American National Standards Institute.

Aniansson, G., Pettersson, K. and Peterson, Y., 1983. Traffic noise annoyance and noise sensitivity in persons with normal and impaired hearing. Journal of Sound and Vibration, 88: 85–97.

ASTM, 1975. Standard recommended procedure for laboratory measurement of airborne sound transmission loss. ASTM 90-75. American standards for testing materials.

Bailey, R.W., 1989. Human Performance Engineering. Prentice Hall, Englewood Cliffs, NJ.

Battman, W. and Schönpflug, W., 1986. Synchrone Belastung durch Schall und Tätigkeiten. Zeitschrift für Lärmbekämpfung, 33: 14–21.

Berglund, B., Berglund, U., Goldstein, M. and Lindvall, T., 1981. Loudness (or annoyance) summation of combined community noises. Journal of the Acoustical Society of America, 70: 1628–1634.

Berglund, B. and Berglund, U., 1987. Measurement and control of annoyance. In: H.S. Koelega (Eds.), Environmental Annoyance: Characterization, Measurement, and Control. Elsevier, Amsterdam, pp. 29–44.

Berrien, F.K., 1946. The effects of noise. Psychological Bulletin, 43: 141–161.

Bhatia, P., Shipra and Muhar, I.S., 1991. Effect of low and high intensity noise on work efficiency. Psychologia, 34: 259–265.

Bienvenue, G.R., Corkery, M.J., Ingrao, B.B. and Rafter, R.V., 1988. Annoyance ratings of office machinery spectra with prominent tone components. In: J.S. Bolton (Eds.), NOISE-CON 88 Proceedings of the 1987 National Conference on Noise Control Engineering. Noise Control Foundation, Poughkeepsie, NY, pp. 421–426.

Björkman, M., 1991. Community noise annoyance: Importance of noise levels and number of noise events. Journal of Sound and Vibration, 151: 497–503.

Björkman, M., Åhrlin, U. and Rylander, R., 1992. Aircraft noise annoyance and average versus peak noise levels. Archives of Environmental Health, 47: 326–329.

Blechman, E.L. and Dannemiller, E.A., 1976. Effects on performance of perceived control over noxious noise. Journal of Consulting and Clinical Psychology, 44: 601–607.

Bohlin, G., 1971. Monotonous stimulation, sleep onset and habituation of the orienting reaction. Electroencephalography and Clinical Neurophysiology, 31: 593–601.

Boyce, P.R., 1974. Users' assessment of a landscaped office. Journal of Architectural Research, 3: 44–62.

Broadbent, D.E., 1957. Effects of noises of high and low frequency on behaviour. Ergonomics, 1: 21–29.

Broadbent, D.E., 1971. Decision and stress. Academic, London.

Broadbent, D.E., 1978. The current state of noise research: reply to Poulton. Psychological Bulletin, 85: 1052–1067.

Broadbent, D.E., 1979. Human performance and noise. In: C.M. Harris (Eds.), Handbook of noise control. McGraw-Hill, New York, pp. 17: 1–17: 20.

Broadbent, D.E. and Little, E.A.J., 1960. Effects of noise reduction in a work situation. Occupational Psychology, 34: 133–140.

Brookes, M.J., 1972. Office landscape: Does it work? Applied Ergonomics, 3: 224–236.

Byström, M., Kjellberg, A. and Landström, U., 1992a. Störningströsklar för kontinuerligt och intermittent bredbandigt buller vid olika uppgifter (Annoyance thresholds for continuous and intermittent noise during the performance of different tasks. Report in Swedish with an English summary). National Institute of Occupational Health, Solna, Sweden, Arbete och Hälsa 1992: 12.

Byström, M., Kjellberg, A. and Landström, U., 1992b. Störningströsklar för olika typer av intermittenta ljud (Annoyance thresholds for different types of intermittent sounds. Report in Swedish with an English summary). National Institute of Occupational Health, Solna, Sweden, Report No. 1992: 39.

Carterette, E.C. and Friedman, M.P., 1978. Handbook of Perception, Volume 4: Hearing. Academic Press, New York.

Cavanaugh, W.J., Farrell, W.R., Hirtle, P.W. and Watters, B.G., 1962. Speech privacy in buildings. Journal of the Acoustical Society of America, 34: 475–492.

Cohen, S., 1980. After-effects of stress on human performance and social behavior: A review of research and theory. Psychological Bulletin, 88: 82–108.

Cohen, S. and Lezak, A., 1977. Noise and inattentiveness to social cues. Environment and Behavior, 9: 559–572.

Cohen, S. and Spacapan, S., 1984. The social psychology of noise. In: D.M. Jones and A.J. Chapman (Eds.), Noise and society. Wiley, Chichester, pp. 221–245.

Costa, G., Apostoli, P. and Peretti, A., 1984. Noise, lighting and climate inside different office work places. In: E. Grandjean (Eds.), Ergonomics and health in modern offices. Taylor & Francis, London, pp. 77–81.

Crocker, M.J., 1978a. Noise of air transportation to nontravellers. In: D.N. May (Eds.), Handbook of Noise Assessment. Van Nostrand, New York, pp. 82–126.

Crocker, M.J., 1978b. Noise of surface transportation to non-travelers. In: D.N. May (Eds.), Handbook of Noise Assessment. Van Nostrand, New York, pp. 39–81.

Davies, D.R. and Jones, D.M., 1975. The effects of noise and incentives upon attention in short-term memory. British Journal of Psychology, 66: 61–68.

Davies, D.R. and Jones, D.M., 1985. Noise and efficiency. In: W. Tempest (Eds.), The Noise Handbook. Academic Press, New York, pp. 87–141.

de Jong, R.G., 1990. Review of research developments in community response to noise. In: B. Berglund and T. Lindvall (Eds.), Proceedings of the 5th International Congress on Noise as a Public Health Problem, Vol. 2. Byggforskningsrådet, Stockholm, pp. 99–113.

deJoy, D.M., 1984. The non-auditory effects of noise: Review and perspectives for research. Journal of Auditory Research, 24: 123–150.

Ericsson Information Systems, 1984. Ergonomic principles in office automation. Ericsson Information Systems, Stockholm.

Glass, D.C. and Singer, J.E., 1972. Urban Stress: Experiments on Noise and Social Stressors. Academic Press, New York.

Goulet, P. and Northwood, T.D., 1973. Subjective rating of broad-band noises containing pure tones. Journal of the Acoustical Society of America, 53: 365–366.

Graham, F.K., 1979. Distinguishing among orienting, defense, and startle reflexes. In: H.D. Kimmel, E.H. van Holst and J.F. Orbleke (Eds.), The Orienting Reflex in Humans. Lawrence Erlbaum, New York, pp. 137–167.

Grandjean, E., 1988. Fitting the Task to the Man. A Textbook of Occupational Ergonomics. Taylor & Francis, London.

Griffiths, I.D. and Delanzun, F.R., 1977. Individual differences in sensitivity to traffic noise: An empirical study. Journal of Sound and Vibration, 55: 93–107.

Griffiths, I.D. and Langdon, F.J., 1968. Subjective response to road traffic noise. Journal of Sound and Vibration, 8: 16–32.

Hamme, R.N. and Huggins, D.N., 1972. The problem of acoustical specifications for office landscape ceilings. Journal of the Acoustical Society of America, 52: 120.

Hartley, L.R., 1973. Similar and opposing effects of noise on performance. In: W.D. Ward (Eds.), Proceedings of the Second International Congress on Noise as a Public Health Problem. Environmental Protection Agency, EPA Report 550/9-73-008, Washington, DC, pp. 379-387.

Hartley, L.R., 1974. Performance during continuous and intermittent noise and wearing ear protection. Journal of Experimental Psychology, 102: 512-516.

Hay, B. and Kemp, M.F., 1972a. Frequency analysis of air conditioning noise in landscaped offices. Journal of Sound and Vibration, 23: 375-381.

Hay, B. and Kemp, M.F., 1972b. Measurements of noise in air conditioned, landscaped offices. Journal of Sound and Vibration, 23: 363-373.

Hedge, A., 1989. Environmental conditions and health in offices. In: D.J. Oborne (Eds.), International Reviews of Ergonomics. Current Trends in Human Factors Research and Practice, Volume 3. Taylor & Francis, London, pp. 87-110.

Hegvold, L.W., 1971. Experimental masking noise installation in an open planned office. Technical Note No. 563. National Research Council of Canada, Division of Building Research.

Hellman, R.P., 1982. Loudness, annoyance, and noisiness produced by single-tone-noise complexes. Journal of the Acoustical Society of America, 72: 62-73.

Herbert, R.K., 1978. Use of the Articulation Index to evaluate acoustical privacy in the open office. Noise Control Engineering Journal, 11: 64-67.

Hockey, G.R.J. and Hamilton, P., 1970. Arousal and information selection in short-term memory. Nature, 226: 866-867.

Hohensee, H., 1990. Die Geräuschbelastung durch Ohrhörer von Diktiergeräten. Zentralblatt für Arbeitsmedizin, Arbeitsschutz und Prophylaxe, 40: 385-390.

Hubert, M., 1978. Lärm und Lärmminderung in Büroräumen. Zentralblatt für Arbeitsmedizin, Arbeitsschutz und Prophylaxe, 28: 85-90.

Hygge, S., 1988. A study of the effects of low-intensity noise and mild heat on cognitive performance and the five-choice serial reaction time task. In: O. Manninen (Eds.), Proceedings of the International Conference on Combined Effects of Environmental Factors. Tampere, Finland, pp. 417-430.

International Standardization Organization, 1974. ISO/TR 3352 Acoustics – Assessment of noise with respect to its effects on the intelligibility of speech. ISO, Geneva.

International Standardization Organization, 1988a. ISO 7779 Acoustics – Measurement of airborne noise emitted by computer and business equipment. ISO, Geneva.

International Standardization Organization, 1988b. ISO 9295 Acoustics – Measurement of high frequency noise emitted by computer and business equipment. ISO, Geneva.

International Standardization Organization, 1988c. ISO 9296 Acoustics – Declared noise emission values from computer and business equipment. ISO, Geneva.

International Standardization Organization, 1990. ISO/DIS 9921-2 Ergonomic assessment of speech communication – Part 2: Assessment of speech communication by means of the modified Articulation Index (MAI method). ISO, Geneva.

International Standardization Organization, 1992a. ISO/DIS 11690-1 Acoustics – Recommended practice for the design of low-noise work-place – Part 1: Noise control strategies. ISO, Geneva.

International Standardization Organization, 1992b. ISO/DIS 11690-2 Acoustics – Recommended practice for the design of low-noise work-place – Part 2: Noise control measures. ISO, Geneva.

Jonah, B.A., Bradley, J.S. and Dawson, N.E., 1981. Predicting individual subjective responses to traffic noise. Journal of Applied Psychology, 66: 490-501.

Jones, D., 1990. Recent advances in the study of human performance in noise. Environment International, 16: 447-458.

Jones, D.M. and Davies, D.R., 1984. Individual and group differences in the response to noise. In: D.M. Jones and A.J. Chapman (Eds.), Noise and Society. Wiley, Chichester, pp. 125-153.

Jones, D.M., Miles, C. and Page, J., 1990. Disruption of proof-reading by irrelevant speech: Effects of attention, arousal or memory? Applied Cognitive Psychology, 4: 89-108.

Jones, D.M. and Morris, N., 1992. Irrelevant speech and serial recall: Implications for theories of attention and working memory. Scandinavian Journal of Psychology, 33: 212-229.

Keighley, E.C., 1970. Acceptability criteria for noise in large offices. Journal of Sound and Vibration, 11: 83-93.

Keighley, E.C. and Parkin, P.H., 1979. Subjective responses to sound conditioning in a landscaped office. Journal of Sound and Vibration, 64: 313-323.

Kjellberg, A. and Goldstein, M., 1985. Loudness assessment of band noise of varying bandwidth and spectral shape. An evaluation of various frequency weighting networks. Journal of Low Frequency Noise and Vibration, 4: 12-26.

Kjellberg, A., Landström, U., Tesarz, M., Söderberg, L. and Åkerlund, E., 1992. Betydelsen av icke-fysikaliska faktorer för bullerstörning i arbetet (The effects of non-physical factors on noise annoyance at work. Report in Swedish with an English summary). National Institute of Occupational Health, Solna, Sweden, Arbete & Hälsa 1992: 37.

Kjellberg, A. and Sköldström, B., 1991. Noise annoyance during the performance of different non-auditory tasks. Perceptual and Motor Skills, 73: 39-49.

Kjellberg, A., Söldström, B., Andersson, P. and Lindberg, L., 1991. Fatigue effects of noise among airplane mechanics. In: A. Lawrence (Eds.), Proceedings of Internoise 91. Australian Acoustic Society, Sydney, pp. 883-886.

Kjellberg, A. and Wide, P., 1988. Effects of simulated ventilation noise on performance of a grammatical reasoning task. In: B. Berglund, U. Berglund, J. Karlsson and T. Lindvall (Eds.), Proceedings of the 5th International Congress on Noise as a Public Health Problem, Vol. 3. Byggforskningsrådet, Stockholm, pp. 31-36.

Klitzman, S. and Stellman, J.M., 1989. The impact of the

physical environment on the psychological well-being of office workers. Social Science and Medicine, 29: 733-742.

Kohfeld, D.L. and Goedecke, D.W., 1978. Intensity and predictability of background noise as determinants of simple reaction time. Bulletin of the Psychonomic Society, 12: 129-132.

Kryter, K.D., 1985. The Effects of Noise on Man. Academic Press, New York.

Kryter, K.D. and Pearson, K.S., 1965. Judged noisiness of a band of random noise containing an audible pure tone. Journal of the Acoustical Society of America, 38: 106-112.

Landström, U., 1987. Laboratory and field studies on infrasound and its effects on humans. Journal of Low Frequency Noise and Vibration, 6: 29-33.

Landström, U. and Byström, M., 1984. Infrasonic threshold levels of physiological effects. Journal of Low Frequency Noise and Vibration, 3: 167-173.

Landström, U., Kjellberg, A. and Byström, M., 1993. Acceptable levels of sounds with different spectral characteristics during the performance of a simple and a complex nonauditory task. Journal of Sound and Vibration, 160, 533-542.

Landström, U., Kjellberg, A. and Söderberg, L., 1991a. Spectral character, exposure levels and adverse effects of ventilation noise in offices. Journal of Low Frequency Noise and Vibration, 10: 83-91.

Landström, U., Kjellberg, A., Söderberg, L. and Nordström, B., 1991b. The effects of broadband, tonal and masked ventilation noise on performance, wakefulness and annoyance. Journal of Low Frequency Noise and Vibration, 10: 112-122.

Landström, U., Kjellberg, A., Tesarz, M. and Åkerlund, E., 1992. Samband mellan exponeringsnivå och störningsgrad för buller i arbetslivet (The relation between exposure level and noise annoyance at work. Report in Swedish with an English summary). National Institute of Occupational Health, Solna, Sweden, Arbete och Hälsa 1992: 42.

Landström, U., Lindblom Häggqvist, S. and Löfstedt, P., 1988. Low frequency noise in lorries and correlated effects on drivers. Journal of Low Frequency Noise and Vibration, 7: 104-109.

Landström, U., Lundström, R. and Byström, M., 1983. Exposure to infrasound - Perception and changes in wakefulness. Journal of Low Frequency Noise and Vibration, 2: 1-11.

Langdon, F.J., 1966. Modern offices: A user survey. National Building Studies Research Paper No 41. London: H. Majesty's Stationery Office. Ministry of Technology, Building Research Station.

Lazarus, H., 1990. New techniques for describing and assessing speech communication under conditions of interference. In: B. Berglund and T. Lindvall (Eds.), Proceedings of the 5th International Congress on Noise as a public Health Problem, Vol 3. Swedish Council for Building Research, Stockholm, pp. 197-235.

Leventhall, H.G., 1988. Low frequency noise in buildings - Internal and external sources. Journal of Low Frequency Noise and Vibration, 7: 74-85.

Lindström, K. and Vuori, J., 1984. Relationship between environmental factors, job satisfaction and mental strain in an open-plan drafting office. In: E. Grandjean (Eds.), Ergonomics and Health in Modern Offices. Taylor and Francis, London, pp. 59-63.

Makley, W.K., 1985. Noise in the office: There are ways to deal with it. The Office, December, pp. 70-72.

Martin, R.C., Wogater, M.S. and Forlano, J.G., 1988. Reading comprehension in the presence of unattended speech and music. Journal of Memory and Language, 27: 382-398.

McKennell, A.C., 1963. Aircraft noise annoyance around London (Heathrow) airport. London: Her Majesty's Stationary Office.

McKennell, A.C., 1980. Annoyance from Concorde flights around Heathrow. In: J.V. Tobias, G. Jansen and W.D. Ward (Eds.), Proceedings of the Third International Congress on Noise as a Public Health Problem. ASHA Reports 10, American Speech-Language-Hearing Association, Rockville, MD, pp. 562-566.

Moreland, J.B., 1988. Role of the screen on speech privacy in open plan offices. Noise Control Engineering Journals. 30(2): 43-56.

Munz, D.C., Ruffner, J.W. and Cross, J.F., 1971. Reduction of noise annoyance through manipulation of stressor relevance. Perceptual and Motor Skills, 32: 55-58.

Nemecek, J., 1984. Music during office work. In: E. Grandjean (Eds.), Ergonomics and Health in Modern Offices. Taylor & Francis, London, pp. 64-69.

Nemecek, J. and Grandjean, E., 1973a. Noise in landscaped offices. Applied Ergonomics, 4: 19-22.

Nemecek, J. and Grandjean, E., 1973b. Results of an ergonomic investigation of large-space offices. Human Factors, 15: 111-124.

Nemecek, J. and Turrian, V., 1978a. Der Bürolärm und seine Wirkungen. Kämpf dem Lärm, 25: 50-57.

Nemecek, J. and Turrian, V., 1978b. Untersuchungen von Störwirkungen durch Lärm in Büros. Zeitschrift für Arbeitswissenschaft, 32: 21-24.

Nemecek, J., Turrian, V. and Sancin, E., 1976. Lärmstörungen in Büros. Sozial- und Präventivmedizin, 21: 133-134.

Niemi, P., Von Wright, J.M. and Koinunen, E., 1977. Arousal and incidental learning: A reappraisal. Report No 44. The Institute of Psychology, University of Turku.

Noweir, M.H., 1984. Noise exposure as related to productivity, disciplinary actions, absenteeism, and accidents among textile workers. Journal of Safety Research, 15: 163-174.

Ollerhead, J.B., 1973. Scaling aircraft noise perception. Journal of Sound and Vibration, 26: 361-388.

Pearsons, K.S., 1990. Review of relevant research on noise and speech communication since 1983. In: B. Berglund and T. Lindvall (Eds.), Proceedings of the 5th International Congress on Noise as a Public Health Problem, Vol. 3. Swedish Council for Building Research, Stockholm, pp. 181-195.

Persson, K., Björkman, M. and Rylander, R., 1985. An experimental evaluation of annoyance due to low frequency noise. Journal of Low Frequency Noise and Vibration, 4: 145-153.

Poulton, E.C., 1976. Continuous noise interferes with work by masking auditory feedback and inner speech. Applied Ergonomics, 7: 79–84.
Poulton, E.C., 1977. Continuous intense noise masks auditory feedback and inner speech. Psychological Bulletin, 84: 977–1001.
Poulton, E.C., 1978. A new look at the effects of noise: A rejoinder. Psychological Bulletin, 85: 1068–1079.
Poulton, E.C., 1979. Composite model for human performance in continuous noise. Psychological Review, 86: 361–375.
Purcell, A.T. and Thorne, R.H., 1977. An alternative method for assessing the psychological effects of noise in the field. Journal of Sound and Vibration, 55: 533–544.
Rabbitt, P., 1966. Recognition memory for words correctly heard in noise. Psychonomic Science, 6: 383–384.
Raw, G.J. and Griffiths, I.D., 1990. Subjective response to changes in road traffic noise: A model. Journal of Sound and Vibration, 141: 43–54.
Scharf, B., 1978. Loudness. In: E.C. Carterette and M.P. Friedman (Eds.), Handbook of Perception. Academic Press, New York, pp. 187–242.
Scharf, B. and Hellman, R., 1979. Comparison of various methods for predicting the loudness and acceptability of noise, Part II: Effects of spectral pattern and tonal components. EPA 550/9-79-102. Environmental Protection Agency, Office of Noise Abatement and Control.
Schultz, T.J., 1978. Synthesis of social surveys on noise annoyance. Journal of the Acoustical Society of America, 64: 377–405.
Schulz, P. and Battmann, M., 1980. Die Auswirkungen von Verkehrslärm auf verschiedenen Tätigkeiten. Zeitschrift für Experimentelle und Angewandte Psychologie, 27: 592–606.
Schulz, P. and Schönpflug, W., 1982. Regulatory activity during states of stress. In: H.W. Krohne and L. Laux (Eds.), Achievement, Stress, and Anxiety. Hemisphere, New York, pp. 51–73.
Schwabe, M., 1982. Eine Studie über den Einfluss von Verkehrslärm auf die Tätigkeiten Korrekturlesen. Zeitschrift für Arbeitswissenschaft, 36: 49–53.
Schönpflug, W., 1982. Aspiration level and causal attribution under noise stimulation. In: H.W. Krohne and H.D. Laux (Eds.), Achievement, Stress and Anxiety. Hemisphere, New York, pp. 291–314.
Schönpflug, W., 1983. Coping efficiency and situational demands. In: R. Hockey (Eds.), Stress and Fatigue in Human Performance. Wiley, New York, pp. 299–326.
Schönpflug, W. and Schulz, P., 1979. Lärmwirkungen bei Tätigkeiten mit komplexer Informationsbearbeitung. Forschungsbericht 79-10501201. Umweltbundesamtes, Berlin.
Smith, A.P., 1983. The effects of noise on strategies of human performance. In: G. Rossi (Eds.), Proceedings of the Fourth International Congress on Noise as a Public Health Problem. Centro Ricerche e Studi Amplifon, Milano, pp. 797–807.

Smith, A.P., 1985. The effects of different types of noise on semantic processing and syntactic reasoning. Acta Psychologica, 58: 263–273.
Smith, A.P., 1991. A review of the non-auditory effects of noise on health. Work and Stress, 5: 49–62.
Smith, A.P. and Broadbent, D.E., 1985. The effects of noise on the naming of colours and reading of colour names. Acta Psychologica, 58: 275–285.
Smith, T.J.B., 1975. Noise control in open-plan offices. Noise Control and Vibration Reduction, 6: 112–117.
Statistics Sweden 1991. Sysselsättning, arbetstider och arbetsmiljö 1986–87 (Employment, working hours and working environment, 1986–87. Report in Swedish with English summary). Statistics Sweden, Stockholm.
Sundstrom, E., 1986. Work Places. The Psychology of the Physical Environment in Offices and Factories. Cambridge University Press, Cambridge.
Sust, C., 1987. Geräusche mittlerer Intensität – Bestandsaufnahme ihrer Auswirkungen. Fb Nr. 497. Bundesanstalt für Arbeitsschutz.
Sörensen, S., 1970. On the possibilities of changing the annoyance reaction to noise by changing the attitudes to the source of annoyance. Nordisk Hygienisk Tidskrift (Suppl 1).
Thompson, S.Z., 1981. Will it hurt less if I can control it? A complex answer to a simple question. Psychological Bulletin, 90: 89–101.
van Cott, H.P. and Kinkade, R.G. (Ed.), 1972. Human engineering guide to equipment design. Wiley, New York.
van Dijk, F.J.H., 1990. Epidemiological research on non-auditory effects of occupational noise exposure since 1983. In: B. Berglund and T. Lindvall (Eds.), Proceedings of the 5th International Congress on Noise as a public Health Problem, Vol. 3. Swedish Council for Building Research, Stockholm, pp. 285–292.
van Heusden, E., Plomp, R. and Pols, L.C.W., 1979. Effects of ambient noise on the vocal output and the preferred listening levels of conversational speech. Applied Acoustics, 12: 31–43.
Vanderhei, S.L. and Loeb, M., 1977. Annoyance and behavioral after-effects following interfering and noninterfering aircraft noise. Journal of Applied Psychology, 62: 719–726.
Verein Deutsche Ingenieure, 1990. Schallschutz und akustische Gestaltung im Büro. VDI 2569 01.90. Verein Deutsche Ingenieure.
von Ising, H., Shenoda, F.B. and Wittke, C., 1980. Zur Wirkung von Infraschall auf den Menschen. Acustica, 44: 173–181.
Völker, E.J., 1978. Privacy und akustische Behaglichkeit im modernen Bürobau. Kampf dem Lärm, 25: 26–31.
Völker, E.J., 1979. Störgeräusche in Grossraumbüros – Pegelverhältnisse und Bewertungen. Kampf dem Lärm, 26: 70–79.
Völker, E.J., 1987. Akustische Behaglichkeit am Büroarbeitsplatz durch schalltechnische Simulierung. Zeitschrift für Lärmbekämpfung, 34: 52–55.
Vos, J., 1992. Annoyance caused by simultaneous impulse,

road-traffic, and aircraft sounds: A quantitative model. Journal of the Acoustical Society of America, 91: 3330–3345.

Warner, H.D. and Heimstra, N.W., 1971. Effects of intermittent noise on visual tasks of varying complexity. Perceptual and Motor Skills, 32: 219–226.

Warnock, A.C.C., 1973. Acoustical privacy in the landscaped office. Journal of the Acoustical Society of America, 53: 1535–1543.

Webster, J.C., 1984. Noise and communication. In: D.M. Jones and A.J. Chapman (Eds.), Noise and Society. John Wiley, New York, pp. 185–220.

Weinstein, N.D., 1974. Effects of noise on intellectual performance. Journal of Applied Psychology, 59: 548–554.

Weinstein, N.D., 1977. Noise and intellectual performance: A confirmation and extension. Journal of Applied Psychology, 62: 104–107.

Weinstein, N.D., 1978. Individual differences in reactions to noise: A longitudinal study in a college dormitory. Journal of Applied Psychology, 63: 458–466.

Weinstein, N.D., 1980. Individual differences in critical tendencies and noise annoyance. Journal of Sound and Vibration, 68: 241–248.

Weinstein, N.D., 1982. Community noise problems: Evidence against adaptation. Journal of Environmental Psychology, 2: 87–97.

West, M., 1973. The sound attenuation in an open-plan office. Applied Acoustics, 6: 35–56.

Wieland, R., 1981. Schwankende Schallpegel, Leistungshandeln und der Wechsel von Arbeit und Erholung. Zeitschrift für Lärmbekämpfung, 28: 117–122.

Wilkins, P.A. and Acton, W.I., 1982. Noise and accidents – A review. Annals of Occupational Hygiene, 25: 249–260.

Wilkinson, R.T., 1974. Individual differences in response to the environment. Ergonomics, 17: 745–756.

Williams, C.E., Stevens, K.N. and Klatt, M., 1969. Judgements of the acceptability of aircraft noise in the presence of speech. Journal of Sound and Vibration, 9: 263–275.

Wisner, A., 1967. Audition et bruit. In: J. Scherrer (Eds.), Physiologie du travail (ergonomie), Tome II. Ambiances physiques travail psycho-sensoriel. Masson, Paris, pp. 3–72.

Ying, S.P., 1988. Noise control of banking business equipment. Sound and Vibration, 22: 28–31.

Zwicker, E., 1987. Meaningful noise measurement and effective noise reduction. Noise Control Engineering Journal, 29(3): 66–76.

Yano, T. and Kobayashi, A., 1990. Disturbance caused by impulsive, fluctuating, and combined noises. In: H.G. Jonasson (Eds.), Proceedings of Internoise 90. Acoustic Society of Sweden, Göteborg, pp. 1189–1192.

Work/rest: Part I – Guidelines for the practitioner[1]

Stephan Konz[a,*]

[a] *Department of Industrial and Manufacturing Systems Engineering, Kansas State University, Manhattan, KS 66506, USA*

1. Audience

The practitioner is assumed to be an ergonomist or a person who use ergonomics as a part of their job responsibilities.

2. Context of use

The Guidelines is a guideline, not a law or regulation. The accompanying knowledge base (Part II, Konz, 1998) gives the literature supporting the guidelines.

3. Glossary

None.

---

* Corresponding author. E-mail: sk@taylor.ie.ksu.edu.

[1] The recommendations provided in this guide are based on numerous published and unpublished scientific studies and are intended to enhance worker safety and productivity. These recommendations are neither intended to replace existing standards, if any, nor should they be treated as standards. Furthermore, this document should not be construed to represent institutional policy.

The following individuals participated in the discussion of the earlier version of this guide. Their suggestions (written or verbal) were incorporated by the authors in this version: A Aaras, Norway; J.E. Fernandez, USA; A. Frelivalds, USA; T. Gallwey, Ireland; M. Jager, Germany; S. Kumar, Canada; H. Krueger, Switzerland; K. Landau, Germany; A. Luttmann, Germany; A. Mital, USA; J.D. Ramsey, USA; M-J. Wang, Taiwan.

4. Problem identification/recognition

The ergonomist will recognize the fatigue problem from the job/task. It is unlikely that illness/injury reports will be the impetus for an investigation. However, there may be a serious accident and then, upon investigation, fatigue and the work/rest situation may be recognized as a problem. Fatigue probably will be related to long daily work hours – especially if there is a lack of sleep. Occasionally, fatigue will be due to long weekly work hours, perhaps combined with shiftwork.

Performance effects of fatigue are more likely to be reflected in errors than in changes in units/h.

5. Data collection/analysis

Data needs to be collected on errors and near errors. It is critical to identify when the error occurred so the connection with fatigue and work/rest can be identified. Subjective evaluation/self-report (if quantifiable) can be useful.

6. Solutions

From a financial viewpoint, some rest time is paid and some is not paid, but, from a fatigue viewpoint, a rest is a rest. In addition to the coffee break, there are three types of at-work breaks: (1) microbreaks (short breaks of a minute or less), (2) informal breaks (work interruptions, training) and (3) working rest (a different task using a different part of the body, such as answering the phone vs. keying data). Most tasks do not require maximum capacity continuously. In particular, the work may be automatic or semi-automatic and the operator may rest during machine time. In addition, there are meal breaks and off-work breaks (evenings/nights, weekends, holidays, vacations).

(1) Ideally, the ergonomics of the job/environment are such that no rest breaks are needed beyond the standard breaks for coffee and meals.

(2) Work breaks are the first choice since they not only provide a break but have output during the break. However, work breaks may not have maximum recovery value.

(3) Improve recovery value of a break by:
 a. Minimizing the fatigue ("dose") before the break.
 b. Maximizing the recovery rate (minimize half-life of fatigue).
 c. Have a sufficient length of a break.

(4) Fatigue can affect different parts of the body: (1) cardio-vascular system, (2) skeletal–muscular system and (3) the brain.

The cardio-vascular system primarily is affected by heavy work such as manual material handling. The skeletal–muscular system is primarily affected by postural static work (e.g. standing), VDT work, and manipulative work. The brain is affected by information overload (concentration and attention) jobs. However, a serious problem is information underload (boredom). Another serious problem for the brain is lack of sleep.

Note that jobs (e.g. nurses, truck drivers, machine operator, VDT operators) usually have a combination of physical and mental fatigue.

The following seven guidelines are divided into (1) fatigue prevention and (2) fatigue reduction.

7. Fatigue prevention

7.1. Guideline 1: Have a work-scheduling policy

The problem is insufficient rest. Two aspects are: (1) too many work hours and (2) work hours at the wrong time.

7.1.1. Too many hours

Count all the hours in "duty time". For example, jobs such as train crews and flight crews often have preparation time required before and after the "primary" job. Watchkeepers at sea often have other duties assigned besides watchkeeping. There may be "shift turnover" time in which the old shift stays to communicate with the new shift. In addition, people may work overtime. Overtime can occur when the entire group works more hours but also can occur when specific individuals have to be replaced (illness, absenteeism). In such cases, very long shifts can occur for individuals. There probably should be restrictions on prolonged overtime, especially over 12 h/day and over 55 h/week.

Lack of sleep can increase if the individual moonlights or has a long commute time.

See Tables 4–8 in Part II for some recommendations.

7.1.2. Work hours at wrong time

Lack of sleep can be due to sleeping at the "wrong time" and due to irregular hours of work. This circadian rhythm problem affects both health and social life. See Tables 11–13 in Part II for some recommendations.

7.2. Guideline 2: Optimize stimulation during work

The problem for the brain is too much stimulation (overload) or too little stimulation (boredom). Stimulation comes from both the task and the environment.

7.2.1. Too much stimulation

The usual solution is to reduce environmental stimulation. For example, for office tasks, increase visual and auditory privacy.

7.2.2. Too little stimulation

Increase stimulation for either or both the task and the environment. Tasks are more stimulating if there is physical activity. For example, if truck drivers are sleepy, have them stop and walk (say 150 m), or have them eat or drink. Add variety within the task; another option is a variety of tasks done by the same person.

Add environmental stimulation by (1) encouraging conversation with others (this may require a two-way radio for those physically isolated), (2) varying the auditory environment (talk radio, stimulating music), (3) varying the visual environment (e.g. windows with a view), (4) varying the physical environment (change temperature, air velocity).

Chemicals (e.g. caffeine) also stimulate the individual. See Table 9 in Part II.

7.3. Guideline 3: Minimize the fatigue dose

The problem is the "dose" of fatigue becomes too great to overcome easily. Two aspects are intensity and work/rest schedule.

7.3.1. Intensity

Good ergonomics practice reduces high stress levels on the person. For example, use machines and devices to reduce hold and carry activities. Static work is especially stressful as specific muscles are activated continuously and there is not the alternation of muscles that occurs in dynamic work.

7.3.2. Work/rest schedule

The "dose" of fatigue increases exponentially (not linearly) with time. Thus, it is important to get rest before the fatigue level becomes too high. (If piece-rate incentives are used, insist that workers take their breaks.)

The normal approach is to schedule a break but another approach is to use part-time workers.

8. Fatigue reduction

8.1. Guideline 4: Use work breaks

The problem with a conventional break is that there is no productivity during the break. A solution is to use a different part of the body to work while resting the fatigued part.

If a machine is semi-automatic, the operator may be able to rest during the automatic part of the cycle (machine time). (Machine time may decrease physical fatigue but increase boredom.) Fatigue recovery is best if the alternative work uses a distinctly different part of the body. For example, loading/unloading a truck could be alternated with driving a truck. Word processing could be alternated with answering a telephone.

Nor quite as good (but still beneficial) is alternating similar work, as there would be differences in body posture, force requirements, mental activity, etc. One example is inspectors inspecting items on a belt conveyor; they shift jobs with other inspectors every hour. An assembly team of 6 rotates jobs every 30 min. In a packaging operation with 14 products on 14 different lines, workers shift lines every 60 min. Checkout clerks could use a left-hand station and then a right-hand station.

Job rotation, in addition to reducing fatigue, reduces the feeling of inequity among workers as everyone shares the good and bad jobs. Job rotation requires cross-trained people (able to do more than one thing); cross-training gives management scheduling flexibility.

8.2. Guideline 5: Use frequent short breaks

The problem is how to divide break time. The key to the solution is that fatigue recovery is exponential; see Table 1 and Figs. 1 and 2 in Part II. If recovery is complete in 60 min, then it takes only 4 min to drop from 100% fatigue to 75% fatigue but it takes 42 min to drop from 25% fatigue to no fatigue. Thus, give break time in small segments. As an additional benefit of frequent breaks, the frequent break may reduce the exponential growth of fatigue (see Guideline 3 above).

Machine-paced work does not allow for individual differences between people (Mary vs. Betty) and within people (Mary on Monday vs. Mary on Tuesday). Operator-controlled breaks are better. However, people may not take enough break time so reminders may be beneficial.

There is some production time lost for each break. Consider minimizing this loss by not turning the machine off and on, by taking the break near the machine, etc.

8.3. Guideline 6: Maximize the recovery rate

The problem is to recover as quickly as possible. In technical terms, reduce the half-life of the fatigue.

For environmental stressors, reduce contact with the stressor. For heat stress, use a cool recovery area; for cold stress, use a warm recovery area. Use a quiet area to recover from noise, no glare to recover from glare, no vibration to recover from vibration.

For muscle stressors, good blood circulation carries away fatigue products and brings nutrients. Athletes use heat (hot showers, hot tubs, saunas, whirlpools) and massage. Workers do not have the time or facilities available during their break for these approaches. Muscles elevated above the heart (e.g. when the arm is raised) recover more slowly then when the muscle is below the heart. In general, blood circulation should be best for prone postures, then sitting, and then standing. Active rest seems better than passive rest. The active rest may be just walking to the coffee area (blood circulation in the leg improves dramatically within 20–30 steps). For exercises done at the workstation (during short breaks), consider their social acceptability (some people may be embarrassed to do some exercises).

It helps to have a good circulation system. That is, a person in good physical shape will recover from muscle fatigue faster than a person in poor shape. People with heat-acclimatization recover from heat faster than people without heat-acclimatization.

8.4. Guideline 7: Increase recovery/work ratio

The problem is insufficient time to recover. The solution is to increase the recovery time or decrease the work time. See Table 1 in Part II. For example, if a specific joint is used 8 h/day, then there are 16 h to recover; 2 h recovery/1 h of work. If the work of the two arms are alternated so one arm is used 4 h/day, then there are 20 h to recover; 5 h recovery/1 h of work. However, overtime or 12 h shifts or moonlighting can cause problems. Working 12 h/day gives 12 h for recovery so there is 1 h recovery/1 h of work.

Consider all break time, both paid and unpaid. In particular, consider machine time and job rotation as well as lunch and coffee breaks. Holidays, weekends and vacations are valuable for reduction of long-term fatigue (long half-life) where there is still a fatigue effect at the start of the day.

References

Konz, S., 1998. Work/rest: Part II – the scientific basis (knowledge base) for the guide. International Journal of Industrial Ergonomics 22: 73–99.

Work/rest: Part II – The scientific basis (knowledge base) for the guide[1]

Stephan Konz[a],*

[a] *Department of Industrial and Manufacturing Systems Engineering, Kansas State University, Manhattan, KS 66506, USA*

1. Problem description

This paper concerns the temporal aspects of ergonomics and fatigue. The goals are to reduce fatigue so workers (1) can maintain/increase productivity and (2) have "optimal" stress.

To optimize these two goals, the design must consider both goals, not just productivity nor just worker stress. For example, when fatigued, workers may maintain productivity by increasing their effort. The "optimal" stress implies fatigue is not accumulated between shifts, that health/safety are not reduced, and that there is reasonable productivity. (After all, the purpose of work is to produce.) Also, in general, resting time is not productive time.

2. Fatigue

Time at work will be divided into working time and resting (recovery) time. Resting time will be divided into "off-work" (evenings, weekends, holidays, vacations), "formal breaks" (lunch, coffee), "informal breaks" (work interruptions, training), "microbreaks" (short pauses of a minute or less) and "working rest" (a different task using a different part of the body, such as answering the phone vs. keying data). From a financial cost viewpoint, it is important that some resting time is paid and some unpaid, but from a fatigue viewpoint, a rest is a rest.

This paper will focus on resting time, rather than on working time.

In this paper, the purpose of resting time is to overcome fatigue.

Fatigue can occur in different parts of the body:
- general body fatigue (cardiovascular system) (physiological),
- muscular fatigue (muscles) (physiological),
- mental fatigue (brain) (psychological).

Jobs will have different combinations of fatigue; the combinations often will vary during the shift. For example, a material handler will have more

---

* E-mail: sk@ksu.edu

[1] The recommendations provided in this guide are based on numerous published and unpublished scientific studies and are intended to enhance worker safety and productivity. These recommendations are neither intended to replace existing standards, if any, nor should they be treated as standards. Furthermore, this document should not be construed to represent institutional policy.

The following individuals participated in the discussion of the earlier version in this guide. Their suggestions (written or verbal) were incorporated by the authors in this version: A. Aaras, Norway; J.E. Fernandez, U.S.A.; A. Freivalds, U.S.A.; T. Gallwey, Ireland; M. Jager, Germany; S. Kumar, Canada; H. Krueger, Switzerland; K. Landau, Germany; A. Luttmann, Germany; A. Mital, U.S.A.; J.D. Ramsey, U.S.A.; M.-J. Wang, Taiwan.

physiological (muscular) stress while an airline pilot will have psychological stress. A truck driver may have psychological fatigue while driving and physiological fatigue while unloading. A VDT operator may have mental fatigue and two different kinds of physiological fatigue (static loading in the back from posture and repetitive strain on the fingers).

A key consideration of a rest is its recovery value (Cakir et al., 1980). The recovery value of a rest is a function of:
- how fatigued the muscle (cardiovascular system, brain) is when the rest begins (the "dose")
- length of the rest (the "response")
- what happens to the muscle (cardiovascular system, brain) during the rest (the "response")

The following wll be considered as axioms (evident without proof):
- Most jobs have peaks and valleys of demand within the shift. i.e., most jobs do not have "constant" loads.
- Fatigue increases exponentially with time.
- Rest is more beneficial if it occurs before the muscle (cardiovascular system, brain) has "too much" fatigue. Thus, "machine-paced" or standardized rests probably are less effective than rests under operator control.
- The value of a rest declines exponentially with time (see Table 1 and Figs. 1 and 2).
- Different parts of the body have different recovery rates.
- Active rest is an alternative to passive rest. For example, heat, exercise and massage can improve blood circulation.
- The general concept is that there is "output" during the work and "no output" during the rest. However, the rest may permit greater output during work (not just prevent decline). During "working rest", the person shifts to another task and so rest and work occur simultaneously. In addition, rest from one job may mean the operator is working for another employer (moonlighting) so rest may not really be rest.

Table 1
Recovery/work ratio (Konz, 1995a,b)

Both the amount of recovery (rest) and the distribution are important.

Amount. Recovery (repair, rest) time can be calculated as a ratio of exposure time; i.e. a recovery/work ratio. For example, in a 24 h day, if a specific joint on a person is used for 8 h, then there are 16 h available for recovery. That is, there is $\frac{16}{8} = 2$ h of recovery for 1 h of exposure. If the joint is used for only 4 h/day (say by alternating with the other arm), then there is $\frac{20}{4} = 5$ h of recovery for 1 h of exposure. Overtime or long shifts can cause considerable reduction of the ratio; 12 h of work gives $\frac{12}{12} = 1$ h of recovery for 1 h of exposure.

Weekends, holidays, and vacations increase the time available for recovery for the total body. Working rest (job variety) allows rest for part of the body.

See Table 2 for a discussion of exposure/recovery for chemicals.

Guideline: Increase recovery time

Distribution. The above assumes that each minute of recovery (repair, rest) time is equally effective. That is, recovery rate vs. time is a horizontal line. But recovery rate declines exponentially with time. See Figs. 1 and 2. That is, the amount of recovery for min 6–10 is much less than for min 0–5. Thus, for the same total length of break, many short breaks are better than occasional long breaks. A single break of 15 min is not as effective as 3 breaks of 5 min. Thus, it is better to rotate jobs within days rather than between days. For example, have Joe work on job A in the morning and have Pete work on it in the afternoon. This is better than Joe working for 8 h on job A on Monday and then on job B for 8 h on Tuesday. Waersted and Westgaard (1991) suggest rotation probably should be after 1 or 2 h rather than after 4 h.

Janaro and Bechtold (1985) point out that there is a time penalty for breaks (e.g. operator turns off equipment, goes to rest area, returns from rest area, powers up equipment, goes to rest area, returns from rest area, powers up equipment); thus, very short breaks may not be cost effective. However, the "microbreak" concept is for the operator to take a break (say 20 s) at the workstation and leave the equipment on, thus, eliminating the travel cost and the on/off cost.

Guideline: Frequent short breaks are better than occasional long breaks

Table 2
Exposure/recovery for chemicals

The Threshold Limit Value-Time-Weighted Average (TLV-TWA) is the concentration, for a normal 8 h workday and 40 h work week, to which nearly all workers may be exposed, day after day, without adverse effects. It primarily recognizes chronic (long-term) effects.
For example, the TLV-TWA for acetone is 750 ppm.
If a person is exposed, for an 8 h day, to over 750 ppm, the exposure is not acceptable.
If a person is exposed for periods different than 8 h, then the permitted exposure is:

TLV Adjustment Factor = 8/h worked. (4)

For a 6 h shift, the TLV Adjustment Factor = $\frac{8}{6}$ = 1.33; for a 12 h shift, it is $\frac{8}{12}$ = 0.67. Thus, the permitted exposure for 6 h is 1.33 (750) = 1000 ppm; for 12 h, it is 500 ppm. Eq. (4) considers only the exposure time, not the recovery time.
Another possibility is to consider both exposure time and the recovery time.

TLV Reduction Factor = (8/h worked) (Hours off − work/16)   (valid only for ⩾ 8 h) (5)

For a 6 h shift, Eq. (5) TLV Reduction Factor is not applicable. For a 12 h shift, it is $\frac{8}{12}\frac{12}{16}$ = 0.67(0.75) = 0.5. Thus, this formula not only permits less exposure for periods over 8 h but also is more restrictive than Eq. (4). Eq. (5) is discussed at length as the "Brief and Scala" model by Paustenback (1994).
For chemicals whose half-life suggests that not all the chemical would be eliminated before returning to work the next day, a weekly formula is used. Examples whose biologic half-life is clearly over 10 h include PCBs, PBBs, mercury, lead, mineral dusts and DDT.

TLV Reduction Factor = 40/h of exposure in one week (6)

Eqs. (4) and (6) are the models used by the US government (Paustenback, 1994).
If the TLV adjustment factor was used alone, very high exposures would be permitted for short work periods. These high exposures are restricted by the Short-term-exposure-limit (TLV-STEL) and Ceiling (TLV-C). The STEL is a 15 min time-weighted exposure which should not be exceeded any time during the day even if the TLV-STEL is met. The TLV-C is the concentration which should not exceed during any part of the working day.
Eqs. (4)–(6) do not consider the exponential decline in toxin concentration and thus, tend to be conservative. See Table 3. If biological half-life (pharmacokinetic approach) is used for adjustment of TLVs, the adjustments are smaller than Eqs. (4) and (6) for longer work periods; and are larger for shorter work periods (Paustenback, 1994, p. 292). All the equations assume no exposure during non-work time; however, this may be violated if a person moonlights.

Table 3
Biologic half-life

The concentration of a compound in the body will decline exponentially with time. Exponential curves with a negative exponent (as is our concern) have a value of $y = 1$ at $x = 0$ and approach $y = 0$ as an asymptote. However, exponential curves also can be approximated as a straight line for log y vs. x. An interesting characteristic of log scales is that the same physical distance along the axis represents a constant *ratio*; i.e. the distance from 100 to 50 is the same as 50 to 25 or 25 to 12.5. See Fig. 2.
This leads to the concept of biologic half-life. The half-life is independent of the concentration for a first-order process (elimination is a function of concentration); it is the time needed to eliminate 50% of the absorbed material. The longer the half-life, the slower the elimination rate.
The potential for the body burden to exceed normal levels during unusually long work periods exists whenever the biologic half-life for the chemical in humans is in the range of 3–200 h (Paustenback, 1994, p. 231). A rule of thumb is steady-state body burden occurs when exposure occurs for a period of greater than 5 biologic half-lives. For moderately volatile substances (e.g. solvents), which have half-lives from 12 to 60 h, and for most work schedules, the steady-state tissue burden will be reached in 2–6 weeks. For volatile chemicals (e.g. low molecular weight solvents) with shorter half-lives, steady state will be reached in 2–4 days (Paustenback, 1994, p. 274).

3. Work hours

It is difficult to generalize about working times as there is an very large variance between countries. In addition, even if a given country is specified, there are large differences by type of work (office vs. factory vs. services vs. agriculture), full vs. part-time work, season of the year, overtime, absenteeism, etc. Nonetheless, I will make some general statements:

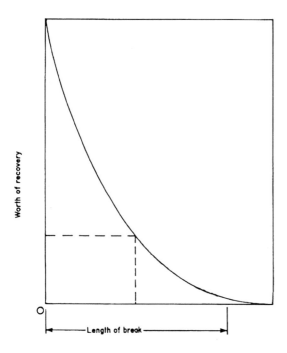

Fig. 1. Since recovery from fatigue is exponential, when time is low, the percent change in "concentration" is greater than the percent change in time. When time is high, the change in concentration is less than the change in time. There is more recovery in the first part of the break than in the latter. See Fig. 2. Fatigue (while working) increases exponentially (curve is reversed). When time is low, a change in concentration is less than a change in time; when time is high, a change in concentration is greater than a change in time. Thus, three breaks of 5 min have more benefit than one break of 15 min as (1) recovery is better and (2) fatigue has not increased as much.

A typical workweek of 40 h has been achieved in most countries. In fact, in many European countries the standard workweek is now less than 40 h (about 38 h); in some German firms, it is 35 h.

The typical hours worked per year also has declined – especially in Europe. In 1992, average annual hours/full-time worker were 2007 in Japan, 1857 in USA, 1646 in France and 1519 in Germany. The decrease has occurred through more and longer holidays and vacations as well as reduction in the hours worked per week. In addition, there seem to be more people working part time. Thierry and Meijman (1994) report that, in the late 1980s, around 15% of the European Community work force was employed part time; there was a great variability between country with 4% in Spain and 24% in Great Britain and Netherlands. (These official figures are underestimates as the underground economy tends to have part-time workers.) In the USA, it was about 20%.

The hours worked per lifetime has declined to about 70 000–85 000. This is due to the reduction in annual hours as well as working less years (entering the workforce later (more education) and leaving it sooner (early retirement)).

Thus, from a health viewpoint, the problem of long hours seems to be decreasing. However, as hours worked decline, some workers opt for income instead of leisure and take a second job – "moonlighting".

However, there seems to be a change in the daily hours worked from the 8 h "standard". There is an economic pressure on employers to get good utilization of their facilities and equipment as well as provide services to customers "around the clock". If a facility is used for 40 h/week for 48 weeks/year, it is used for 1920 h/year. If 10 holidays are subtracted, then it is available for 80 h less or 1840 h/year. But the year has 8760 h! 1840/8760 = 21%. Thus, employers tend to want the equipment to be used for more hours. (Of course, some organizations such as hospitals, police, fire, etc. have always worked 8760 h/yr.) Thus, there is an "uncoupling" or "decoupling" of hours worked by a specific worker and the hours the firm operates.

In addition, there has been increasing recognition of the problem of variations in customer demand (seasonal, etc). Traditionally, firms built to stock; fluctuations were in inventory rather than worker hours. However, firms increasingly have tried to have some of the fluctuation occur also in worker hours by using overtime, short schedules during part of the year, etc. For example, a firm may work 35 h/week for a period of slow demand and 40 h/week for a period of higher demand.

As a result, there have been many changes to the 8 h "standard". (About 20% of American workers work a non-standard or altered work shift (Office of Technology Assessment, 1991).) One change is more part-time workers. A specific type of part-time work is job sharing (where 2 people share 1 full-time job). There also has been increased use of

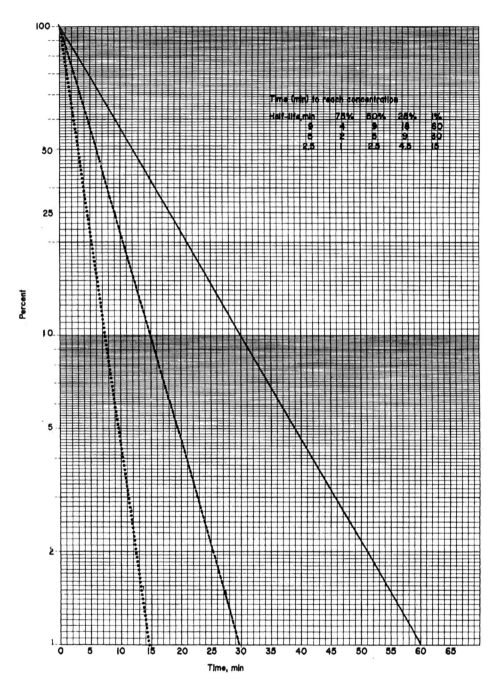

Fig. 2. As discussed in Table 3, an exponential curve (such as Fig. 1) can be approximated by a straight line on semi-log paper. In theory, an exponential curve approaches 0 concentration asymptotically (i.e. never reaches zero). An approximation of the end point is the time at which concentration reaches 1%. If concentration is 1% at 60 min, the 50 level was reached in 9 min. The concentration took only 4 min to drop from 100% to 75% but took 42 min to drop from 25% to 1%.

shiftwork (both second and third shift) and even weekend work. Another change is longer hours on some days. One possibility is a compressed work week of 4 days of 10 h/day; another choice is 12 h/day for 3 days one week and 4 days the next; compressed work weeks tend to in non-manufacturing settings. (Although popular with workers, long shifts present potential problems of cumulative trauma and chemical exposure due to the reduced recovery/work ratio. Long shifts also present overtime problems.) There has been relatively little study of compressed work (Duchon and Smith, 1993). Duchon et al. (1995) reported mining workers "decreased personal levels of effort" for 12 h shifts (vs. 8 h shifts).

In summary, although weekly, annual and lifetime hours are decreasing and thus present less health problems, there may be some health/safety problems for daily hours – especially for people working over 8 h/day and for people working without adequate sleep. In addition, longer hours may present productivity problems.

4. Rest hours (Allowances)

Engineered time standards have a normal time (the time for an experienced normal worker) developed either from time studies or predetermined time standards (either from microsystems such as Methods Time Measurement or Work-Factor or from collections of elements of time studies). This normal time is adjusted by allowances (often subdivided into personal, delay and fatigue) to yield standard time.

Note that allowances only consider the *duration* of the rest and not what happens during the rest. In addition, each minute of rest is equally valuable. The "dose" of fatigue is considered by the table entry. Allowances are based on an 8 h day and 5 day week and do not consider recovery due to weekends, holidays, and vacations.

Allowances can be added to normal time in two ways because allowances can be defined in two ways: (1) work allowances (allowances as a percent of work time) and (2) shift allowances (allowances as a percent of shift time), i.e.

$$STTIME = NOTIME (1 + WORKAL), \qquad (1)$$

or

$$STTIME = NOTIME/(1 - SHFTAL), \qquad (2)$$

where STTIME is the standard time for a task over an entire shift, h/unit, NOTIME the normal time, h/unit, WORKAL the allowance (allowance based on percent of time worked), (%), and SHFTAL the shift allowance (allowances based on percent of shift time) (%).

For example, if NOTIME for a task was 0.05 h and WORKAL was 10%, then STTIME = 0.05 + (0.10 × 0.05) = 0.055 h. If a SHFTAL of 10% was used, then STTIME = 0.05/(0.9) = 0.0556 h. From a small informal survey, it seems most American firms define their allowances as work allowances.

4.1. Personal allowances

Personal allowances are given for such things as blowing your nose, going to the toilet, getting a drink of water, etc. They are paid time. They do not vary with the task but are the same for all tasks in the organization. There is no scientific or engineering basis for the percent to give.

In many organizations, there are standardized break periods (coffee/tea breaks) – e.g., 15 min in the first part of the shift and the same in the second part. Some firms (especially in offices) allow workers the 15 min break but it is taken as the work permits rather than at a standardized time. If the person is paid for an 8 h shift and takes 2 breaks of 15 min, then the breaks are 30/450 = 6.67% of the work or 30/480 = 6.25% of the shift. In some firms, there is a personal allowance added to the normal time in addition to the "coffee break" allowance. (If the firm wants to maintain production during the breaks (generally giving about 10% more output/shift), it can stagger the breaks and have "tag relief operators" for those on break. A tag relief worker tends to step in for 6–8 workers/shift. However, some workers will have their break after a short time of work and others after a long time.

Some firms grant an additional break for shifts longer than 8 h. For example, for a 12 h shift, there is an additional break of 15 min at the end of the 9th hour. This would maintain the allowance at 6.67% of the work or 6.25% of the shift.

In addition, for 8 h shifts, there generally is an unpaid meal break ("lunch") in the middle of the shift. Depending on the firm, the type of work, and the culture, the break is from 20 to 60 min. Assuming 30 min and an 8 h shift, this is an additional 6.67% of the work or 6.25% of the shift. However, if the 24 h day is divided into two shifts of 12 h, then what happens to lunch? One possibility is to have it be unpaid and have 11.5 h of paid work; another possibility is to have it be paid and pay 12 h for 11.5 h of work. Another possibility is to have people eat while working; i.e., 12 h pay for 12 h work.

In addition, some firms give special allowances for certain jobs. For example, there may be a cleanup allowance (either to clean up the machine or the person) or a travel allowance. For work in mines, there may be portal-to-portal pay; this means pay begins when the worker crosses the mine portal, even though the worker may not arrive at the working surface until some time later.

The rational of personal allowances is to compensate the person for "necessary" personal activities.

4.2. Delay allowances

Delay times should vary with the task but not with the operator. Delay allowances are meant to compensate the operator for short delays beyond the control of the operator. Four examples are machine breakdowns, interrupted material flow, conversations with supervisors, and conversations with inspectors.

If the delay is "long" (e.g. 30 min), the operator clocks out (records the start and stop time of the delay on a form) and works on something else during the clocked-out time.

The rationale of delay allowances is to compensate the person for unavoidable delays. Avoidable delays, when the operator deliberately works slowly, should not be compensated for by delay allowances.

4.2.1. Machine time
In many situations, people work with automatic or semi-automatic machines. For an automatic machine, the operator acts as a supervisor/inspector/maintainer. For an example of semi-automatic, a person may load/unload a machine which runs automatically once loaded or a person may operate a machine which has automatic loading/unloading. The person may work with multiple machines (1 operator for 2 or 3 machines) or a team may work with multiple machines (2 operators for 5 machines).

"Inside work" is work which is done while the machine is operating automatically – i.e. work can be done "inside" the automatic time. For example, previous parts could be inspected while the machine is making new parts. (If no work is done by the operator during inside work time, it becomes "unoccupied" time.) "Outside work" is work which must be done when the machine is not operating automatically. An example is loading or unloading the machine.

4.2.2. Fatigue recovery
Note that personal time and delay time may occur at the same time as machine time. For example, while the machine is operating automatically, the operator drinks some coffee (personal allowance) or talks to the supervisor (delay allowance). Personal time, delay time and machine time all allow the person to recover from fatigue.

4.3. Fatigue allowances

The rationale of fatigue allowances is to compensate the person for the time lost due to fatigue. That is, if there is no fatigue, then there need be no fatigue allowance. If the person spends 6% of the time recovering from fatigue, then 6% additional time is included in the work standard. Although it may be obvious, the concept is to reduce fatigue while working; the breaks should not be accumulated and taken by leaving work early at the end of the day!

Four systems of fatigue allowances have been published by Page (1964), Williams (1973), Cornman (1970) and the International Labor Office (ILO, 1979, 1989). They are discussed in detail by Konz (1995a,b). All four systems divide fatigue allowances into three major categories: physical, mental and environmental.

4.3.1. Physical

Physical fatigue allowances are divided into (1) material handling (considers both local muscle load and whole-body (metabolic) load, (2) short cycle (consider lack of recovery time of local muscle groups for short (< 0.2 min) cycles), (3) posture or static load (standing, squatting, crouching) and (4) restrictive clothing.

Physical fatigue allowances must consider whole body (metabolic) load and local muscle load. To dig deeper, read Freivalds and Goldberg (1988b), Price (1990a,b) and Mital et al. (1991).

4.3.1.1. Metabolic.
Murrell (1965) gave the following formula for metabolic load:

$$PREST = (MET - LTMET)/(MET - RSTMET), \quad (3)$$

where PREST is the percent rest time, MET the metabolic rate of the task, W, LTMET the long-term metabolic rate, W (e.g. 350 W or 5 kcal for males and 4.2 kcal for females, assuming 1/3 aerobic capacity for each), and RSTMET the resting metabolic rate, W (e.g. 100 W or 1.5 kcal).

The formula implies no metabolic allowance is needed if MET < LTMET. A problem is that the formula depends upon the metabolic rate of the individual (both long-term and resting); these values vary greatly. Another problem is the assumption that the job is done continuously for an entire day. Price (1990a) gives an approach which considers varying tasks for various lengths of time. He explicitly considers the recovery which takes place during various idle times during the shift. Also see Freivalds and Goldberg (1988a).

4.3.1.2. Local muscle fatigue.
Price (1990b) gives formulas for fatigue for static muscle fatigue, abnormal postures, and dynamic muscle fatigue.

4.3.2. Mental

Mental fatigue allowances considers discipline, concentration, mental and visual demand, and mental and visual strain. Mental fatigue may be influenced by boredom. See Goldberg and Freivalds (1988c) and Mital and Genaidy (1990).

4.3.3. Environmental

Environmental fatigue allowances consider climate, noise, vibration and illumination. See Freivalds and Goldberg (1988c).

4.3.3.1. Climate.
Climate will be divided into heat stress and cold stress.

Heat stress is indicated primarily by an increase in central body temperature. The body temperature is affected by both the metabolic rate and the environment. Fig. 3 shows how the threshold limit value of environmental temperature declines as metabolic rate increases. The solid line is for working 8 h continuously. When there is 25%, 50% or 75% rest in each hour, the metabolic rate will decrease so the environmental threshold-limit temperature increases.

There do not seem to be any studies on recovery time from cold. The amount of cold stress depends not only on the temperature and air velocity, but also the clothing, metabolic rate of the person, and duration of exposure.

4.3.3.2. Noise.
The correction for duration of noise exposure (Burns and Robinson, 1970) is D, db = $10 \log_{10} T$ where T is the time of exposure. The 10 is based on the equal energy principle. However, Kryter (1985), summarizing many studies by many authors, says it should be 20 (the equal-pressure principle) for noise exposure but 10 for recovery. If the noise was for 4 h/day instead of 8 h/day, Kryter would subtract 6 db while Burns and Robinson would subtract 3 db. The US Occupational Safety and Health Administration used a political compromise of 5; the US Environmental Protection Agency and some other countries use 3.

4.3.3.3. Vibration.
For hand-arm vibration, ACGIH (1995) recommends 4 m/s² for 4 to < 8 h, 6 m/s² for 2 to < 4 h, 8 m/s² for 1 to < 2 h, and 12 m/s² for < 1 h. They recommend a cessation of vibration exposure for about 10 min/continuous vibration hour.

For whole-body vibration, ACGIH (1995) proposes (based on ISO 2631) two figures (one for the vertical axis and one for the transverse X and Y axes). Due to resonance of the human body, the limiting frequency for the vertical axis is

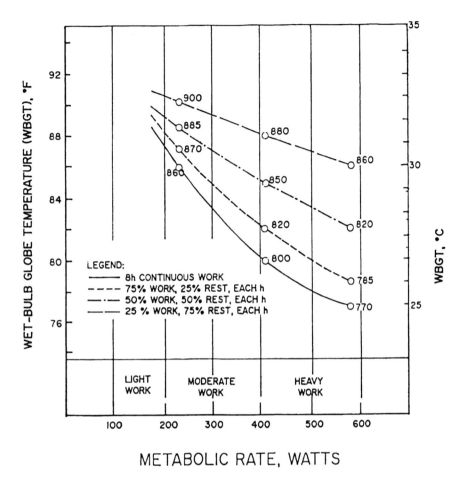

Fig. 3. Threshold limit values for heat-(ASHRAE, 1991) for heat acclimitized people. The rest area is assumed to have the same wet bulb globe temperature (WBGT) as the work area. If the rest area WBGT is below 24°C, reduce resting time by 25%. However, Konz et al. (1983) found no benefit of a cool rest area unless the work environment temperature was over 35°C Effective Temperature. For non-permeable protective clothing, Constable et al. (1994) recommended microclimate cooling garments during the rest period.

4000–8000 Hz while for the X and Y axes it is 1000–2000 Hz. The figures give permitted acceleration as a function of frequency and exposure duration. For example, at 4000 Hz, vertical acceleration of 2.8 m/s^2 is allowed for 1 min, 1.8 m/s^2 for 25 min and 1.2 m/s^2 for 60 min.

4.3.3.4. Illumination. The Illuminating Engineering Society (1987) gives recommended illumination levels (lux) as a function of type of task, with corrections for the occupant's age, speed/accuracy tradeoff and room surface reflectance. Goldberg and Freivalds (1988a,b) recommend 2% additional time if actual lighting is one category lower than the recommendation and 5% if actual lighting is two categories lower.

5. Body parts: Cardio-vascular system

The cardio-vascular system is fatigued during "heavy" work. The cardio-vascular system has five responses to exercise: (1) heart rate, (2) stroke volume, (3) artery–vein differential, (4) blood distribution and (5) going into debt (anaerobic metabolism). If anaerobic metabolism is used, then the debt must be repaid – with "interest". The most common task stressing the cardio-vascular system is manual material handling.

The NIOSH Lifting Guideline (Waters et al., 1993, 1994) adjusts the lifting frequency multiplier by a "lifting duration/session". It has 3 categories:
- short = 0.001 h to ⩽ 1 h, with recovery time of at least 1.2 (duration),
- moderate = ⩾ 1 h but ⩽ 2 h, with recovery time of at least 0.3 (duration),
- long = > 2 h but ⩽ 8 h.

Mital (1984a,b) determined that male material handlers could sustain, without overexertion, for 8 h workdays, 29% of their maximal oxygen uptake (bicycle aerobic capacity); the female value was 28%. For 12 h workdays, the values decline to 23% and 24%. However, Mital et al. (1994) found workers in an air-cargo firm's package-handling area working, for a 2 h shift, at 40–53% of their treadmill aerobic capacity! Note that if aerobic capacity is used as a predictor of when rest is needed, the calculation would have to be made for individuals on a job, rather than for the job itself.

Hagberg et al. (1980) tested men at 50, 65 and 80% of their maximal aerobic power for 5 and 20 min on a bicycle ergometer. They found oxygen recovery had two components: (1) an initial rapid component (half-life of 0.5 min) and (2) a subsequent slower component (half-life of 30 min). The rapid component was proportional to exercise intensity but not affected by exercise duration. The slow component was not affected by either intensity or duration except that, at 20 min exercise at 80% of maximal aerobic power, half-life was about 150 min.

6. Body parts: Skeletal–muscular system

6.1. Static work (including posture)

Vollestad and Sejersted (1988) speculate that decreased Ca^{2+} availability for release from the sarcoplasmic reticulum might contribute to fatigue during all types of exercise. However, Sjogaard (1990) feels potassium loss from the intracellular space is the fatigue mechanism. Sjogaard states that muscle fatigue is primarily within the muscle (as opposed to blood or brain). Kahn and Monod (1989) feel local muscle fatigue is due to lack of elimination of byproducts rather than lack of new energy or oxygen supplies. They report fatigue from isometric contraction occurs sooner in muscles with type II (fast-twitch) fibers. (Eye muscles are fast-twitch fibers while the soleus muscle of the leg (a postural muscle) is slow-twitch.) Kadefors et al. (1996), using whole day EMG, report that some parts of muscles (low-threshold motor units, also called C units) work almost continuously (i.e. rarely rest) even in low-metabolic rate tasks. The picturesque label for such units is "Cinderella units" (first up and last to bed).

Dul (1991) developed a work/rest model for static postures. This model was challenged by Mathiassen and Winkel (1992). Miedema et al. (1997) compared maximum holding time (MHT) for 19 different standing postures; MHT varied from 2 to 35 min. If the hand position was less than 50% of shoulder height, posture was terminated by lower back and leg pain; between 50 and 100% of shoulder height, posture was terminated by pain in the shoulders and arms.

Fig. 4 shows MHT for which subjects could hold a forward bent posture. For low stress, recovery is rapid; for high stress, recovery is slow. Kilbom et al.

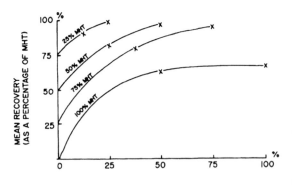

Fig. 4. Maximum Holding Time (MHT) for a forward-stooped posture is plotted for 25, 50, 75 and 100% MHT duration (Milner et al., 1986). When the stress is low (e.g. 25% MHT), even 0% rest gives 75% recovery. When rest = 25% (i.e. rest = MHT), recovery = 100%. When stress is high (e.g. 100% MHT), 0% rest has 0% recovery. When rest = MHT, recovery = 50%.

(1983) suggest it takes several days for recovery from maximal effort. Bystrom et al. (1991) report maximum voluntary contraction was significantly reduced 24 h after a continuous handgrip exertion. Bigland-Ritchie (1986) point out that there is great differences in the fatigue resistance of various muscles; e.g. using the same protocol, endurance times of the soleus was 7 times greater than the quadriceps.

When standing with a sit–stand seat, the lower leg and foot are at an obtuse angle; this problem can be overcome by occasional walking or shifting posture (e.g. using a bar rail).

Van Dieen and Vrielink (1996a,b) reported subjective comfort in the legs and back for poultry inspectors was worst for 60 min standing working followed by 15 min sitting; 45–15, 30–15 and 30–30 work/rest schedules did not differ significantly.

Wood et al. (in press) developed a work/rest model for isometric grip strength; fatigue increases (vs. 32% MVC for 19 s work in a 60 s work/rest cycle) for high force for a short time (48% MVC for 13 s) or low force for a long time (16% MVC for 37 s).

Pure static work is not very common in most tasks. Usually there are small movements or the body parts may be partially supported.

6.2. Dynamic work

In dynamic work, the muscles automatically create "micropauses" (in contrast to the constant load of static work). For example, when reaching out with the arm, the set of muscles for reaching out work while the set of muscles for reaching in have a rest. Thus, dynamic work inherently has some recovery "built in".

Dynamic work is divided into (1) VDT and (2) Non-VDT work.

6.2.1. VDT

After an introduction, various VDT rest studies are discussed.

6.2.1.1. Introduction. VDT, the abbreviation for Video Display Terminals, also could stand for Very Demanding Task. It is demanding physically due to the sedentary, minimum-activity posture for most of the body combined with highly repetitive motions by the fingers. It is demanding mentally due to the constant attention required. It is demanding visually due to the poor-quality character reproduction on screens.

VDT tasks include:
- data entry (document intensive because the operator routinely enters data from documents),
- data acquisition (screen-intensive because information from a form is matched with information from the screen),
- word processing (document- and screen-intensive because attention is focused on both places),
- interactive (occasional user of screen such as travel agent, CAD/CAM user, programmer).

The number of keystroke repetitions can be very high.

Assume a person keys words at a rate of 50 words/min. Assuming that a typical word (including spaces and punctuation) has 7 strokes, then 50 words/min $\times$ 7 strokes/word = 350 strokes/min. Keying 400 min/shift (allowing some time for breaks and other work) then results in $350 \times 400 = 140\,000$ strokes/day. In a month, this is 21 days $\times$ 140 000 = 2 940 000 strokes; in a year, this is 235 days $\times$ 140 000 = 32 900 000 strokes. In 10 yr, this is 329 000 000 strokes! It takes about 30 yr to reach 1 billion strokes. Yet we want people to work for 40–45 yr.

We would be very impressed with any machine which operated over 1 billion cycles. We need to reduce stress on the "human machine". Stress can be reduced by two primary ways: reduce exposure and increase worker ability to endure the stress. This paper focuses on increasing worker ability to endure the stress.

Table 1 discusses the recovery/work ratio.

6.2.1.2. VDT rest studies. Four aspects are time before a break, break length, microbreaks, and active vs. passive rest.

Time before a break. Zwahlen et al. (1984) concluded that a 15 min break after 90 min was insufficient to control musculoskeletal discomfort of VDT operators.

Misawa et al. (1984) studied word processing on a VDT for 120 min with three conditions: (a) 30 min with a 5 min break, (b) 60 min with a 10 min break and (c) 120 min continuously. They suggested VDT work should be 60 min or less before a break.

Floru et al. (1985) studied simulated data entry for 120 min. Performance declined with time until a bottom (about 50% of initial level) around 60 min. Then, however, performance climbed back to its initial level. Performance was correlated with EEG activity.

Break length. Horie (1987) concluded 10 min rest in 60 min work was best; however, if the VDT work was for more than 2 h in a day, then rest should be 15 min for every 60 min work. Yoshimura and Tomoda (1994) studied 6 combinations: working 40, 50 or 60 min with breaks of either 10 or 20 min. The best combination was 50 min work with 20 min rest. Yoshimura and Tomoda (1995) studied a work period of 50 min with rest periods of 5, 10, 15, 20 and 25 min; they recommended the 15 min rest period.

Kopardekar and Mital (1994) studied VDT work with 30 min work followed by 5 min of rest (30–5), 60 min with 10 min of rest (60–10) and 120 min with 0 rest (120–0). The 120–0 condition was significantly worse than the 30–5 and the 60–10, which did not differ significantly. They recommended the 60–10 as it caused fewer interruptions, although 30–5 would be recommended if errors were important.

The rest break standard that has been adopted in the majority of German work places is 10 min of non-working rest after 60 min of continuous computer work; $10/60 = 16.7\%$ (Boucsein and Thum, 1995). Boucsein and Thum (1995) compared rest breaks of 7.5 min after 50 min of computer work vs. 15 min after 100 min ($15/100 = 15\%$). The day was from 0900 to 1730 (8.5 h or 510 min); they had 45 min of rest breaks plus 37.5 min for lunch so total rest time was 82.5 min; $82.5/427.5 = 19\%$. They reported responses (heart-rate variability, neck EMG, subjective opinions) were better for short breaks often during the 1100 and 1500 tests but long breaks occasionally were better for the 1700 test.

Microbreaks. Henning et al. (1989) studied microbreak ("concealed" breaks) length. Subjects keyed for 40 min with a discretionary length microbreak after 20 min. The breaks had a log-normal distribution with a mean length of 27 s. Henning et al. feel the subjects should have taken longer microbreaks to ensure complete recovery from fatigue.

Henning et al. (1994) studied two types of microbreaks: (1) regimented (20 s every 300 s) or (2) compensatory (20 s every 300 s, if spontaneous pauses totalled less than 17 s). They concluded compensatory breaks could be used as they were as effective as regimented breaks and eliminated unnecessary task interruptions. Henning et al. (1995, 1996) gave a 30 s break every 8 min for 55 min to VDT operators but with two conditions. Condition (1) had feedback of the actual rest vs. the goal; Condition (2) did not. They concluded feedback was desirable. Asundi (1995) describes a number of software packages available to remind VDT operators about rest breaks and exercises. An example program would count time (or keystrokes) from the last break and then indicate when a break should be taken; the programs usually recommend various exercises.

Active vs. passive rest. What is done during the break? "Active" rest varies from going to another area to doing gymnastics; "passive" tends to be sitting at the workstation. Asmussen and Mazin (1978) found active rest better than passive rest. Winkel and Jorgensen (1986) studied foot swelling in office work for (1) inactive sitting (chair with no wheels), (2) semi-active sitting (chair with wheels, but stay at desk), (and) (3) active sitting (chair with wheels, rolled 175 cm and back every 10 min). Foot swelling decreased with activity; heart rate was highest for inactive sitting. Sundelin and Hagberg (1989) studied three types of rest: (1) passive, (2) active (seated gymnastics) and (3) active (walk in corridor). The 20 s breaks occurred every 6 min; operators preferred active over passive rest. Thompson (1990) had exercise breaks of 5 min added twice a day for 85 data-entry operators; lost time due to injury decreased and productivity increased. Swanson and Sauter (1993) found "inconspicuous exercises taken at the workplace" may help prevent a decline in productivity. Henning et al. (in press) found breaks beneficial for VDT work but breaks with non-computer work were not significantly better than breaks with passive rest.

6.2.2. Non-VDT work

Pollack and Wood (1949) reported mean venous pressure at the ankle during sitting was 56 mmHg and during standing was 87 mmHg i.e. approximately equal to hydrostatic pressure from the right auricle. After 8 steps of walking, venous pressure dropped to 23 mmHg. The immediate fall in pressure occurs as the person contracts the calf muscles in taking the next step before venous filling has been completed; thus, additional blood is pumped out of the leg, causing a further drop in pressure when the calf muscles relax. This drop stabilizes in about 8 steps when blood flow into the vein from the capillaries equals flow pumped out.

Bhatia and Murrell (1969) studied industrial workers with 6 breaks of 10 min vs. 4 breaks of 15 min. The 6 breaks of 10 min was preferred. Ramsey et al. (1974) reported that inspectors did better with short (0-24 min) duration sessions than with long (50-74 min) sessions. Genaidy et al. (1995) had meat packers use microbreaks of prolonged static stretching when they perceived discomfort. The perceived discomfort was lower when microbreaks were used. The break frequency was 2/shift with a mean duration of 48 s.

Davis and Konz (1995) reported that, after 1 min of pushups, hand steadiness had not returned to basal levels within 6 h.

Nakamura et al. (1996) reported that, after pedalling, recovery was fastest in a 30°C bath, then 38°C bath and then in air. Fatigue sensation (1 = not tired, 2 = slightly tired, 3 = a little tired, 4 = tired and 5 = tired) had $r^2 = 55\%$ vs. lactate concentration, mmol/l.

Recent developments with EMG may permit the separation of the fatigue and force aspects (Luttman et al., 1996; Van Dieen and Vrielink, 1996a,b; Spaepen and Hermans, 1996; Kumar and Mital, 1996); this would give an "on-line physiological index" of fatigue instead of depending upon subjective opinions.

7. Body parts: Brain

The divisions are (1) optimum stimulation, (2) concentration and attention, (3) sleep/biological clock and (4) shiftwork.

7.1. Optimum stimulation

It has been pointed out many times that fatigue is not all physiological; there is a strong psychological (lack of motivation) component. For a recent example, Finkleman (1994) analyzed 3700 people who reported fatigue in their work. Physically demanding jobs had *less* fatigue reported than low physically demanding jobs! Significant predictors of fatigue included job pay, job control, and supervisor quality- emphasizing the importance of lack of motivation. Finkleman concluded an important predicator of fatigue was processing too much information (overload and thus, fatigue) or too little information (underload and thus, boredom).

To avoid too much information or too little information, consider that information comes both from the task and from the environment. If the task is high stimulation, then reduce stimulation from the environment. For example, in office tasks, increase visual and auditory privacy.

If the task is low stimulation, then add stimulation to the task or from the environment. Adding task stimulation often is done by adding physical movement. Smith (1981) reviewed 50 references on boredom and found that a change in task was as good as a rest. Adding environmental stimulation can be (1) encouraging conversation with others, (2) background music, (3) windows with a view, (4) a varied visual environment and (5) a varied physical environment (temperature, air velocity).

7.2. Concentration and attention

Tasks with concentration and attention are sedentary but with considerable mental activity. Example activities with mental overload are simultaneous translation and education; activities with mental underload are monitoring in control rooms and watchkeeping at sea, and driving vehicles (cars, trucks, trains, airplanes).

Meijman (1995) points out people can maintain mental performance by exerting more effort; he used the 0.1 Hz component in the heart beat as an index of mental effort.

For work with concentration and attention, a number of studies in the 1960s and 1970s showed that machine-paced work needs additional rest vs. operator-paced work (Konz, 1979).

Megaw (1995) reviews 95 articles on visual fatigue and concludes we do not know much about visual fatigue.

Table 4
Limits of hours of work (including overtime) in nuclear power plant control rooms (Lewis, 1985). The limits exclude shift turnover time (typically 30 min/turnover). (Shift turnover time is the time when the old shift communicates with the new shift.)

Period, days[a]	Maximum h unless unusual circumstances[b]	Maximum h unless very unusual circumstances[c]
1	12[d]	–
2	24	–
7	60	72
14	112	132
28	192	228
365	2260	2300

[a]A "day" is any period of 24 consecutive hours.
[b]Deviations from this column must be approved by the plant manager; the authorization must be documented and available for Nuclear Regulatory Commission (NRC) inspection. An extended shutdown shall not be considered unusual circumstances.
[c]Deviations from this column shall be authorized, up to specific limits, by the NRC.
[d]In the case of a problem during operation (such as the unexpected absence of an operator), overtime may be worked on an individual basis. No individual should be allowed to work more than 16 h straight, more than 1 period of 16 hours in a 7 day period, or more than 2 periods of 16 h in a 28 day period.

7.2.1. Translation

Simultaneous translators work in pairs. One translates while the other rests. In Quebec (French/English) they switch every 30 min; in Japan (Japanese/English) they switch every 20 min. Sign language translators for the hearing impaired also work in pairs; 30 min on and 30 min rest.

7.2.2. Education

The typical schedule at universities in the USA is 50 min of lecture with a 10 min break before the next class. However the typical student schedule is 15 class hours/week so there is considerable recovery time between classes. Professors tend to have 6 to 12 teaching "hours" (each of 50 min) each week; typically they teach for 32 weeks/yr.

High school students (age 14–18) in USA typically have 6 or 7 periods/day; each period is 60 min, which includes a 5 min break. In addition to lunch, one period typically is for physical education. Typically, they go to school 186 days/yr.

Bennett et al. (1974) studied college students doing arithmetic for 180 min. The three conditions were no rest, passive rest (sit quietly) and active rest (go to another room and have a soft drink). Active rest was superior to passive and no rest. In a second experiment, performance with two 10 min changes of task was superior to working without any change.

Henning (1987) had subjects make choice reaction time responses; they could take a variable length "microbreak" between the 4 min trials. Subjects took a mean moicrobreak of only 10 s. Henning concluded this break length was insufficient to recover from fatigue and that operators either needed to have feedback concerning the break or have a longer break that was externally scheduled.

Translation and education are examples of overload; monitoring and driving are examples of underload.

7.2.3. Monitoring/vigilance

Example tasks would be process-control monitoring, hospital patient monitoring, radar monitoring and industrial inspection.

Table 4 gives the recommended work hours for control operators in nuclear power plants. Table 5 compares the Nuclear Regulatory Commission (NRC) values vs. the US Air Force and US regulatory values for truck, railroads and airlines. A concern was the use of large annual amounts of overtime (say 400 h) in nuclear power plants. The primary criterion was operator alertness/safety. The recommendations in Table 4 were not adopted by the NRC.

For control operators in nuclear power plants, Table 6 gives the recommendations for 8 h shifts and Table 7 gives the recommendations for 12 h shifts.

Table 5
Comparison of limits of hours of work (Adapted from Lewis, 1985).

Period, days	NRC policy	Lewis recommendation		US air force	Non-nuclear industries		
		unusual	very unusual		Truck	RR	Airlines
1	16	12	–	12[a]	10	12	8
2	24	24	–	24[b]	(20)	(24)	(16)
7	72	60	72	–	60	–	30
14	(144)	112	132	–	(120)	–	(60)
28[c]	–	192	228	125[a]	–	–	100
91	–	(626)	(734)	330	–	–	300
365	–	2260	2300	(1320)	–	–	1000

[a]During Desert Shield and Desert Storm, duty days of 16 h for unaugmented crews often were extended to 20 h; duty days of augmented crews were 24 h. Because some pre-flight activity was not counted as part of a duty day, time continuously awake at the end of a duty day (i.e. at landing) was as much as 29 h for unaugmented crews and 33 h for augmented crews. The 125 h/month was increased to 150 (Neville et al., 1994). Neville et al. report the critical variable was the sleep in the previous 48 h; the increase from 125 to 150 for the month had no effect on fatigue.
[b]Numbers in brackets are extrapolations from one time period to the next longer time period.
[c]For US Air Force and airline pilots and crew, the time period is 30 days.

Table 6
Recommendation for routine 8 h schedules for control rooms (Lewis, 1985)

1. The schedule should be limited to a maximum of 7 consecutive days of work.
2. In any 4 week period, the schedule should not exceed 21 days of work (including training).
3. In any period of 9 consecutive days, the schedule should include at least 2 consevutive full days off.
4. A series of night shifts should be followed by at least 2 full days off.
5. The schedule should rotate forward, not backward.

Table 7
Recommendation for routine 12 h schedules for control rooms (Lewis, 1985)

1. The schedule should contain a maximum of 4 consecutive 12 h work days.
2. Four consecutive 12 h work days should be followed by no fewer than 4 days off.
3. The basic 12 h day schedule should be "2 on, 2 off", "3 on, 3 off", "4 on, 4 off" or a systematic combination of these such as the "every other weekend off" schedule (which combines the "2 on, 2 off" with "3 on, 3 off").
4. The general safety record of the plant should be satisfactory.
5. The plant should have the capability to cover unexpected absences satisfactorily without having any individual work more than 12 h/day.
6. The round trip commute time for the operators should not exceed 2.5 h. (Commute time reduces time available for sleep).

Table 8 gives the recommendations for watchkeeping at sea. Note that watchkeeping implies the use of permanent shifts. At sea, this may be reasonable since there are less social pressures outside work and thus circadian inversion may occur. Colquhoun et al. (1988) surveyed 97 cases of merchant shipping from 30 countries; 52 followed a 4 h on/8 h off system, 27 a 6 h on/6 h off system and 3 a 12 h on/12 h off system.

Flight attendants in the US must be provided at least 9 h of scheduled rest within a 24 h time span – if they have been on duty for up to 14 h. Attend-

Table 8
Recommendations for watchkeeping at sea (Buck et al., 1995)

1. A duty schedule must be published before sailing. This schedule must be observed during the voyage.
2. The cycle of repetition should be 24 h.
3. The duty schedule could be varied from day to day. Hours on duty and off duty could begin no more than 1 h before the scheduled starting time; they could be extended up to the limits imposed by maximal hours on duty and minimal hours off duty.
4. Watchkeepers should not be permitted to spend more than 18 h on duty in any period of 24 h and not more than 24 h in any period of 48 h. All hours spent working should be counted as being on duty and not just hours spent watchkeeping.
5. Watchkeepers should spend not less than 6 consecutive hours off duty in any one period of 24 h and not less than 24 h off duty in any one period of 48 h.

ants must have a 24 h rest every 7 calendar days (Phillips, 1994). The Federal Aviation Administration has proposed flight crews could work 14 h/day instead of the previous 16 h/day (Pasztor and Gruley, 1995).

Craig (1985) summarizes some field studies on vigilance. Changes which reduce boredom are beneficial; one example is loading/unloading as well as inspecting coins so that inspection took 14 min out of each 24 min. Another example is 30 min of inspection followed by 60 min of other tasks. This "task variety" can be considered as "active rest".

7.2.4. Driving/piloting

Buck and Lamonde (1993), in their survey of fatigue among locomotive drivers, report that locomotive drivers tended to have irregular schedules (interfering with sleep) and also failures to respond were more likely to occur close to 0300 and 1500 h. Young and Hashemi (1996) propose a two-tier fatigue model; some drivers arrive at work fatigued and have accidents early in the work period while other drivers start rested and become fatigued during the task. Fatigue may be more of a problem when the driver is "externally scheduled"; for example, when tired, a bus driver or a pilot cannot stop and take a break while a car or truck driver generally can stop since they are self-paced. On some limited-access highways, there may be a long distance between safe places for a nap. Japanese taxi drivers typically work a 16 h day and then have the following 1 or 2 days off.

Wedderburn (1987) reports that, to overcome fatigue, many truck drivers recommend eating (while still driving). Other temporary solutions are having a drink with caffeine, adjusting the ventilation, and listening to the radio.

Meijman et al. (1992) studied workload of driving examiners. The question was how many drivers should be examined/shift. Recovery time (including lunch) was 26% for examining 9 drivers/shift, 20% for 10 drivers/shift and 15% for 11 drivers/shift. They recommended (and the organization implemented) the 9 drivers/shift due to worker stress during the 10 and 11 divers/shift. Perhaps most interesting is that this stress (blood pressure, adrenaline) was higher *after* the shift as well as during the shift. This points out that stress effects may continue after work.

There seems to be more recognition by scientists (but not necessarily by regulators) that fatigue is probably more affected by lack of sleep than hours spent working (Brown, 1994). Note that all the regulations assume the hours of the day are equal and interchangeable.

7.3. Sleep/biological clock

Two areas are sleep and drugs.

7.3.1. Sleep

Sleep restores the functions of the brain, rather than the body. Only sleep allows some form of cerebral shutdown. Sleep has two functions: (1) obligatory sleep and (2) facultative sleep (Horne, 1985). Obligatory sleep is the first 5 h of sleep. If sleep is restricted, people lose the later part of sleep (the facultative portion, such as REM (dreaming) and stage 2 (basic sleep), but retain deep sleep (stages 3 and 4). Sleep length follows a normal distribution with young adults having a mean of 7.5 h and standard deviation of 1 h. Night shift workers tend to sleep about 1 h less than those on the other shifts.

Sleep deprivation of 24–48 h primarily affects motivation to perform rather than ability to perform; thus uninteresting, undemanding, simple tasks are most affected. Longer sleep deprivation begins to affect more "cognitive" tasks. However, "ranger type" soldiers could perform their duties even with only 6 h sleep in a 72 h mission (Ball et al., 1984).

The sleepiness/alertness of a person can be tested with the multiple sleep latency test (MSLT). It measures the amount of time between trying to fall asleep and falling asleep. The scale is from 0–20 with Maximum Sleepy = 0 min and Maximum Alert = 20 min. The maximum sleepiness is around 0300 with a secondary dip around 1400 (post-lunch dip). The "post-lunch dip" is not related to food consumption since it does not occur after breakfast or dinner. Incidents of falling asleep while driving peak at 0200 with a secondary peak at 1400 (Schwing, 1990). Hamelin (1987) show truck driver risk rate vs. time of day has two peaks; the higher one (as expected) is from midnight to 0400 but a secondary peak occurs around noon. Summala and Mikkola (1994) report drivers over age 56 had many fatigue fatalities in the late afternoon.

Many people have mild but chronic sleep deprivation. A typical MSLT score after 8 h of sleep is 15 min. However if the 2 previous nights have 10 h of nighttime sleep, the MSLT = 20 (i.e. maximum alertness). Conversely, restricting sleep to 4 h/night for 2 nights gives a daytime MSLT = 5 min (Coleman, 1985, p. 165).

Naps typically occur between 1430 and 1700 h and last about 90 min (Gillberg, 1985). Although naps taken at 0300 are of some benefit, people tend to think of them as "inadequate opportunities to sleep"; Minors and Waterhouse (1987) recommend they be presented instead as a bonus (an opportunity for partial recuperation).

Fig. 5 (Coleman, 1986) shows circadian timing and five physiological functions over 2 days. (It is called "circadian" rhythm from the Latin circa dies (about a day); without external clues, the rhythm is about 25 h rather than the 24 h we get from time givers (zietgebers).) Outdoor light (and darkness) is the primary zietgeber. When shifting phases with coherent zietgebers (such as time zone shifting or going from night work to day work), the phase shift is about 2 h/day. When shifting phases with incoherent zietgebers (e.g. time of dawn is not changed but time of going to bed changes), the phase shift is about 1 h/day (Moog, 1987). Melatonin seems to reduce jet lag (Arendt et al., 1987).

Alertness peaks in the late afternoon and bottoms during sleep (2300 to 0600). Internal temperature and potassium also peak during the day and bottom at night. "Morning type" people (larks) have about a 0.9°C difference between their maximum and minimum body temperature while "evening types" (owls) have a difference about 1.0°C. A number of studies show morning types are less tolerant to shiftwork (Moog, 1987; Gander et al., 1993). Cortisol peaks in the early morning and then declines; growth hormone is low during the day but peaks at night.

7.3.2. Drugs

The physiological effect of some compounds is affected by circadian rhythm. For example, ethanol (alcohol) affects a person more in the evening than the day (Coleman, 1986 p.27); 0.4 g/kg of ethanol had the same effect after 5 h of sleep as 0.8 g/kg produced after 8 h of sleep (Roehrs et al., 1989). Ethanol has less effect on daytime alertness/sleepiness for fully rested subjects than ethanol does for people with sleep deficits (Lumley et al., 1987).

The most common drug to decrease sleepiness/increase alertness is caffeine (see Tables 9 and 10); amphetamines work well. Zwyghuizen-Doorenbos (1990) demonstrated caffeine increased daytime alertness although people began to develop a tolerance for caffeine after 4 administrations of caffeine. Johnson and Merullo (1996) report 200 mg caffeine attenuates the vigilance decrement during simulated sentry duty. Bonnet and Arand (1994) advocate use of an "investment" nap (taken before the period of sleep loss) in combination with caffeine. For example, in hospitals a secure afternoon nap and the availability of caffeine at night is

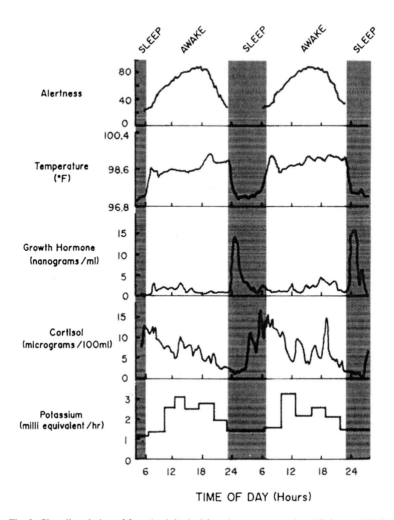

Fig. 5. Circadian timing of five physiological functions over two days (Coleman, 1986).

recommended. Under conventional scheduling, where doctors take naps during the night if workload permits, doctors cannot take caffeine as it would interfere with potential sleep. If doctors take a nap at night and are awakened during the "deep sleep" phases for an emergency, they may experience sleep inertia (poor performance for about 15-30 min after being wakened) (Dinges et al., 1985). Although a common recommendation is to avoid caffeine during the last 2 h of the shift (to allow good sleep after the shift), real shiftworkers, especially if they have little problem sleeping, may prefer to drink caffeine to maintain work alertness and alertness on the drive home.

To improve daytime sleep, while not interfering with nighttime alertness, Walsh (1990) recommends a short-acting benzodiazepine (trizolam) rather than a long-acting one (flurazepam). Note that if the person may work immediately upon awaking, then a sedative is not recommended as the drug effect outlasts the sleep period. Also see Table 11.

Some people use alcohol or some of the older antihistamines (e.g. diphenhydramine) which has the effect of increasing sleepiness/decreasing

Table 9
Caffeine

Caffeine, a drug with primarily stimulating effects, has a metabolic half-life of 3–7 h; some factors influencing the half life are pregnancy (increases half life to as much as 18 h) and smoking (decreases half life) (Evaluation of Caffeine Safety, 1987) (Grice and Murray, 1987). Caffeine has no day-to-day accumulations as it almost completely disappears from the body overnight (Ensminger, 1994).

A pharmacological active dose is about 200 mg (about 3 mg/kg), depending upon the individual (body weight, body tolerance). Rosenthal et al. (1991) reported increased sleep latency with doses as low as 75 mg.

Caffeine:
1. Stimulates the central nervous system (brain), thereby prolonging wakefulness with alert intellectual facilities.
2. Stimulates the heart action.
3. Relaxes smooth muscle (digestive tract, blood vessels).
4. Increases urine flow (a diuretic).
5. Stimulates stomach acid secretion.
6. Increases muscle strength and the amount of time a person can perform physically exhausting work.

A dose of 1000 mg generally will produce adverse effects (insomnia, restlessness, excitement, trembling, rapid heart beat with extra heart beats, increased breathing, desire to urinate, ringing in the ears, heartburn). Daily consumption of about 800 mg will cause user dependence on caffeine.

Table 10 gives the caffeine content in some foods and drugs.

Table 10
Caffeine content

Item	Measure	Caffeine (mg)
Alertness tablets		
NoDoz	Tablet	100
Vivarin	Tablet	200
Pain relievers		
Excedrin	Tablet	65
Anacin	Tablet	32
Cold allergy relief	Tablet	15–32
Weight-control		
Dexatrim	Tablet	200
Dietac	Tablet	200
Coffee (percolated)[a]	240 mL	21–148[b]
Coffee (instant)	240 mL	66
Tea[a]	240 mL	20–130
Soft drinks with caffeine	375 mL	36–54
Chocolate, baking	30 mL	70
Chocolate, sweet or dark	30 mL	40
Chocolate, milk	30 mL	12

[a] Longer brewing increases caffeine content.
[b] 58 different people prepared coffee ranging from 21 to 148 mg/cup (Eisenberg, 1989).

alertness. Alcohol will have more effect on alertness in the early afternoon or early morning as the alcohol then will reinforce the body's natural sleep tendencies. Walsh (1990) says "alcohol produces significant sleep maintenance difficulty due to the rapid metabolism of alcohol and subsequent sympathetic arousal". At high blood-alcohol concentrations, the enzyme dehydrogenase is saturated and the half-life of alcohol in the blood is increased (Paustenback, 1994, p. 244). Walsh et al. (1991) postulate that alcohol-related vehicle accidents at night may be "as likely due to reduced alertness as to impaired motor function or reaction time". Schwing (1990) reports vehicle accidents in the USA are risky for after-midnight hours, especially on weekends; 4.5% of the travel yields 32% of fatalities. The combination of alcohol and fatigue are powerful indicators for death on the highway.

7.3.3. Shiftwork

There are many different criteria upon which to design shiftwork systems. Knauth (1993) gives the general guidelines of Table 12. Schonfelder and Knauth (1993) give the math model given in Table 13. One of the virtues of a math model is the explicit statement of the variables and the coefficients. Thus you can select the variables you want to emphasize and their importance. Wedderburn and Scholarios (1993) demonstrated that shiftworkers do not agree with all the recommendations of shiftwork experts.

Table 11
Tips for day sleeping (adapted from Konz, 1995a,b)

---

1. Develop a good sleeping environment (dark, quiet, cool, bed). Have it *dark* (e.g. opaque curtains). Have it *quiet* since it is difficult to go back to sleep when daytime sleep is interrupted. Minimize changes in noise volume. Consider earplugs, unplugging bedroom phones, turning down phone volume in other rooms, reducing TV volume in other rooms. Train your children. Have the sleeping area *cool*. *Bed* (normally OK but may be poor if the sleeper is not sleeping at home; e.g. is part of an "augmented crew" for trucks, aricraft). Then provide a good mattress and enough space.
2. Plan your sleeping time. Tell others your schedule (minimize interruptions). Consider sleeping in two periods (5–6 h during the day and 1–2 h in the late evening before returning to work). Less daytime sleep and more late evening sleep not only makes it easier to sleep but also may give a better fit with family/social activities. Morning to noon bedtimes are the most unsuitable times to sleep (Akerstedt, 1985).
3. Have a light (not zero or heavy) meal before sleep. Liquid consumption increases the need to urinate (which wakes you up). Avoid caffeine (see Table 10). A warm drink before your bedtime (perhaps with family members starting their day) may help your social needs. Avoid foods which upset your stomach – and thus, wake you up.
4. If under emotional stress, relax before going to bed. One possibility is light exercise.

Table 12
Shift system design recommendations (Knauth, 1993)

---

1. Permanent nightwork does not seem to be advisable for the majority of shiftworkers. Full entrainment of physiological functions to night work is difficult. Even permanent night workers have problems due to readapting to day cycles during weekends, holidays and vacations. If shifts rotate, rapid rotation is preferable to slow (weekly) rotation.
2. Shift durations of 12 h have advantages and disadvantages. Some potential problems are: fatigue, covering absentees, overtime, limitation of toxic exposure, and possible moonlighting when workers have large blocks of leisure time.
3. Avoid an early (before 7:00) start for the morning shift.
4. Distribution of leisure time is important. Have sufficient time to sleep between shifts (e.g. during shift changeovers). Limit the number of consecutive working days to 5–7. For every shift system, have some non-working weekends with at least two successive full days off.
5. Shifts should rotate forward (day, evening, night).
6. Keep the schedule simple and predictable. People want to be able to plan their personal lives. Make work schedules understandable. Publically post them in advance so people can plan; 30 days in advance is a good policy.

Shiftwork seems to become less tolerable about age 50 (Monk and Folkard, 1985; Gander et al., 1993). "Level of commitment" is an important individual variable in adjusting to shiftwork; that is, how willing is the worker willing or able to structure life about the need to work at unusual hours? Physical fitness increases tolerance to shiftwork – probably due to improved sleep (Harma, 1993). Moog (1996) emphasizes how poorly we are able to predict individual responses to changing shifts. Monk and Folkard (1992) tell how to make shiftwork tolerable.

When people work a permanent night shift, they gradually (about 20 days) adjust their circadian rhythms – if they follow the same schedule on days off as work days. However, most people comform to the rest of society during days off and, since reentrainment is very fast, they never really adjust. Thus, shiftwork is characterized by internal temporal disorder.

A recent approach is to have very bright lights in the control room at night (Czeisler et al., 1990; Bovian et al., 1994). This gives a quick adaptation of the circadian rhythm with better nighttime alertness and better day sleep. At the beginning of a block of shifts, light (level 4) is 4000 to 5500 lux. During the block of shifts, the light is decreased to level 3 (2000 to 3000 lux), to level 2 (800 to 1200 lux) and then to level 1 (default) with 50 lux. (Detailed attention needs to be paid to glare.)

Table 13
Shiftwork criteria (Schonfelder and Knauth, 1993)
The following criteria for evaluating shiftwork are based on "health" and "social life". The larger the number the worse the shift-work system

Factor 1: Consecutive night shifts

Night shifts	Cost	Night shifts	Cost
0	0	4	44.1
1	7.4	5	51.4
2	22.0	6	61.2
3	29.4	≥ 7	73.5

Factor 2: Quota of night shifts

$F_2 = n_n/n_s$,

where
  n_n = Number of night shifts,
  n_s = Number of days in the shift cycle.

Factor 2	Cost	Factor 2	Cost	Factor 2	Cost
0	0	$0.21 < F2 \leq 0.35$	68.6	$0.56 < F2 \leq 0.63$	137.2
≤ 0.07	17.2	$0.35 < F2 \leq 0.42$	85.7	$0.63 < F2 \leq 0.70$	154.4
$0.07 < F2 \leq 0.14$	34.3	$0.42 < F2 \leq 0.49$	102.9	$0.70 < F2 \leq 1$	171.5
$0.14 < F2 \leq 0.21$	52.4	$0.49 < F2 \leq 0.56$	120.0		
$0.21 < F2 \leq 0.35$	68.6				

Factor 3: Number of consecutive working days

Factor	Cost	Factor	Cost	Factor	Cost
1	0	5	35	9	70
2	8.8	6	43.8	10	78.8
3	17.5	7	52.5	> 10	87.5
4	26.2	8	61.2		

Factor 4: Weeks with > 40 h working time

$F4 = N_{>40}/n_w$,

where
  $n_{>40}$ = Number of weeks with > 40 h working time,
  n_w = Number of weeks in the shift cycle.

Factor	Cost	Factor	Cost	Factor	Cost
0	0	$0.3 < F4 \leq 0.4$	35	$0.7 < F4 \leq 0.8$	70.0
$F4 \leq 0.1$	8.8	$0.4 < F4 \leq 0.5$	43.8	$0.8 < F4 \leq 0.9$	78.8
$0.1 < F4 \leq 0.2$	17.5	$0.5 < F4 \leq 0.6$	52.5	$0.9 < F4 \leq 1.0$	87.5
$0.2 < F4 \leq 0.3$	26.2	$0.6 < F4 \leq 0.7$	61.2		

Factor 5: Ratio of unfavorable shift sequence

$F5 = n_u/n_w$,

where
  n_u = Number of forbidden or unfavourable shift sequences:
  Forbidden = N/M, M/A, A/N, N/A, A/M, M/N e.g. N/M means night shift is followed by morning shift N = night, M = morning, A = afternoon.
  Unfavourable = N/-/M, N/-/N
  n_w = Number of weeks in the shift cycle.

Factor	Cost	Factor	Cost	Factor	Cost
0	0	$0.3 < F5 \leq 0.4$	84	$1.5 < F5 \leq 2.0$	168
$F5 \leq 0.1$	21	$0.4 < F5 \leq 0.5$	105	$2.0 < F5 \leq 2.5$	189
$0.1 < F5 \leq 0.2$	42	$0.5 < F5 \leq 1.0$	126	$2.5 < F5 \leq 7$	210
$0.2 < F5 \leq 0.3$	63	$1.0 < F5 \leq 1.5$	147		

Table 13 (Continued)

Factor 6: Index of shift rotation order

$F6 = (F - B)/(F + B)$,

where
F = Sum of forward rotation (M/A/N)
B = Sum of backward rotation (N/A/M)

Factor	Cost	Factor	Cost	Factor	Cost
$F6 = 1$	0	$-0.2 \leqslant F6 < 0.2$	17.5	$-0.8 \leqslant F6 < -0.6$	28
$0.8 \leqslant F6 < 1.$	3.5	$-0.4 \leqslant F6 < -0.2$	21	$-1.0 \leqslant F6 < -0.8$	31.5
$0.6 \leqslant F6 < 0.8$	7.0	$-0.6 \leqslant F6 < -0.4$	24.5	$F6 = -1.0$	35
$0.4 \leqslant F6 < 0.6$	10.5				

Factor 7: Start time of morning shift

Time	Cost	Time	Cost	Time	Cost
After 8:00	0	6:31–7:00	21	5:01–5:30	31.5
7:31–8:00	3.5	6:01–6:30	21	Before 5:01	35
7:01–7:30	7	5:31–6:00	28		

Factor 8: Weekend leisure time index

$F8 = (n_{ss} + 0.43 n_s)/n_w$,
where
n_{ss} = Number of weeks without free Saturday and Sunday,
n_s = Number of weeks with free Saturday or free Sunday,
n_w = Number of weeks in the shift cycle.

Factor	Cost	Factor	Cost	Factor	Cost
$F8 = 0$	0	$0.3 < F8 \leqslant 0.4$	50.4	$0.7 < F8 \leqslant 0.8$	100.8
$0 < F8 \leqslant 0.1$	12.6	$0.4 < F8 \leqslant 0.5$	63	$0.8 < F8 \leqslant 0.9$	113.4
$0.1 < F8 \leqslant 0.2$	25.2	$0.5 < F8 \leqslant 0.6$	75.6	$0.9 < F8 \leqslant 1.0$	126
$0.2 < F8 \leqslant 0.3$	37.8	$0.6 < F8 \leqslant 0.7$	88.2		

Factor 9: Index of working time adjustment

$F9 = (n_z + 0.25 n_p)/n_w$,
where
n_z = Number of additional shifts beyond collective agreement,
n_p = Number of paid days off,
n_w = Number of weeks in the shift cycle.

Factor	Cost	Factor	Cost	Factor	Cost
$F9 = 0$	0	$0.8 < F9 \leqslant 1.0$	12	$1.6 < F9 \leqslant 1.8$	21.6
$F9 \leqslant 0.2$	2.4	$1.0 < F9 \leqslant 1.2$	14.4	$1.8 < F9 \leqslant 7$	24.0
$0.2 < F9 \leqslant 0.4$	4.8	$1.2 < F9 \leqslant 1.4$	16.8		
$0.4 < F9 \leqslant 0.6$	7.2	$1.4 < F9 \leqslant 1.6$	19.2		

Factor 10: Ratio of weeks with evening leisure time

$F10 = n_{we}/n_w$,

where
n_{we} = Number of weeks without free evening,
n_w = Number of weeks in the shift cycle.

Factor	Cost	Factor	Cost	Factor	Cost
$F10 = 0$	0	$0.3 < F10 \leqslant 0.4$	50.4	$0.7 < F10 \leqslant 0.8$	100.8
$F10 \leqslant 0.1$	12.6	$0.4 < F10 \leqslant 0.5$	63	$0.8 < F10 \leqslant 0.9$	113.4
$0.1 < F10 \leqslant 0.2$	25.2	$0.5 < F10 \leqslant 0.6$	75.6	$0.9 < F10 \leqslant 1.0$	126
$0.2 < F10 \leqslant 0.3$	37.8	$0.6 < F10 \leqslant 0.7$	88.2		

Table 13 (Continued)

Factor 11: Weeks in shift rota

Weeks	Cost	Weeks	Cost	Weeks	Cost
1	0	5	2.4	9	4.8
2	0.6	6	3.0	10	5.4
3	1.2	7	3.6	> 10	6
4	1.8	8	4.2		

Factor 12: Number of changes between working days and days off in the basic pattern

Changes	Cost	Changes	Cost	Changes	Cost
1	0	3	2.4	5	4.8
2	1.2	4	3.6	≥ 6	6

Factor 13: Number of different types of shifts

Number	Cost	Number	Cost	Number	Cost
1	0	3	2.4	5	4.8
2	1.2	4	3.6	≥ 6	6

Factor 14: Shift sequence index
L_b = Length of shift blocks for each type of shift, maximum minus minimum,
I_b = Interval between shift blocks for each type of shift, maximum minus minimum.
Determine $F14$ from the following matrix:

		L_b				
		0	1	2	3	≥ 4
I_b	0	1	1.5	2	2.5	3
	1	1.5	2	2.5	3	3.5
	2	2	2.5	3	3.5	4
	3	2.5	3	3.5	4	4.5
	≥ 4	3	3.5	4	4.5	5

Factor	Cost	Factor	Cost	Factor	Cost
$F14 = 1$	0	$2.2 < F14 \leq 2.6$	2.4	$3.8 < F14 \leq 4.2$	4.8
$1 < F14 \leq 1.4$	0.6	$2.6 < F14 \leq 3.0$	3.0	$4.2 < F14 \leq 4.6$	5.4
$1.4 < F14 \leq 1.8$	1.2	$3.0 < F14 \leq 3.4$	3.6	$4.6 < F14 \leq 5$	6.0

References

ILO, 1989. Working Time Issues in Industrialized Countries SWT1/1988/6, International Labour office, Geneva.

American Congress of Governmental Industrial Hygienists (ACGIH) 1995. Threshold Limit Values, Cincinnati.

Akerstedt, T., 1985. Adjustment of physiological circadian rhythms and the sleep-wake cycle to shiftwork. In: Folkard, S., Monk, T. (Eds.), Hours of Work. (ch. 15.) Wiley, New York.

Arendt, J., Aldhous, M., English, J., Marks, V., Arendt, J.H., Marks, M., Folkard, S., 1987. Some effects of jet-lag and their alleviation by melatonin. Ergonomics 30 (9), 1379-1393.

ASHRAE, 1991. HVAC applications. Refrigeration and Air Conditioning, ch. 25, American Society of Heating, Atlanta.

Asmussen, E., Mazin, B., 1978. Recuperation after muscular fatigue by 'diverting activities'. European Journal of Applied Physiology 38, 1-7.

Asundi, P., 1995. Rest reminders and multimedia. Ergonomics 38 (10), 2131-2133.

Ball, C., Funk, T., Noonan, D., Velasquez, J., Konz, S., 1984. Degradation of performance due to sleep deprivation: a field study. Proceedings of the Human Factors Society, pp. 570-574.

Bennett, C., Marcellus, F., Reynolds, J., 1974. Counteracting phychological fatigue effects by stimulus changes. Proceedings of the Human Factors Society, pp. 219-224.

Bhatia, N., Murrell, K., 1969. An industrial experiment in organized rest pauses. Human Factors 11 (2), 167–174.

Bigland-Ritchie, B., Furbush, F., Woods, J., 1986. Fatigue of intermittent submaximal voluntary contractions: central and peripheral factors. Journal of Applied Physiology 61 (2), 421–429.

Bonnet, M., Arand, D., 1994. The use of prophylactic naps and caffeine to maintain performance during a continuous operation. Ergonomics 37 (6), 1009–1020.

Boucsein, W., Thum, M., 1995. Recovery from strain under different work/rest schedules. Proceedings of the Human Factors and Ergonomics Society, pp. 785–788.

Bovian, D., Duffy, J., Kronauer, R., Czeisler, C., 1994. Sensitivity of the human circadian pacemaker to light. Proceedings of the APSS, Boston, MA.

Buck, L., Lamonde, F., 1993. Critical incidents and fatigue among locomotive engineers. Safety Science 16, 1–18.

Buck, L., Greenley, M., Loughnane, D., Webb, R., 1995. Statutory regulations for optimizing work schedules. 27th Annual Conference of the Human Factors Association of Canada, pp. 245–250.

Burns, W., Robinson, D., 1970. Hearing and Noise in Industry. London: Her Majesty's Stationery Office.

Bystrom, S., Mathiassen, S., Fransson-Hall, C., 1991. Physiological effects of micropauses in isometric handgrip exercise. European Journal of Applied Physiology 63, 405–411.

Cakir, A., Hart, D., Stewart, T., 1980. VDT Terminals. Wiley, New York, pp. 247–253.

Coleman, R., 1985. Wide Awake at 3:00 A.M. Freeman, New York.

Colquhoun, W., Rudenfranz, J., Goethe, H., Neidhart, B., Condon, R., Plett, R., Knauth, P., 1988. Work at sea. International Archives of Occupational and Environmental Health 60, 321–329.

Constable, S., Bishop, P., Nunnely, S., Chen, T., 1994. Intermittent microclimate cooling during rest increases work capacity and reduces heat stress. Ergonomics 37 (2), 277–285.

Cornman, G., 1970. Fatigue allowances – a systematic method. Industrial Engineering 2 (4), 10–16.

Craig, A., 1985. Field studies of Human Inspection: the application of vigilance research. In: Folkard, S., Monk, T. (Eds.), Hours of Work. Wiley, New York.

Czeisler, C., Johnson, M., Duffy, J., Brown, E., Ronda, J., Kronauer, R., 1990. Exposure to bright light and darkness to treat physiologic maladaptation to night work. New England Journal of Medicine 322, 1253–1259.

Davis, R., Konz, S., 1995. Effects of muscular exertion on hand steadiness. In: Bittner, A., Champney, P. (Eds.), Advances in Industrial Ergonomics and Safety VII. Taylor & Francis, London.

Dinges, D., Orne, M., Orne, E., 1985. Assessing performance upon abrupt awakening from naps during quasi-continuous operations. Behavior Research Methods, Instruments and Computers 17 (1), 37–45.

Duchon, J., Smith, T., 1993. Extended workdays and safety. International Journal of Industrial Ergonomics 11, 37–49.

Duchon, J., Smith, T., Keran, C., Koehler, E., 1995. Psychophysiological effects of extended workshifts. Proceedings of Human Factors and Ergonomics Society pp. 794–798.

Dul, J., Douwes, M., Smitt, P., 1991. A work–rest model for static postures. In: Queinnec, Y., Daniellou, F. (Eds.), Design for Everyone, vol. 1, Taylor & Francis, London, pp. 93–95.

Eisenberg, S., 1989. Looking for the perfect brew. Food Technology 43, 42–45.

Ensminger, A. et al., 1994. Foods and Nutrition Encyclopedia, 2nd ed., vol. 1 CRC Press, Bocca Raton, 289–291.

Evaluation of Caffeine Safety, 1987. Food Technology 41, 105–113.

Finkleman, J., 1994. A large database study of the factors associated with work-induced fatigue. Human Factors 36 (2), 232–243.

Floru, R., Cail, F., Elias, R., 1985. Psychophysiological changes during a VDU repetitive task. Ergonomics 28(10), 1455–1468.

Freivalds, A., Goldberg, J., 1988a. A methodology for assigning variable relaxation allowances: manual work and environmental conditions. In: Aghazadeh, F. (Ed.), Trends in Ergonomics/Human Factors V. Elsevier, New York, pp. 457–464.

Freivalds, A., Goldberg, J., 1988b. Specification of bases for variable relaxation allowances: 1 standing, abnormal positions and use of force or muscular energy. The Journal of Methods Time Measurement XIV, 2–8.

Freivalds, A., Goldberg, J., 1988c. Specification of bases for variable relaxation allowances: 3. environmental conditions. The Journal of Methods Time Measurement XIV, 16–23.

Gander, P., Nguyen, D., Rosekind, M., Connell, L., 1993. Age, circadian rhythms, and sleep loss in flight crews. Aviation, Space and Environmental Medicine 64 (3), 189–195.

Genaidy, A., Delgado, E., Bustos, T., 1995. Active microbreak effects on musculoskeletal comfort ratings in meatpacking plants. Ergonomics 38 (2), 326–336.

Gillberg, M., 1985. Effects of naps on performance. In: Folkard, S., Monk, T. (Eds.), Hours of Work. Wiley, New york.

Goldberg, J., Freivalds, A., 1988a. A methodology for assigning variable relaxation allowances: visual strain, illumination and mental strain. In: Adhazadeh, F. (Ed.), Trends in Ergonomics/Human Factors V. Elsevier, New York, pp. 161–168.

Goldberg, J., Freivalds, A., 1988b. Specification of bases for variable relaxation allowances: 2. visual strain ad low lighting. The Journal of Methods Time Measurement XIV, 9–15.

Goldberg, J., Freivalds, A., 1988c. Specification of bases for variable relaxation allowances: 4. mental strain, monotony, and tediousness. The Journal of Methods Time Measurement XIV, 24–29.

Grice, H., Murray, T., 1987. Caffeine: a perspective on current concerns. Nutrition Today 22, July–August, 36–38.

Hagberg, J., Mullin, J., Nagle, F., 1980. Effect of work intensity and duration on recovery O_2. Journal of Applied Physiology 48, 540–544.

Hamelin, P., 1987. Lorry driver's time habits in work and their involvement in traffic accidents. Ergonomics 30 (9), 1323–1333.

Harma, M., 1993. Individual differences in tolerances to shiftwork. Ergonomics 36 (1–3), 101–109.

Henning, R., 1987. Worker-terminated micro-breaks and perceptual-motor performance. In: Proceedings of the Fourth Mid-Central Ergonomics/Human Factors Conference. Springer, New York, pp. 374–380.

Henning, R., Sauter, S., Salvendy, G., Krieg, E., 1989. Microbreak length, performance, and stress in a data entry task. Ergonomics 32(7), 855–864.

Henning, R., Jacques, P., Kissel, G. Sullivan, A., Alteras-Webb, S. Frequent, short breaks from computer work: effects on productivity and well-being at two field sites. Ergonomics, in press.

Henning, R., Kissel, G., Maynard, D., 1994. Compensatory rest breaks for VDT operators. International Journal of Industrial Ergonomics 14, 243–249.

Henning, R., Callaghan, E., Guttman, J., Braun, H., 1995. Evaluation of two self-managed rest break systems for VDT users. Proceedings of Human Factors and Ergonomics Society, pp. 780–784.

Henning, R., Callaghan, E., Ortega, A., Kissel, G., Guttman, J., Braun, H., 1996. Continuous feedback to promote self-management of rest breaks during computer use. International Journal of Industrial Ergonomics 18, 71–82.

Horie, Y., 1987. A study on optimum term of work hour with rest pause for VDT workers. Japanese Journal of Ergonomics 23, 373–383.

Horne, J., 1985. Sleep loss: underlying mechanisms and tiredness. In: Folkard, S., Monk, T. (Eds.), Hours of Work. ch. 5. Wiley, New York.

Illuminating Engineering Society, 1987. IES Lighting Handbook: Application Volume. New York.

International Labour Office, 1979. Introduction to Work Study. 3rd ed., ILO, Geneva, Switzerland.

Janaro, R., Bechtold, S., 1985. A study of the reduction of fatigue impact on productivity through optimal rest break scheduling. Human Factors 27 (4), 459–466.

Johnson, R., Merullo, D., 1996. Effects of cafffeine and gender on vigilance and marksmanship. Proceedings of Human Factors and Ergonomic Society, pp. 1217–1221.

Kadefors, R., Sandsjo, L., Oberg, T., 1996. Evaluation of pause distribution patterns in the trapezius muscle. In: Mital, A., Krugger, H., Kumar, S., Menozzi, M., Fernandez, J. (Eds.), Advances in Occcupational Ergonomics and Safety . International Society for Occupational Ergonomics and Safety, Cincinatti, OH, pp. 545–550.

Kahn, J., Monod, H., 1989. Fatigue induced by static work. Ergonomics 32 (7), 839–846.

Kilbom, A., Gamberale, F., Persson, J., Annwall, G., 1983. Physiological and psychological indices of fatigue during static contractions. European Journal of Applied Physiology 36, 7–17.

Knauth, P., 1993. The design of shift systems. Ergonomics 36 (1–3), 15–28.

Konz, S., 1979. Endurance and rest for work with concentration and attention. Proceedings of Human Factors Society, pp. 210–213.

Konz, S., Rohles, F., McCullough, E., 1983. Male responses to intermittent heat. ASHRAE Transactions Part 1B, 79–100.

Konz, S., 1995a. Environmental fatigue allowances. International Journal of Industrial Engineering 2 (1), 5–13.

Konz, S., 1995b. Work Design: Industrial Ergonomics. Publishing Horizons, Scottsdale, AZ.

Kopardekar, P., Mital, A., 1994. The effect of different work-rest schedules on fatigue and performance of a simulated directory assistance operator's task. Ergonomics 37 (10), 1697–1707.

Kryter, K., 1985. Effects of Noise on Man. 2nd ed., Academic Press, New York.

Kumar, S., Mital, A., 1996. Electomyography in Ergonomics. Taylor & Francis, London.

Lewis, P., 1985. Recommendations for NRC policy on shift scheduling and overtime at nuclear power plants. NUREG/CR-4248, US Nuclear Regulatory Commission, Washington DC 20555.

Lumley, M., Roehrs, T., Asker, D., Zorick, F., Roth, T., 1987. Ethanol and caffeine effects on daytime sleepiness/alertness. Sleep 10 (4), 306–312.

Luttman, A., Jager, M., Sokeland, J., Laurig, W., 1996. Joint analysis of spectrum and amplitude (JASA) of electromyograms applied for the indication of muscular fatigue among surgeons in urology. In: Mital et al. (Ed.), Advances in Occupational Ergonomics and Safety. International Social for Occupational Ergonomics and Safety, Cincinatti, OH, pp. 523–528.

Mathiassen, S., Winkel, J., 1992. Can occupational guidelines for work-rest schedules be based on endurance time data?. Ergonomics 35 (3), 253–259.

Megaw, E., 1995. The definition and measurement of visual fatigue. In: Wilson, J., Corlett, N. (Eds.), Evaluation of Human Work. 2nd ed., Taylor & Francis, London.

Meijman, T., Mulder, G., van Dormolen, M., Cremer, R., 1992. Workload of driving examiners: a psychophysiological field study. In: Kragt, H. (Ed.), Enhancing Industrial Performance. Taylor and Francis, London, pp. 245–258.

Meijman, T., 1995. Mental fatigue and the temporal structuring of working times. Proceedings of the Human Factors and Ergonomic Society. Santa Monica. CA, pp. 789–793.

Milner, N., Corlett, N., O'Brien, 1986. A model to predict recovery from maximal and submaximal isometric exercise. In: Corlett, N. (Ed.), The Ergonomics of Working Postures. Taylor & Francis, London, pp. 126–135.

Miedema, M., Douwes, M., Dul, J., 1997. Recommended maximum holding times for prevention of discomfort of static standing postures. International Journal of Industrial Ergonomics 19, 9–18.

Minors, D., Waterhouse, J., 1987. The role of naps in alleviating sleepiness during an irregular sleep-wake schedule. Ergonomics 30 (9), 1261–1273.

Misawa, T., Yoshina, K., Shigeta, S., 1984. An experimental study of the duration of a single spell of work on VDT performance. Japanese Journal of Industrial Health. 26, 296–302.

Mital, A., 1984a. Comprehensive maximum acceptable weight of lift database for regular 8 h work shifts. Ergonomics 27, 1127–1138.

Mital, A., 1984b. Maximum weights of lift acceptable to male and female industrial workers for extended work shifts. Ergonomics 27, 1115–1126.

Mital, A., Genaidy, A., 1990. Review and evaluaton of techniques for determining work-rest periods. IE Ergonomics News 24 (2), 1–7.

Mital, A., Bishu, R., Manjunath, S., 1991. Review and evaluation of techniques for determining fatigue allowances. International Journal of Industrial Ergonomics 8, 165–178.

Mital, A., Hamid, F., Brown, M., 1994. Physical fatigue in high and very high frequency manual materials handling: percieved exertion and physiological factors. Human Factors 36 (2), 219–231.

Monk, T., Folkard, S., 1985. Individual differences in shiftwork adjustment. In: Folkard, S., Monk, T. (Eds.), Hours of Work. Wiley, New York.

Monk, T., Folkard, S., 1992. Making Shiftwork Tolerable. Taylor & Francis, London.

Moog, R., 1987. Optimization of shift work: physiological contributions. Ergonomics 30 (9), 1249–1259.

Moog, R., 1996. Problems in determining the practical importance of light to ease circadian phase adaptation to night work. In: Mital et al. (Ed.), Advances in Occupational Ergonomics and Safety. International Society for Occupational Ergonomics and Safety, Cincinnati, OH, pp. 205–208.

Murrell, K., 1965. Human Performance in Industry. Reinhold, New York.

Nakamura, K., Takahaski, H., Shima, S., Tanaka, M., 1996. Effects of immersion in tepid bath water on recovery from fatigue after submaximal exercise in man. Ergonomics 39 (2), 257–266.

Neville, K., Bisson, R., French, J., Boll, P., Storm, W., 1994. Subjective fatigue of C-141 aircrews during Operation Desert Storm. Human Factors 36 (2), 339–349.

Office of Technology Assessment, 1991. Biological Rhythms: Implications for the Worker (OTA-BA-463). US Govt. Printing Office, Washington, DC.

Page, E., 1964. Determining fatigue allowances. Industrial Management 1–3 and 14 February.

Paustenback, D., 1994. Occupational exposure limits, pharmacokinetics and unusual work schedules. In: Harris, R., Cralley, L.J., and Cralley, L.V., (Eds.), Patty's Industrial Hygiene and Toxicology, 3rd ed., vol. 3A, Ch. 7. Wiley, New York, pp. 222–348.

Pasztor, A., Gruley, B., 1995. FAA, in safety move, to seek to toughen limits on shifts of commercial pilots. Wall Street Journal December 14.

Phillips, E., 1994. FAA mandates rest for cabin crews. Aviation Week and Space Technology 29, 22nd August.

Pollack, A., Wood, E., 1949. Venous pressure in the saphenous vein at the ankle in man during exercise and change in posture. Journal of Applied Physiology 1, 649–662.

Price, E., 1990a. Calculating relaxation allowances for construction operatives – Part 1: metabolic cost. Applied Ergonomics 21 (4), 311–317.

Price, E., 1990b. Calculating relaxation allowances for construction operatives – Part 2: local muscle fatigue. Applied Ergonomics 21 (4), 318–324.

Ramsey, J., Halcomb, C., Mortagy, A., 1974. Self-determined work/rest cycles in hot environments. International Journal of Production Research 12 (5), 623–631.

Roehrs, T., Zwyghuize-Doorenbos, A., Timms, V., Zorick, F., Roth, T., 1989. Sleep extension, enhanced alertness and the sedating effects of ethanol. Pharmacology Biochemistry and Behavior 34, 321–324.

Schwing, R., 1990. Exposure-controlled highway fatality rates: temporal patterns compared to some explanatory variables. Alcohol, Drugs and Driving 5/4, 275–285.

Sjogaard, G., 1990. Exercise-induced muscle fatigue: the significance of potassium. Acta Physiological Scandinavica (Suppl.) 593, Copenhagen.

Smith, R., 1981. Boredom: a review. Human Factors 23 (3), 329–340.

Shiftwork Systems Inc., 1995. Circadian lighting for headquarters operations officers. NRC contract NRC-04-93-076, Shiftwork Systems Inc. 1 Kendell Square, Cambridge, MA.

Schonfelder, E., Knauth, P., 1993. A procedure to assess shift systems based on ergonomic criteria. Ergonomics 36 (1–3), 65–76.

Spaepen, A., Hermans, V., 1996. EMG in occupational settings: measurement of muscle load or muscle fatigue. In: Mital et al. (Ed.), Advances in Occupational Ergonomics and Safety. International Society for Occupational Ergonomics and Safety, Cincinnati, OH, pp. 557–561.

Summala, H., Mikkola, T., 1994. Fatal accidents among car and truck drivers: effects of fatigue, age and alcohol consumption. Human Factors 36 (2), 315–326.

Sundelin, G., Hagberg, M., 1989. The effect of different pause types on neck and shoulder EMG activity during VDU work. Ergonomics 32(5), 527–537.

Swanson, N., Sauter, S., 1993. The effects of exercise on the health and performance of data entry operators. In: Luczak, H., Cakir, A., Cakir, G. (Eds.), Work with Display Units. Elsevier, Amsterdam, pp. 288–291.

Thierry, H., Meijman, T., 1994. In: Triandis, H., Dunnett, M. and Hough, L. (Eds.), Handbook of Industrial and Organizational Psychology, 2nd ed., vol. 4. Consulting Psychologists Press, Palo Alto, p. 393.

Thompson, D., 1990. Effect of exercise breaks on musculoskeletal strain among data-entry operators: a case study. In: Sauters et al. (Ed.), Promoting Health and Productivity in Computerized Offices. Taylor & Francis, London.

Vollestad, N., Sejersted, O., 1988. Biochemical correlates of fatigue. European Journal of Applied Physiology 57, 336–347.

Van Dieen, J., Vrielink, H., 1996a. Toward an optimal sampling strategy of EMG and EMG spectral parameters, when using test contractions to monitor muscle fatigue. In: Mital et al. (Ed.), Advances in Occupational Ergonomics and Safety. International Society for Occupational Ergonomics and Safety, Cincinnati, OH, pp. 534–539.

Van Dieen, J., Verielink, H., 1996b. Evaluation of work-rest schedules with respect to postural workload in standing work. In: Mital et al. (Ed.), Advances in Occupational Ergonomics and Safety. International Society for Occupational Ergonomics and Safety, Cincinatti, OH, pp. 394–399.

Waersted, M., Westgaard, R., 1991. Shoulder muscle tension introduced by two VDT-based tasks of different complexity. Ergonomics 34 (2), 265–276.

Walsh, J., 1990. Using pharmacological aids to improve waking function and sleep while working at night. Work and Stress 4 (3), 237–243.

Walsh, J., Humm, T., Muehlback, M., Sugerman, J., Schweitzer, P., 1991. Sedative effects of alcohol at night. Journal of Studies on Alcohol 52 (6), 597–600.

Waters, T., Putz-Anderson, V., Garg, A., Fine, L., 1993. Revised NIOSH equation for the design and evaluation of manual lifting tasks. Ergonomics 36 (7), 749–776.

Waters, T., Putz-Anderson, V., Garg, A., 1994. Applications Manual for the Revised NIOSH Lifting Equation. DHHS (NIOSH) Publication No PB94-176930.

Wedderburn, A., Scholarios, D., 1993. Guidelines for shift-workers: trials and errors? Ergonomics 36 (1–3), 211–217.

Williams, H., 1973. Developing a table of relaxation allowances. Industrial Engineering 5 (12), 18–22.

Winkel, J., Jorgensen, K., 1986. Evaluation of foot swelling and lower-leg temperatures in relation to leg activity during long-term seated office work. Ergonomics 29 (2), 313–328.

Wood, D., Fisher, D., Andres, R., in press. Minimizing fatigue during repetitive jobs; optimal work-rest schedules. Human Factors.

Young, S., Hashemi, L., 1996. Fatigue and trucking accidents: two modes of accident causation. Proceedings of Human Factors and Ergonomic Society, pp. 952–956.

Yoshimura, I., Tomoda, Y., 1994. A study of fatigue estimation by integrated analysis of psychophysiological function-relating to continuous working time and rest pause for VDT work. Japanese Journal of Ergonomics 30 (2), 85–97.

Yoshimura, I., Tomoda, Y., 1995. A study of fatigue grade estimation for VDT work – Investigation of the rest pause. Japanese Journal of Ergonomics 31 (3), 215–223.

Zwahlen, H., Hartman, A., Rangarajulu, S., 1984. Effects of rest breaks in continuous VDT work on visual and musculoskeletal comfort/discomfort and on performance. In: Salvendy, G. (Ed.), Human Computer Interaction. Elsevier, Amsterdam, pp. 315–319.

Zwyghuizen, A., Roehrs, T., Lipschutz, L., Timms, V., Roth, T., 1990. Effects of caffeine on alertness. Psychopharmacology 100, 36–39.

Managing stress in the workplace:
Part I – Guidelines for the practitioner [*]

Ann M. Williamson

National Institute for Occupational Health and Safety (Worksafe Australia), P.O. Box 58, Sydney, Australia 2001

1. Audience

Stress is ubiquitous in all workplaces and can have negative effects on work performance and worker health and well-being. Strategies to help control or manage stress when it is a problem in the workplace include changing the work environment and assisting individual workers to control their responses to stress experiences.

This guideline is aimed therefore at practitioners in the workplace who are responsible for ensuring the healthy, safe and productive conduct of work. In these terms a practitioner could include both those who have direct responsibility for the worker's performance and well-being, and also those who are more concerned with the management and organisation of the work and the way it is done.

2. Context for use of this guideline

The guideline is intended to provide some assistance in reducing the negative consequences of stress in all workplaces. The guideline, as a consequence, is quite general. There are too many potential sources of stress in workplaces to cover them separately. Discussion about specific sources of stress will be limited to only a few examples. The real aim of the guideline is to provide an overview of the approaches that could be adopted to reduce the negative effects of stress from all sources in the workplace.

The guideline is aimed to be useful both in providing suggestions for prevention of stress problems and for suggested actions to identify that a problem exists and ways of solving it.

---

[*] The recommendations provided in this guide are based on numerous published and unpublished scientific studies and are intended to enhance worker safety and productivity. These recommendations are neither intended to replace existing standards, if any, nor should be treated as standards. Furthermore, this document should not be construed to represent institutional policy.

The following individuals participated in the discussion of the earlier version of this guide. Their suggestions (written or verbal) were incorporated by the authors in this version: Arne Aaras, *Norway*; Fred Aghazadeh, *USA*; Roland Andersson, *Sweden*; Jan Dul, *The Netherlands*; Jeffrey Fernandez, *USA*; Ingvar Holmér, *Sweden*; Matthias Jäger, *Germany*; Åsa Kilbom, *Sweden*; Anders Kjellberg, *Sweden*; Olli Korhonen, *Finland*; Helmut Krueger, *Switzerland*; Shrawan Kumar, *Canada*; Ulf Landström, *Sweden*; Tom Leamon, *USA*; Anil Mital, *USA*; Ruth Nielsen, *Denmark*; Jerry Ramsey, *USA*; Murray Sinclair, *UK*; Rolf Westgaard, *Norway*; Jørgen Winkel, *Sweden*; Pia Zätterström, *Sweden*. The guide was also reviewed in depth by several anonymous reviewers.

3. Glossary

A number of terms will be used in this guideline which need clarification. These are as follows:

Stress response: The individual's reaction to stressors which involves a natural increase in the state of arousal or preparedness for action that includes physiological, psychological and/or behavioural change which is uncontrolled or uncontrollable. It is also known as strain. The stress response can be short term or acute in nature and has main effects on performance. It can also occur over longer periods which can produce health problems as well.

Stressor: The factor or situation that causes a stress response. A number of factors exist as potential stressors, having positive effects on individuals such as motivation at some levels of exposure, but at other levels stressors will induce negative effects. For example, for most people there is an optimum work pace which they find challenging and motivating, at which they perform best. If the work pace is much higher or much lower than this, however, they will respond negatively and the level of performance will be poorer.

Stress: The process or relationship between stressors and the stress response

4. Data collection

Various forms of data can be used to obtain estimates of the sorts of stress indicators listed in Table 1 in establishing whether a stress problem exists. They include both objective and subjective indicators that stress is occurring and come from a number of sources:

(1) Personnel records – Collection of records of both certificated and uncertificated illnesses. In addition, the pattern of use of recreation leave can be informative, as taking small amounts of time off is a method commonly used to help deal with unpleasant working conditions.
(2) Medical records – Collection of information about the reasons for time off due to illness or injury. In conjunction with the absolute numbers of days off for medical problems,

Table 1
Indicators of stress problems in the work group and in individuals

Indicators of stress problems in the work group	Indicators of stress problems in individuals
Personnel: Minor or major illness especially stress-related unjustified absenteeism	*Physiological:* Increased blood pressure Increased heart rate Increased muscle tone Headaches Hypertension, etc.
Safety: High or increased incident/accident rates	*Psychological:* increased anxiety depression agression confusion
Performance: Poor or reduced work output	*Behavioural:*
Interpersonal relations: Dissatisfaction with work Industrial action	increased drinking increased smoking irritability obsessive concern with otherwise trivial issues poor work performance

analysis of the nature of the problems can provide insight into whether they may be stress-related.

(3) First aid books – Information should be collected about accidents and incidents, regardless of whether or not injury results, as the pattern of small and large incidents can provide insights into a stress problem.

(4) Compensation records – These can provide good evidence of stress problems, but only about problems that have been pre-existing for some time. The other sources of information are more useful for detecting stress problems at an earlier stage, which have the advantage of being considerably easier to overcome.

(5) Industrial relations records – Collection of this information will provide insights into the level of job dissatisfaction in the work place. Major industrial relations incidents like strikes are indicators of stress problems.

(6) Worker's perceptions of the work environment – This is collected usually by questionnaires and is important for assessing the perceived sources of stress in the work environment.

(7) Psychological state – This is usually collected by questionnaire and is aimed at assessing the level of psychological distress in individual workers and in the work group as a whole.

(8) Behavioural state – This is, again, usually collected by questionnaire and focuses on the way that individual workers cope with the stressors that they experience, particularly in relation to behavioural coping.

(9) Physiological/biochemical state – This can be assessed in a number of ways. Physiological measures include heart rate, blood pressure, and muscle tension. Biochemical measures include tests for catecholamines and corticosteroids (stress hormones).

5. Data analysis

As for many other issues in occupational health and safety, preventive action is the best management approach, but it is still important to keep up a constant vigil for any problems due to stress. There are four main reasons for analysis of data related to stress. These include the establishing of the existence of a problem, identifying the stressors, determining which workers are most vulnerable to stressors and deciding what solutions to pursue.

Different types of information are needed to answer each of these questions. All of the information required is potentially available in workplaces. Some is often collected as a matter of routine for other reasons, but other information needs to be collected for the purpose. The following section discusses the analysis of data for each question in turn.

5.1. Establishing that a problem exists

A decision can be made that a problem exists from the statistical types of data sources described above. To some extent, analysis of this type of information may only suggest that a problem exists, not necessarily that stress is the cause of the problem. There are many reasons for increased absenteeism or higher accident rates, for example, other than stress-related problems.

Closer examination of information from statistical data sources should provide more information about the origin of the problem. If the reasons can be linked to stress, this provides better evidence that increases in, or higher-than-acceptable, levels of these statistics indicate a stress problem. Linkage to stress can be inferred when, for example, the absence is for stress-related illness like ulcers, musculoskeletal disorders or tension headaches or the report of the accident or incident includes well-known stressors, like prolonged periods of night work or an argument in the workplace. Better definition of the problem can be achieved by using measures designed to assess the existence of the stress response in members of the work group. This includes the final three types of measures listed above, psychological, behavioural and physiological indicators of stress. The decision that there is a stress problem can be made much more firmly if this sort of data is available.

Many of the measures used to assess individual

responses to stress are subjective, that is they relate to aspects of the person that are unobservable. As a consequence these measures are often dismissed as unreliable. Such a view is unjustified, however, as subjective measures of stress play an important role in identifying that individuals have a problem. Most importantly, subjective or self-report measures provide information about the person's state that cannot be obtained by any means other than asking them. In addition to information about the nature of the person's current state, they can also inform about the quality of the individual's current state, whether positive or negative, and also about the individual's perceived need for help.

The combination of objective and subjective indicators of stress will provide the best evidence on which to identify stress problems in individuals. A combination of indicators like subjectively measured increased anxiety levels and objectively measured hypertension suggests very strongly that the individual is suffering problems due to stress.

Identifying that the level of stress in an organisation is enough to constitute a problem is often not a clear-cut decision. It needs to take into account whether the level of any of the stress indicators is sufficiently high or has deteriorated significantly to indicate a problem. A possible rule-of-thumb is that a problem might be said to exist in the work group, and action is required where there are two or more individuals experiencing stress-related symptoms. An alternative approach is to make the judgement when the proportion of workers experiencing stress-related symptoms is higher than the background incidence of chronic symptoms due to sources other than stress. Estimates of this proportion have been set at about five percent of the population, so an incidence of symptoms higher than five percent should be regarded as a problem.

For many of the indicators, especially measures of psychological, behavioural and physiological well-being, clinical ranges of normal are available to help establish that a problem exists. Ranges of normal are available for most physiological and biochemical measures and for most well-developed questionnaires so it is possible to interpret an individual's response in this light.

In identifying that stress is a problem for individual workers, we must be able to distinguish between the experience of potential stressors which can motivate the individual and enhance their performance, and the experience of stressors which have negative consequences for the individual. Only the latter would be regarded to be a problem, even though the same factor is causing the stress response. In addition we need to recognise that stress may be experienced as a short term phenomenon or as a long term problem for the person and the indicators of stress may vary depending on the length of time that the person has experienced it. For example, short term exposure to a stressor may produce increased muscle tension; however, if the stressor persists the response may develop into tension headaches.

In identifying stress problems in the workplace, it must be recognised that the workplace may not be the only or even the main source of the problem. Stress due to life events and experiences outside work can cause a response to stress that carries over into the workplace. When this occurs, it could be argued that it is not a workplace problem and therefore no action is necessary. The argument fails, however, since the worker's experience of stress can affect their work no matter what is the origin of the stress response. This means that the cause of the stress response is not relevant to identification that stress problems exist. It is relevant, however, to the solution of the problem.

5.2. The nature of the problem

Once it is established that a problem exists, the nature of the problem can be established quite readily through survey of the workforce using a workplace perception questionnaire. This will allow analysis of what are the most common problems in the view of the workforce. Surveys of perceptions of the workplace can also be effective in attempting to predict what the sources of stress might become.

In addition, questionnaires that look at how workers deal with workplace stressors will help analysis of the contribution of maladaptive coping behaviours to the problem. This last data source should not be interpreted on its own, however, as what may be interpreted as poor coping behaviours may be behaviours that workers adopt to protect themselves from poor or unreasonable work environments. In this case, they are a response to stress rather than a source of stress, although their ineffectiveness may lead to increases in stress levels.

5.3. Identifying workers who are most vulnerable

Analysis of the information from the statistical sources such as absenteeism, sickness and accident records is useful for identifying individuals who have already experienced significant consequences of stress. This sort of analysis is not useful, however, for preventing serious problems in these individuals, although it does indicate that there may be problems for others in the work setting.

For preventive purposes, it is more productive to determine who is most likely to experience health and safety problems as a result of their exposure to stressors. Increased vulnerability can be picked up by investigating which workers have been experiencing adverse stress responses, but not necessarily adverse health and safety consequences. The most direct way of analysing this is through physiological and biochemical and psychological state indicators of the stress response.

Increased vulnerability can also be inferred where workers are currently using poor or maladaptive methods to cope with stressors. These workers may also be more likely to progress to health and safety problems if coping methods are not enhanced. It is detected best through analysis of questionnaires on behavioural adaptation to stressors.

5.4. Deciding what interventions / solutions to pursue

In deciding what to do about a stress problem, it is important to get an overview of the problem by looking at all the measures that have been taken in conjunction with one another. It is essential to focus on both aspects of the problem, the workplace and the worker, as they are intrinsically related. This should allow identification of both what needs to be changed in the workplace and whether and how workers themselves may be assisted in coping with problems they encounter in the workplace. Particular attention should be paid to measures of worker perceptions of the sources of stress in the workplace and to measures of how workers are coping with stressors in the workplace.

To some extent, using these sources of information makes the decision about targets for intervention quite easy. It is possible to identify the target as simply the one or more aspects of the work setting that are identified by most workers as the source of the problem. Similarly, the pattern of maladaptive coping behaviours across workers can be used to establish what type of intervention would be suitable to assist most workers in improving their coping skills.

Where the decision becomes more complicated, however, is in assisting individual workers and smaller subgroups. It must be realised that an intervention that suits most workers may not suit some individuals, and, in fact may increase their stress levels. The decision about what and how to intervene must look at who may be negatively affected and at alternative solutions that are available for these individuals. This also points out the need for any intervention to be evaluated at least 6 to 12 months after its implementation.

6. Solutions

Solutions to the problem of stress can be put into two main categories, interventions that focus on changing the work environment and work organisation, and those focussing on assisting and improving the individual worker's coping skills. These will be discussed separately in the sections below.

6.1. Changing the wider work environment

The number of factors in the work environment and work organisation that might emerge as stressors is almost limitless, but they can be grouped as follows:

(1) stressors due to aspects of the physical work environment like the level of noise, lighting, heat and other physical stressors,
(2) stressors due to the nature of the job like pace of work, level of complexity, hours of work and work load,
(3) stressors due to the person's role in the organisation like role conflict and ambiguity and level of responsibility,
(4) stressors due to poor relationships within the work group including poor delegation,
(5) stressors due to career structures in the organisation like under or over promotion,
(6) stressors due to the structure and climate of the organisation including the organisation of the flow of work and information and little worker participation in decision-making,
(7) stressors due to the relationship between work and home or social life.

The most suitable type of intervention depends on the nature of the stressor identified as a problem, however almost all interventions that deal with problems in the wider work environment are related to management of the workplace. In short, good management of stressors from the work environment can be achieved through good people management. Typical strategies may involve improving communication channels in the work setting, allowing workers a level of control or decision-making power over their work, or changing the organisation of working hours to suit physiological and social needs of the workforce.

It is not possible in this guide to cover each possible stressor in detail. Many potential stressors will be dealt with as occupational hazards in their own right (e.g.: work-rest allowances, noise in the office, exposure to hazardous chemicals, etc.). Nevertheless, there are some general principles that should be followed in undertaking any intervention in this area. These are:

(1) The form of the intervention should be developed in consultation with the work group.
(2) The intervention should be tried in the workplace for a long enough period for workers to become familiar with it (not likely to be less than 3 to 6 months minimum).
(3) The intervention should be evaluated at the end of the trial period, preferably on an ongoing basis using the measures employed to demonstrate that a problem existed so that any change can be detected.

6.2. Improving workers' ability to manage potential stressors

There are a number of types of interventions that aim at helping the individual worker's ability to manage potential stressors. The first is ensuring that the workers have been selected for the right job and that they have adequate work skills to cope with the demands of the task. This includes vocational training of any type that is related to the job, including advanced vocational courses and such training as dealing with customers, safety training for hazards encountered in the particular job and the use of computers. This type of intervention is usually seen to be part of the human resources function in organisations, when performed successfully, it can significantly reduce the exposure to workplace stressors.

The second type of individual-based intervention focusses on attempts to change the individual's perception of an experience as stressful. These include methods to help the worker develop alternative ways of thinking about a situation that they may find stressful, such as rational emotive therapy and other cognitive therapies. These techniques are usually conducted on a one-to-one basis by trained professionals in a clinical setting.

The third type of intervention aims at improving the individual's ability to cope with stressors by controlling their stress responses. This is the largest group of interventions for individuals and includes most of the commonly used techniques. Coping interventions include techniques that concentrate on the physiological responses to stress, like biofeedback, psychological responses like meditation and a very wide range of methods for changing the behavioural response to stressors.

Many of the methods for dealing with the behavioural response would also be found in management training courses and could be regarded as good management skills. These include time management, improving interpersonal skills like assertiveness, leadership and delegation and enhancing communication skills.

The coping techniques differ in the way they are applied in the workplace. Physiological coping methods are conducted on an individual basis, usually in clinical settings. Psychological and behavioural techniques on the other hand are normally conducted in groups, often at the worksite.

The fourth group of techniques aim to provide individuals with information on maintaining and improving their lifestyle. The premise behind the use of these techniques for stress management is that healthy, fit workers are more able to withstand work stressors than workers who have unhealthy lifestyles. Included in this group are education programmes on nutrition, physical fitness and other facets of healthy lifestyle like smoking cessation. This type of technique is usually conducted in groups in the workplace and is often conducted as part of a general health promotion campaign.

Individual stress management is usually run though the health service attached to the workplace, through Employee Assistance Programmes in the workplace and/or by independent consultants who specialise in these sorts of interventions.

6.3. Choosing the best intervention strategies

Many of the individual stress management techniques are useful for managing stress in general, not just that generated in the workplace, and therefore have added benefits for the worker. This is especially so for the coping techniques and for the lifestyle change techniques. Similar to other methods for improving the individual's health or quality of life, all individual stress management techniques require a degree of motivation from the person to be successful. This is probably the most common reason for failure for any of the individual management techniques described above. In addition, almost all individual stress management operates at a palliative level. Apart from the human resources or personnel techniques and to a certain extent the management skills methods, all other techniques attempt to help the person to deal with stressors that already exist. They do not change the stressor itself, only how the person sees the potential stressor. Consequently, for prevention, organisational and environmental strategies are needed. The difficulty with organisational stress management is that it requires commitment and motivation from workplace decision-makers for it to be implemented, but is likely to have long-lasting effects.

On balance, for effective stress management, both types of methods are needed, organisational techniques to reduce or even remove the stressors that affect most workers in the workplace and individual stress management to assist workers to cope with situations that they find stressful.

Disclaimer

The views expressed in this paper are those of the author and do not necessarily reflect those of the National Institute of Occupational Health and Safety (Worksafe Australia).

Reference

Williamson, A.M., 1994. Managing stress in the workplace: Part II – The scientific basis (knowledge base) for the guide. International Journal of Industrial Ergonomics 14(1-2): 171–196 (this issue).

Managing stress in the workplace: Part II – The scientific basis (knowledge base) for the guide

Ann M. Williamson

National Institute for Occupational Health and Safety (Worksafe Australia), P.O. Box 58, Sydney, Australia 2001

1. Problem description

The stress response is the body's natural reaction to situations that are perceived to require more than the normal or comfortable amount of effort to achieve required performance. Situations that are unfamiliar to the individual or that challenge or threaten them in some way will produce a stress response. One of the pioneers of the area of stress, Hans Selye, described stress in terms of the nonspecific result of any demand made on the body (Selye, 1936, 1982). This could include situations that are well-recognised to produce a stress response in individuals not used to them, like having to speak in public, taking on a new or difficult task, or undergoing examination or inspection. In all of these situations the physiological, psychological and behavioural changes that form the stress response have beneficial effects on performance. Biochemical and other changes help to, amongst other things, increase the level of energy and muscle tension of the individual and improve their ability to concentrate on the task at hand, thereby helping to enhance their performance. A certain level of stress that brings about these sorts of changes is a necessary aspect of maintaining productivity.

A very large amount has been written about when stress occurs, and many theories and models have been developed (Eichler et al., 1986; Kahn and Byosiere, 1992). This is an important issue because the model which describes when stress occurs will provide the framework for all other work on the problem, including how it is prevented and managed. The earlier models of stress emphasised the biological processes involved and described stress as occurring when the body state changed in response to external demands (Cannon, 1935; Selye, 1936, 1982). Later models emphasised the interaction between the environment or situation (stressors) and the person, for example the Person-Environment Fit model (French et al., 1982) or the job demands and control model (Karasek, 1979). More recent models have been based on the same person–environment relationship but have focussed on different features. For example some models have emphasised the transactional or process-oriented nature of the relationship between the person and their environment (Lazarus and Folkman, 1984; Cox, 1978), and others have included the role of moderating or mediating factors like age, education and the amount of social support in the person–environment relationship in addition to the other acknowledged factors (Ivancevich and Matteson, 1980; Kahn and Byosiere, 1992).

The often-expressed view in the stress literature is that there is little agreement on the concept of stress (Bailey and Bhagat, 1987; Kasl, 1983) and, partly as a consequence, too much emphasis has been placed on the person part of the relationship and not enough on the environmental causes (Baker, 1985). On examination, however, it can be seen that the basic structure of most stress models developed over the last 15 years is remarkably consistent, most incorporating a relationship between the person and the environment which can vary depending on a number of factors that will change the individual's perception of the situation as stressful. While the concept of stress still needs further resolution, enough is known currently to provide a degree of guidance for controlling and managing stress in the workplace.

It is generally agreed that stress becomes a problem when the response to it is uncontrolled or uncontrollable (Quick et al., Quick, 1987; Fisher, 1989). That is, where the changes that are part of the stress response become so pronounced that they hamper rather than facilitate performance and where the individual does not have the resources to reduce these uncontrolled changes. For example the person who is asked to address their work group for the first time and finds that their muscle tension is so great that their voice quivers and quakes will find it difficult to produce good or even adequate performance. Their problem is made more difficult if they are not able to bring their level of arousal down to a manageable level.

There are a number of difficulties that may be encountered in identifying that a stress problem exists.

1.1. Establishing when stress exerts negative effects

One of the most commonly expressed problems in the area is that the types of situation or experiences that can cause the negative outcomes of a stress response can also cause beneficial responses in different people or even in the same person when experienced under different conditions or at different levels of exposure. This has led a number of authors to distinguish "good" or beneficial stress from "bad" or negative stress (Selye, 1982; Quick et al., 1987). This, however can make the identification of problems due to stress somewhat difficult. For most hazards, such as toxic chemicals or other physical hazards, problems only occur during exposure to high levels and there is really no level that could be regarded as beneficial. There are some hazards, however, which are similar to stress, for example noise can have beneficial effects of stimulating arousal for workers doing monotonous work, but will also have negative effects on performance and on worker health at higher levels of exposure. This issue will be discussed in more detail below.

1.2. The subjective nature of stress

Problems arise because of the subjective nature of stress and the question is often raised whether we can rely only on subjective measures to determine whether a person is experiencing stress. Fleming and Baum (1987) put a case for the role of subjective assessment in identifying stress problems. They argue that if an individual says he or she is experiencing stress, it must be accepted that it is so. This difficulty can be overcome to a large extent by combining subjective assessments with objective symptoms which suggest lack of control over stress responses, such as increased drinking, increased aggression or irritability or poor job performance (Kasl, 1986).

1.3. Establishing the relative importance of acute and chronic stressors

Problems due to the issue of the relative importance of acute and chronic stressors have not been resolved although it is argued that the effects differ (Pratt and Barling, 1988). Since both forms of stressors can constitute problems, for the purposes of this review the distinction will not be made.

1.4. Distinguishing work and nonwork causes of stress

The issue of separation of the causes of stress between work and nonwork factors needs to be addressed so that stress problems due to work can be identified and attempts can be made to control the workplace sources. It must be recognised, however, that stressed workers, no matter what the cause, present a problem in the workplace which needs to be dealt with.

2. Scope of the problem

2.1. Workers' compensation statistics

Estimates of the size of the problem can be gained most easily by examining workers' compensation figures. This source is likely to underestimate the problem, however, due to difficulties in coding such as the linking of specific illnesses with stress and also since compensation is usually only for longer illnesses.

In the USA, a study by the National Council on Compensation Insurance (1985) showed that claims for "gradual mental stress" accounted for 11% of all occupational disease claims. Furthermore, the costs for such claims were increasing while those for all other disabling injuries were decreasing. In 1980, Ivancevitch and Matteson reported that the cost associated with stress in the US was estimated to be between $75 and $90 billion annually. Statistics such as these have led to the inclusion of psychological disorders in the list of the ten leading work-related diseases and injuries suggested by US-NIOSH (Millar, 1984) and led the Director of NIOSH to assert that "stress-related conditions may be among the most important problems, occupational or nonoccupational of the 1990s and beyond" (Millar, 1990).

Similar findings are evident from other countries. In Australia, recent reports of stress claims for compensation (coded as mental distress) have showed marked increases in the rates for stress-related problems compared to those for other disabilities. For example, in a number of states, the workers' compensation claims for stress nearly doubled over three years (Queensland Workers' Compensation Board, 1990/91; Victorian Accident Compensation Commission, 1989/1990).

2.2. Work-related measures

The scope of the problem can also be seen in effects on productivity and safety in addition to the effects of time-off work due to illness. There have been many reports that work performance can suffer due to work stress (Corlett and Richardson, 1981; Hockey, 1983), but it is much harder to quantify this effect than the effects of stress on health. It is generally acknowledged that productivity costs are much greater than health costs; however, it is often overlooked that there are two sources of effects of poor productivity. One is the indirect effect of poor health and the other is poor work performance resulting from the direct effect of stressors in the work environment. Some attempts have been made to quantify the former source. Matteson and Ivancevitch (1987) included the costs of absenteeism, extra employees to do the job of stress-affected employees, turnover and sabotage and estimated that these indirect aspects of work stress cost the US economy more than $300 billion per year.

The impact of poor performance produced by stress is much harder to quantify outside individual workplaces. There are many studies of the effects of poor work environments on work performance. For example, studies show a positive relationship between high workload and error rates due to pressures on the individual's capacity to perform (Kahneman, 1973; Wickens 1984). Many studies of work organisation have also shown that work performance can be improved by the way work is arranged (Gardell, 1987).

This brief discussion shows some of the evidence that stress can result in negative effects on health, safety and work performance. It shows that the problems due to stress have been increasing significantly over recent years and that the proportionately greater increases in costs associated with these problems present cause for concern in the workplace.

3. Background

There is a very large literature on stress in the workplace. In the review which follows, no attempt was made to be comprehensive. The aim of this review is to provide enough background detail to support the Guideline. With this in mind, the issues addressed in this section included:
- The evidence that stress can be regarded as a problem.

- What causes stress in most people?
- Why don't all people experience stress when faced with the same stressors?
- What can be done about both the workplace causes and vulnerability in individuals?

3.1. Evidence that stress constitutes a problem

Most of our view of stress as a problem comes from research on its health effects (Quick et al., 1987; Marmot and Madge, 1987). The research in this area includes effects of stress on physical, mental and social well-being in addition to disease or injury (Levi, 1987). A number of studies have shown a positive relationship between high stress levels and the incidence of physical diseases like cardiovascular problems (Cobb and Rose, 1973; Kawakami et al., 1989; Murphy, 1991) and gastrointestinal trouble (Dunn and Cobb, 1962; Cobb and Rose, 1973). A similar relationship has also been found for psychological and psychiatric problems (French and Caplan, 1972; Cooper and Melihuish, 1980; Keenan and Newton, 1987; Phelan et al., 1991) and some evidence for the same relationship with infectious diseases (Cohen and Williamson, 1991).

Stress can also be seen as a problem in terms of its effects on health-related behaviours. Work on absenteeism, for example, suggests that stress in the form of social, economic and company policy pressures plays an important role in determining whether or not a worker will decide to attend work or not (Wood, 1986). Studies have demonstrated that high stress levels can predict subsequent absence from work (Gupta and Beehr, 1979; Jamal, 1984), suggesting that stress is a reason for absenteeism.

A recent survey by Manning and Osland (1989) has challenged this idea. This survey showed small but significant correlations between stress measures and prior work absence, but not subsequent absence, which the authors hypothesised was due to absenteeism causing stress. Some support for this idea comes from the findings of a survey of personnel practices reported by the US Bureau of National Affairs (1981), which showed that about 50 percent of worker absences could be avoided by the companies dealing with the physical and emotional needs of their employees (quoted in Seamonds, 1986)

Again, however, it seems clear that there is a relationship between work absences and stress, but the nature of the relationship is not clear.

Other health-related behaviours that have been implicated with stress include alcohol consumption (Gorman, 1988), cigarette smoking and drug abuse (Quick et al., 1987).

Despite the large literature on the health effects of stress, there is still a degree of dispute about their interpretation. Kasl's widely quoted view is that research has failed to provide the explanation for the relationship between stress and illness (Kasl, 1983). Leventhal and Tomarken (1987), however, detail a number of specific difficulties with much of the research, including the lack of random assignment to high and low stress conditions, the lack of a model or theory on which to base hypotheses and therefore selection of measures, lack of control over the measurement of the different levels of response to stress and the lack of longitudinal research designs.

There is no doubt that these problems need to be addressed before we can understand the nature of the relationship between high stress levels and ill health. Nevertheless, on the weight of evidence alone, including animal studies, it is possible to conclude that a relationship does exist (Sharit and Salvendy, 1982). This view is strengthened by the findings that changes have also been detected in high stress individuals that may mediate some of the health effects. For example, Siegrist and Klein (1990) found increased cardiovascular reactivity in workers experiencing high levels of stress, indicating, the authors argued, a possible predictor of cardiovascular risk. Endresen et al., (1991) found immunological change in individuals experiencing stress, changes which may mediate the body's responses to infection when experiencing stress.

3.2. Measures to detect the presence of a stress problem

Clearly the appropriate measures for assessing whether or not a stress problem exists, are those which are most likely to be affected by stress.

Table 1
Measures to detect the presence of a stress problem

Type of information	Measures
Absences/Turnover	Sickness leave, Recreation leave – with or without certificates – especially for very long periods or for many short periods – for stress-related symptoms Attrition rates Industrial disputes
Accidents/Incidents	Compensation statistics, first aid statistics – especially where there is a pattern of incidents over time – especially where there are clusters in particular areas
Productivity/Performance	Product outputs, product quality estimates, amount of down time, employee participation – especially where level is poor or highly variable

From the discussion above, it can be seen that these will include collection of information on absences from work of all types, information on accidents and information on productivity and work performance. Many work places collect these sources of information in some form as a matter of course (see Table 1); however, the usefulness of this information for detecting that a stress problem exists depends on how the information is collected.

Absences

Absence information is usually collected in the form of the number of hours absent per unit time; however, alternative ways of measuring absence may reveal more about stress influences. Folger and Belew (1985) suggested that one-day absences are the best conceptual measure for indicating stress effects as they are usually voluntary. Manning and Osland (1989) suggest that longer absences can also indicate stress effects since they occur due to actual illnesses which may be stress-induced. In this case, it is important to examine the reason for the absence in order to determine whether the illness could be regarded as stress-related.

Accidents

Accident information is usually collected as first aid records and compensation statistics. Information about stress can be gained from these sources by examining the relative distribution of accidents under different conditions which might be predicted to cause stress. For example, a study of pilots and crew revealed that those who had been involved in aircraft accidents had been exposed to more stressful life events than flight crew who had not been involved in accidents (Alkov, 1981)

Performance and productivity

Productivity and performance effects can be measured in a range of ways depending on the type of industry. Currently these factors are the source of a great deal of attention, due to innovations like the interest in management for quality (Samson, 1991), industry "Best Practice" (Feigenbaum, 1983) and the introduction of new organisational forms (ILO, 1982). As for accident statistics the judgement that a work productivity problem exists is usually made by comparing the measures under different situations or conditions. For example, Folkard and Monk (1985) reviewed research on the relationship between poor work performance (near misses) and time of day and showed strong circadian influences on performance. In this way the distribution of performance effects on accidents or incidents can also help to identify the source of stress.

Establishing a level of concern

For all of these measures, it is necessary to establish the level at which they indicate a problem, usually by comparing them to some standard. It is sometimes possible to take a time-based or longitudinal approach by looking at changes in the measures over time, preferably in relation to changes in exposure to potential stressor(s). An alternative, but weaker approach is to compare the measures for groups that are distinguished by their exposure to a specific stressor, where one group is exposed to high levels of stress and the other to low levels or no exposure.

The most common approach is probably to compare these measures against some industry standard, like the compensation record for similar industries. The technique of "benchmarking" or using industry-specific information to set goals for productivity and quality standards (Camp, 1989) is very relevant to this discussion. One difficulty with assessing against a standard, though, is achieving a common basis for comparison. Often the measurement standards and criteria differ between workplaces, industries and national databases, making comparisons difficult. In addition, it is necessary to examine the pattern of difference or change in the measures in order to determine that the source of the problem is stress. Information about the type of illness, the timing of the accident or the circumstances in which poor performance occurs will help in this respect.

In reality, however, these statistical types of measures may be too gross and insensitive to be used in single workplaces to differentiate a problem due to a particular hazard (Kalimo, 1985). When this is the case, the focus must shift to measures that indicate adverse stress responses in individual workers. The nature of these measures will be discussed in a later section; however, they are also important for determining that a stress problem exists in a workplace.

Unfortunately, there has been surprisingly little discussion in the literature about what should be regarded as the level of concern for stress symptoms in the work group. A large number of studies have established that stress symptoms are common in industry (Fletcher, 1988), yet there have been very few attempts to determine their origin. Surveys of stress symptoms will pick up individuals who have chronic constitutional conditions unrelated to stress as well as those who are responding to stress. It is consequently very difficult to estimate just how large is the contribution of stress.

One solution to this difficulty is to take a pragmatic approach and acknowledge that a clustering of two or more individuals with symptoms in a work group, suggests that a problem exists. The decision can be strengthened if the cluster occurs in the presence of a known stressor, or where the individuals have little previous history of symptoms. This approach assumes nothing about the distribution of individuals with constitutional symptoms in the work group, but responds only to the existence of the problem.

Tinning and Spry (1981) attempted to make some quantitative estimates of when a stress problem could be said to exist. Based on their own survey and other surveys of the distribution of symptoms, they estimated that a fairly constant approximately five percent of the population have chronic constitutional symptoms unrelated to stress. At any time, they argue, a further five percent will be added to the symptom group due to stress-related causes. In their view it is this second group who will be amenable to stress reduction and control strategies. From this analysis, it can be argued that a level of concern exists where more than five percent of the work group show stress symptoms.

4. What causes stress in the workplace? What are the stressors?

The most common method of examining the sources of stress in the work place is by survey. Studies have focussed on groups like white collar occupations (Cooper and Marshall, 1976; Turnage and Spielberger, 1991), blue collar occupations (Poulton, 1978; Cooper and Smith, 1985) and occupations in which VDUs are used (Kahn and Cooper, 1986; Amick and Smith, 1992; Pickett and Lees, 1991). Specific occupations that have been covered include medical practitioners and specialists (Groenewegen and Hutten, 1991; Arnetz, 1991; Cooper et al., 1989) nurses (McLaney and Hurrell, 1988; Revicki and May, 1989; Estryn-Behar et al., 1990) teachers (Innes and Kitto, 1989; Bain and Conn, 1990) police (Brown and Campbell, 1990), prison officers (Kalimo, 1980), drivers (Carrere et al., 1991; Raggatt, 1991) and farmers and other farm workers (Giesen et al., 1989; Ellis and Gordon, 1991). In addition, there have been a few studies of such diverse occupations as financial market dealers (Kahn and Cooper, 1986) and hazardous waste workers (Fiedler et al., 1989).

Most of these studies demonstrated that workers doing these occupations are likely to experience significant stressors at work, but the studies are difficult to compare since the stressors studied differed greatly and even when the same stressors were studied, measurement methods often differed. Apart from demonstrating that stress should be considered as a factor in all of these jobs, these studies do not contribute a great deal to our knowledge of what the general sources of stress are likely to be.

More useful are studies that attempt to look at the effects of specific potential stressors. From these studies it has been established that a range of experiences and factors encountered in the workplace can be stressors. For example, unemployment has been shown to be related to increased psychophysiological symptoms, lower self esteem and increased hostility (Stokes and Cochrane, 1984) and job insecurity to decreased personal well-being and deterioration of work performance and attitudes (Roskies and Louis-Guerin, 1990). A number of studies have also shown that exposure to a traumatic experience at work can be a potent stressor which has long-lasting effects on individuals who are exposed to it (McGuire, 1990).

Another factor that has been studied extensively is stress due to the role in the organisation, specifically role conflict (where the demands of the job are not in concert with what the worker believes they should be doing) and role ambiguity (where the worker does not have enough information about his/her work role). Since the original study by Kahn et al. (1964) showing negative effects of role conflict and role ambiguity, there has been a considerable amount of research on the effects of these factors. Jackson and Schuler (1985) performed a meta-analysis on this research and showed significant negative correlations between role conflict and ambiguity and a number of stress-related responses like job satisfaction, participation, commitment and involvement.

The role stress plays in the relationship between poor physical aspects of the work environment and work performance is less clear. While there is abundant evidence that work performance is affected by such conditions as poor lighting, noise, heat, cold, glare, etc., stress is not commonly included in the explanation of the mechanism of the effect. Stress can, however, play a secondary role in this relationship (Poulton, 1978; Levi, 1987). For example, poor lighting is well-recognised to produce poor work performance, because it hampers the worker's ability to gain enough information from the work environment to complete the task successfully. This then requires the individual to increase the amount of effort required to do the job, thereby increasing their level of stress and affecting work performance (Kahneman, 1973; Eysenck, 1983).

A number of authors have attempted to develop lists of workplace risk factors for stress. One of the earliest was developed by Cooper and Marshall (1976), who grouped the sources of managerial job stress into six main groups shown in Table 2.

The general models of stress in the workplace can also provide a framework for making predictions about which workplace factors are risk factors for stress. On the basis of the demand/control model (Karasek et al., 1981), for example, the most likely work stressors are factors that create or increase demand (or perceived demand) on

Table 2
Sources of managerial job stress (from Cooper and Marshall, 1979)

1. *Factors intrinsic to the job*
   e.g.: working conditions such as those that demand fast paced work or high levels of physical energy as well as jobs which have high work load.

2. *Role in the organisation*
   e.g.: role ambiguity, role conflict and responsibility for staff or equipment.

3. *Relationships at work*
   e.g.: interpersonal relationships between workers and their supervisors, management and subordinates and colleagues in the work group.

4. *Career development*
   e.g.: job insecurity and status incongruity, and under or over-promotion.

5. *Organisational structure and climate*
   e.g.: little worker participation in decision making and restrictions on behaviour.

6. *Extra-organisational sources of stress*
   e.g.: competition between the demands of work and home or social life.

the individual worker. Similarly, the person–environment fit model of French et al. (1982) would predict that the factors that create high stress risk in the workplace are those involving high demands on the individual or which offer little to the individual.

Most of these frameworks for thinking about the workplace stressors are not founded on a strong theoretical basis (Baker, 1985). In fact at this point in the development of the literature on stress, there is no overall theoretical framework for describing potential workplace stressors. In fact one may never be developed because of the huge array of possible risk factors for stress that may be encountered at work. By this reasoning, it should be necessary to evaluate each workplace separately to establish what are the risk factors for stress in that work setting. It is, however, very costly in time and effort.

Frese and Zapf (1988) suggested that a group of stressors exist which could be regarded as objective stressors. They use the term objective to mean not being related to one specific individual's perception. In this sense, an objective stressor is one which would cause the average worker to experience a stress response. The example the authors use is of a job requiring the average worker to react very quickly to a danger signal which would be stressful for the majority of workers. The concept of objective stressors is important as they could provide the foundations of a framework for conceptualising work stressors. This has direct implications for how stress is measured and managed.

It seems reasonable to argue that some idea of what are objective stressors may be gained by reviewing the stressors that show up as stress risk factors across a number of studies. Sauter et al., (1990) used this approach in their proposed National Strategy for the Prevention of Work-Related Psychological Disorders. The risk factors that qualified were workload and work pace, the temporal scheduling of work, role stressors, career security factors, interpersonal relations with other members of the work group including supervisors, colleagues and subordinates, job content or the nature of the task and intervening variables or those situation and personal variables that exist outside work, but affect the worker's experience with work.

At the present time, there are a number of problems with attempting to decide what are the workplace stressors. One major problem is that most studies are correlational and therefore reveal nothing about the direction of causation. Broadbent (1985) tackled this issue, however, and argued that work factors cause symptoms and not vice versa. This view is also held by others (Kelloway and Barling, 1991). Other problems that also need to be considered are the inability for any

Table 3
Measurement of work stressors

Type of stressor	Measurement method	Source
General work stressors	Work environment scale	Moos, 1974
	Job diagnostic survey	Hackman et al., 1975
	Stress diagnostic survey	Ivancevitch et al., 1980
	Job content questionnaire	Karasek, 1985
	Occupational stress questionnaire	Elo et al., 1992
	Occupational stress inventory	Cooper et al., 1988
Specific work stressors	Teacher stress survey	Kyriacou et al., 1978
	Prison personnel stress	Kalimo, 1980
Specific stressors	Role conflict/ambiguity	Rizzo, 1970
	Organisational change inventory	Sarason et al., 1979
	Assessment of daily experiences	Stone et al., 1982
	Organisational conflict	Rahim, 1983

single study to assess all possible sources of stress, as discussed earlier, and the problem that sources of stress are dynamic in most workplaces and may be changing constantly (Landy, 1992).

Despite these problems, it is striking that there is so much overlap between the studies and reviews on what the risk factors for stress might be. Considering the huge number of factors that could be stressors this may be surprising. It could be explained in terms of simply limited scope of measurement, but it is also likely that these are the major stressors in many work settings. If the latter is the case, it may yet be possible to develop an overall framework for stressor identification.

4.1. Measurement of stressors

Measurement of stressors depends again on making predictions about what the stressors might be. Measures for general workplace stressors exist as well as ones that look at specific stressors or groups of stressors (see Table 3). Instruments that attempt to assess general work stressors include the Work Environment Scale (Moos, 1974), the Stress Diagnostic Survey (Ivancevitch and Matteson, 1980), the Job Diagnostic Survey (Hackman and Oldman, 1975) and the Job Content Questionnaire (Karasek, 1985). In addition, there are a number of general stressor measures that have been developed for specific occupations. For example, Kyriacou and Sutcliffe (1978) developed an instrument for assessing the sources of stress in teachers.

Instruments that focus on specific groups of stressors include a scale designed to measure role conflict and role ambiguity by Rizzo (1970) an instrument to measure organisational conflict (Rahim, 1983) and instruments that assess acute stressors such as the Organisational Change Inventory (Sarason and Johnson, 1979) and the group of life events scales (e.g.: Assessment of Daily Experiences (Stone and Neale, 1982).

One of the criticisms of many measures of the sources of stress, is that they only ask the respondent to report the existence of the factor in the workplace, not whether this constitutes a problem for them (Dohrenwend and Stout, 1985; Shirom, 1988). Reporting the existence of a factor in the workplace does require some limited subjective judgement, but it does not require the respondent to consider how they actually feel about the factor being in the workplace. Consequently these measures may not actually reflect stressors, only that the characteristics are present. Dewe (1991) argues that this approach denies the transactional nature of stress and also that the individual's appraisal of the characteristic plays a role. This has led Dewe and others (Bailey and Bhagat, 1987; Frese and Zapf, 1988) to argue that we need new approaches to the assessment of work stressors. Williamson et al., (1993) made an attempt to take this criticism into account by modifying the Work Environment Scale to include a response for each item requiring the respondent to report whether the characteristic *should* be present (yes, no or don't care) in addition to the usual requirement to indicate whether it existed at all. This approach reveals which of the characteristics respondents regard as a problem and which they see as stressors.

A second difficulty with the measurement of sources of stress is that it is done almost invariably by subjective methods (usually questionnaires or interviews). Fleming and Baum (1987) argue that the easiest and most direct way of finding out whether people are stressed is to ask them. A similar argument could be advanced for measures of what people think causes stress. Nevertheless, a number of authors have criticised the purely subjective approach to stress measures, arguing that it weakens and limits the possible conclusions that can be made about stress (Kasl, 1983; Contrada and Krantz, 1987; Leventhal and Tomarken, 1987; Frese and Zapf, 1988). The main difficulties raised about subjective measures are that they may be affected by biases in the identification of events and factors (i.e., that they exist) and in their evaluation (i.e., that they are a problem).

Objective measures of workplace stressors do exist. For example, Carrere et al. (1991) used an unobtrusive behavioural index of stress and Murphy (1991) used the Position Analysis Questionnaire (PAQ), a copyrighted job analysis inventory, to gain an objective measure of job characteristics. Bailey and Bhagat (1987) advocate the use of archival reports like absenteeism as non-reactive or objective measures. It should be recognised, however, as Frese and Zapf point out, that many so-called objective measures, particularly those relying on observer ratings, still involve subjective elements but due to the observer rather than the worker.

A third criticism of the work stress measures currently in use is that they do not cover all possible stressors. As discussed above, the array of stressors is potentially huge, so that this criticism of current work stressors may be simply a reality. It is notable, though, that instruments like the Job Content Questionnaire are obviously viewed by their designers as dynamic since new sections have been added to cover such areas as the affect of new technology (Turner and Karasek, 1984). As a general rule, work stress measures should be as broad as possible, especially if they are to be used for screening for stressors and should always include measures of objective stressors.

In the final analysis, there is agreement in the literature that many of the problems discussed in this section will be overcome by the use of a multiple method approach to assessing stressors (Folger and Belew, 1985; Kasl, 1983; Fleming and Baum, 1987). In this way objective and subjective methods can be used to strengthen the interpretation of potential stress problems.

5. The stress response

Not everyone exposed to stress will respond in exactly the same way to the same stressors. Even so, we do know that there are some basic changes that occur as part of the stress response. The work of Selye (Selye, 1936, 1982) demonstrated that the natural response of any organism to a demanding or threatening situation or experience (stressor) was a predictable series of nonspecific changes in physiological, psychological and behavioural functions which he named the General Adaptation Syndrome. This syndrome was shown to be a temporally-arranged sequence of changes called the alarm, resistance and exhaustion stages respectively, which Selye demonstrated could result ultimately in illness.

The initial response to perceived stressors involves all body systems. Overall this initial response acts to heighten the level of arousal and mobilise the person's resources for physical action. This occurs by increasing heart rate, blood pressure, muscle tension and general physical and mental alertness. Mital and Mital (1984), for example, showed that exposure to a stressor for a short period, such as an examination, produced significant changes in blood pressure, pulse rate and oral body temperature.

If the stressor persists, the resistance stage follows in which the person attempts to manage or cope with the demand or threat through psychological, physiological and behavioural means. At this stage, the initial physiological changes such as increased hormone levels and increased heart rate return to normal. If the attempts to control or cope with the stressor are unsuccessful, the exhaustion stage is reached in which psychological and physiological capacities become overstretched and the individual becomes susceptible to illness and disease. An increasing number of diseases have been identified as having stress-related origins (Quick et al., 1987).

If the stress response is not managed in the initial stages, a number of adverse outcomes are likely to occur. The behavioural consequences that have been identified include increased smoking (Lindenthal et al., 1972; Conway et al., 1981) alcohol and drug abuse (Jones and Boye, 1992) and poor work performance (Hockey, 1983). The adverse psychological consequences that have been identified include increased anxiety, depression, irritability and aggression and sleep disturbances (Quick and Quick, 1984).

Why the experience of work stressors will only produce an acute stress reaction in some workers, go on to produce a long term, or chronic response in others and yet produce little response in another group of workers is not understood. There are a number of views about why individuals differ in their reaction to stressors, the most common being vulnerability and other personal characteristics, modifying and buffering factors and differences in appraisal and coping processes. These are discussed in more detail below.

Vulnerability

The concept of vulnerability to stressors has been used by a number of writers (McLean, 1986; Dohrenwend and Dohrenwend, 1981; Rabkin, 1982) to explain individual variability in response to stressors. Vulnerability implies characteristics or reactions of the individual that increase or decrease the likelihood that they will experience a stress response in the face of a specific stressor. Wolff (1952) argued that individuals have characteristic physiological responses to managing stress which are responsible for the variation among individuals in illness patterns. There is some evidence to support this view. For example, individuals with cardiovascular disease showed greater variability in heart rate and respiration than healthy individuals (Fleming and Baum, 1987). Hypertensive patients and healthy individuals with a family history of hypertension also showed increased blood pressure in response to a number of stressors (Shapiro, 1961).

Other personal characteristics

There is evidence also that personal characteristics like trait anxiety or the tendency to be anxious (Brief et al., 1988; Depue and Monroe, 1986), neuroticism (Costa and McCrae, 1985; Innes and Kitto, 1989) and the Type A behaviour pattern (Matteson et al., 1984; Kelly and Houston, 1985) have also been linked positively to stress responses.

Other personality characteristics have been identified as affecting the response to stress. The most influential of these include the characteristic of resilience or psychological hardiness (Kobasa, 1979), which has been associated in a number of studies with effective handling of stress (Maddi and Kobasa, 1984). The level of self-esteem or the degree of positive self-regard (Rosen et al., 1982) and the level of self-focused attention or the tendency to focus attention on aspects of the self like thoughts, feelings, attitudes and body sensations (Carver and Scheier, 1981) have also been shown to be important in changing the stressor–stress response relationship. Finally, the concept of locus of control (Rotter, 1966), which was developed to describe the characteristic degree to which individuals believe they have control or mastery over their environments, has been related to coping with stress, since those who see themselves as having greater control appear to cope better with stress (Rotter, 1990).

In addition, a range of obvious personal characteristics like sex (Cox and Mackay, 1979) and age (Beale and Nethercott, 1986; Lindstrom, 1988) have been shown to be positively related to experiences of the stress response (Payne, 1988). It is thought that these individual characteristics can affect the way the person reacts emotionally, cognitively, physiologically or behaviourally to a stressor (Cooper and Marshall, 1976; Levi, 1984).

Modifying or buffering factors

In contrast to personal characteristics factors, some authors argue that individual differences in stress responses are due to the effect of a number of factors which change the relationship between stressor and the individual's response (Williams and House, 1985; Cohen and Wills, 1985). Modifying factors imply a broader set of variables including aspects of the situation and personal context in which the stressor occurs and which buffer the individual against the influence of stressors, diminishing their impact (LaRocco, et al., 1980).

Social support or the resources provided by other people, work mates, family or friends is the most frequently identified stress moderator. Cohen and Wills (1985) reviewed the literature on social support and concluded that there is reasonable evidence for its stress-buffering role. They also concluded that the most effective social supports were those that increased the person's self-esteem or that provided information-type support by helping the person to understand or respond to the demands of the stressor. Unfortunately, a number of studies have failed to demonstrate consistent findings regarding the buffering role of social support (Barrera, 1988), leading Greller et al. (1992) to conclude that overall there is evidence for social buffering, but the nature and circumstances under which social supports moderate the stress response are not clear.

Appraisal and coping

The importance of the concepts of vulnerability and moderation lies in their effects on the way individuals think about, perceive and respond to stressors. In the response to potential stressors, Lazarus and Folkman (1984) distinguish the judgement or appraisal process from the management or coping process. Thus individuals first judge a particular situation as demanding or threatening (known as primary appraisal). Secondary appraisal then occurs which involves the person determining whether they have the skills, abilities and resources to deal with the perceived demands or threat that the stressor presents. This needs to occur before coping can begin. Folkman and Lazarus (1980) define coping as all cognitive and behavioural efforts to master, reduce or tolerate demands.

A number of models have been developed to explain the role coping plays in reducing stress levels. Edwards (1988) reviewed critically the major theoretical approaches to coping and argued that there are drawbacks to each of them. He put forward an alternative in which coping is viewed as part of the stress process in which individual perceptions of a potential stressor are central and affect coping depending on how they fit in with the individual's desires. Any discrepancy between perceptions and desires will be judged in terms of its importance for the individual and efforts to cope will be adjusted accordingly. Thus greater discrepancy between perception and desires means greater importance for the individual and increased coping efforts. This approach has a number of advantages over previous ones, particularly since it acknowledges a decision-making component to the coping process in which individuals make judgements about the importance of the stressor which then influences the amount of effort they put into coping. The theory specifies a number of basic pathways through which coping moderates stress. These are summarised in Table 4.

Folkman et al. (1986) argue that success of coping depends more on choosing the right coping approaches for the perceived stressor than on the inherent usefulness of any particular coping strategies. The results of a study by Dewe (1991) support the importance of primary appraisal and to some extent secondary appraisal as important influences in successful coping. What coping measures are used depends on the individual's available resources for coping. This is influenced by a range of factors, including the individual's skills, abilities, experiences and personality or typical ways of behaving. For example, there is evidence that individuals who exhibit the Type A behaviour pattern respond differently to the demands of their job than do Type B individuals (Ganster, 1987). In addition, the impact of particular coping strategies will be influenced by the physical and social situation in which they are used. Research on the use of avoidance as a coping style, for example, suggests that it is most beneficial when work and life events are out of the individual's direct control (Roth and Cohen, 1986).

Our understanding of the use of coping is further complicated by the findings that individuals rarely adopt a rational approach to selecting coping strategies (Simon, 1976). Even though an individual may have the resources to cope with a particular stressor, they may not use them when needed. There appear, however, to be some patterns to the use of coping strategies. Under low levels of stress, for example, individuals seem to

Table 4
Pathways for moderating stress through coping (from Edwards, 1988)

Method of coping	Interpretation
Changing perceptions of stressor	I think I'll take a break
Changing a characteristic of the person	I am doing assertiveness training
Changing or challenging existing information	That can't be right, I'll ask someone else
Changing the cognitive basis of perceptions of stressor	I didn't hear what you said
Reducing the discrepancy between desire and importance of the stressor	I didn't really want it

use tried and true methods, whereas under high or prolonged stress it seems that individuals will try new strategies (Weiss et al., 1982; Janis and Mann, 1977). Under time pressures the selected coping strategies tend to be the obvious ones, only requiring very simple decisions (Janis and Mann, 1977). Individuals tend to use strategies that they have used successfully in the past and avoid those that have been unsuccessful (Mac-Crimmon and Taylor, 1976). These research findings show that in order to help individuals to improve their ability to cope, it is important not just to train them in new coping strategies, but also to provide assistance in knowing when and where to apply them.

5.1. Measurement of the stress response

Table 5 summarises some of the wide variety of methods available to measure the stress response. The aim of all methods is to detect individuals who are experiencing short or long term changes due to stress. They fall most readily into three groups, those that assess physical state, including health, physiological and biochemical status, those assessing psychological state and behavioural state measures.

Physical state

Assessment of physical state includes a range of different types of measures, from symptom

Table 5
Measures of the stress response

Aspect of stress response	Measurement method	Source
Physical state	*Symptom checklists:* Cornell Medical Index	Brodman et al., 1949
	Physiological: Heart rate/variability Blood pressure Muscle tension	Fried, 1988
	Biochemical: Cortisol Adrenaline/noradrenaline	Fleming et al., 1987
Psychological state	*Trait or dispositional:* Jenkins Activity Survey Locus of Control Scale Cognitive Hardiness Scale Self-Consciousness Scale	Jenkins et al., 1965 Rotter 1966 Kobasa, 1979 Scheier et al., 1985
	Current state: General Health Questionnaire, Mental Health Inventory, General Well-Being Schedule, State-Trait Anxiety Inventory	Goldberg, 1978 Veit et al., 1983 McDowell et al., 1987 Spielberger et al., 1970
Behavioural state	*Various behaviours e.g.:* Alcohol use (MAST, Mortimer-Filkens Questionnaire)	Schlosser et al., 1984
	Work performance	Wilson et al., 1991
	Appraisal and coping: Defence-mechanism inventory Ways of Coping Scale	Gleser et al., 1969 Folkman et al., 1980

checklists to direct measures of body state. Symptom checklists are designed to elicit medical symptoms from which the health effects of the stress experience can be inferred. The Cornell Medical Index, one of the best known, is a self-report questionnaire in which the respondent is asked questions about the presence of symptoms in each of 18 areas, including the nervous system, fatigability, depression, anxiety and tension (Brodman et al., 1949). These types of measures suffer from the same problems of subjectivity as discussed earlier. In addition, however, they present problems of distinguishing transient, perhaps minor symptoms from more serious ones.

Objective assessment of the body's response to stress can be made using measures of physiological and biochemical state. Physiological measures assess the response to stress of the sympathetic nervous system through its affects on specific organs. This includes cardiovascular indicators like heart rate, heart variability, blood pressure, gastrointestinal indicators, like peptic ulcer and musculoskeletal indicators like muscle tension. Biochemical measures assess the hormonal response to stress through indicators such as the catecholamines, adrenaline and noradrenaline, cholesterol and corticosteriods like cortisol.

Fried (1988) and Fleming and Baum (1987) reviewed the area of physiological and biochemical assessment in the workplace. Both reviews point to the usefulness of these types of indicators; however, they both also point to the difficulties in applying them in the workplace. The major problems identified were the difficulty in accounting for the effects of confounding factors that promote stable characteristics like familial susceptibility, sex or dietary differences, confounding factors that are transient like the effects of temperature, time of day and consumption of substances like alcohol and coffee before the measurement session and confounding factors due to problems in the procedure used to obtain the measurement, such as the number of repeats of the measurement, the instructions to the subject and how the sample is stored.

The conclusion from both these reviews and others (Sharit and Salvendy, 1982) was that physiological and biochemical indicators can be taken in the workplace, provided that attention is paid to the problems they raised. Fried argued that the problems can be reduced further by limiting their use only to those measures that match the expected effects from acute or chronic exposure to the stressors of interest, and by adopting a multidimensional approach to evaluate the complexities of the stress response.

Psychological state

The assessment of psychological indicators of the stress response is usually by self-report questionnaire or interview. A very large number of instruments are available to measure psychological morbidity and psychological well-being. Some focus on trait or dispositional characteristics of the person which tend to be relatively stable over time and will predispose the individual to behave in certain ways. Others focus on the current state of the individual.

The first group of trait or dispositional measures assess characteristics of individuals which increase their vulnerability to stress. For example, the Locus of Control Questionnaire (Rotter, 1966) assesses the degree to which the individual views the world as being within their personal control (internal locus of control) or outside their own control (external locus of control) and the Self-Consciousness scale (Scheier and Carver, 1985) assesses the individual's level of self-focussed attention.

In the same vein, measures have also been developed for psychological hardiness, self-esteem and Type A behaviour pattern. Kobasa (1979) devised the Cognitive Hardiness scale to measure psychological resilience and a revised scale has been developed by Nowack (1991). The characteristic of self-esteem or positive self-regard can be assessed by a number of instruments including that devised by Rosenberg (1965). The Jenkins Activity Survey (Jenkins et al., 1965) is one of a number of questionnaire-type approaches to measuring the presence of Type A characteristics such as self-imposed time pressure, impatience, competitiveness, and aggression. In addition the Type A behaviour pattern can be assessed using the Structured Interview technique which is a standard interview method designed to elicit specific types of behaviours from Type A individuals (Chesney et al., 1980).

Of the second group of psychological instruments which focus on the current mental state of the individual, one of the most well-known is the General Health Questionnaire (GHQ, Goldberg, 1978). The GHQ assesses the individual's current level of psychological distress. In McDowell and Newell's (1987) review of six psychological well-being scales, they concluded that the GHQ is a very well-developed scale which has many advantages for use as a screening tool for mental disorders in the general population.

Other similar measures of current psychological state include the General Well-Being Schedule (Dupuy, 1977, reviewed in McDowell and Newell, 1987) and the Mental Health Inventory (Veit and Ware, 1983). Both of these instruments have been reviewed by McDowell and Newell (1987). They concluded that both are useful indicators of subjective well-being in general populations, but the Mental Health Inventory is more useful due to its broader scope and because there is more published material about it.

A few psychological instruments focus on both dispositional characteristics and current state. The State-Trait Anxiety Inventory (Spielberger et al., 1970) is an example since it focuses on current levels of anxiety as well as more enduring traits or stable individual proneness for anxiety.

Behavioural state

Measures of behavioural change due to the stress response are not nearly as numerous as self-report psychological measures. Included in this group of measures are instruments that look at the way people cope with stress or the way stress affects their performance.

Behaviours like smoking and alcohol and drug consumption are likely to be affected in the stress response. Measures of these health-related behaviours are usually by questionnaire or interview for groups, which are recognised to have problems of under-reporting (Pill, 1991). Questionnaires like the Michigan Alcoholism Screening Test (MAST) and the Mortimer-Filkins Questionnaire have been used in a number of studies to identify workers whose drinking behaviour may require some sort of intervention (Schlosser and McBride, 1984).

On an individual basis these behaviours can be assessed by observation of direct or indirect measures. Fleming (1986) for example, listed a number of behavioural signs of alcohol abuse that can be used in the workplace.

There are a considerable number of measures of coping, many of which may be regarded as psychological rather than behavioural, but, for the purposes of this review, are included in this section as it is assumed that all will have impact on stress-related behaviours. The majority of measures of this type again use self-report questionnaire or interview methods. Cohen (1987) reviewed these types of measures of coping and concluded that all commonly used scales have a number of deficiencies, including poor construct and content validity. She argues that choice between such measures of coping style as the defence-mechanism inventory (Gleser and Ihilevich, 1969) or measures of the coping methods used in particular situations such as the Ways of Coping scale (Folkman and Lazarus, 1980) depends on the type of situation to be studied, the view of the investigator about what type of coping methods could be used and the type of interpretations that the investigator is wanting to make. Overall, however, we need better, more comprehensive measures of the individual's coping methods.

Indicators of the effect of the stress response on work performance have not been well-documented, except in studies of particular stressors. Research on the effects of shiftwork on performance, for example, have used a wide range of sensory, psychomotor and cognitive tests and also specific workplace measures of performance (Folkard and Monk, 1979).

Summary

In summary, there are many alternative measures from which to choose indicators of the stress response. The choice of particular measures should depend on the reason for needing them and the type of group to be studied, the setting in which the assessments are to be made, the number of assessments needed and the time over which the assessments are to be made. As discussed earlier, it is advantageous to include a range of different methods in order to improve the quality of interpretations about the existence of a stress response. For example, where two or more different types of indicators show higher than normal levels, the conclusion that a stress problem exists may be drawn with much more confidence.

In a great many studies, the approach to the selection of psychological measures is to extract sections of existing questionnaires and put them together to form a new one (Sharit and Salvendy, 1982). This creates problems in comparing the results with existing data on the measures and, if the format of the new questionnaire is not considered carefully, questions from each section may confound one another.

In addition, any alterations to questions in existing instruments need to be made with some caution. When changes are made, even small ones, they may affect the sense of the question, thus influencing interpretation of the results and making impossible any comparisons between existing data using the instrument.

Lastly, psychological and behavioural measures are invariably liable to the same set of problems due to subjectivity as discussed above in the discussion on identifying stressors. For psychological measures, however, the argument is slightly different since there is really no other way of determining how a person feels, except by asking them.

6. Research on successful interventions to reduce the problem

The aim of interventions for stress problems is to reduce stress, not eliminate it. This is based on the acknowledgement that stress is a necessary and functional response to the world, which only becomes a problem when the relationship between the world and the individual who must operate in it becomes unbalanced.

In this vein, there are two kinds of approaches for intervening to reduce the problem of stress: those which focus on the work environment and organisation part of the relationship and those focussing on the individual. Considerably more studies in this area have focussed on interventions for the individual rather than for the wider environment although this state of affairs is changing. It has been argued that the lag between these two types of studies has occurred because historically the concept of stress tended to over-emphasise the individual's role in causing the stress problem (Murphy and Hurrell, 1987; Karasek and Theorell, 1990). Furthermore, Neale et al. (1982) argued that individual approaches are easier to implement, do not disrupt work as much, can be evaluated more easily and are readily accepted by employers and managers as they imply that the problem rests with the worker. Nevertheless, these authors deplored the apparent mutual selective ignoring by both management and unions of the two approaches to stress management, environmental strategies by management and individual strategies by unions. They argued that to deny the existence of one side of the stress relationship is not constructive. Both are needed. The next sections, therefore, examine the work that has been done on each type of approach.

6.1. Changing the wider environment

Interventions in the wider work environment aim at reducing the action of stressors. This is usually done by altering working conditions through organisational change and job redesign. A number of reviews have been published on the management of stressors (Murphy and Hurrell, 1987; Murphy, 1988; Landy, 1992; MacLennan, 1992; Luczak, 1992); however, the relevant literature comes from a wide variety of sources, especially the management and organisational psychology and sociology areas.

In essence, the major strategies to manage or control workplace stressors can be summarised as good management practice. Much of the work on new organisational forms and better management techniques has enormous relevance to stress management because they often focus on the same problems, the same causes of the problems and aim for the same basic outcomes. For example, the move to greater industrial democracy and participative management in workplaces was started to improve productivity through improving decision making by shopfloor employees (ILO, 1982). There is evidence that these changes also had an impact of reducing stress levels (Gardell, 1991) and that the degree of control allowed in the workplace is inversely related to the amount of stress experienced in the workplace (Karasek and Theorell, 1990).

Another example is the recent emphasis on establishing quality culture in organisations. This has led managements towards adopting improvements in the workplace that are likely to reduce or eliminate many stressors. Principles integral to management for quality are that all members of an organisation should contribute to the improvement culture and process and that most problems are due to poor system design rather than worker error (Samson, 1991). The former principle encourages worker participation and control over aspects of their job, a commonly advocated strategy for reducing stress. The latter principle encourages the development of work systems which reduce or eliminate stressors through job and task redesign, which are again regarded to be useful stressor reduction strategies. Other new forms of management that might reduce or eliminate stressors from the work environment include flexible working hours (Tepas, 1985; Prieto and Martin, 1990), and the introduction of autonomous work groups (Gardell, 1982)

Unfortunately, apart from the work by Gardell and his associates (Johnson and Johansson, 1991) there has been little empirical research on the impact of new management approaches on reducing stress. Conclusions about their worth as strategies for stressor reduction can therefore be gained only through inference. Alderfer (1976) cautions, however, that organisational change can be successful at one level, but produce unwanted and unforeseen changes at another level. If this is the case, it seems reasonable to assume that new management systems which take an organisation-wide approach are likely to be most useful in reducing stressors. This caution also points to the need to evaluate the effects of any organisational change.

Stressor reduction can also be achieved through job and task redesign using ergonomic and engineering approaches (Luczak, 1992). Not all stressors are psychosocial in nature. Many workplace conditions can constitute stress hazards which can be overcome or improved using ergonomic approaches. For example, Jankovsky (1981) described a number of ergonomic solutions to stress problems in a light engineering factory. There are, however, relatively few studies that examine the use of ergonomic methods to respond to stress or which measure the stress-related consequences of ergonomic interventions. Luczak (1992) described a conceptual approach to good work design through ergonomic and industrial engineering methods. This approach would provide a useful basis for further research on the issue.

6.2. Individual stress management

Individual stress management strategies concentrate mainly on three basic forms, namely, assisting workers to reconstruct their views about particular situations or experiences, assisting workers to cope with particular stressors in the short or long term and to adopt lifestyles that improve their ability to withstand the pressure of stressors.

Changing perceptions of potential stressors

Strategies that have been used to change perceptions of stressors or cognitive restructuring techniques can range from simple "commonsense" methods like concentrating on positive rather than negative experiences to clinical approaches like rational-emotive therapy (Ellis, 1962), stress inoculation training (Meichenbaum, 1977) and systematic desensitisation (Thomas et al., 1979). All of these strategies attempt to help the individual change the way they conceptualise and appraise potentially stressful experiences. Mostly these strategies use logical rather than emotional approaches which are the basis of the stress response. Since individual perceptions play such an integral role in the stress response, these strategies are seen to be most useful for reducing and eliminating the stress response. In the main, however, they have not been evaluated widely

(Ivancevich and Matteson, 1988) especially in the context of work stressors.

Coping techniques

Many methods have been developed for improving coping with the response to stressors, including methods that attempt to reduce the physiological, psychological and behavioural response to stress (Rice, 1992). By far the most commonly used physiological strategies are muscle relaxation and biofeedback. In his review of individual stress management programmes, Murphy (1984) found that muscle relaxation was common to all programmes. The aim of this strategy is to reduce the level of general body arousal. While there is evidence of significant reductions in physiological and psychological stress indicators when relaxation training is used (Murphy, 1984), there are a number of studies using relaxation techniques which have conflicting findings. For example, Fiedler et al. (1989) found that relaxation training had little effect on the blood pressure of asymptomatic employees.

Biofeedback training has been used for a wide variety of problems including stress management. It attempts to provide the individual with information about the current state of particular physiological functions so that the person can make efforts to control them. The commonly used function in stress management is electromyography (EMG), in which feedback is provided about muscle activity. For this reason EMG biofeedback is often used in conjunction with relaxation training. Ivancevich and Matteson (1988) cite research which they argue casts doubts on the effectiveness of EMG biofeedback for treating stress-related problems.

In the workplace, meditation is the most commonly used strategy for reducing psychological responses to stress. Meditation is, again, aimed at reducing arousal levels, but by concentrating on mental rather than physical state. Reviews of the use of meditation as a stress management technique conclude overall that meditation appears to be effective (Murphy, 1984; Delmonte, 1985). In fact, a study by Kabat-Zinn et al. (1992) reported reductions in anxiety and depression for three years following meditation training in a group of patients diagnosed with anxiety disorders.

In addition, there are a very large number of behavioural methods that have been used for coping with stress, for example, time management, improved interpersonal skills like assertiveness and communication skills. These are techniques that attempt to change the way individuals deal with their environment by encouraging behaviours that will reduce the stress. These techniques are also used in management training as they are seen to be important skills for managers (Dunphy, 1985). Hence emphasis on good management practice can again be useful for reducing stress problems, both in terms of reducing stressors for others in the work group as discussed above and in terms of controlling stress responses of managers themselves. The utility of management skills type training for stress management has not been assessed effectively, mainly because they are hardly ever used alone in workplace programmes and there are very few comparative evaluations of stress reduction techniques (Murphy, 1984).

Lifestyle techniques

The final group of strategies that are common to stress management training are general health promotion or lifestyle strategies. These include training in nutrition, exercise and other healthy lifestyle behaviours. Evidence suggests that positive or healthy changes in these behaviours can change the individual's response to stressors (Spring et al., 1986) and may reduce stress (Crews and Landers, 1987).

Comparing types of individual interventions

Unfortunately, there has been little evaluation of the effectiveness of any type of intervention to assist the individual (Murphy, 1984). Many of the studies that have attempted evaluation suffer from deficiencies on a number of aspects. These included issues like whether the technique will generalise beyond training to other situations, whether long-term benefits can be gained from the technique, whether the programme focusses on workers with apparent stress problems or on the general work group and whether the technique is cost-effective. These need to be addressed in judging whether a stress management technique is successful for stress management (Murphy, 1985; Ivancevich and Matteson, 1988).

Despite these difficulties there is evidence that stress management training can result in such positive changes as reduced anxiety (Murphy, 1985, 1988) reductions in accident rates (Murphy et al., 1986) and lowered rates of absenteeism (Bertera, 1990) and possibly in job satisfaction (Murphy, 1988).

There is less evidence, however, about the relative success of the various stress management techniques. One study by Bruning and Frew (1985) compared the effectiveness of management skills training, meditation, and physical exercise alone and in combination with each other for reducing physiological indicators of stress. They found that all strategies reduced blood pressure and pulse rate but the biggest decrease occurred when management skills training was used.

An alternative approach to discovering which strategies might be most effective has been to investigate the strategies used by people who manage stress successfully. For example, Howard et al. (1975) determined that the most successful strategies for coping with stress used by managers were changing to a non-work activity, building better resistance through regular sleep and good health habits, separating work and non-work life, talking with coworkers and engaging in physical exercise. Many of these techniques have been associated with lower stress levels in intervention studies, for example Fremont and Craighead (1987) found that exercise programmes helped to control the level of emotional stress.

It should be pointed out, however, that the behaviours shown to be used by individuals who cope successfully with stress tend to be emotion-based strategies which have little effect on minimising actual exposure to work stressors (Pearlin and Schooler, 1978). Problem-based strategies are needed to cope with the source of stress. This suggests that the strategies that people spontaneously generate to cope do not tackle the real causes of their stress problems, but only help to tolerate them. It can be argued, therefore, that workers need training to help deal with work stressors, but that this training must emphasise problem-based strategies, and it must be linked to other organisational strategies to reduce their exposure to stressors.

Implementing individual interventions

Individual stress management in organisations normally takes one or both of two forms, training programmes conducted usually by a consultant expert and Employee Assistance Programmes. The individual stress management strategies described in the preceding discussion would typically be conducted as part of a training programme. Techniques like muscle relaxation, meditation and time management are taught in workplaces often in a general health promotion context rather than in response to a known problem. Employee Assistance Programmes, in contrast, are used typically to solve the existing problems of individual workers which may or may not be due to stress.

Employee Assistance Programmes (EAPs), usually operating in the context of company policy, can play a relatively large role in workplace attempts to manage stress. Mostly they are used to respond to employees' personal and emotional problems which interfere with job performance either directly or indirectly and to organise the rehabilitation of these troubled employees. Masi (1986), on reviewing the cost/effectiveness of EAPs, concluded that they could reduce productivity losses in terms of disability benefit payments to employees, absenteeism and lost time.

Mostly EAPs do not take an active role in stress prevention. This can be seen in two of the tenets of EAP philosophy put forward by Phillips and Mushinski (1992), that EAPs facilitate but do not make changes in organisations and that they offer opinions to management but do not intervene in management decisions. It has been argued, however, that EAPs should widen their scope to cover more preventive activities like health education and health promotion and focus less on productivity and personnel problems (MacLeod, 1985).

Martin (1992) advanced a very lucid model for improving stress management training for work-

Table 6
Model for stress management training (From Martin, 1992)

Stress management training should:
1. be relevant to the audience,
2. be in a context of organisational efforts to eliminate or remove stressors,
3. be based on a theoretical model of stress,
4. use the worker's own resources to assist in making the change,
5. acknowledge that interventions will not always be successful because individuals differ,
6. be reoccurring and preferably sustained by a support group.

ers which is detailed in Table 6. He maintains that stress training should aim at more than just awareness of stress and achieving lifestyle change. Stress training should, he argues, help individual workers to increase their skills in controlling their own experiences, particularly those relating to their workplace.

7. What intervention should be chosen?

The decision about how to intervene should be based on the results of stress measures. Identifying what are the stressors for the majority of workers will provide information about the most useful strategies to pursue for the majority of workers. Usually measures of the stress response only reveal that stress levels are high and need to be lowered. They signal the existence of a problem. They do not in themselves suggest specific strategies for reducing stress levels. Comparing individuals who are exposed to different stressors in different parts of the organisation can also provide information about what the main stressors might be, and therefore help guide stress reduction and management strategies.

In deciding on the best strategies for interventions, there are some distinct advantages to tackling sources of stress in preference to individual sources. First, as has been discussed, there are likely to be groups of objective stressors in every workplace which are most likely to cause problems for all workers who encounter them. Removing objective stressors will therefore reduce stress levels for the majority of workers (Frese and Zapf, 1988). In contrast, fewer workers are likely to be assisted by interventions which deal only with individual sources of stress.

Second, tackling environmental sources of stress, once achieved, is more likely to produce lasting effects since the sources of the stress are often removed. Individual interventions, on the other hand, are much harder to achieve in the long term as they require time, effort and commitment from the person to be successful. Bellingham (1990) takes this further by arguing that it is a myth that individual health promotion programmes work, alleging that they are successful only in organisations that are healthy.

On the other hand, it is often argued that organisational approaches present problems since they usually require direct attention and effort from managers and are therefore much harder to implement (Karasek and Theorell, 1990). Individual-focussed approaches, in contrast, are much easier to implement, do not challenge the organisation's hierarchy as they concentrate on changing the individual worker, and are much easier to evaluate (Murphy and Hurrell, 1987).

Despite these issues, the nature of stress requires that when stress is a problem both sides of the environment–person relationship need to be addressed. On balance, while the most wide-reaching and lasting solutions are likely to be achieved through organisational and environmental approaches, it is important that individuals have access to strategies which will help them to enhance and reinforce their own coping skills. Most importantly, however, any strategies that are implemented to reduce or manage stress must be monitored and evaluated, and if found to be ineffective, must be modified appropriately.

Disclaimer

The views expressed in this paper are those of the author and do not necessarily reflect those of the National Institute of Occupational Health and Safety (Worksafe Australia).

References

Alderfer, C.P., 1976. Change processes in organisations. In: M.D. Dunnette (Ed.), Handbook of Industrial and Organisational Psychology. Rand McNally, Chicago.

Alkov, S., 1981. Psychosocial stress, health and human error. Professional Safety, 5: 12–14.

Amick, B.C. and Smith, M.J., 1992. Stress, computer-based work monitoring and measurement systems: A conceptual overview. Applied Ergonomics, 23: 6–15.

Arnetz, B.B., 1991. White collar stress: What studies of physicians can teach us. Psychotherapy and Psychosomatics, 55: 197–200.

Bailey, J.M. and Bhagat, R.S., 1987. Meaning and measurement of stressors in the work environment: An evaluation. In: S.V. Kasl and C.L. Cooper (Eds.), Stress and Health: Issues in Research Methodology. John Wiley, New York.

Bain, C., Conn, M., 1990. Occupational stress among government school teachers: Causes and preventive measures. Journal of Occupational Health and Safety Australia and New Zealand, 6: 381–386.

Baker, D.B., 1985. The study of stress at work. Annual Review of Public Health, 6: 367–381.

Barrera, M. Jr., 1988. Models of social support and life stress: Beyond the buffering hypothesis. In: L.H. Cohen (Ed.), Life events and psychological functioning: Theoretical and methodological issues. Sage, Newbury Park, CA.

Beale, N. and Nethercott, S., 1986. Job loss and health – The influence of age and previous morbidity. Journal of the Royal College of General Practitioners, 36: 261–264.

Beck, A.T., Ward, C.H., Mendelson, M. Mock, J. and Erbaugh, J., 1961. An inventory for measuring depression. Archives of General Psychiatry, 4: 561–571.

Bellingham, R., 1990. Debunking the myth of individual health promotion. Occupational Medicine: State of the Art Reviews, 5: 665–675.

Bertera, R.L., 1990. The effects of workplace health promotion on absenteeism and employment costs in a large industrial population. American Journal of Public Health, 80: 1101–1105.

Brief, A.P., Burke, M.J., George, J.M., Robinson, B.S. et al., 1988. Should negative affectivity remain an unmeasured variable in the study of job stress? Journal of Applied Psychology, 73: 193–198.

Broadbent, D.E., 1985. The clinical impact of job design. British Journal of Clinical Psychology, 24: 33–44.

Brodman, K., Erdmann, A.J., Lorge, I. and Wolff, H.G., 1949. The Cornell Medical Index. Journal of the American Medical Association, 140: 531.

Brown, J.M. and Campbell, E.A., 1990. Sources of occupational stress in the police. Work and Stress, 4: 305–318.

Bruning, N.S. and Frew, D.R., 1985. The impact of various stress management training strategies: A longitudinal experiment. In: R.B. Robinson and J.A. Pearce (Eds.), Academy of Management Proceedings. Academy of Management, San Diego, CA.

Camp, R.C., 1989. Benchmarking. The search for industry best practices that lead to superior performance. Quality Press, Milwaukee, WI.

Cannon, W.B., 1935. Stresses and strains of homoeostasis. American Journal of Medical Science, 189: 1.

Carrere, S., Evans, G.W., Palsane, M.N. and Rivas, M., 1991. Job strain and occupational stress among urban public transit operators. Journal of Occupational Psychology, 64: 305–316.

Carver, C.S. and Scheier, M.F., 1981. Attention and self-regulation: A control theory approach to human behaviour. Springer-Verlag, New York.

Chesney, M.A., Eagleston, J.R. and Rosenman, R., 1980. The Type A structure interview: A behavioural assessment in the rough. Journal of Behaviour Assessment, 2: 255–272.

Cobb, S. and Rose, R.M., 1973. Hypertension, peptic ulcer and diabetes in air traffic controllers. Journal of American Medical Association, 224: 489–494.

Cohen, F., 1987. Measurement of coping. In: S. Kasl and C.L. Cooper (Eds.), Stress and Health: Issues in Research Methodology. John Wiley, Chichester.

Cohen, F. and Lazarus, R., 1973. Active coping processes, coping dispositions and recovery from surgery. Psychosomatic Medicine, 35: 375–389.

Cohen, S. and Wills, T.A., 1985. Stress, social support and the buffering hypothesis. Psychological Bulletin, 98: 310–357.

Cohen, S. and Williamson, G.M., 1991. Stress and infectious disease in humans. Psychological Bulletin, 109: 5–24.

Cohen, S. and Wills, T.A., 1985. Stress, social support and the buffering hypothesis. Psychological Bulletin, 98: 310–357.

Contrada, R.J. and Krantz, D.S., 1987. Measurement bias in health psychology research designs. In: S.V. Kasl and C.L. Cooper (Eds.), Stress and Health: Issues in Research Methodology. Wiley, Chichester.

Conway, T.H., Vickers, R.T., Ward, H.W. and Rahe, R.H., 1981. Occupational stress and Variation in cigarette, coffee and alcohol consumption. Journal of Health and Social Behaviour, 22: 155–165.

Cooper C.L. and Melihuish, A., 1980. Occupational stress and the manager. Journal of Occupational Medicine, 22: 588–592.

Cooper, C.L. and Marshall, J., 1976. Occupational sources of stress: A review of the literature relating to coronary heart disease and mental health. Journal of Occupational Psychology, 49: 11–29.

Cooper, C.L. and Smith M.J., 1985. Job stress and blue collar work. John Wiley, New York.

Cooper, C.L., Rout, U. and Faragher, B., 1989. Mental health, job satisfaction, and job stress among general practitioners. British Medical Journal, 298, 366–370.

Cooper, C.L., Sloan, S.J. and Williams, S., 1988. Occupational stress indicator: Management guided. NFER-Nelson, Windsor.

Corlett E.N. and Richardson, J., 1981. Stress Work and Productivity. John Wiley, Chichester.

Costa, P.T. and McCrae, R.R., 1985. Hypochondriasis, neuroticism and aging: When are somatic complaints unfounded? American Psychologist, 40: 19–28.

Cox, T., 1978. Stress. Macmillan, London.
Cox, T. and Mackay, C.J., 1979. The impact of repetitive work. In: R. Sell and P. Shipley (Eds.), Satisfactions in job design. Taylor and Francis, London.
Crews, D.J. and Lunders, D.M., 1987. A meta-analytic review of aerobic fitness and reactivity to psychosocial stressors. Medicine and Science in Sports and Exercise, 19: S114–S120.
Delmonte, M.M., 1985. Meditation and anxiety reduction: A literature review. Clinical Psychology Reviews, 5: 91–102.
Depue, R.A. and Monroe, S.M., 1986. Conceptualisation and measurement of human disorder in life stress research. Psychological Bulletin, 99: 35–51.
Dewe, P., 1991. Measuring work stressors: The role of frequency, duration, and demand. Work and Stress, 5: 77–91.
Dohrenwend, B.S. and Dohrenwend, B.P., 1981. Stressful life events and their contexts. Rutgers University Presss, New Brunswick, NJ.
Dohrenwend, B.P. and Stout, P.E., 1985. Hassles in the conceptualisation and measurement of life stress variables. American Psychologist, 40: 780–785.
Dunn, J.P. and Cobb, S., 1962. Frequency of peptic ulcer among executives, craftsmen and foremen. Journal of Occupational Medicine, 4: 343–348.
Dunphy, D.C., 1985. Organizational change by choice. McGraw-Hill, Sydney.
Edwards, J.R., 1988. The determinants and consequences of coping with stress. In: C.L. Cooper and R. Payne (Eds.), Causes, Coping and Consequences of Stress at Work. John Wiley, Chichester.
Eichler, A., Silverman, M.M. and Pratt, D.M., 1986. How to define and research stress. American Psychiatric Press Inc., Washington.
Ellis, A., 1962. Reason and Emotion in Psychotherapy. Stuart, New York.
Ellis, J.L. and Gordon, P.R., 1991. Farm family mental health issues. Occupational Medicine: State of the Art Reviews. 6: 493–502.
Elo, A.-L., Leppanen, A., Lindstrom, K. and Ropponen, T., 1992. Occupational Stress Questionnaire: Users Instructions. Institute of Occupational Health, Helsinki.
Endresen, I.M., Ellertsen, B., Endresen, C., Hjelmen, A.M., Matre, R. and Ursin, H., 1991. Stress at work and psychological and immunological parameters in a group of Norwegian female bank employees. Work and Stress, 5: 217–227.
Estryn-Behar, M., Kaminski, M., Peigne, E., Bonnet, N., Vaichere, E., Gozlan, C., Azoulay, S., and Giorgi, M., 1990. Stress at work and mental health status among female hospital workers. British Journal of Industrial Medicine, 47: 20–28.
Eysenck, M., 1983. Anxiety and individual differences. In: R. Hockey (Ed.), Stress and Fatigue in Human Performance. Wiley, Chichester.
Fiedler, N. Vivona-Vaughan, E. and Gochfeld, M., 1989. Evaluation of a work-site relaxation training program using ambulatory blood pressure monitoring. Journal of Occupational Medicine, 31: 565–602.
Feigenbaum, A.V., 1983. Total Quality Control. McGraw-Hill, New York.
Fisher, S., 1989, Stress, control, worry prescriptions and the implications for health at work: A psychobiological model. In: S.L. Sauter, J.J. Hurrell and C.L. Cooper (Eds.), Job Control and Worker Health. John Wiley, Chichester.
Fleming, T.C., 1986. Alcohol and other mood-changing drugs. In: S. Wolf and A.J. Finestone (Eds.), Occupational Stress. Health and Performance at Work. PSG Publishing Co., Littleton, MA.
Fleming, I. and Baum, A., 1987. Stress: Psychobiological assessment. In: J.M. Ivancevich and D.C. Ganster (Eds.), Job Stress; From Theory to Suggestion. Haworth Press, New York.
Fletcher, B.C., 1988. The epidemiology of occupational stress. In: C.L. Cooper and R. Payne (Eds.), Causes, Coping and Consequences of Stress at Work. John Wiley, Chichester.
Folger, R. and Belew, J., 1985. Nonreactive measurement: A focus for research on absenteeism and occupational stress, Research in Organisational Behaviour, 7: 129–170.
Folkard, S. and Monk, T.H., 1979. Shiftwork and performance. Human Factors, 21: 483–492.
Folkard, S. and Monk, T.H., 1985. Hours of Work. John Wiley, Chichester.
Folkman, S. and Lazarus, R., 1980. An analysis of coping in a middle-aged community sample. Journal of Health and Social Behaviour, 21: 219–239.
Folkman, S., Lazarus, R., Dunkel-Schetter, C., DeLongis, A. and Gruen, R., 1986. Dynamics of a stressful encounter: Cognitive appraisal, coping and encounter outcomes. Journal of Personality and Social Psychology, 50: 992–1003.
Fremont, J. and Craighead, L.W., 1987. Aerobic exercise and cognitive therapy in the treatment of dysphoric moods. Cognitive Therapy and Research, 11: 241–251.
French, J.R.P. and Caplan, R.D., 1972. Occupational stress and individual strain. In: A.J. Morrow (Ed.), The Failure of Success. AMACOM, New York.
French, J.R.P., Caplan, R.D. and Van Harrison, R., 1982. The mechanisms of job stress and strain. John Wiley, New York.
Frese, M. and Zapf, D., 1988. Methodological issues in the study of work stress: Objective and subjective measurement of work stress and the question of longitudinal studies. In: C.L. Cooper and R. Payne (Eds.), Causes, Coping and Consequences of Stress at Work. John Wiley, Chichester.
Fried, Y., 1988. The future of physiological assessments in work situations. In: C.L. Cooper and R. Payne (Eds.), Causes, Coping and Consequences of Stress at Work. John Wiley, Chichester.
Ganster, D.C., 1987. Type A behaviour and occupational stress. In: J.M. Ivancevich and D.C. Ganster (Eds.), Job Stress: From Theory to Suggestion. Haworth Press, New York.

Gardell, B., 1982. Work participation and autonomy; A multilevel approach to democracy at the workplace. International Journal of Health Services, 12: 527–558.

Gardell, B., 1987. Work Organisation and Human Nature. The Swedish Work Environment Fund, Stockholm.

Gardell, B., 1991. Worker participation and autonomy: A multilevel approach to democracy in the workplace. In: J.V. Johnson and G. Johansson (Eds.), The Psychosocial Work Environment: Work Organisation, Democratization and health. Baywood, New York.

Giesen, C., Maas, A. and Vriens, M., 1989. Stress among farm women: A structural model approach. Behavior Medicine, 15: 53–62.

Gleser, G.C. and Ihilevich, D., 1969. An objective instrument for measuring defence mechanisms. Journal of Consulting and Clinical Psychology, 33; 51–60.

Goldberg, D.P., 1978. Manual of the General Health Questionnaire. NFER Publishing, Windsor. UK.

Gorman, D.M., 1988. Employment, stressful life events and the development of alcohol dependence. Drug and Alcohol Dependence, 22: 151–159.

Greller, M.M., Parsons, C.K. and Mitchell, D.R.D., 1992. Additive effects and beyond: Occupational stressors and social buffers in police organisation. In: J.C. Quick, L.R. Murphy and J.J. Hurrell (Eds), Stress and Well-being at Work. American Psychological Association, Washington.

Groenewegen, P.P. and Hutten, J.B.F., 1991. Workload and job satisfaction among general practitioners: A review of the literature. Social Science Medicine, 32; 1111–1119.

Gupta, N. and Beehr, T.A., 1979. Job stress and employee behaviours. Organisational Behaviour and Human Performance, 23: 373–387.

Hackman, J.R. and Oldman, G.R., 1975. Development of the Job Diagnostic Survey. Journal of Applied Psychology, 60: 159–170.

Hockey, R., 1983. Stress and Fatigue in Human Performance. Wiley, Chichester.

Howard, J.H., Reichnitzer, R.A. and Cunningham, D.A., 1975. Coping with job tension – Effective and ineffective methods. Public Personnel Managment, 4: 317–326.

International Labour Organisation, 1982. New Forms of Work organisation 1. ILO, Geneva.

Ingram, R.E., 1990. Self-focused attention in clinical disorders: Review and conceptual model. Psychological Bulletin, 107: 156–176.

Innes, J.M. and Kitto, S., 1989. Neuroticism, self-consciousness and coping strategies and occupational stress in high school teachers. Personality and Individual Differences, 10, 303–312.

Ivancevitch, J.T. and Matteson, M.T., 1980. Stress and Work: A Managerial Perspective. Scott Foreman and Co., Glenview, IL.

Ivancevitch, J.T. and Matteson, M.T., 1988. Promoting the individual's health and well-being. In: C.L. Cooper and R. Payne (Eds.), Causes, Coping and Consequences of Stress at Work. John Wiley, Chichester.

Jackson, S.E. and Schuler, R.S., 1985. A meta-analysis and conceptual critique of research on role ambiguity and role conflict in work settings. Organisational Behaviour and Human Decision Processes, 36: 16–78.

Jamal, M., 1984. Job stress and job performance controversy: An empirical assessment. Organisational Behaviour and Human Performance, 33: 1–21.

Janis, I.L. and Mann, L., 1977. Decision Making. Free Press, New York.

Jankovsky, F., 1981. An ergonomics study of a press workship with the objective of improving working conditions in a new factory. In: E.N. Corlett and J. Richardson (Eds.), Stress, Work Design and Productivity. John Wiley, Chichester.

Jenkins, C.D., Zyzanski, S.J. and Rosenman, R.H., 1965. Jenkins Activity Survey. The Psychological Corporation, New York.

Johnson, J.V. and G. Johansson (Eds.), 1991. The Psychosocial Work Environment: Work organisation, Democratization and Health. Baywood, New York.

Jones, J.W. and Boye, M.W., 1992. Job stress and employee counterproductivity. In: J.C. Quick, L.R. Murphy and J.J. Hurrell (Eds.), Stress and Well-Being at Work. American Psychological Association, Washington.

Kabat-Zinn, J., Massion, A.O., Kristeller, J., Peterson, L.G., Fletcher, K.E., Pbert, L., Lenderking, W.R. and Santorelli, S.F., 1992. Effectiveness of a meditation-based stress reduction program in the treatment of anxiety disorders. American Journal of Psychiatry, 149: 936–943.

Kahn, R.L., Wolfe, D.M., Quinn, R.P., Snoek, J.R. and Rosenthal, R.A., 1964. Organisational stress: Studies in role conflict and ambiguity. Wiley, New York.

Kahn, H. and Cooper, C.L., 1986. Computing stress. Current Psychological Research and Reviews, Summer: 148–162.

Kahn, R. and Byosiere, P., 1992. Stress in organisations. In: M.D. Dunnette and L.M. Hough (Eds.), Handbook of Industrial and Organizational Psychology. Consulting Psychologists Press, Inc. Palo Alto, CA.

Kahneman, D., 1973. Attention and Effort. Prentice Hall, Englewood Cliffs, NJ.

Kalimo, R., 1980. Stress in work. Conceptual analysis and study on prison personnel. Scandinavian Journal of Work, Environment and Health, 6 (Suppl. 3): 1–148.

Kalimo, R., 1985. Objectives, scope and current approaches in monitoring of psychosocial stressors and strain at work. An overview. In: J.J. Sanchez-Sosa (Ed.), Health and Clinical Psychology. Elsevier, Amsterdam.

Karasek, R.A., 1979. Job demands, job decision latitude and mental strain: Implications for job redesign. Administrative Science Quarterly, 24: 285–307.

Karasek, R.A., 1985. Job content questionnaire. Department of Industrial and Systems Engineering, University of Southern California, Los Angeles.

Karasek, R. and Theorell, T., 1990. Healthy work: Stress, productivity, and the reconstruction of working life. Basic Books, New York.

Karasek, R.A., Baker, D., Marxer, F., Ahlbom, A. and Theorell, T., 1981. Job decision latitude, job demands and

cardiovascular disease: A prospective study of Swedish men, American Journal of Public Health, 71: 694–705.
Kasl, S.V., 1983. Methodological issues in stress research. In: C.L. Cooper (Ed.), Stress research: Issues for the eighties. Wiley, Chichester.
Kawakami, N., Haratani, T., Kaneko, T. and Araki, S., 1989. Perceived job-stress and blood pressure increase among Japanese blue collar workers: One year follow-up study. Industrial Health, 27: 71–81.
Keenan, A. and Newton, T.J., 1987. Work difficulties and stress in young professional engineers. Journal of Occupational Psychology, 60; 133–145.
Kelloway, E.K. and Barling, J., 1991. Job characteristics, role stress and mental health. Journal of Occupational Psychology, 64: 291–304.
Kelly, K.E. and Houston, B.K., 1985. Type A behaviour in employed women: Relation to work, marital and leisure variables, social support, stress, tension, and health. Journal of Personality and Social Psychology, 46, 1067–1079.
Kobasa, S.C., 1979. Stressful life events, personality and health: An enquiry into hardiness. Journal of Personality and Social Psychology, 37: 1–11.
Kobasa, S.C., 1979. Personality and resistance to illness. American Journal of Community Psychology, 7; 413–423.
Kyriacou, C. and Sutcliffe, J., 1978. Teacher stress: Prevalence sources and symptoms. British Journal of Educational Psychology, 48: 159–167.
Landy, F.J., 1992. Work design and stress panel. In: G.P. Keita and S.L. Sauter (Eds.), Work and Well-being. An Agenda for the 1990s. American Psychological Association, Washington.
LaRocco, J.M., House, J.S. and French, J.R.P., 1980. Social support, occupational stress and health. Journal of Health and Social Behaviour, 21: 202–218.
Lazarus, R.S. and Folkman, S., 1984. Stress, appraisal and coping. Springer, New York.
Leventhal, H. and Tomarken, A., 1987. Stress and illness: Perspectives from health psychology. In: S.V. Kasl and C.L. Cooper (Eds.), Stress and Health: Issues in Research Methodology. John Wiley, Chichester.
Levi, L., 1984. Work, stress and health. Scandinavian Journal of Work and Environmental Health, 10: 495–500.
Levi, L., 1987. Definitions and the conceptual aspects of health in relation to work. In: R. Kalimo, M. El-Batawi and C.L. Cooper (Eds.), Psychosocial Factors at Work. WHO, Geneva.
Lindenthal, J.J., Myers, J.K. and Pepper, M.P., 1972. Smoking, psychological status and stress. Social Science Medicine, 6: 583–591.
Lindstrom, K., 1988. Age-related differences in job characteristics and in their relation to job satisfaction. Scandinavian Journal of Work Environmental Health, 14: 24–26.
Luczak, H., 1992. "Good work" design: An ergonomic, industrial engineering perspective. In: J.C. Quick, L.R. Murphy and J.J. Hurrell (Eds.), Stress and Well-Being at Work. American Psychological Association, Washington, DC.
MacCrimmon, K.R. and Taylor, R.N., 1976. Decision making and problem solving. In: M.D. Dunnette (Ed.), Handbook of Industrial and Organisational Psychology. Rand McNally, Chicago, IL.
McDowell, I. and Newell, C., 1987. Measuring Health: A Guide to Rating Scales and Questionnaires. Oxford University Press, Oxford.
McGuire, B., 1990. Post-traumatic stress disorder: A review. The Irish Journal of Psychology, 11, 1–23.
McLaney, M.A. and Hurrell, J.J., 1988. Control, stress, and job satisfaction in Canadian nurses. Work and Stress, 2: 217–224.
McLean, A.A., 1986. Ages, stages and vulnerability to stressors. Occupational Medicine: State of the Art Reviews, 1: 569–581.
McLennan, B.W., 1992. Stressor reduction: An organisational alternative to individual stress management. In: J.C. Quick, L.R. Murphy and J.J. Hurrell (Eds.), Stress and Well-being at Work. American Psychological Association, Washington.
MacLeod, A.G.S., 1985. EAPs and blue collar stress. In: C.L. Cooper and M.J. Smith (Eds.), Job Stress and Blue Collar Work. John Wiley, Chichester.
Maddi, S.R. and Kobasa, S.C., 1984. The Hardy Executive: Health under Pressure. Dow Jones-Irwin, Homewood, IL.
Manning, M.R. and Osland, J.S., 1989. The relationship between absenteeism and stress. Work and Stress, 3: 223–235.
Marmot, M.G. and Madge, N., 1987. An epidemiological perspective on stress and health. In: S. Kasl and C.L. Cooper (Eds.), Stress and Health: Issues in Research Methodology. John Wiley, Chichester.
Martin, E.V., 1992. Designing stress training. In: J.C. Quick, L.R. Murphy and J.J. Hurrell (Eds.), Stress and Well-Being at Work. American Psychological Association, Washington.
Masi, D.A., 1986. Employee assistance programs. Occupational Medicine: State of the Art Review Reviews, 1: 653–665.
Matteson, M.T. and Ivancevitch, J.M., 1987. Controlling work stress: Effective human resources and management strategies. Jossey-Bass, San Francisco, CA.
Matteson, M.T., Ivancevich, J.M. and Smith, S.V., 1984. Relation of Type A behaviour to performance and satisfaction among sales personnel. Journal of Vocational Behaviour, 25: 203–214.
Meichenbaum, D., 1977. Cognitive-Behaviour Modification: An Integrative Approach. Plenum Press, New York.
Millar, D., 1984. The NIOSH-suggested list of the ten leading work-related diseases and injuries. Journal of Occupational Medicine, 26: 340–341.
Millar, D., 1990. Mental health and the workplace. An interchangeable partnership. American Psychologist, 45: 1165–1166.
Mital, A. and Mital, C., 1984. Mental stress and physiological responses. In: A. Mital (Ed.), Trends in Ergonomics/Human Factors 1. Elsevier Science, Amsterdam, pp. 353–357.
Moos, R., 1974. The Work Environment Scale. Consulting Psychologists Press, Palo Alto, CA.
Murphy, L.R., 1984. Occupational stress management: A re-

view and appraisal. Journal of Occupational Psychology, 57: 1–15.
Murphy, L.R., 1985. Individual coping strategies. In: C.L. Cooper and M.J. Smith (Eds.), Job Stress and Blue Collar Work. John Wiley and Sons, Chichester.
Murphy, L.R., 1985. Evaluation of worksite stress management. Corporate Commentary, 1: 24–32.
Murphy, L.R., 1988. Workplace interventions for stress reduction and prevention. In: C.L. Cooper and R. Payne (Eds.), Causes, Coping and Consequences of Stress at Work. John Wiley, Chichester.
Murphy, L.R., 1991. Job dimensions associated with severe disability due to cardiovascular disease. Journal of Clinical Epidemiology, 44: 155–166.
Murphy, L.R. and Hurrell, J.J., 1987. Stress measurement and management in organisations: Development and current status. In: A.W. Riley and S.J. Zaccaro (Eds.), Occupational Stress and Organisational Effectiveness. Praeger, New York.
Murphy, L.R., DuBois, D. and Hurrell, J.J., 1986. Accident reduction throught stress management. Journal of Business and Psychology, 1: 5–18.
National Council on Compensation Insurance, 1985. Emotional stress in the workplace – New legal rights in the eighties. NCCI, New York.
Neale, M.S., Singer, J.A., Schwartz, G.A. and Schwartz, J., 1982. Conflicting perspectives on stress reduction in occupational settings: A systems approach to their resolution. Report to NIOSH on P.O. 82-1058, Cincinnati.
Nowack, K.M., 1991. Psychosocial predictors of health status. Work and Stress, 5: 117–131.
Payne, R., 1988. Individual differences in the study of occupational stress. In: C.L. Cooper and R. Payne (eds.), Causes Coping and Consequences of Stress at Work. Chichester, John Wiley, pp. 209–232.
Pearlin, L.I. and Schooler, C., 1978. The structure of coping. Journal of Health and Social Behaviour, 19: 2–21.
Phelan, J., Schwartz, J.E., Bromet, E.J., Dew, M.A., Parkinson, D.K., Schulberg, H.C., Dunn, L.O. Blane, H. and Curtis, E.C., 1991. Work stress, family stress and depression in professional and managerial employees. Psychological Medicine, 21: 999–1012.
Phillips, A.B. and Mushinski, M.H., 1992. Configuring an employee assistance program to fit the corporation's structure: One company's design. In: J.C. Quick, L.R. Murphy and J.J. Hurrell (Eds.), Stress and Well-Being at Work. American Psychological Association, Washington.
Pickett, C.W.L. and Lees, R.E.M., 1991. A cross-sectional study of health complaints among 79 data entry operators using video display terminals. Journal of the Society of Occupational Medicine, 41: 113–116.
Pill, R., 1991. Issues in lifestyles and health: Lay meanings of health and health behaviour. In: B. Badura and I. Kickbusch (Eds.), Health Promotion Research. WHO Regional Publications European Series No. 37, Geneva.
Poulton, E.C., 1978. Stress and blue collar work. In: C. Cooper and R. Payne (Eds.), Stress at Work. John Wiley, New York.

Pratt, L.I. and Barling, J., 1988. Differentiating between daily events, acute and chronic stressors: A framework and its implications. In: J.J. Hurrell, Jr, L.R. Murphy and S.L. Sauter (Eds.), Occupational Stress: Issues and Developments in Research. Taylor and Francis, New York.
Prieto, J.M. and Martin, J., 1990. New forms of work organisation. Irish Journal of Psychology, 11: 170–185.
Quick, J.C. and Quick, J.D., 1984. Organizational stress and preventive management. McGraw-Hill, New York.
Quick, J.D., Horn, R.S. and Quick, J.C., 1987. Health consequences of stress. In: J.M. Ivancevich and D.C. Ganster (Eds.), Job Stress: From Theory to Suggestion. Haworth Press, New York.
Rabkin, J., 1982. Stress and psychiatric disorders. In: L. Goldberger and S. Breznitz (Eds.), Handbook of Stress. Free Press, New York.
Raggatt, P.T.F., 1991. Work stress among long-distance coach drivers: A survey and correlational study. Journal of Organizational Behavior, 12: 565–579.
Rahim, M., 1983. Measurement of organisational conflict. Journal of General Psychology, 109: 189–199.
Revicki, D.A. and May, H.J., 1989. Organizational characteristics, occupational stress and mental health in nurses. Behavioural Medicine, 15: 30–36.
Revicki, D.A., May, H.J. and Whitley, T.W., 1991. Reliability and validity of the Work-Related Strain Inventory among health professionals. Behavioral Medicine, 17: 111–120.
Rice, P.L., 1992. Stress and Health. Brooks/Cole, Pacific Grove, CA.
Rizzo, J., House, R. and Lirtzman, S., 1970. Role conflict and ambiguity in complex organisations. Administrative Science Quarterly, 15: 150–163.
Rosen, T.J., Terry, N.S. and Leventhall, H., 1982. The role of esteem and coping in response to a threat communication. Journal of Research in Personality, 16: 90–107.
Rosenberg, M., 1965. Society and the adolescent self-image. Princeton University Press, Princeton, NJ.
Roskies, E. and Louis-Guerin, C., 1990. Job insecurity in managers: Antecedents and consequences. Journal of Organizational Behavior, 11: 345–359.
Roth, S. and Cohen, L., 1986. Approach, avoidance and coping with stress. American Psychologist, 41: 813–819.
Rotter, J.B., 1966. Generalised expectancies for internal versus external control of reinforcement. Psychological Monographs, 80.
Rotter, J.B., 1990. Internal versus external control of reinforcement: A case history of a variable. American Psychologist, 45: 489–493.
Samson, D., 1991. Manufacturing and Operations Strategy. Prentice Hall, Sydney.
Sanders, A.F., 1981. Stress and human performance: A working model and some applications. In: G. Salvendy and M.J. Smith (Eds.), Machine Pacing and Occupational Stress. Taylor and Francis, London.
Sarason, I.G. and Johnson, J.H., 1979. Life stress and job satisfaction. Psychological Reports, 44: 75–77.
Sauter, S.L., Murphy, L.R. and Hurrell, J.J., 1990. Prevention of work-related psychological disorders: A national strat-

egy proposed by the National Institute for Occupational Safety and Health (NIOSH). American Psychologist, 45: 1146-1158.

Scheier, M.F. and Carver, C.S., 1985. The Self-Consciousness Scale: A revised version for use with general populations. Journal of Applied Social Psychology, 15: 687-699.

Schlosser, D. and McBride, J.W., 1984. Estimating the prevalence of alcohol and other drug related problems in industry. Report to the New South Wales Drug and Alcohol Authority (B 84/3), Sydney.

Seamonds, B.C., 1986. The control of absenteeism. In: S. Wolf and A.J. Firesone (Eds.), Occupational Stress. Health and Performance at Work. PSG Publishing Co., Littleton, MA.

Selye, H., 1936. A syndrome produced by diverse noxious agents. Nature, 138: 32.

Selye, H., 1982. History and present status of the stress concept. In: L. Goldberger and S. Breznitz (Eds.), Handbook of Stress. Free Press, New York.

Shapiro, A.P., 1961. An experimental study of comparative responses of blood pressure to different noxious stimuli. Journal of Chronic Disease, 13: 293-311.

Sharit, J. and Salvendy, G., 1982. Occupational stress: Review and reappraisal. Human Factors, 24: 129-162.

Shirom, A., 1988. Situationally anchored stress scales for the measurement of work-related stress. In: J.J. Hurrell, L.R. Murphy, S.L. Sauter and C.L. Cooper (Eds.), Occupational Stress: Issues and Developments in Research. Taylor and Francis, London.

Siegrist, J. and Klein, D., 1990. Occupational stress and cardiovascular reactivity in blue-collar workers. Work and Stress, 4: 295-304.

Simon, H.A., 1976. Administrative Behaviour. Free Press, New York.

Spielberger, C., Gorsuch, R. and Lushene, R., 1970. State-Trait Anxiety Inventory. Consulting Psychologists Press, Palo Alto, CA.

Spring, B.J., Lieberman, H.R., Swope, G. and Garfield, G.S., 1986. Effects of carbohydrates on mood and behaviour. Nutrition Reviews: Diet and Behaviour, 44: 51-61.

Stokes, G. and Cochrane, R., 1984. A study of the psychological effects of redundancy and unemployment. Journal of Occupational Psychology, 57: 309-322.

Stone, A.A. and Neale, J.M., 1982. Development of a methodology for assessing daily experiences. In: A. Baum and J.E. Singer (Eds.), Advances in Environmental Psychology - Environmental and Health, Lawrence Erlbaum, Englewood Cliffs, NJ, pp. 49-83.

Tepas, D.I., 1985. Flexitime, compressed workweeks and other alternative work schedules. In: S. Folkard and T.H. Monk (Eds.), Hours of Work. John Wiley, Chichester.

Thomas, M.R., Rapp, R.S. and Gentles, W.M., 1979. An inexpensive automated desensitisation procedure for clinical application. Journal of Behaviour Therapy and Experimental Psychiatry, 10: 317-321.

Tinning, R.J. and Spry, W.B., 1981. The extent and significance of stress symptoms in industry - with examples from the steel industry. In: E.N. Corlett and J. Richardson (Eds.), Stress Work Design and Productivity. Wiley, Chichester.

Turnage, J.J. and Spielberger, C.D., 1991. Job stress in managers, professionals and clerical workers. Work and Stress, 5: 165-176.

Turner, J.A. and Karasek, R.A., 1984. Software ergonomics: Effects of computer application design parameters on operator task performance and health. Ergonomics, 27: 663-690.

Veit, C.T. and Ware, J.E. Jr., 1983. The structure of psychological distress and well-being in general populations. Journal of Consulting and Clinical Psychology, 51: 730-742.

Weiss, H.M., Ilgren, D.R. and Sharbaugh, M.E., 1982. Effects of life and job stress on information search behaviours of organisational members. Journal of Applied Psychology, 67: 60-66.

Wickens, C.D., 1984. Engineering Psychology and Human Performances Charles Merrill, Columbus, OH.

Williams, D.R. and House, J.S., 1985. Social support and stress reduction. In: C.L. Cooper and M. Smith (Eds.), Job Stress and Blue Collar Work. Wiley, London pp. 207-224.

Williamson, A.M., Gower, C.G.I. and Clarke, B., 1993. Changing the hours of shiftwork: Comparing 8 and 12 hour rosters. Ergonomics (in press).

Wilson, J.R. and Corlett, E.N., 1991. Evaluation of Human Work. A Practical Methodology. Taylor and Francis, London.

Wolff, H.G., 1952. Stress and Disease. Charles C. Thomas, Springfield, MA.

Wood, L.W., 1986. Absenteeism. Occupational Medicine: State of the Art Reviews, 1: 591-607.

Economic evaluation of ergonomic solutions: Part I – Guidelines for the practitioner *

E. Roland Andersson

Department of Work Science, The Royal Institute of Technology, S-10044 Stockholm, Sweden

A. The practitioner

These guidelines are aimed at practitioners who want to examine the profitability of 'ergonomic solutions', or to increase the reliability of investment appraisal with respect to ergonomics. The practitioner can be anyone in the production system (ergonomist, engineer, occupational health and safety professional, manager, supervisor, etc.).

B. Scope of the guidelines

The guidelines mainly focus on different methods and criteria for capital expenditure evaluation, and their relative merits and limitations with respect to ergonomics. Evaluation methods and criteria can be useful at various stages in the evaluation of an ergonomic solution:
- in the formal accept/reject decision of an ergonomic project (a solution alternative) with respect to profitability,
- in the choice between several ergonomic options (alternative solutions) that meet the goal, i.e., solve the same problem,
- in ranking independent projects, in case the capital required exceeds available funds,
- in choosing between an ergonomic project and other investments, and
- in controlling costs of implemented solutions (actual costs versus budget).

In each case, costs and revenues must be identified, quantified, and related to each other. The data needed and the format in which this relationship can be expressed and used will be considered in this guide. A 'project' is a proposal that requires capital expenditures to solve an ergonomic problem.

Correspondence to: E. Roland Andersson, Department of Work Science, The Royal Institute of Technology, S-10044 Stockholm, Sweden.

* The recommendations provided in this guide are based on numerous published and unpublished scientific studies and are intended to enhance worker safety and productivity. These recommendations are neither intended to replace existing standards, if any, nor should be treated as standards. Furthermore, this document should not be construed to represent institutional policy.

The following individuals participated in the discussion of the earlier version of this guide. Their suggestions (written or verbal) were incorporated by the authors in this version: Alvah Bittner, *USA*; Peter Buckle, *UK*; Jan Dul, *The Netherlands*; Bahador Ghahramani, *USA*; Juhani Ilmarinen, *Finland*; Sheik Imrhan, *USA*; Asa Kilbom, *Sweden*; Shrawan Kumar, *Canada*; Tom Leamon, *USA*; Mark Lehto, *USA*; William Marras, *USA*; Barbara McPhee, *Australia*; James Miller, *USA*; Anil Mital, *USA*; Don Morelli, *USA*; Maurice Oxenburgh, *Australia*; Jerry Purswell, *USA*; Jorma Saari, *Finland*; W. (Tom) Singleton, *UK*; Juhani Smolander, *Finland*; Terry Stobbe, *USA*; Rolf Westgaard, *Norway*; Jørgen Winkel, *Sweden*. The guide was also reviewed in depth by several anonymous reviewers.

C. Concepts and definitions

The profitability of capital expenditures in ergonomic projects is influenced by:
- initial capital outlay and salvage value after life-time,
- annual expenses and revenues,
- life-time, and
- discount-rate.

Capital expenditures may be defined as any act involving 'the sacrifice of an immediate and certain satisfaction in exchange for a future expectation'. The distinguishing feature of a capital expenditure is that the costs that it incurs and the revenue it generates arise over a period of time. Most investment decisions are taken on the basis of the relationship between the cost of the project and the incremental profit expected to accrue to the firm as a result of undertaking the project. The goal is to evaluate uncertain future cash flow and to relate it to outlays in the immediate and near future.

The *life-time* is the project's technical life or the duration over which the project remains economically worthwhile.

The *initial capital outlay* is the original investment required to initiate a project, such as purchase, development and implementation costs. It is the actual cash expenditure incurred to acquire or replace a resource. It is the cash that must be given up. The *salvage value* is the remaining value of the project after its life-time.

The *annual expenses* are the cash outflows due to the project, over the project's life-time. In investment appraisal only *specific* cash outlays and expenses of different project alternatives are taken into consideration, other things being equal.

A *revenue* is either a cash inflow or a cost reduction (a saving) resulting from the project, for example, savings resulting from increased productivity, and a reduction of costs connected with absenteeism and labor turn-over.

In cost calculations the costs are normally divided into direct and indirect costs (overheads). The 'direct costs' are those costs that can be easily identified with the production of a unit, normally direct labor (salary or wage) and direct material. 'Overhead' includes all other costs. The costs of absenteeism and labor turnover are normally part of the factory overhead. In cost calculations the factory overhead is normally added to direct labor and direct material costs, and is a certain percentage. This percentage (rate) is normally based on total overhead divided by production volume or total hours worked. In cost calculation predicted costs (standards) per unit of various cost factors are used. Those 'standard costs' are based on historical costs (from accounting records and reports), present or expected market values, and time studies of production. 'Standards' are normally developed and used for various cost factors, for example absenteeism and labor turnover. In human resource calculation we should also distinguish between actual cash expenditures (outlay costs), and imputed costs. 'Imputed costs', for example the difference in productivity between a trainee and a skilled employee, do not require actual cash outlays and are thus not shown in the accounting records and reports as the 'outlay costs'.

The *discount-rate* expresses the time preference for payments at different times. It converts payments in the future to their base year equivalents and makes payments at different times comparable. Normally, the discount-rate expresses the demands for return on investment. The minimum rate or return specified by management is commonly referred to as the *hurdle rate* or *minimum attractive rate of return* (MIRR). This rate is based on the firm's cost of capital. Estimating a firm's cost of capital is an extremely difficult task. In general, one can define the cost of capital as the rate of return required by those who provide capital to the firm. The rates of return required by the different suppliers of capital – debt holders, preferred stockholders and common stockholders – are determined in the market by the actions of investors competing against one another.

The assessment of the investment criteria is based upon the assumption that the objective of the decision maker is to maximize profit.

D. Problem identification

The principal problem of capital budgeting in most companies is allocation of available funds to the most worthwhile projects. Therefore, *quantitative* evaluation methods and criteria are important in ranking projects, and for formal accept/reject decisions. In investment appraisal of capital expenditures in ergonomic projects the relationship between cost of the project and expected savings from four key-factors, (1) labor turnover, (2) absenteeism, (3) spoiled and defective goods and (4) productivity, is of special interest (other savings are also possible). The steps described in sections E–G should be taken (see also table 4).

E. Data collection

Annual savings in key-factors
(see also Part II, section 3)

(1) Labor turnover. Determine the total annual number of labor turnover and total 'replacement costs' that can be eliminated by the project.

Note: 'Replacement costs' include acquisition costs, development costs, and separation costs. Many companies have developed 'standard costs' for the acquisition, development, and separation related to various company jobs. If not, 'standard costs', to be used in cost calculation, have to be developed. Determine replacement costs per employee from actual replacement costs and total number of turnover per year for various jobs.

Empirical work has shown that replacement costs for workers can be predicted to be between, 10,000–60,000 SEK/employee (1989), depending on supply and demand of labor and qualifications needed.

Costs to be included in turnover standards

Acquisition costs are the costs incurred in the recruitment, selection, hiring, and induction of employees. 'Recruitment costs' would include job advertisements, agency fees, recruiters' salaries and benefits, travel and entertainment, materials such as recruitment brochures, administrative costs, etc. The principal costs of 'selection' are interviewing, testing, evaluating and any other processing cost. 'Hiring costs' are incurred in bringing the employee into the enterprise.

Development costs are the costs to train a person to the level needed or to enhance the individual's skills. The three components of development costs are orientation costs (general information about the job and the company), on-the-job training cost and off-the-job training costs. One of the principal costs during 'on-the-job training' is the imputed cost, that is, the difference between the productivity of a trainee and of an experienced employee. Various types of learning curves have been developed to show graphically the reduction in learning cost as the training progresses and productivity increases. Specific costs for 'orientation' and 'off-the-job training' include salaries for persons providing the training and information, materials, travels, etc.

Separation costs include severance pay costs, low-performance costs, and vacant position costs. 'Low-performance cost' represents the loss of productivity before the employee leaves the company. 'Vacant position costs' may include imputed costs, due to losses from definite sales, etc., but may also cause cash outlays, due to overtime for others. 'Severance pay cost' is the amount of prepaid compensation given to the employee upon leaving the company.

Example

Savings in imputed costs. Training time is estimated to average 16 weeks per individual. During training it is estimated that 40% production is achieved by each individual. This means that the company loses 60% in 16 weeks or 384 hours of production (work week = 40 hours). For each turnover saved, the savings in imputed costs will be direct labor costs per hour (salary incl. benefits and taxes) $\times$ 384 hours. Payroll taxes and fringe benefits are normally treated as direct labor and applied to gross wages by a fixed percentage.

(2) Absenteeism. Determine the total cost for absenteeism resulting from accidents and musculoskeletal disorders that can be eliminated by the project.

Note: Costs for absenteeism include compensation costs, medical costs and replacement costs. If 'standard costs' are not available in the company for various jobs they must first be developed. Determine absenteeism costs per hour from total actual costs for absenteeism and total actual hours of absenteeism per year for various positions. Rate and length of absenteeism for injuries (accidents, illness and diseases) can often be determined from health and safety records.

Empirical work has shown that the total cost/hour is generally between 2 and 10 times the direct compensation costs (sickness wages, payroll taxes and fringe benefits that have to be paid by the company to the injured person during absence).

Costs to be included in absenteeism standards

Compensation costs are the costs that have to be paid by the company to the injured person during his or her absence. They include sickness wages, payroll taxes and fringe benefits such as pension and vacation. These costs are today often insured (or taxed or funded) and paid jointly by the employer and the employee. The cash outlay therefore varies due to actual insurance systems, agreements, terms of absenteeism, tax regulations, etc.

Medical costs include costs for medical care and hospitalization paid by the company.

If the work is not assumed to be done by another person during the employee's absence or if the absent employee will not take care of it upon returning, appropriate production *replacement costs* must also be taken into consideration. See *(1) Labor turnover* for appropriate cost factors (e.g. over-employment, productivity-losses, overtime, etc.). Also include costs for retraining.

Example
- Compensation costs: 30 SEK/hour (sickness wages, payroll taxes and fringe benefits during absence)
- Medical and replacement costs: 150 SEK/hour (estimated by experience to 5 times the compensation costs)
- Total costs/hour: 30 + 150 = 180 SEK/hour
- Average length of absenteeism: 20 days (from occupational health and safety records)
- Injuries reduced: 10 injuries/year (estimated from occupational health and safety records)

Total savings: 180 SEK/hour × 8 hours × 20 days × 10 injuries = *288,000 SEK/year*.

(3) Spoiled and defective goods. Determine the total annual number of units and costs for spoiled and defective goods that can be eliminated by the project.

Note: The difference between spoiled goods and defective goods is that defective goods are reworked to be sold with good units, while spoiled goods are sold for their salvage value or just discarded without additional work being performed on them. Costs for spoiled and defective goods are often available from accounting records and reports. Determine the imputed costs for rework per unit defective goods (direct wages incl. payroll taxes and benefits) and the losses (income losses or productive time losses incl. direct wasted material, whichever is relevant) from the spoiled goods. Actual numbers of spoiled and defective goods that can be eliminated by the project can be determined from accounting records and reports and from interviews with manufacturing personnel.

(4) Productivity. Determine possible changes in labor efficiency from time and motion studies with prototypes, etc. (if possible) or estimate possible changes from experience.

Note: Experimental and empirical findings have shown an increase in productivity up to 20% due to ergonomic measures. The possible outcome is the difference between the actual number of hours needed for a job in question, and the estimated number of hours needed due to the 'solution'.

Example

The time needed to complete a job is estimated to be 1.5 hours/unit. As a result of ergonomic measures a 10% increase in productivity is expected (from time studies). The saving/unit is then direct labor costs (including benefits and payroll taxes) × 10% of 1.5 hours. Total savings depend on the actual number of units manufactured.

Costs of the project

Initial outlay. Determine the original investment in the project (purchase, investigation, development and implementation costs).

Expenses. Determine the annual cash outlay for the project for every year under its life-time (service, materials, etc.).

Life

Determine the project's *life* (technical and economic, whichever is smaller).

Determine the *salvage value* of the investment after the life of the project.

Discount rate

Determine the 'hurdle rate' (or MIRR) for the project to be profitable. The hurdle rate is often easily available in the company. If not, use actual market interest rate as a starting point.

F. Data analysis
(see also Part II, section 4)

F1

Perform capital expenditure analysis from the data collected by using capital expenditure evaluation methods. Typically the following four different methods are used in evaluating the profitability of different projects: (*i*) Net present value (NPV), (*ii*) internal rate of return (IRR), (*iii*) annuity method (AM) and (*iv*) the payback period method (PPM).

A project example

To illustrate the different methods for investment appraisal, their criteria and relative merits, an example will be given. The project example, which is based on data from earlier work (Andersson, 1988, 1990a,b) is about the decision whether to buy a buttress for a hand-held chisel hammer for a building site or not. Hand-held chisel hammers are used on construction sites for cracking concrete blocks, bricks etc., and weigh between 6 and 20 kg. The use of chisel hammers entails exposure to demanding and unsuitable handling, especially when the machines are held high. The statistics show that such chiseling causes 96 injuries (accidents and diseases) per 1 million chiseling hours. The injuries are primarily due to strenuous movements and working positions. The average absenteeism is 20 days per injury. The compensation costs for absenteeism are 30 SEK/hour (insured salary, fringes and taxes), and the medical and replacement costs are estimated from experience to be 10 times the compensation costs. Labor turnover and quality of work are not affected by the proposed solution. The economic life of the buttress is estimated at 3 years (the construction time of the building). The initial capital outlay (purchase price) is 4,000 SEK, with no salvage value after the construction. According to a time and motion study, the buttress reduces the work-load from 50 to 40% of maximal aerobic power, and makes possible an increase of 15% in productivity. The annual use of the buttress is estimated to be 100 hours and the direct labour cost is 150 SEK/hour (incl. fringes and taxes). The discount rate for capital investment is established by the company to be 20%.

Table 1
Annual savings.

Savings from increased productivity:
100h × 150SEK/h × 15% = 2,250 SEK

Reduced risk of injuries:
96/1000000 × 100h × (30 + 300) × 30days × 8h = 510 SEK

Total annual savings
2,250SEK + 510SEK = *2,760 SEK*

From these data an annual savings of 2,760 SEK was calculated (table 1). Two types of questions must now be answered: the formal accept/reject question, and how the project should be ranked in comparison with the other projects (alternative solutions to the same problem and other independent projects).

(i) Net Present Value (NPV)

The NPV is defined as the present value equivalent of all cash inflows less all cash outlays associated with a project. If the NPV is greater than zero, the project is worthwhile from an economic standpoint. If a choice has to be made between projects, the project with the greatest NPV should be selected. The NPV method converts payments in the future to present values and makes them comparable. The NPV method can be expressed mathematically as:

$$\text{NPV} = \sum_{i=1}^{i=n} \frac{S_i}{(1+r)^i} - C \quad (1)$$

where i = the actual year, S = annual savings, $1/(1+r)^i$ = the discounting factor, r = the discount (interest) rate, and C = the initial outlay (investment).

As shown by the calculations in table 2 the NPV is +1,814 SEK for the buttress. Hence, a purchase is acceptable according to the above criterion because the NPV is positive, and the investment is thereby profitable. If a choice among various projects has to be made, because projects are competitive or investment funds are limited, the projects should be ranked for selection purposes in order of their NPVs. That is, the greater the NPV, the higher the ranking of a competing project. If the competing projects differ in capital outlay, the relationship between the absolute dif-

Table 2

Calculation of the NPV with $r = 20\%$.

Year, i	Outlay C (SEK)	Savings S (SEK)	Discount factor $1/(1+r)^i$	Discounted value $S \times 1/(1+r)^i$ (SEK)
0	−4,000			
1		+2,760	0.8333	+2,300
2		+2,760	0.6944	+1,917
3		+2,760	0.5787	+1,597
=	−4,000			+5,814

NPV = +5,814 − 4,000 = +1,814 SEK

ference in outlay and the expected improvement in NPV must be examined.

Note: Advantage: The NPV takes the long-term effect of the project into account. Disadvantage: The NPV criterion can discriminate 'small' ergonomic projects in ranking. That is, the projects that require 'small' capital outlays.

(ii) Internal Rate or Return (IRR)

This method is also known as the 'solution' rate of interest. It is defined as the rate of interest, r, that equates the discounted present value of expected future receipts to the present value of the stream of cash outlays; that is, it solves the equation for the discount (interest) rate that makes NPV = 0. Thus, in our example, the IRR is the solution for r in

$$4,000 = \sum_{i=1}^{i=3} \frac{2,760}{(1+r)^i} \quad (2)$$

The project is acceptable if the internal (solution) rate of return equals or exceeds the required rate of return (20%). If $r = 50$ is substituted into the equation, the NPV is +3,883 SEK. The solution rate is found to be 48%, and the project is thereby acceptable. In ranking projects the one with the highest IRR over the cost of capital is selected according to the profit-maximising criterion.

Note: Advantage: Like the NPV criterion the IRR takes long term effects of the project into account. Disadvantage: The IRR criterion can discriminate the 'biggest' projects. That is, the projects that require 'big' capital outlays.

(iii) Annuity Method (AM)

The AM converts an initial capital outlay (investment) to equal annual outlays. The conversion is made with the help of an annuity factor,

$$\frac{r(1+r)^n}{(1+r)^n - 1} \quad (3)$$

where r = the discount (interest) rate and n = number of years.

A project is acceptable if the annual savings are equal to or greater than the annuity of the expenditures. In our example the annuity of the initial capital outlay is −1,900 SEK (−4,000 SEK (initial outlay) × 0.4747 (the annuity factor)). That makes an annual average surplus of +860 SEK (2,760 SEK (annual savings) −1,900 SEK (annual outlay)). In accordance with the investment criterion a purchase is acceptable. If a choice between projects must be made, the one with the biggest annual surplus should be selected.

Note: Advantage: Same as the NPV and the IRR methods. Has also some merits in actual cost calculation. Disadvantage: Same as the NPV method.

(iv) The Payback Period Method (PPM)

The payback period is the time, usually expressed in years, required to generate sufficient savings to recover the initial capital outlay of the project. The shorter the payback period, the more favourably a project is regarded. The computation of the payback method is as follows:

$$\text{Payback period } (P) = \frac{\text{Initial outlay } (C)}{\text{Annual savings } (S)}$$

$$= \text{years} \quad (4)$$

The payback period for our example is 1.5 year (4,000 SEK (C)/2,760 SEK (S)).

Note: Advantage: A simple method that may be used to screen alternative projects. Disadvantage: Does not take long-term effects into account, such as interest rate, which could lead to a discrimination of ergonomic projects, compared to others in ranking.

Table 3

Calculation of the Discounted Value (DV) for the annual savings (+ 2,700 SEK) and the Net Present Value (NPV) for the example at different discount (interest) rates (r). Initial outlay (C) = −4,000 SEK.

Year	Discount (interest) rate (r)		
	20%	30%	40%
1	2,300	2,123	1,970
2	1,917	1,633	1,408
3	1,597	1,256	1,005
DV	+5,812	+5,012	+4,383
NPV (DV−C)	+1,814 SEK	+1,012 SEK	+383 SEK

F2

Perform *sensitivity analysis* of the result by varying the prerequisites for the data.

Note: The calculations so far ignore the effects of risk and uncertainty on decision making. *Sensitivity analysis* is essentially a determination of how a change in the key factors in the analysis will affect the decision.

Example

The simplest approach to accounting for uncertainty is to increase the discount (interest) rate of return. For our example table 3 shows how sensitive the NPV is with respect to the required rate of return.

Because of the large uncertainties in estimating the actual costs for labor turnover and absenteeism primarily from present accounting records and reports, also 'low' and 'high' values can be used in capital expenditure evaluation. As mentioned above, a 'low' value for absenteeism can be estimated to be 'compensation costs' × 2, and a 'high' value as 'compensation costs' × 10. In our example, also the 'minimum' outcome for the project to be profitable can be calculated. This corresponds to a raise in productivity of 9% or to a reduction of annual absenteeism of approximately a day (given no raise in productivity at all), provided the annuity of the initial capital outlay (1,900 SEK).

It should also be noted that the calculus boundaries will affect the outcome and, thereby, the ranking of the project. In our example the calculus boundaries were placed 'around' the building site. But if the boundaries had been placed 'around' the construction company, as a whole, the reduced risk for injuries would have motivated a purchase of 54 buttresses (given 200 machines and the same prerequisites).

F3

Make reject/accept decisions of the project on the basis of the investment criterion chosen (and other material).

Note:
NPV criterion: If the discounted value of all cash revenues equals or is greater than the initial capital outlay the project is acceptable.
IRR criterion: If the internal rate of return equals or exceeds the 'hurdle rate' (MIRR) the project is acceptable.
AM criterion: If the annual saving for the project equals or is greater than the annuity of the initial capital outlay the project is worthwhile.
PPM criterion: Time required to generate savings to recover the initial capital outlay.

F4

Rank the acceptable project to alternative projects if a choice among acceptable projects must be made. Either because the projects meet the same goal (solve the same problem) or if the capital required exceeds available funds (independent projects).

Note:
NPV criterion: If a choice has to be made between projects, the project with the greatest NPV should be selected.
IRR criterion: In ranking projects the one with the highest IRR is selected.
AM criterion: The one with the highest surplus should be selected.
PPM criterion: The shorter the payback period, the more favourably a project is regarded.

G. Selected projects

The selected projects will affect the inflow and outflow of cash in the company during its life. The change in the flow of money are taken into consideration in the budget process. Normally two kinds of budgets exist: the capital expenditure budget and the operating annual budget.

The figures within parentheses refer to the project example.

1. Make the necessary changes in capital expenditure budget for the job in question.
   + initial capital outlay   (+ 4,000 SEK)
2. Make the necessary changes in the operating (annual) budget for the job in question.

Table 4
Capital expenditure evaluation form for ergonomic projects.

Project: _____

Initial outlay:
_____: – _____
_____: – _____   Total = – _____

Annual expenses:
_____: – _____
_____: – _____   Total = – _____

Annual revenues/savings:

(1) *Labor turnover*
    Acquisition   + _____ /employee
    Development   + _____ /employee
    Separation   + _____ /employee
    Total cost   + _____ /employee   × Total number saved _____ = + _____

(2) *Absenteeism*
    Compensation + _____ /hour
    Medical care + _____ /hour
    Replacement   + _____ /hour
    Total cost   = _____ /hour   × Total hours saved   _____ = + _____

(3) *Spoiled and defective goods*
    Rework   + _____ /unit   × Total units saved   _____ = + _____
    Losses   + _____ /unit   × Total units saved   _____ = + _____

(4) *Labor efficiency*
    Direct labor: _____/hour
    × Total productive hours: _____
    = Total direct labor costs: _____ × change in %   _____ = + _____

(5) *Other savings* (see enclosure)   = + _____

                                               Total (1 + 2 + 3 + 4 + 5)   = + _____

Salvage value: + _____

Discount rate: _____ %   Life-time: _____ years

RESULTS:
NPV: + / – _____   IRR: _____ %
AM: + / – _____   PPM: _____ years

+ annual expenses
  − costs for absenteeism   (−510 SEK)
  − costs for labor turnover
  − costs for spoiled and defective products
  − costs for direct labor
    due to productivity   (−2,250 SEK)
3. Establish control procedures in the accounting and the statistics departments.
   − actual outcome relative to budget

Reference

Andersson, E.R., 1992. Economic evaluation of ergonomic solutions: Part II – The scientific basis. International Journal of Industrial Ergonomics, 10: 173–178 (this issue).

Economic evaluation of ergonomic solutions: Part II – The scientific basis

E. Roland Andersson
Department of Work Science, The Royal Institute of Technology, S-10044 Stockholm, Sweden

1. Problem description

Capital expenditure justification can be made before the actual investment or after. In recent years the need to study capital expenditures before rather than after the commitment is made has been emphasized. The purpose of capital budgeting is the process of ensuring that capital expenditures planned represent the most profitable outlays of funds, that these expenditures are in accordance with company policy and that such expenditures do not jeopardize the financial well-being of the company. The major goal of capital budgeting in most companies is the allocation of available funds to the most worthwhile projects. Therefore, quantitative evaluation methods and criteria are important in ranking projects in order of priority and for making formal accept/reject decisions. Procedures for funding allocation vary widely among companies; from 'rule of thumb' to the use of scientific methods.

Several recent studies have shown that capital expenditures in technical and organizational 'solutions' to ergonomic problems can yield profitable outcomes through increased productivity and reduced absenteeism (see Spilling et al., 1986, Liukkonen, 1987; Andersson, 1988; Simpson and Mason, 1990). It has also been demonstrated that the choice of investment methods and criteria may have a decisive effect on the selection and outcome of expenditures in ergonomic 'solutions' (Andersson, 1988; Westlin, 1989; Simpson and Mason, 1990). Westlin (1989) has also indicated, from a study in Swedish companies, that capital expenditures in ergonomic 'solutions' are often discriminated, because companies lack knowledge about actual costs and possible outcomes of ergonomic projects in capital planning and control.

The purpose of these guidelines is to discuss different quantitative methods for capital expenditure evaluation, their relative merits and limitations from an ergonomic viewpoint. That is, how to justify, economically, ergonomic solutions and how to choose the most economical one among alternatives. The ergonomic investment 'project' is a proposed solution to a problem that requires capital expenditures. This might involve redesigning existing machines, jobs, tools, etc. or developing entirely new ergonomic tools, designs, and concepts.

2. Scope of the problem

Quantitative evaluation methods and criteria are needed at several stages in the project evaluation process:
- in the formal accept/reject decision of an ergonomic project (a solution proposal), with respect to profitability.
- in choosing between several ergonomic projects (alternative solutions) that accomplish the same goal, i.e., effectively solve the same problem.
- in ranking independent ergonomic projects, because the amount requested for capital ex-

---

Correspondence to: E. Roland Andersson, Department of Work Science, The Royal Institute of Technology, S-10044 Stockholm, Sweden.

penditures may be greater than the available funds.
- in choosing between an ergonomic project and another non-ergonomic investment project, and
- in controlling cost of implemented solutions.

In each case, costs and revenues must be identified, quantified and related to each other. The data needed and the forms in which this relationship can be expressed and used will be considered in detail in this guide. The emphasis will be on ergonomic projects.

3. Background and need for further research

3.1. Cost data relevant to ergonomics

Cost accounting is primarily concerned with gathering and analysing cost information, for internal use by managers for planning, control and decision making. In investment appraisal of capital expenditures, the relationship between the cost of the project and expected savings in four key-areas, absenteeism, labor turnover, spoiled and defective goods and productivity, is of special interest. The data needed for capital expenditure evaluation of a proposed ergonomic solution may, therefore, include the following:

(1) A breakdown of machine, job or process operating hours;
(2) Hourly wage rate, classified by worker function;
(3) Total labor hours and total labor costs;
(4) A breakdown of labor hours and costs for absenteeism resulting from accidents and musculoskeletal disorders, by machine, job or process;
(5) Total labor hours and costs for absenteeism resulting from accidents and musculoskeletal disorders,
(6) A breakdown of labor turnover and costs by machine, job or process;
(7) Total labor turnover and costs;
(8) A breakdown of units and costs for spoiled and defective goods by machine, job or process;
(9) Total costs and units for spoiled and defective goods;
(10) Efficiency standards for labor, and
(11) Variances in data.

The costs elements of a product, or its integral components, are materials, labor and overhead. Material and labor are normally divided into direct and indirect costs. The direct costs are those costs that can be easily identified with the production of a finished product. Overhead costs are all other costs – other than direct labor and direct material – of manufacturing a product (Cashin and Polimeni, 1985). The definition is thereby dependent on how the costs are defined and accounted. In cost calculations the overhead is added to direct labor and direct material costs and is a certain percentage. This percentage (rate) is normally based on total overhead divided by production volume or total hours worked. The costs of absenteeism and labor turnover are normally part of factory overhead.

In calculation of *absenteeism* costs, there may be compensation costs such as sickness wages (normally an insured cost), fringe benefits such as pension and vacation plus payroll taxes that the company has to pay for the injured employee during absence. These costs can easily be traced to the injured and thereby be treated like direct costs. Overhead, in absenteeism cost calculation, is all other costs such as medical costs and production replacement costs resulting from the injury but not easily traced to it. A number of studies have also pointed out the difficulties in estimating especially the replacement costs caused by accidents and musculoskeletal disorders (Kjellen, 1986; Westlin, 1989; Harms-Ringdahl, 1990; Söderquist et al., 1990). Therefore, accounting practices have to be developed within companies, if capital expenditure evaluation and later control procedures of ergonomic expenditures are to be reliable. The problems of estimating the replacement costs have been discussed by Hjort (1978), Samuelson and Mauro (1983), Markkanen (1986), Liukkonen (1987) and Brody et al. (1990). In general, these studies estimate the total costs for accidents and injuries to be between 2 and 10 times the direct compensation costs, depending on the differences in the insurance systems in each country and trade, the definition of direct and indirect costs and the accounting methods used.

Also the costs for *labor turnover* (acquisition, development and separation) are hard to estimate from the present cost accounting practices.

However, Cashin and Polimeni (1985) have suggested a cost structure, and Liukkonen (1989) has, from empirical studies in Swedish companies, estimated the costs for labor turnover to be between 10,000–60,000 SEK per worker, depending on supply and demand for labor and qualifications needed.

The information about *spoiled and defective goods* is often available from accounting reports. It is, however, not easily traced to ergonomic problems. Especially the possibility to address costs for spoiled and defective goods to ergonomic problems needs to be established.

Because of the difficulties in obtaining reliable information about costs for absenteeism, labor turnover, and quality from present accounting practices, Liukkonen (1987) recommends the firms to determine 'standard costs' based on actual hourly rates, time estimations and statistics available. The 'standards' are to be used in capital planning and control.

3.2. The use of statistics

From health and safety records (at national, branch or company level) it is often possible to determine (or estimate) both the rate and the duration of absenteeism that can be eliminated by the proposed solution. Also, labor turnover rate is often available within companies. However, the quality of data varies widely and must be improved, if the cost calculations (and later cost control procedures) are to be reliable. Especially the relationship between the type and nature of absenteeism and turn-over and certain machines, jobs and processes needs to be established. Injury-rates for accidents, diseases and illness are normally determined from numbers of actual injuries in relation to hours worked. This relationship (the injury-rate) is often expressed as 'injuries per 1 million hours worked'. By using data obtained from statistics, in combination with costs estimated on hourly or other unit bases (see above), the total costs for absenteeism and labor turnover that can be eliminated by the proposed ergonomic solution can be calculated.

3.3. The use of efficiency standards

Planning is defined as the formulation of objectives as well as plans of actions to achieve these objectives. Objectives and plans are prepared on a short- and long-term basis to provide guidelines for daily operations as well as future activities. The information provided by a cost accounting system is combined with other data and analysed. Based on these findings the management makes decisions and formulates strategies for the future, affecting areas such as employees' welfare and health.

The planning often also results in *efficiency standards* which can be used in capital expenditure evaluation. 'Efficiency standards' are primarily used by management to determine the effectiveness of operations by comparing them with actual costs. 'Efficiency standards' for labor are often developed with the help of time and motion studies. In these studies an analysis is made of the procedures to be followed by workers and the conditions under which the worker must perform assigned tasks. Several time-studies have shown that ergonomic measures that lead to improved postures and loadings often also lead to increased productivity. For example, experimental findings by Andersson et al. (1980), Andersson (1988), and Glimskär and Höglund (1978) have shown an increase in productivity as high as 20%. Andersson (1988) has also shown that time-studies with simplified models, or early prototypes, can be used as reliable predictors of the final outcome. Results from time and motion studies are, therefore, important in capital expenditure evaluation and cost control of proposed ergonomic solutions. A widely used measure for productivity is output per man-hour of labor. The saving is the difference between actual hours required for a specified job and the estimated hours needed, when the proposed solution is implemented.

4. Methods for capital expenditure evaluation

Typically, the following different methods are used in evaluating the profitability of proposed solutions: (*i*) Net Present Value (NPV), (*ii*) Internal Rate of Return (IRR), (*iii*) Annuity Method (AM) and (*iv*) the Payback Period Method (PPM).

4.1. Summary of the methods

(i) NPV criterion:
If the discounted value of all cash savings (or revenues) equals or is greater than the initial

capital outlay, the proposed solution is acceptable.

(ii) IIR criterion:
If the internal rate of return exceeds the Minimum attractive Rate of Return (MIRR), the proposed solution is acceptable.

(iii) AM criterion:
If the annual savings for the proposed solution equals or is greater than the annuity of the initial capital outlay, the proposed solution is worthwhile.

(iv) PPM criterion:
Time required to generate savings to recover the initial capital outlay.

4.2. Comparison of the methods

The *payback method* has the virtue of simplicity. The calculation is admittedly simple, but it is less apparent that this is a virtue since there are very serious deficiencies associated with its use. First, it does not consider the time pattern of receipts. Second, it takes no account of any earnings that may accrue to a project after the expiration of the payback period. It favors, therefore, short-term projects and ignores the possibility of long-term growth in profits. In fact it is not a profitability criterion but the embodiment of a liquidity concept; that is, it is concerned with the rapid recovery of outlays. However, where liquidity, rather than profitability, is an important business consideration the method has merit. It may also have some merit as an initial screening method. However, the most important deficiency of the method compared to the discounting methods (NPV, IRR, AM) is that it favors projects that yield immediate profits.

Concerning the differences between the various *discounting methods*, we have on the one hand the NPV and the AM methods, which consider the total savings, and on the other hand the IRR method that provides the return in relation to initial outlays. In the evaluation of investment opportunities two types of decisions must be made: the formal accept/reject decision and the ranking of alternative projects. In the first case the two discounting criteria produce identical answers. All projects that have yield in excess of the cost of capital must have a positive NPV when discounted at the marginal cost of capital. In other words, all projects accepted on an NPV basis have a discount rate higher than that given by the marginal cost of capital to reduce their NPV to zero. When the decision is one of ranking, however, they may not give the same results. It has frequently been argued that the IRR method does not rank projects in their true order of profitability. The reason for this is that the NPV method and the AM show absolute figures, whereas the IRR is related primarily to the amount of capital involved and the duration of the investment period. That is, a project may have a high rate of return in relation to initial investment but might yield a low absolute amount of profit. It may be argued that most decisions will be of the accept/reject type, in which the additional information generated by the NPV method will be irrelevant. However, there are situations in which the ranking projects become important, for instance, when capital rationing prevails, or when mutually exclusive choices are encountered. The term 'capital rationing' is used to describe a situation in which a firm cannot raise capital beyond a fixed limit. In this case the true opportunity cost of capital is clearly in excess of the market rate. Under these conditions it is impossible to calculate a well-defined cost of capital with which to evaluate projects on an IRR basis. The majority of economists also consider the NPV method to be superior. The reason for this preference is profit maximization for the investment. Therefore, the projects that produce the highest NPV, which are not necessarily those with the highest IRR, should be ranked higher.

Concerning the relative merits of the different methods from an ergonomic standpoint, there is insufficient information. Westlin (1989) suggests that the payback method should be limited to projects with a short economic life. This is because ergonomic measures normally take a long time to yield results. Therefore, he suggests using the NPV method. However, the higher the discount-rate, the more the long-term solutions discounted by the NPV method are discriminated (see Part 1, table 3). There is also a lack of studies investigating the influence of ergonomic measures on accidents and diseases as a function of time. In comparison with the IRR method, the NPV method (and the Annuity Method) favor projects that produce the highest absolute amount of money. These projects are not necessarily those with the highest IRR. Since ergonomic solutions are often unique the IRR method seems preferable to other methods, at least in ranking with other types of project. However, most economists recommend the use of several quantitative methods in investment appraisal.

4.3. Risk and uncertainty

In economics a distinction has long been drawn between 'risk' and 'uncertainty'. 'Risk' describes situations in which a plurality of outcomes is possible and where objective probabilities of occurrence can be assigned to each possible outcome. 'Uncertainty' describes a state of nature in which such objective probabilities cannot be determined. Most business decisions are characterised as involving uncertainty because they are unique decisions. The existence of uncertainty has a number of different implications for the economic analysis. In a narrow sense, it poses problems in formulating techniques to appraise investments. However, in another context, uncertainty is relevant to the development of more broadly based theories of corporate policy making and the role of quality analysis.

The simplest approach to the accounting for presence of uncertainty is to increase the discount rate of return. The discount rate of return used to discount the cash flow from operations in the NPV analysis or used as the 'hurdle rate' in the IRR analysis assumes that projects have the same risk. Management can require a premium over the required rate of return to account for additional risk. Such an addition reduces the value of the discount factor. Therefore, the greater the uncertainty, the more the expected return is reduced because expected cash flows are multiplied by a smaller fraction. However, in addition to the difficulty of determining the appropriate premium, this approach implies that uncertainty is a simple increasing function of time, which is frequently not the case.

The probabilities relating to a possible outcome may also be used as weights that may be applied to estimated cash-flow for each year of a project's life.

Alternatively, present value equivalents for each possible outcome may be calculated and then weighted according to the probability of their likely occurrence. The result is a weighted present value of the outcome.

Given that probabilities cannot be determined on an objective basis, we must use other strategies to appraise investment. In this regard, the available literature recognises a number of different investment strategies based upon different attitudes to the presence of uncertainty. In most business decisions, the factors that affect the decisions are subject to variations from the expected value. *Sensitivity analysis* is essentially a determination of how a change in the key factors in the analysis will affect the decision.

The firm may also choose to perform the analysis based on the most optimistic and the most pessimistic conditions for the cash flow from operations and the initial outlay. If the investment project is profitable under the most pessimistic conditions, the decision makers will have more confidence in their decision. Because of the large uncertainties in estimating the actual costs for accidents and injuries in the first place, Harms-Ringdahl (1990) suggests the use of both 'low' and 'high' values in capital expenditure evaluation.

The cost-effectiveness analysis has also been suggested due to the uncertainty in estimating the real costs for absenteeism and labor turn-over from the present accounting practices. Cost-effectiveness analysis relates the cost of the project to possible effects on absenteeism and labor turnover rates (Schaapveld et al., 1990).

References

Andersson, E.R., 1988. The use of system groups in product development: An experiment from the perspective of ergonomics. Royal Institute of Technology, doctoral thesis, Report Trita-AAV-1022, ISSN 0280-7521, Stockholm.

Andersson, E.R., 1990a. A systems approach to product design and development: An ergonomic perspective. International Journal of Industrial Ergonomics, 6: 1–8.

Andersson, E.R., 1990b. Adoptionsprocessen för en arbetsmiljöinnovation (The adoption process for an ergonomic innovation). The Swedish Council for Building Research, Report R76: 1990, Stockholm.

Andersson, E.R., Hellsten, M. and Westerdahl, B., 1980. Ergonomi vid arbete i ledningsgravar (Ergonomics while working in pipeline trenches). The Swedish Council for Building Research, Report R68: 1980, Stockholm.

Brody, B., Rohan, P.C., Letourneau, Y. and Poirier, A., 1990. Real indirect costs of work accidents: Results from our new model. Abstract, Journal of Occupational Accidents, 12: 99.

Cashin, J.A. and Polimeni, R.S., 1985 (3rd ed.). Cost accounting. McGraw-Hill, London.

Glimskär, B. and Höglund, P.-E., 1978. Montering av TRP-plåt på tak (Erecting corrugated metal sheets in roof construction). Royal Institute of Technology, Report Trita-BEL-004, Stockholm.

Harms-Ringdahl, L., 1990. On the evaluation of systematic safety work at companies. Journal of Occupational Accidents, 12: 89-98.

Hjort, L., 1978. Yrkesskador og sykefravaer: hva koster det? (Occupational accidents and absenteeism: what is the cost?) Norak Produktivitetsinstitutt, Oslo.

Kjellen, U., 1986. Att beräkna kostnader för olycksfall inom företag: förstudie inom nordisk möbelindustri (Cost accounting of occupational accidents: A preliminary study in Nordic furniture industry). Swedish Council for Building Research, Report G3: 1986: 48-57, Stockholm.

Liukkonen, P., 1989. Vad kostar frånvaron? (What's the cost of absenteeism?). Swedish Employers Association (SAF), Stockholm.

Liukkonen, P., 1988. En företagsekonomisk utvärdering av arbetsmiljöförbättringar i smedjan (Profitability analysis of ergonomic measures in a workshop). Swedish Work Environment Fund, Abstract 1237, Stockholm.

Markkanen, J., 1986. Kostnader för arbetsolyckor och försäkringspremier inom byggnadsindustrin (Costs of occupational accidents and insurances in the building industry) Swedish Council for Building Research, report G3: 1986 98-117.

Samuelson, N.C. and Mauro, S.O., 1983. Benefits of safety management in construction. Reprints from American Society of Civil Engineers congress in Philadelphia, May 16-19, 1983, Pennsylvania.

Schaapveld, K., Bergsma, E.W., van Ginneken, J.K.S. and van der Water, H.P.S., 1990. Setting priorities in prevention, NIPG-TNO, no. 90.099, Leiden, Netherlands.

Simpson, G. and Mason, S., 1990. Economic analysis in ergonomics. In: R. Wilson and N. Corlett (Eds.), Evaluation of human work. A practical ergonomics methodology. Taylor and Francis, London.

Spilling, S., Eitrheim, I. and Aarå, A., 1986. Cost-benefit analysis of work environment, investment at STK's telephone plant at Kongsvinger. In: N. Corlett, J. Wilson and I. Manenica (Eds.), The ergonomics of working postures, models, methods and cases. Proceedings of the first international occupational ergonomics symposium, Zadar, Yugoslavia, 15-17 April, 1985. Taylor and Francis, London.

Söderqvist, A., Rundmo, T. and Aaltonen, M., 1990, Costs of occupational accidents in the Nordic furniture industry (Sweden, Norway, Finland). Journal of Occupational Accidents, 12: 79-88.

Westlin, A., 1989, Arbetsmiljöåtgärder, bakgrund och konsekvenser, Beräkningsmodeller. (Measures in working environment, background and consequencies. Methods for evaluation). Swedish Employers Federation (SAF), Stockholm.

Author Index

Andersson, E.R. 463, 473

Bernard, T.E. 329, 337
Brauchler, R. 1, 9

Dukes-Dobos, F.N. 329, 337

Gallwey, T.J. 301, 313

Holmér, I. 347, 357

Ilmarinen, J. 189, 199
Imrhan, S.N. 233, 241

Kilbom, Å. 145, 151, 213
Kilborn, Å. 217
Kjellberg, A. 367, 371
Konz, S. 281, 285, 397, 401
Kulkarni, M. 33, 61
Kumar, S. 103, 123

Landau, K. 1, 9
Landström, U. 367, 371
Leamon, T.B. 179, 183
Lehto, M.R. 249, 257

Mital, A. 33, 61, 103, 123, 213, 217
Motorwala, A. 33, 61

Ramsey, J.D. 329, 337
Rohmert, W. 1

Siemieniuch, C. 33, 61
Sinclair, M. 33, 61

Westgaard, R. 79, 83
Williamson, A.M. 429, 437
Winkel, J. 79, 83